6th International Conference of Geography and Disaster Management 2024 (6th ICGDM2024)

IOP Conference Series: Earth and Environmental Science
Volume 1503

Online
12-13 December 2024

Editors:

Jumadi
Vidya Nadhiyatul Fikriyah
Christopher Gomez

ISBN: 979-8-3313-2246-5
ISSN: 1755-1307

Printed from e-media with permission by:

Curran Associates, Inc.
57 Morehouse Lane
Red Hook, NY 12571

Some format issues inherent in the e-media version may also appear in this print version.

Institute of Physics
Dirac House, Temple Back
Bristol BS1 6BE UK

Phone: 44 1 17 929 7481
Fax: 44 1 17 920 0979

techtracking@iop.org

Additional copies of this publication are available from:

Curran Associates, Inc.
57 Morehouse Lane
Red Hook, NY 12571 USA
Phone: 845-758-0400
Fax: 845-758-2633
Email: curran@proceedings.com
Web: www.proceedings.com

TABLE OF CONTENTS

IOP Conf. Series: Earth and Environmental Science **1503** (2025) 011001 doi:10.1088/1755-1315/1503/1/011001

PREFACE

The 6th International Conference of Geography and Disaster Management 2024 (ICGDM 2024) was successfully held by the Faculty of Geography, Universitas Muhammadiyah Surakarta. The conference was held via online Zoom meeting on 12th-13th December 2024. ICGDM 2024 had a relevant theme for the tropical regions which is "*Interdisciplinary of Geography on Nature, Technology, and Disaster Management for Sustainable Development Goals*". It aimed to be a scientific platform for research dissemination, ideas sharing, and collaboration related to topics of environmental studies, socio-economic, regional development, disaster management, remote sensing, and GIS applications. ICGDM 2024 invited speakers from different backgrounds who are experts in the field of either physical or socio-economic geography. The keynote speakers were Prof. Nilanchal Patel (Birla Institute of Technology Mesra Ranchi, India), Dr. Farha Sattar (Charles Darwin University, Australia), Dr. Lam Kuok Choy (Universiti Kebangsaan Malaysia, Malaysia), Prof. Taieb Boumeaza (Universite Hassan II de Cassablanca, Morocco), and Dr. Danang Sri Hadmoko (Universitas Gadjah Mada, Indonesia). This ICGDM 2024 was attended by 116 participants and presenters. The committee has received 129 paper submissions, and after the review process, 59 papers were selected for publication in the IOP Conference Series: Earth and Environmental Science. We would like to thank all the authors who have actively participated in the conference. We also highly appreciate the full support from Universitas Muhammadiyah Surakarta.

Best regards, Aziz Akbar Mukasyaf, Ph.D

Chairman of the 6th ICGDM 2024

IOP Conf. Series: Earth and Environmental Science **1503** (2025) 011001 doi:10.1088/1755-1315/1503/1/011001

Organising Committee

Conference Chair

Aziz Akbar Mukasyaf, Ph.D (Universitas Muhammadiyah Surakarta, Indonesia)

Advisory Board

Prof. Josaphat Tetuko Sri Sumantyo, M.Eng., Ph.D.
(Center for Environmental Remote Sensing, Chiba University, Japan)
Prof. Dr. Muhammad Dimyati, M.Sc.
(Geography, Universitas Indonesia, Indonesia)
Prof. Dr. Kuswaji Dwi Priyono
(Geography, Universitas Muhammadiyah Surakarta, Indonesia)
Prof. Christopher Gomez
(Kobe University, Japan)
Prof. dr. hab. Mariusz Lamentowicz
(Adam Mickiewicz University, Poznań, Poland)
Dr. Nawaz M Mian
(National University of Singapore, Singapore)

Scientific Committee

Prof. Dr. Sanusi Adeeyo Yekeen (Federal University of Technology Minna, Nigeria)
Dr. Mohd Hairy bin Ibrahim (Universiti Pendidikan Sultan Idris, Malaysia)
Dr. Balazs Bradak (Graduate School of Maritime Sciences, Kobe University, Japan)
Dr. Sharif Shofirun (Univeristi Utara Malaysia, Malaysia)
Asaba Joyfred, M.A. (Kyambogo University, Uganda)
Jumadi, Ph.D. (Universitas Muhammadiyah Surakarta, Indonesia)
Drs. Yuli Priyana (Universitas Muhammadiyah Surakarta, Indonesia)
Dr. Choirul Amin (Universitas Muhammadiyah Surakarta, Indonesia)
Dr. Rohman Hakim (Universitas Muhammadiyah Surakarta, Indonesia)
Aditya Saputra, Ph.D (Universitas Muhammadiyah Surakarta, Indonesia)
Afif Ari Wibowo, M.Sc. (Universitas Muhammadiyah Surakarta, Indonesia)
Wahyu Tyas Pramono, M.Sc. (Universitas Muhammadiyah Surakarta, Indonesia)
M. Iqbal Taufiqurrahman Sunariya, S.Si., M.Sc., M.URP (Universitas Muhammadiyah Surakarta, Indonesia)

Management Committee

Umar El Izzudin Kiat, M.P.W.K. (Universitas Muhammadiyah Surakarta, Indonesia)
Basyar Ihsan Arijuddin, M.Sc. (Universitas Muhammadiyah Surakarta, Indonesia)
Danardono, M.Sc (Universitas Muhammadiyah Surakarta, Indonesia)
Hamim Zaky Hadibasyir, M.GIS. (Universitas Muhammadiyah Surakarta, Indonesia)
Vidya Nadhiyatul Fikriyah, M.Sc. (University of Twente, The Netherlands)

Secretariat

Dewi Novita Sari, M.Sc. (Universitas Muhammadiyah Surakarta, Indonesia)
Afifah Nurlaila, S. Kom. (Universitas Muhammadiyah Surakarta, Indonesia)

IOP Conf. Series: Earth and Environmental Science **1503** (2025) 011001 doi:10.1088/1755-1315/1503/1/011001

Treasurer
Dr. Annisa Trisnia Sasmi (Universitas Muhammadiyah Surakarta, Indonesia)

Equipment, IT and Documentation
Rosin Novaditya Nur Wahid, M.Pd (Universitas Muhammadiyah Surakarta, Indonesia)
Rudiyanto, S.Si (Universitas Muhammadiyah Surakarta, Indonesia)
Dyah Widowati (Universitas Muhammadiyah Surakarta, Indonesia)

Logistic
Muryadi (Universitas Muhammadiyah Surakarta, Indonesia)
Rohmat Hidayat (Universitas Muhammadiyah Surakarta, Indonesia)

Editor
Jumadi, Ph.D. (Universitas Muhammadiyah Surakarta, Indonesia)
Vidya Nadhiyatul Fikriyah, M.Sc. (University of Twente, Netherlands)

Guest Editor
Prof. Christopher Gomez (Kobe University, Japan)

Track Director
Track 1: Physical Geography
Aditya Saputra, Ph.D (Universitas Muhammadiyah Surakarta, Indonesia)
Basyar Ihsan Arijuddin, M.Sc. (Universitas Muhammadiyah Surakarta, Indonesia)

Track 2: Human Geography
Dr. Choirul Amin S,Si., M.M (Universitas Muhammadiyah Surakarta, Indonesia)
Wahyu Tyas Pramono, M.Sc. (Universitas Muhammadiyah Surakarta, Indonesia)

Track 3: Information Technology Application on Geography
Dr. Rohman Hakim (Universitas Muhammadiyah Surakarta, Indonesia)
Afif Ari Wibowo, M.Sc. (Universitas Muhammadiyah Surakarta, Indonesia)

Track 4: Natural Disaster
Dr. Annisa Trisnia Sasmi (Universitas Muhammadiyah Surakarta, Indonesia)
Umar El Izzudin Kiat, M.P.W.K. (Universitas Muhammadiyah Surakarta, Indonesia)

IOP Conf. Series: Earth and Environmental Science **1503** (2025) 011001 doi:10.1088/1755-1315/1503/1/011001

The 6th International Conference of Geography and Disaster Management (ICGDM) 2024 was successfully held virtually, ensuring the participation of a diverse range of academics, researchers, practitioners, students, and policymakers from various countries. The decision to conduct the conference online was primarily driven by the need to accommodate participants facing travel restrictions and to enhance accessibility for a wider global audience.

The conference was organized from Faculty of Geography, Universitas Muhammadiyah Surakarta, where the organizers managed the event coordination and technical aspects. The virtual model of the conference included plenary sessions featuring distinguished keynote and invited speakers, as well as parallel sessions for contributed papers. Each speaker was allocated a specific time slot for their presentation, followed by interactive Q&A sessions to foster engagement and discussions among participants.

The event successfully attracted 129 submited papers representing 8 countries with diverse academic and professional background in geography, technology, and disaster management. The virtual platform allowed a seamless exchange of knowledge and interdisciplinary discussions, ensuring that all attendees could contribute effectively.

To facilitate the conference, we utilized zoom meeting application, providing a stable and interactive environment for presentations and discussions. Although the virtual format posed certain challenges, such as potential connectivity issues for some participants and the lack of in-person networking opportunities, the overall engagement remained high. We acknowledge that the shift to a virtual format impacted the traditional conference experience but also opened new avenues for wider participation and collaboration beyond geographical limitations.

Despite these challenges, the 6th ICGDM 2024 successfully achieved its objectives of promoting interdisciplinary discussions, knowledge sharing, and collaboration in the fields of geography, technology, and disaster management. We extend our sincere gratitude to all participants, speakers, reviewers, and organizers for their invaluable contributions in making this conference a success.

IOP Conf. Series: Earth and Environmental Science **1503** (2025) 011002 doi:10.1088/1755-1315/1503/1/011002

Peer Review Statement

All papers published in this volume have been reviewed through processes administered by the Editors. Reviews were conducted by expert referees to the professional and scientific standards expected of a proceedings journal published by IOP Publishing.

- **Type of peer review:** Double Anonymous
- **Conference submission management system:** Morressier
- **Number of submissions received:** 129
- **Number of submissions sent for review:** 68
- **Number of submissions accepted:** 59
- **Acceptance Rate (Submissions Accepted / Submissions Received × 100):** 45.7
- **Average number of reviews per paper:** 2
- **Total number of reviewers involved:** 14
- **Contact person for queries:**
  Name: Aziz Akbar Mukasyaf
  Email: icgdm@ums.ac.id
  Affiliation: Universitas Muhammadiyah Surakarta

Stakeholder Analysis in the Development of the Kendal Special Economic Zone

Bagus Ramadhan[1,2], Muhammad Sani Roychansyah[1] and Tri Mulyani Sunarharum[1]

[1] Department of Architecture and Planning, Gadjah Mada University, Yogyakarta, Indonesia
[2] Department of Urban and Regional Planning, Faculty of Science and Technology, Universitas Amikom Yogyakarta, Indonesia

*E-mail: bagusramadhan@amikom.ac.id

Abstract. In 2009, a policy was made for developing Special Economic Zone (SEZ) in Indonesia through the Special Economic Zone Law Number 39/2009. This policy aimed to increase export value, develop regional economies, and attract Foreign Direct Investment (FDI). However, many designated locations have yet to operate according to schedule due to a lack of infrastructure support and land acquisition. On the opposite, the Kendal SEZ, established in 2019, was successfully operated in 2021 after the National SEZ Council had prepared an action plan that prompted infrastructure development with relevant stakeholders. Therefore, this research aimed to analyze stakeholder participation in the development of Kendal SEZ. Data obtained from interviews and questionnaires were then analyzed using stakeholder analysis and Social Network Analysis (SNA). The results showed that SEZ development comprised complex stakeholders, such as state-owned enterprises, the private sector, and the central, provincial, district, and village governments. Important stakeholders are the Central Java Special Economic Zone Council and the SEZ Development and Construction Agency (BUPP) because both have extensive networks of relationships and become the key stakeholders in developing special economic zone in the context of Indonesia.

Keywords: special economic zone, stakeholder, stakeholder analysis, social network analysis

1. Introduction

Economic zone is an area characterized by certain geographical boundaries, developed industrial spaces and incentives, focused on exports, and relevant managers [1]. It had been developed since the 16th century in Leghorn Port and various European cities as free ports [2]. Meanwhile, in 1958, a special economic zone centered on industrial activities was developed in the Shannon EPZ (Export Processing Zone), Ireland. The zone was perceived as an important instrument for spreading the SEZ concept throughout the world [2]. There are various forms of economic zones such as General, Special, and Hybrid [1]. General economic zones are industrial areas, constituting various enterprises and technologics. Special economic zones are exclusive areas characterized by diverse regulations and institutions from other regions to overcome institutional barriers posed by lower-middle income countries. Additionally, hybrid economic zones are common areas open to all types of industries and SEZ focused on export-oriented activities [1]. This definition shows that SEZ is a delineation of different regulations from other countries.

A typical example is Shenzhen, one of the rapidly developed SEZs in China. The establishment of Shenzhen as an SEZ in a riverine village bordering Hong Kong in 1980 changed the physical form of the area into a metropolitan city within 20 years [3]. This was due to the economic openness compared to other parts of China, focusing on the market economy which motivated regional cooperation, including both domestic and international collaboration [4]. The success of Shenzhen was also considered a milestone in the country, with the SEZ policy later expanded to other cities [5].

Many lower-middle income countries had adopted SEZs as a catalyst for economic development in respective regions [6]. Several successful examples such as Shenzhen had also motivated other countries to establish SEZs with the intention of transforming the diverse economies. However, the outcome of the process in lower-middle income countries tend to vary due to various factors [7]. However, some of the factors that limit the development of SEZs are high competition, lack of infrastructure, and connectivity with the economy in the surrounding region [8].

In Indonesia, SEZ development is characterized by geoeconomic and geostrategic advantages, as well as functions to accommodate industrial activities, exports, imports, and other related events that are internationally competitive, with high economic value [9]. The objectives include labor absorption, economic growth, and acceleration of regional development. Compared to other countries, SEZs in Indonesia have many facilities and incentives such as regional tax relief and those stated in SEZ Law 39/2009. However, within 2009 and 2019, there had been no significant development due to several problems namely land acquisition, infrastructure support, overlapping regulations, and the professionalism of SEZ Management Business Entities [10].

Based on this perspective, there were 19 SEZs in the country within 2009 to 2021, but only four had optimal performance, including the Kendal SEZ [11]. It was established in 2019 through Government Regulation Number 85/2019 concerning the Kendal SEZ. The SEZ started operational since May 2021, relatively two years since it was established in 2019. Meanwhile, in the 2021 report, the Kendal SEZ also had the highest workforce absorption and the second utmost investment realization compared to other SEZs.

The actors should play a significant role during the development process. As with regional development in general, collaboration between these actors was essential, resulting in the discovery of unequal power levels [12; 13]. In accordance with Law 39/2009 concerning SEZs, the relevant actors include the National SEZ, and Zone Councils, Development and Construction Business Entity (BUPP), Zone Administrator, and tenant. The National SEZ Council was tasked with evaluating proposals and monitoring performance annually. The Zone Council functioned as an extension of the National SEZ Council, forming and supervising administrators, as well as determining strategic steps in overcoming related problems. The council consisted of various stakeholders from both vertical and several other agencies in the provincial government. The regional administrator was tasked with providing business permits and other licences for actors in the SEZ consisting of vertical and related agencies in the regency or city government. The BUPP was responsible for organizing business activities in the SEZ. In addition to these stakeholders, the National Special Economic Zone Council prepared an Action Plan for the development process.

The development of the Kendal SEZ in accordance with the Action Plan prepared by the National Special Economic Zone Council in 2019 lacked infrastructure. Some basic ones such as water, gas, and port required the support of various actors who played a significant role in the realization of the Kendal SEZ covered in the Action Plan.

Rapid and excellent performance in the development process does not only include BUPP (as a business entity that builds and manages the SEZ), because it is tasked with the arrangement procedure in the immediate area. The supporting infrastructure outside the SEZ is the responsibility of other governments based on the type. Some of these infrastructures are provided by the central, provincial, and local government, while others are developed by the private sector. Cooperation between stakeholders in the development of the Kendal SEZ is essential to map the network for learning purposes, considering that a significant problem in Indonesia is the duration of infrastructure development.

With many stakeholders playing a major role in the development of the Kendal ZEZ which became operational in less than two years since the establishment, there is need to conduct an analysis of the entire process. Stakeholder analysis is a method adopted to impact knowledge of actors, both individuals and organizations. This was aimed to understand behavior, relationships, and interests, including evaluating the resources used in the implementation of relevant policies [14]. Stakeholder interest and power dynamics mapping is a tool that enabled the understanding of potential conflicts, including determining appropriate engagement methods [15]. Furthermore, conducting the stakeholder analysis was intended to increase the success rate of the project because it provided information regarding the planning, implementation, and evaluation processes to all relevant stakeholders [14]. The analysis can also determine the relevant policies in the development of other SEZ, for instance cooperation between stakeholders in the Kendal SEZ effectively improved performance.

2. Literature Review

The literature review discussed the challenges encountered during SEZ development and how these were resolved with stakeholder participation.

2.1. Challenges of SEZ development

The development of SEZ worldwide had become attractive due to the ability to improve the economy of a country. This led to an increase of 5,400 zones in 147 countries [16], although the real impact on the economy remains unclear [17]. However, the economic development was carried out by providing incentives or facilitation for actors who developed own businesses in the area, enabling SEZ to improve spillover effects in the form of FDI and technology transfer, as well as motivating economic transformation throughout the country [16]. In East Asian countries, SEZ provided several benefits to the economy such as investment, employment, exports, and economic growth. Although, the reverse was the case in other countries, because SEZ are developed in an enclave, having fewer connection with the surrounding economy [17].

During the development process, many challenges occurred based on adopted practices in various parts of the world. These can be categorized at the initiation or early and maturity stages. In Africa, SEZ development, especially during the early stages, was hampered by lack of infrastructure and funding, resulting in the need for China to partake in the process [17]. Rodríguez-Pose et al., [18], also stated that these challenges occur in Africa because the focus of the SEZ developed differed from the comparative advantages of the country, regulatory design similar to competitors, less attractive locations, lack of quality infrastructure, political commitment, cross-sectoral and institutional policies. The development process adopted by China in four sub-Saharan African countries, encountered the following challenges land access, regulatory barriers, resettlement, and lack of infrastructure [19].

In Pakistan, the participation of the government in promoting and attracting investment to SEZ had been a barrier to the growth process, resulting in the adoption of strategies such as PPP to motivate potential investors [6]. Additionally, in neighboring India, these challenges were related to land acquisition issues which led to anti-SEZ resistance [20].

Some of the challenges described in the previous section were encountered during the mature stage. However, for effectively developed SEZ such as those in China, the issue was to maintain success [21]. These included the need for increased innovation in order to ascend the global value chain, ensuring the sustainability of export-oriented sectors against crises affecting foreign demand, and environmental viable issues. Meanwhile, some SEZ do not provide adequate social services for employees, causing it to become less attractive to skilled employees [21]. Another challenge is resilience in the face of crises or shocks such as the Covid-19 pandemic [22].

2.2. Stakeholders in SEZ development

Stakeholder analysis is essential in the development of a region because it helps identify the main actors respective roles and relationships in complex urban projects [23; 24]. Moreover, in regional development, stakeholders include government agencies, citizens, academics, investors, and NGOs [25; 26]. In the case of SEZ, it helps identify the main actors, resolve conflicts, and prioritize common interests [27]. Mapping stakeholder interests and power dynamics is also essential in understanding potential conflicts and determining appropriate engagement techniques [15]. Systematic methods, including social network analysis (SNA), provide valuable insights into sustainable development [28]. In the context of foreign direct investment, which is the focus of many SEZ developments, stakeholder analysis aided in the identification of environmental and social challenges, namely air quality and water consumption, including opportunities such as energy reserves and socio-economic growth [29].

In the development, SEZ has various stakeholders that depend on the respective government system of the country where it is located [17]. Although there is no specific model, from general practice, some stakeholders included the government, SEZ authorities, area developers, operators, and tenants.

Stakeholder theory was formulated in response to an increasingly globalized world, the development of information technology, progressively liberal countries, and awareness of the impact of a business on society or the country [30]. Freeman defined stakeholders as groups or individuals who can impact or be influenced by the objecives of an organization, including playing a critical role to the sustainability and success of the company [30]. The various definitions possess the connecting verb impacted, as well as outlined the relationship between organizations and stakeholders [31].

The development of SEZ requires support from various stakeholders, for example the government provided assistance through commitments in the form of infrastructure and policies. This was because several obstacles related to minimal infrastructure, implementation of policies, land provision, and compliance with environmental regulations were encountered [32]. Furthermore, success also depended on connectivity and national competitiveness, where connectivity included the provision of relevant infrastructure namely ports, roads, and energy [32] which can be developed by other stakeholders asides the government. More dynamic private sector participation in regional development in various countries refers to a source of expertise and risk management. The government tend to focus on developing infrastructure that supports the private sector, especially ports and quality road networks to the zone [32].

In respect to the development of sustainable SEZ, UNCTAD [17] stated that cooperation between stakeholders helped to identify opportunities and develop action plans. These stakeholders were tasked with developing policies, infrastructure, and administrative support [17]. With the development process comprising various stakeholders both from the government at various levels and the private sector, led to the realization of efficiency during strategic execution [17].

During SEZ development, in Indonesia, stakeholders encountered several problems. The challenges related to governance were viewed from the diverse interests of stakeholders resulting in the inability to execute an effective role in the development process. This led to the adoption of a top-down method, to ensure the local governments and SEZ developers did not ignore the local context and needs [33]. In the annual report of the National SEZ Council, these problems include land acquisition, provision of infrastructure and utilities [11], resulting in the need from local stakeholders to enlist in the Zone Council motivating community participation in the development and operation of SEZ [33].

The major stakeholders that played a significant role in the SEZ structure, included the National SEZ Council, a collection of ministries from the central government, the Zone Council, consisting of various provincial-level agencies and related shareholders, BUPP, and management enterprises. This also comprised the Area Administrator, a representative of the central, provincial and regency or city governments tasked with assisting the Zone Council in supporting SEZ operations [9].

The research conducted by Kismartini et.al [27] in Bangka Belitung stated that mapping the roles of stakeholders is important to identify those who play a major part in implementing related policies. The National SEZ Council is the actor with the highest level of power and interest in the development process. However, the central government lacked commitment to SEZ development. As a tourism SEZ, local community participation also played minimal support in Bangka Belitung [27].

3. Methodology

3.1. Research Area

Kendal SEZ is located in the Central Java Province, approximately 20 km from Semarang City, the capital. The SEZ is precisely situated on the North Coast of Java and the Trans Java Toll Roads connecting major cities and ports. It is strategically positioned due to the connection with the Trans Java toll road, adjacent to Tanjung Emas Port in Semarang City.

Prior to becoming an SEZ, it was already an Industrial Area managed by Jababeka, a development company with decades of experience in the country. For the development of this area to be more optimal in 2018, the status was submitted to become designated as an SEZ in 2019 based on Government Regulation Number 85/2019 concerning the Kendal SEZ.

This was the first industrial SEZ in Java Island, established in 2019. Java is the center of growth in Indonesia and the development process focused on areas outside Java that still needed new growth generators. Therefore, the development of Kendal SEZ tends to be new with a higher target burden compared to other industrial SEZ outside Java Island.

3.2. A combination of Stakeholder and Social Network Analyses

This research uses mixed methods: the qualitative approach is used to analyse the stakeholder through stakeholder analysis, and the quantitative approach is used to analyse the stakeholder network analysis. The combination of these two evaluations were used to conceal the lack of quality and consistency of stakeholder analysis only used to understand the participating parties. Moreover, the combination with SNA provides an overview of multi-governmental arrangements and collaborative efforts adopted [34]. SNA also presented an outline of stakeholder characteristics, including exploring the relationship structure [35]. However, stakeholder analysis is perceived as a managerial tool, seeking opportunities to gain support for specific objectives [14].

SNA is a method based on the assumption that relationships between related individuals are important to the system, showing how the interactive nature can be analyzed. From the results of SNA using Gephi software, several stakeholders were obtained as labels for various interactive communities with strong ties, depicted in different colors [36].

This research started with carrying out a stakeholder analysis that explored participation in the development of Kendal SEZ. The stakeholders were identified by respective roles in the preparation and implementation of the Kendal SEZ Action Plan. This initiative was supported by various roles of stakeholders in the development process. In addition to the Action Plan, the Secretariat of the Central Java Zone Council was requested to reveal other stakeholders. The following 29 actors were realized from the Action Plan and information from the Secretariat of the Central Java Zone

Table 1. Stakeholder Group based on authority

Group	Actors	Code
Village Government	Wonorejo Village Government	V1
	Brangsong Village Government	V2
	Mororejo Village Government	V3
	Turunrejo Village Government	V4
	Rejosari Village Government	V5
	Purwokerto Village Government	V6
	Kumpulrejo Village Government	V7
	Karangtengah Village Government	V8
Regency Government	Kendal Regency Investment and One-Stop Integrated Service Department	R1
	Kendal Regency Public Works and Spatial Planning Department	R2
	Kendal Regency Transportation Department	R3
Provincial Government	Secretariat of the Central Java Zone Council	P1
	Central Java Province Transportation Department	P2
	Central Java Province Energy and Mineral Resource Department	P3
	Central Java Province Regional Development Planning Agency	P4
	Central Java Province Public Works, Water Resources and Spatial Planning Department	P5
	Central Java Province Regional Secretariat	P6
Central Government	1st Class of Harbor Master and Port Authority Office Tanjung Emas	C1
	Pemali-Juana River Basin Center	C2
	Secretariat General of the National Council for Special Economic Zones	C3

Group	Actors	Code
	Zone Administrator	C4
	Ministry of Public Works	C5
	Ministry of Energy and Mineral Resource	C6
	Ministry of Transportation	C7
State-Owned Enterprise	Pelindo	S1
	PLN	S2
	PGN	S3
Privat Enterprise	PT. KIK (BUPP)	PE1
	Jababeka	PE2

The questionaires were distributed to the 29 stakeholders interviewed for Stakeholder and SNA analyses. The criteria for the selected stakeholders included participation and contribution to the development of the Kendal SEZ. From the 29 stakeholders, 27 were willing to fill out the questionnaire and be interviewed. The two stakeholders who were unwilling to fill out the questionnaire were PT. KIK (BUPP) and Jababeka (as the parent company of PT. KIK). During the stakeholder analysis, these two were excluded for not filling out the questionnaire (level of power, interest, and knowledge) but were included in the SNA analysis due to the relationship with other actors. The results of the stakeholder analysis, was shown in a diagram representing the position of each actor. This was grouped into four quadrants, namely players who had great power and interest, context setters characterized by great power but low interest, subjects possess low power but great interest, while crowds possessed low power and interest [37; 38]. During stakeholder analysis data collection, a questionnaire calculation was performed using a Likert scale (none or minimal, very low, low, high, and very high) to measure knowledge, power, and interest.

The other part of the questionnaire was was used to analyze the SNA, namely which actors collaborated with each stakeholder in the development of the Kendal SEZ. Then to show the relationship between stakeholders, Gephi software was used to describe the map, including analyzing the network. The data from the questionnaire that was inputted to Gephi Software are the interactions between stakeholders. This data was input into a spreadsheet, which formed a nodes sheet and an edges sheet. Nodes are described as stakeholders, and edges are the interactions between stakeholders. Based on the results of the Gephi software analysis, the network form, degree distribution, modularity, weighted degree, closeness, and betweenness centralities were then described.

4. Result

4.1. Analysis of Stakeholder Characteristics

In accordance with the results of the questionnaire, it was discovered that the stakeholders who had high levels of power and interest consisted of the central and provincial government as shown in Figure 1. These two groups are the main actors in the process of establishing SEZ, especially in providing support related to infrastructure and other facilities. The stakeholders included in the high-power but low-interest quadrant comprised those from the central government and state-owned enterprises. Not all ministries at the central government level were interested in developing SEZ, despite being highly powerful in providing infrastructure. Similarly, state-owned enterprises also had the power to provide infrastructure but were necessarily not

IOP Conf. Series: Earth and Environmental Science **1503** (2025) 012001 doi:10.1088/1755-1315/1503/1/012001

interested in developing it to support SEZ. This was because the consumers of own infrastructure were not clear at the beginning of the development process. The next stakeholder enlisted was the subject section, characterized by low power level but high interest. This section focused on the following stakeholder groups such as the Central, Regency, and Village Governments interested in the development process, but rarely intervened during the implementation of the policies.

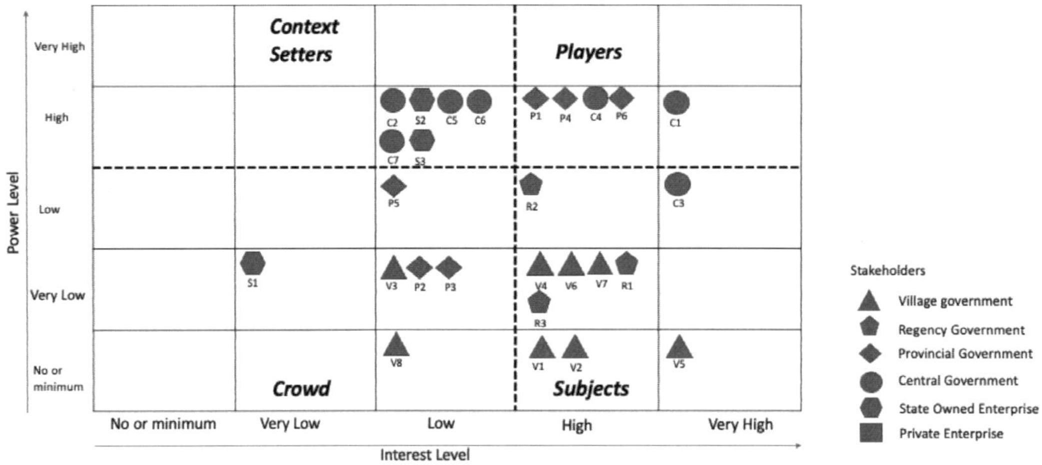

Figure 1. Stakeholder Group Map

The next figure showed stakeholders overlaid with respective knowledge level in SEZ development or supporting infrastructure. In this section, players in the C1 quadrant possessed very high knowledge due to respective regulatory positions in the development of ports around the Kendal SEZ. Several stakeholders in this quadrant who were highly knowledgable comprised P1, the coordinator at the provincial level in the development process. This also included P4 tasked with organizing related programs, and C4 responsible for the provision of facilities for tenants in the SEZ. In the setters quadrant, all stakeholders were highly knowledgeable in respective fields. In the subject quadrant, two stakeholders possessed the highest knowledge, namely C3 who is the main stakeholder at the central government level, and R2 specifically concerned in the development of clean water infrastructure in the Kendal SEZ.

IOP Conf. Series: Earth and Environmental Science **1503** (2025) 012001 doi:10.1088/1755-1315/1503/1/012001

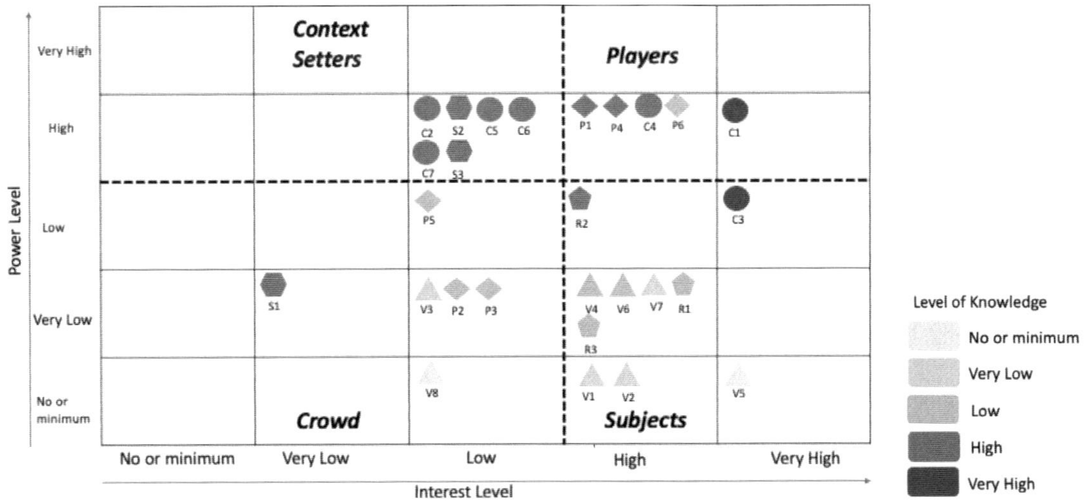

Figure 2. Map of stakeholder groups and knowledge levels

4.2. Stakeholder Network Structure

The results of the distributed questionnaire, including interactions with each stakeholder are shown in Figure 3. The middle section contains the Secretariat of the Central Java Zone Council, the stakeholder with the highest level. Furthermore, the provincial level focused on the liaison between various stakeholders in the development of the Kendal SEZ.

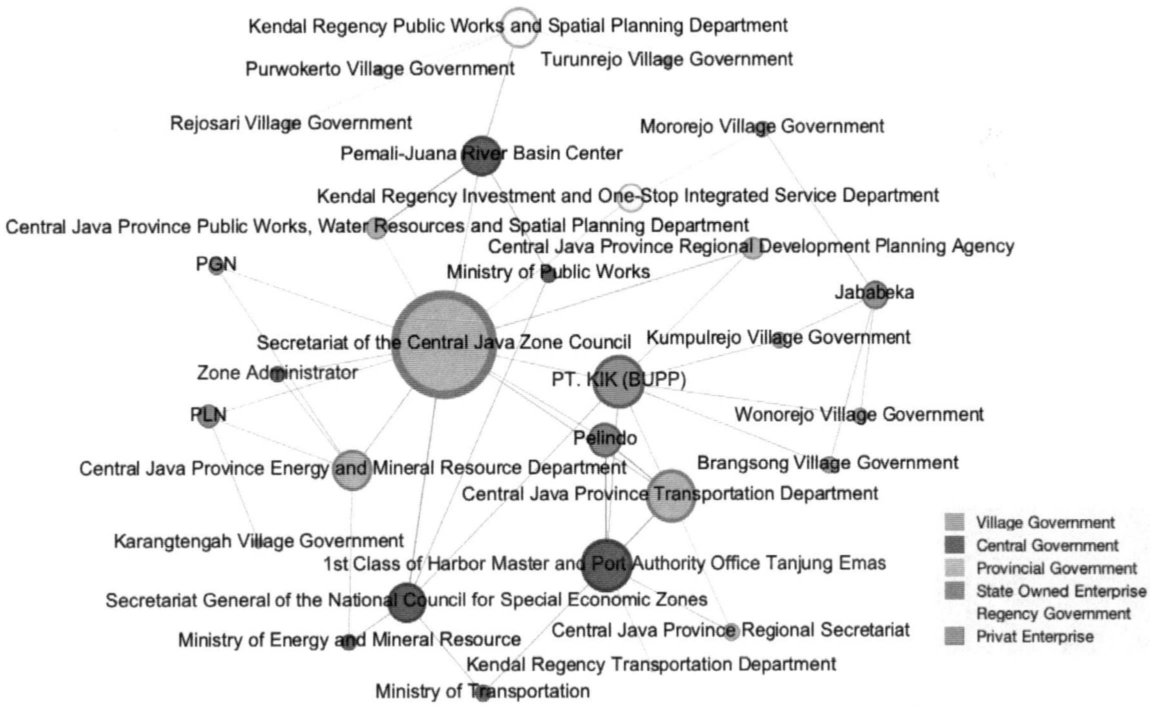

Figure 3. Map of stakeholder networks and groups

Based on the degree of distribution, the most influential actor had 12 relationships, namely the Secretariat of Central Java Zone Council. In addition, another actor namely PT. KIK (BUPP) had eight relationships, with the value of the degree distribution equivalent to 3.17. From the results of the interview, the Secretariat of Central Java Zone Council coordinates other actors in the development of the Kendal SEZ. This council does not possess the power to influence other actors, rather when obstacles are encountered, it tends to find the needed solutions in collaboration with relevant actors. Meanwhile, PT. KIK (BUPP) responsible for managing the development of the Kendal SEZ is limited to related activities within the area. This actor is definitely connected to other stakeholders due to the need for various infrastructure outside the area to function optimally. Figure 4 shows the modularity analysis where the stakeholders were divided into several clusters according to the intensity of respective interactions.

Figure 4. Stakeholder network map based on modularity analysis

Based on the results of the modularity analysis, a value of 0.46 was obtained with five clusters of interrelated stakeholders. These were divided into clusters according to respective fields, depicted by different colors in the graph

1. Cluster 1 (purple) is an energy infrastructure development, characterized by the presence of actors such as PGN, PLN, and the Central Java Provincial ESDM Office. In this cluster, the major stakeholder, namely the Secretariat of the Central Java Zone Council, was also included and Central Java Province Regional Development Planning Agency and Kendal Regency Investment and One Stop Integrated Service Department.

2. Cluster 2 (dark blue) is a port infrastructure development, with several concerned stakeholders, namely Pelindo, 1st Class of Harbour Master and Port Authority Office Tanjung Emas, the Central Java Transportation Office, and the Ministry of Transportation.
3. Cluster 3 (light blue) is a clean water infrastructure development for the Kendal SEZ, with several stakeholders playing a role in the clean water network, including the Pemali-Juana River Basin Center, Central Java Province Public Works, Water Resource and Spatial Planning Development, Kendal Regency Public Works and Spatial Planning Department, as well as several villages passed by the clean water network, namely Rejosari, Purwokerto and Turunrejo.
4. Cluster 4 (gray) played a role in the provision of land for the development process, driven by the following main stakeholders, PT. KIK (BUPP), Jababeka (as the developer of the Kendal Industrial Area before it was termed the Kendal SEZ), and several villages cleared for the development of the Kendal SEZ, including Wonorejo, Mororejo, Brangsong and Kumpulrejo.
5. Cluster 5 (green) is the stakeholder at the central level, namely the Ministries of Energy and Mineral Resources, Transportation, Public Works. Additionally, the Secretariat General of the National SEZ Council also coordinates with other stakeholders.

Table 2. Centrality Calculation of Various Stakeholders

Stakeholder	**Weighted Degree**	**Rank**	**Closeness Centrality**	**Rank**	**Betweenness Centrality**	**Rank**
Secretariat of the Central Java Zone Council	17,00	**1**	0,57	**1**	0,57	**1**
PT. KIK (BUPP)	8,00	**2**	0,49	**2**	0,26	**2**
Central Java Province Transportation Department	8,00	**2**	0,44	**3**	0,07	8
1st Class of Harbor Master and Port Authority Office Tanjung Emas	8,00	**2**	0,37	7	0,10	7
Secretariat General of the National Council for Special Economic Zones	6,00	**3**	0,44	**3**	0,12	6
Pemali-Juana River Basin Center	6,00	**3**	0,43	**4**	0,13	**5**
Kendal Regency Public Works and Spatial Planning Department	6,00	**3**	0,35	8	0,21	**3**
Central Java Province Energy and Mineral Resource Department	6,00	**3**	0,40	**5**	0,03	9
Pelindo	5,00	**4**	0,40	**5**	0,02	10
Kendal Regency Investment and One-Stop Integrated Service Department	4,00	**5**	0,43	**4**	0,17	**4**
Jababeka	4,00	**5**	0,30	12	0,03	9
Central Java Province Regional Development Planning Agency	3,00	6	0,40	**5**	0,00	12

Stakeholder	Weighted Degree	Rank	Closeness Centrality	Rank	Betweenness Centrality	Rank
Central Java Province Public Works, Water Resources and Spatial Planning Department	3,00	6	0,40	**5**	0,00	12
PLN	3,00	6	0,38	6	0,07	8
Zone Administrator	2,00	7	0,37	7	0,00	12
PGN	2,00	7	0,37	7	0,00	12
Central Java Province Regional Secretariat	2,00	7	0,32	11	0,00	12
Ministry of Public Works	2,00	7	0,35	8	0,01	11
Ministry of Transportation	2,00	7	0,34	9	0,01	11
Ministry of Energy and Mineral Resource	2,00	7	0,33	10	0,00	12
Mororejo Village Government	2,00	7	0,33	10	0,03	9
Wonorejo Village Government	2,00	7	0,35	8	0,01	11
Brangsong Village Government	2,00	7	0,35	8	0,01	11
Kumpulrejo Village Government	2,00	7	0,34	9	0,01	11
Karangtengah Village Government	1,00	8	0,28	13	0,00	12
Kendal Regency Transportation Department	1,00	8	0,27	14	0,00	12
Turunrejo Village Government	1,00	8	0,26	15	0,00	12
Rejosari Village Government	1,00	8	0,26	15	0,00	12
Purwokerto Village Government	1,00	8	0,26	15	0,00	12

Source: Analysis, 2024

The table shows the results of the analyzed degree centrality, closeness, and betweenness centralities based on calculations performed with the Gephi software. Based on the analysis, the most influential actors in terms of weight were the Secretariat of the Central Java Zone Council, PT. KIK (BUPP), the Central Java Provincial Transportation Agency, including the Office of the Harbor Master and Class I Port Authority of Tanjung Emas. The most influential actor was the Secretariat of the Central Java Zone Council perceived as the center of interaction because it has a relationship with virtually all stakeholders. This is in accordance with the duties to manage and develop the SEZ, especially in the context of infrastructure support by connecting the various relevant actors. Therefore, in this analysis, the Secretariat of the Central Java Zone Council was the most influential actor in the development of the Kendal SEZ.

In closeness centrality, the node or actor that can best disseminate information on this graph is the Secretariat of the Central Java Zone Council, PT. KIK (BUPP), and the Central Java Provincial Transportation Agency. This is particularly appropriate for the Secretariat of the Central Java Zone Council because the position enables the coordination of actors, especially those within the scope of the Central Java Provincial Government and according to respective needs. The problems faced in the development of the Kendal SEZ were often discussed in forums organized by the Central Java Zone Council leading to the closeness with various other actors. While PT. KIK (BUPP) also relates with other actors according to the interests of BUPP or tenants in the SEZ.

The next order for closeness centrality comprised the Central Java Provincial Transportation Agency and the Secretariat General of the National Council for SEZ. The Central Java Provincial Transportation Agency is quite interesting to explore, ranked third, even though it only participated in the development of port, and also continued to interact with other actors. Meanwhile, the Secretariat General of the National Council for SEZ is an actor at the central government level that grants SEZ status, prepares action plans until the SEZ becomes operational, and participates in solving problems in the realization process, especially when it requires intervention from the central government. This role strengthened the network between stakeholders, thereby reaching out to others.

The Betweeness Centrality measured crucial actors, namely determining important nodes or actors in this network flow that can provide the shortest path in the dissemination of information. This is also included the Secretariat of the Central Java Zone Council and PT. KIK (BUPP). The two actors are in the middle and have relationships with several others. In addition, the Public Works and Spatial Planning Agency of Kendal Regency plays a role in connecting actors in certain villages passed by the clean water network. PT. KIK (BUPP) and actors both at the provincial and national levels played a major role in developing this network.

5. Discussion

5.1. *Action Plan to form Interaction Clusters*

Every SEZ developed in Indonesia has an Action Plan supported by various stakeholders, including infrastructure and non-infrastructure. Similarly, during the development of the Kendal SEZ, an Action Plan was prepared by the National SEZ Council. This Action Plan was supported by various stakeholders who provided infrastructure support for the Kendal SEZ. The support comprised the development of ports, gas networks, and clean water. This was in accordance with the cluster that materialized in the modularity analysis which stated certain stakeholders played a role in the development of ports, gas or energy networks, and clean water. Although actors who play a role in the process are more than those tasked with preparing the Action Plan due to additional needs. In the modularity analysis, the central government formed a separate cluster. The Action Plan united several ministries with respective development fields, and this differed from the modularity analysis. This action plan also in line with recommendation from Aggarwal [16], action plan is needed as set of objective to be developed and contributed for strategic role proposed for SEZ.

5.2. *Major stakeholders in SEZ development*

Based on the results of the weighted degree analysis, closeness, and betweenness centralities, the Secretariat of the Central Java Zone Council and PT. KIK (BUPP), were ranked first and second, respectively. Both actors played crucial roles in developing the Kendal SEZ. The Secretariat of the Central Java Zone Council coordinated support for the development process, especially those outside the region and not under the authority of BUPP. Meanwhile, PT. KIK (BUPP) developed infrastructure, as well as the region within the Kendal SEZ, and continued to interact with various actors. The initiative was aimed to motivate the development of supporting infrastructure that required access from outside. This is a lesson for the development of other SEZ that in the Indonesian context, the Zone Council and BUPP played important roles, including determining whether the SEZ can develop optimally. Both actors must possess adequate coordination and capacity to determine the appropriate steps and facilitation. This result inline with Pratama, et.al. [39] that local government plays a crucial role in developing SEZ as mediating the aspiration and potential conflict between stakeholders. Central Java Zone Council as one of the stakeholder from

IOP Conf. Series: Earth and Environmental Science **1503** (2025) 012001 doi:10.1088/1755-1315/1503/1/012001

local government in Central Java Province become the most influential stakeholder. This slightly differed from the results of Kismartini [27] in the Bangka Belitung SEZ which stated the main actor was the National SEZ Council, followed by the primary and secondary stakeholders namely the local government and tin mining entrepreneurs assumed to be BUPP, respectively. The findings also differed from the results of the present research analysis where the Zone Council and BUPP were regarded as the main actors in SEZ development.

5.3. Complexity of SEZ Development

The stakeholder identification conducted in the initial section showed that actors who played a role in the development of SEZ belonged to diverse groups namely village, regency, provincial, and central government. In addition, there are also State-Owned and Private Enterprises stakeholders who also played a significant role. The large number of stakeholders is a challenge in the development process because, from the results of stakeholder analysis, these actors have varying interest. Therefore, SEZ needs to be understood as an instrument for developing both regional and national economy. The participation of other stakeholders who may not be interested in the development of SEZ, but indirectly contributed to increasing FDI, labor absorption, and export value considered the main objectives of the entire process was required. Thus the development of SEZ needs a good stakeholder management, because the success of SEZ in Indonesia depends on careful planning, effective stakeholder management, and strategy execution [40].

6. Conclusion

In conclusion, several interesting attributes were discovered in relation to the practice of SEZ development in the future. Moreover, in the case of Kendal SEZ, actors who had great interests and power were dominated by those from different levels of government. This implied that the government was the main actor or driver in the development process. The finding was in line with the focus of the research related to infrastructure support which in Indonesia should be provided by both the government and state-owned enterprises.

During the SNA analysis, it was found that the central actors were dominated by the Secretariat of the Central Java Zone Council and PT. KIK (BUPP). Both actors were the driving force in the development of the Kendal SEZ. The Secretariat of the Central Java Zone Council coordinated actors who could support the Kendal SEZ infrastructure despite the inability to influence the decisions of one another. PT. KIK (BUPP) had a great interest in developing the SEZ optimally by building relationships with various actors even though the decision on the contribution to the development process depended on each.

This practice showed that the Zone Council and BUPP are two relevant entities in SEZ development. The Zone Council was perceived as a tactical entity in accepting other actors to support SEZ development. This also needed to be equipped with a professional and cooperative BUPP, and other relevant actors to ensure continuous activeness in obtaining infrastructure support and increased competitiveness. In the future, the Zone Council and BUPP should possess a cordial relationship, cooperating to establish a successful SEZ such as the Kendal SEZ.

Funding: This research was funded by the Final Project Recognition (RTA) of Gadjah Mada University, Yogyakarta, Indonesia, number 5075/UN1.P.II/Dit-Lit/PT.01.01/2023.

References

[1] Aggarwal, A. (2022) *'Special Economic Zones in The Indonesia–Malaysia–Thailand Growth Triangle'*. Asian Development Bank

[2] Aggarwal, A. (2010) 'Economic impacts of SEZs: Theoretical approaches and analysis of newly notified SEZs in India'. Munich Personal RePec Archive Paper no. 20902.

[3] Zacharias, J. and Tang, Y. (2010) 'Restructuring and repositioning Shenzhen, China's new mega city', *Progress in Planning*, 73(4), pp. 209–249. Available at: https://doi.org/10.1016/j.progress.2010.01.002.

[4] Robinson, B. (2022) 'African Special Economic Zones', *Springer Books* [Preprint].

[5] Liu, K. (2020) *Shenzhen: A Model of The China Model? If Yes, Is It Replicable?* Available at: https://ssrn.com/abstract=4470326.

[6] Naeem, S., Waheed, A. and Khan, M.N. (2020) 'Drivers and barriers for successful special economic zones (SEZs): Case of SEZs under china Pakistan economic corridor', *Sustainability (Switzerland)*, 12(11). Available at: https://doi.org/10.3390/su12114675.

[7] Frick, S.A., Rodríguez-Pose, A. and Wong, M.D. (2019) 'Toward Economically Dynamic Special Economic Zones in Emerging Countries', *Economic Geography*, 95(1), pp. 30–64. Available at: https://doi.org/10.1080/00130095.2018.1467732.

[8] Tugushev, A. (2022) 'Factors Affecting Performance of Special Economic Zones', in *Handbook of Research on Special Economic Zones as Regional Development Enablers*. IGI Global, pp. 58–72.

[9] Indonesian Government. (2009). Law of the Republic of Indonesia Number 39 Year 2009 Regarding Special Economic Zones.

[10] SEZ National Council. (2020). SEZ Annual Report Year 2020. Retrieved from https://kek.go.id/id/service/report

[11] SEZ National Council. (2021). SEZ Annual Report Year 2021. Retrieved from https://kek.go.id/id/service/report

[12] Ansell, C. and Gash, A. (2008) 'Collaborative governance in theory and practice', *Journal of Public Administration Research and Theory*, 18(4), pp. 543–571. Available at: https://doi.org/10.1093/jopart/mum032.

[13] Healey, P. (1997) *Collaborative Planning: Shaping Places in Fragmented Societies*. UBC Press.

[14] Varvasovszky, Z. and Brugha, R. (2000) 'A stakeholder analysis'. *Health policy and planning*, 15(3), 338-345.

[15] Wang, Q., Cheong, K.C. and Li, Y. (2020) 'Who benefits from development? Analyzing the stakeholder contestations in a traditional settlement of Malaysia', *Housing Policy Debate*, 30(5), pp. 861–877.

[16] Aggarwal, A. (2019) *SEZs and economic transformation: towards a developmental approach*. Available at: https://ssrn.com/abstract=3623039.

[17] Mugano, G. (2021) *Special economic zones*. Springer International Publishing.

[18] Rodríguez-Pose, A. *et al.* (2022) 'The challenge of developing special economic zones in Africa: Evidence and lessons learnt', *Regional Science Policy and Practice*, 14(2), pp. 456–481. Available at: https://doi.org/10.1111/rsp3.12535.

[19] Zeng, D.Z. (2015) Global Experiences with Special Economic Zones – With a Focus on China and Africa. *World Bank policy Research working paper*, (7240).

[20] Bedi, H.P. (2013) 'Special Economic Zones: National land challenges, localized protest', *Contemporary South Asia*, 21(1), pp. 38–51. Available at: https://doi.org/10.1080/09584935.2012.757582.

[21] Zeng, D.Z. (2012) *China's Special Economic Zones and Industrial Clusters: Success and Challenges*. Lincoln Institute of Land Policy.

[22] Zeng, D.Z. (2021) 'The Past, Present, and Future of Special Economic Zones and Their Impact', *Journal of International Economic Law*, 24(2), pp. 259–275. Available at: https://doi.org/10.1093/jiel/jgab014.
[23] Long, Z. *et al.* (2021) 'Stakeholder Analysis in Urban Community Regeneration in China: The Case of Shenzhen', in *ICCREM 2021*, pp. 810–817.
[24] Semenova, A.M. *et al.* (2021) 'Stakeholder analysis as a tool for improving strategic planning of the urban environment'. Vestn. St. Petersburg Univ. Manag., vol. 20, no. 3, pp. 391-409
[25] Mathur, V.N. *et al.* (2007) 'Defining, identifying and mapping stakeholders in the assessment of urban sustainability. IN: Horner, et. al.', in *Proceedings: SUE-MoT Conference 2007: International Conference on Whole Life Sustainability and its Assessment, Glasgow, Scotland, 27th-29th June*. Citeseer.
[26] Wojewnik-Filipkowska, A. and Węgrzyn, J. (2019) 'Understanding of public–private partnership stakeholders as a condition of sustainable development', *Sustainability*, 11(4), p. 1194.
[27] Kismartini *et al.* (2020) 'Policy Of Special Economic Zones And Environmental Policies In Bangka Belitung: A Stakeholder Analysis', *Humanities & Social Sciences Reviews*, 8(4), pp. 1460–1470. Available at: https://doi.org/10.18510/hssr.2020.84134.
[28] André, K. *et al.* (2012) 'Method development for identifying and analysing stakeholders in climate change adaptation processes', *Journal of Environmental Policy & Planning*, 14(3), pp. 243–261.
[29] Huang, Y., Fischer, T.B. and Xu, H. (2017) 'The stakeholder analysis for SEA of Chinese foreign direct investment: the case of 'One Belt, One Road'initiative in Pakistan', *Impact Assessment and Project Appraisal*, 35(2), pp. 158–171.
[30] Freeman, R.E. *et al.* (2010) 'Stakeholder theory: The state of the art'. Cambridge University Press.
[31] Friedman, A.L. and Miles, S. (2006) *Stakeholders: Theory and practice*. OUP oxford.
[32] Farole, T. and Akinci, G. (2011) *Special economic zones: progress, emerging challenges, and future directions*. World Bank Publications.
[33] Hidayat, S. and Negara, S.D. (2020) 'Special economic zones and the need for proper governance: empirical evidence from indonesia', *Contemporary Southeast Asia*, 42(2), pp. 251–275. Available at: https://doi.org/10.1355/cs42-2e.
[34] Lienert, J., Schnetzer, F. and Ingold, K. (2013) 'Stakeholder analysis combined with social network analysis provides fine-grained insights into water infrastructure planning processes', *Journal of environmental management*, 125, pp. 134–148.
[35] Wu, W. *et al.* (2020) 'Stakeholder analysis and social network analysis in the decision-making of industrial land redevelopment in China: The case of Shanghai', *International Journal of Environmental Research and Public Health*, 17(24), p. 9206.
[36] He, J. *et al.* (2024) 'Measuring social network influence on power relations in collaborative planning: A case study of Beijing City, China', *Cities*, 148, p. 104866.
[37] Patton, M.Q. (2011) *Essentials of utilization-focused evaluation*. Sage Publications.
[38] Ackermann, F. and Eden, C. (2011) 'Strategic management of stakeholders: Theory and practice', *Long range planning*, 44(3), pp. 179–196
[39] Pratama, I. P., Winarso, H., Hudalah, D., & Syabri, I. (2023). Social relations in megaproject planning: exploring the roles of local government in the development of Indonesia's special economic zones. *Town Planning Review*, *94*(5), 513-533.
[40] Affandi, R. A., Purnama, I. F., Noor, M. M. M., Permana, A., & Kembaren, E. E. (2025). Sustainable and equitable development through SEZ in Indonesia. In *IOP Conference Series: Earth and Environmental Science* (Vol. 1441, No. 1, p. 012001). IOP Publishing.

IOP Conf. Series: Earth and Environmental Science **1503** (2025) 012002 doi:10.1088/1755-1315/1503/1/012002

Asymmetric Power Among Actors on Landscape Transformation

(a Study of Ecosystem Degradation)

Heru Purwandari[1*], Arya Hadi Dharmawan[2], Ekawati Sri Wahyuni[3], and Satyawan Sunito[4]

[1234] Department Communication Sciences and Community Development, Bogor Agricultural University, Bogor, Indonesia

*E-mail: heru_purwandari@apps.ipb.ac.id

Abstract. Resource contestation leads to the dominance of a certain actor's power over another as a consequence of asymmetric power relations. Rural areas are arenas where various powers take place and have an impact on the transformation of rural landscapes. Landscape transformation is often preceded by an actor accessing certain resources along with the process of exclusion for other groups. This exclusion process involves various power mechanisms, such as regulation, force, the market, and legitimation. Actors interested in certain resources will apply one or more powers (a bundle of powers). This study focuses on how power relations contribute to landscape transformation. This study aims to analyze the power that transforms agricultural villages into mining villages and its impact on ecological functions. This study began by comparing landscape maps from 2005 to 2023 to analyze the land-use changes. The method applied using GIS by performing geo-spatial mapping, in-depth interviews, and organizing FGDs with the local community to gain power relations descriptions among actors. The results show that land transfer occurs through various powers depending on the period and circumstances that encourage people to release their land. This process provides opportunities for specific actors to utilize and transform the landscape according to their interests. Rice fields receive indirect and direct impacts from mining activities; therefore, the function of this ecosystem is collapsing, and the ecological function decreases dramatically.

Keywords: access, asymmetric power, bundle of power, landscape transformation

1. Introduction

The transformation of landscapes has been a focal point of environmental studies because it encapsulates the dynamic interplay between natural ecosystems and human activities [1]. Within this domain, the concept of asymmetric power among actors [2] emerges as a critical factor influencing the trajectory of landscape change. This study investigates the complex mechanisms through which varying degrees of influence and authority among stakeholders drive ecosystem degradation [3].

Landscape transformation often results from complex interactions among diverse actors, including governments, corporations, local communities, and non-governmental organizations [4]. Thus, actor mapping plays a crucial role in identifying the specific interests of each actor.

Regarding the socio-ecological context, this research examines several actors involved in the land transfer and mining extraction process. Rinku *et al.* explain a multi-scalar approach that simultaneously examines processes at different scales to explain transformations in urban migration landscapes [5]. These entities possess different levels of power, resources, and interests, which shape their capacity to affect environmental outcomes [6]. The asymmetry in power relations leads to uneven impacts on ecosystems, as those with greater influence can steer decisions and policies that favour their objectives, often at the expense of environmental sustainability and the well-being of less powerful groups [7]. The difference in access and power among actors has two consequences: inclusion for powerful groups and exclusion for the less powerful groups especially farmers [8].

This study explores the manifestations and consequences of such power imbalances on landscape transformation. It examines case studies of ecosystem degradation, highlighting how powerful actors exploit natural resources, create land-use changes, and influence environmental governance [9]. By analysing these dynamics, this study aims to uncover the underlying causes of ecological decline and identify potential pathways for more equitable and sustainable landscape management.

Understanding the role of asymmetric power is crucial for developing effective strategies to mitigate ecosystem degradation in the future. Ecosystem degradation often follows the decline in ecosystem services. Land that previously had complex ecosystem functions suddenly lost its function due to large-scale land clearing. Tania Li's article on Upland shows the disruption in the mountainous area caused by the entry of new actors [10]. A similar phenomenon occurred at the study location, where the land-clearing process for quarries caused the river to dry up and the destruction of the rice field area. For years, intensive mining exploitation has not only reduced the productivity of rice field areas but also significantly changed the entire local landscape.

This research not only contributes to the academic discourse on environmental governance and power dynamics but also provides insights for policymakers, conservationists, and community leaders striving to foster more balanced and inclusive approaches to landscape stewardship in the region. Through a comprehensive analysis of power asymmetries, this study seeks to illuminate the path towards healthier and more resilient ecosystems in the face of ongoing environmental challenges. Therefore, this research intends to answer several research questions: How does landscape transformation occur? To what extent do landscape transformations involve various actors? How do actors build relationships with others to transform the landscape?

2. Theoritical Framework

The transformation of the landscape has consequences for shifting actor interests. These changes in interest are reflected in the land-use patterns. Borras and Franco suggest that landscape changes are related to shifting land-use patterns. His research shows that land transformation consists of four types: land intended for food changes to non-food use, land originally used for food changes to fuel purposes, land used for non-food uses changes to food use, and non-food land changes to fuel sources[11].

The following literature highlights the importance of understanding the role of asymmetric power relationships among actors in shaping landscape transformation and ecosystem degradation. Integrating social-ecological perspectives, incorporating stakeholder values, and adopting adaptive management approaches can help address these complex dynamics and support more sustainable ecosystem management.

IOP Conf. Series: Earth and Environmental Science **1503** (2025) 012002 doi:10.1088/1755-1315/1503/1/012002

Ecosystem degradation is a pressing global issue, with over 60% of ecosystems being managed unsustainably[12]. Understanding the complex dynamics and drivers of ecosystem degradation is crucial for effective management and restoration efforts [13]. One key factor that has been identified is the role of asymmetric power relationships among different actors in shaping landscape transformation. Power dynamics can influence the access and distribution of ecosystem services, leading to trade-offs between different stakeholders. Empowered stakeholders may have the ability to regulate access to ecosystem services, negatively impacting less powerful groups [14].

Ecosystem-based approaches have been proposed to address these social-ecological dynamics, but they often lack explicit consideration of power dynamics. Integrating insights from social-ecological systems research can help us better understand how power relationships shape the provision and distribution of ecosystem services [15]. Several studies have highlighted the importance of incorporating societal preferences and values in the valuation of ecosystem service [16], [17]. This can help to identify relevant services for different stakeholders, potential social conflicts, and trade-offs among ecosystem services. Visualizing these trade-offs can serve as a tool for decision-making and identifying the impacts of different management options [16].

However, there are limited discussions of asymmetric power in examining the small scale of the mining industry. Conversely, asymmetric power relations play a significant role in shaping landscape transformations. Power imbalances between different stakeholders, such as the state, developers, and local communities, can lead to conflicts over the use and management of shared landscape resources[18]–[20].

Furthermore, the literature suggests that asymmetric power relations can also manifest in the negotiation of access rights to land and natural resources, such as recreational access to inland waterways in England. Legal and institutional frameworks governing land tenure and property rights can create power imbalances that disadvantage certain groups, such as anglers and boaters [21]. In addition, the literature explores how asymmetric power relations can shape the transformation of urban landscapes, such as through processes of gentrification and the displacement of marginalized communities. The cultural and economic forces that drive these landscape changes are often rooted in power imbalances and uneven distribution of resources and influence [2].

Landscape transformation, followed by the degradation of ecosystem quality, is also close to the concept of the right to access versus the power to access. In contrast to the right to access, which relies on formal property issues and authority [22], access is defined as the ability of an individual or group to derive benefits from thing [23] and rely on the bundle of power. By activating their power, individuals or institutions can exclude other actors through regulation, force, the market, and legitimation [24]. Meanwhile, research on the impact of landscape changes on livelihood systems has been widely studied in sociology. Livelihood resources (natural, financial, human, and social capital) are the starting point for assessing a household's ability to activate its assets to respond to landscape change [25]. In short, the livelihood system developed after landscape transformation is a reflection of the household's adaptive capacity,

According to the discussion above, it is very important to seek an explanation of the emergence of asymmetric power through the use of livelihood capital owned by each actor. Thus, the orchestration of the concepts of power and livelihood assets is expected to answer the process of landscape transformation in rural areas.

IOP Conf. Series: Earth and Environmental Science **1503** (2025) 012002 doi:10.1088/1755-1315/1503/1/012002

3. Method and materials

Batujajar is an andesite mining located in Cigudeg District, Bogor Regency, West Java, approximately 30 km south of Jakarta, the capital city of Indonesia (as shown in Figure 1). The type of mining is an open pit and exploit andesite 30 m until under the surface (as shown in Figure 1). Within almost forty years, Batujajar Villages experienced the expansion of the andesite mining industry that transformed agriculture into a mining landscape.

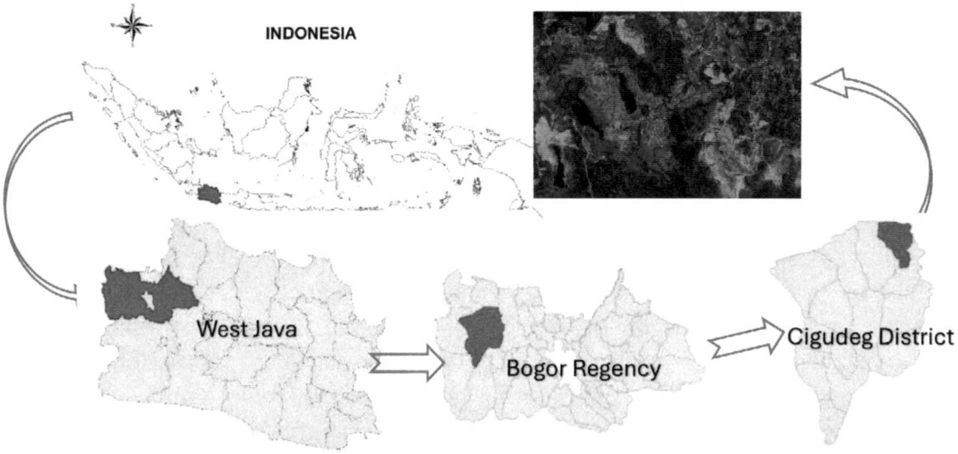

Figure 1. Location of Batujajar Mine (map resources from Indonesia-geospatial; satellite image from Google Earth).

To answer the research questions that have been formulated, a qualitative method was conducted by focusing on a specific strategy to obtain the data. Primary data were obtained from in-depth interviews, especially to acquire deep information regarding actors and power relations among actors. Socio-ecological mapping information was collected using transect and map analysis methods. Secondary data from the Central Bureau of Statistics (Indonesia-BPS) and Google Earth were also obtained to understand the comprehensive landscape transformation. To gain comprehensive information, the research was conducted over 3 months from June 20 to September 17, 2024.

The research was conducted in two stages. First, to obtain information about landscape changes, map analysis was conducted by analyzing Indonesian geospatial maps combined with Google Earth to visualize changes in the landscape at the study location. Transect strategies were also used to gain depth into the landscape changes and the process of how they occurred. The second stage focused on qualitative methods: FGD and in-depth interview activities to understand how the landscape transformation process is the result of unequal relations between actors. Key persons who are the main source of information are dry land owners whose land has been transferred to companies, formal and informal leaders, mining actors, middlemen involved in the process of land transfer and marketing of mining material, and households that rely on mining activities, agricultural activities, or a combination of both (as shown in Table 1).

Table 1. Types and source of data, key persons, and information.

Information	Types and source of data	Key persons
Land use change/landscape transformation	Primary: Transect Secondary: Indonesian-geospatial and Google Earth	Formal and informal leader, local people
Ecosystem degradation	Primary: observation, in-depth interviews, FGD	Formal-informal leader, local people
Actor analysis	Primary: observation, in-depth interview, FGD	Land owner, government, mining actor, farmer, middle man
Power relations	Primary: in-depth interviews, FGD	Mining actors, middle man

4. Results and discussions

The research findings focus on three aspects: landscape transformation and how it affects ecosystem services, land transfer process, and the mining extraction process, particularly emphasizing the involvement of actors, including power relations among them in both processes.

4.1 Landscape transformation and ecosystem services

Significant changes have occurred in the last 20 years. Land-use changes in 2005 were compared with the data from 2022. A visual representation of the landscape transformation is shown in Figure 2. The expansion of extractive mining activities continues at 14 quarry locations (black shading on the map) and converts the 200ha agricultural landscape that was initially utilized for agriculture.

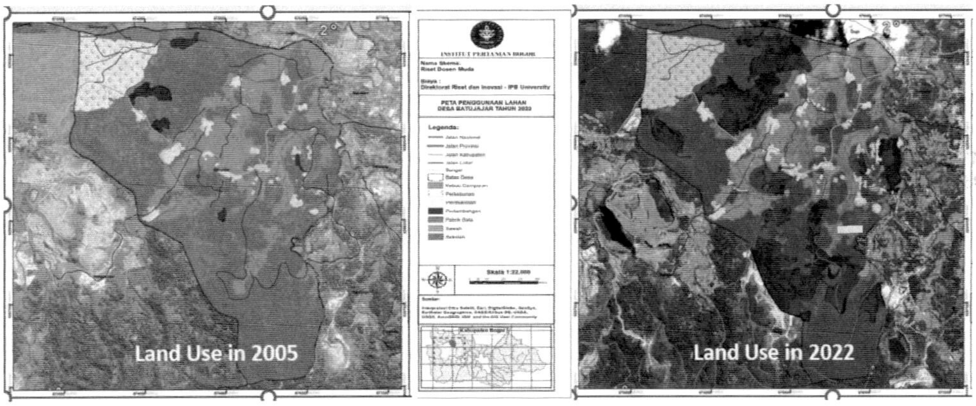

Figure 2. Land-use changes in Batujajar Mining Village between 2005 and 2022. Source: Secondary data, processed, 2023.

As land conversion continues, the environmental quality in Batujajar is deteriorating, particularly in the agricultural sector. There are socio-economic and ecological impacts of landscape transformation. The land-clearing process for opening a quarry significantly transforms the characteristics of the local landscape. In mining industrial activities, land is utilized for processing material into various sizes. Loading activities also take place on the site, and the material is distributed from the quarry to the consumers. This condition creates an environmental crisis, especially related to high air and noise pollution due to passing trucks.

Another problem is flooding for residents living along the road. Trucks carrying stone materials require strong concrete road conditions. Therefore, the road always increases in height and ultimately exceeds the height of the houses.

The opening of mountainous areas causes critical issues in agricultural land. Water flows are limited, along with the disruption of agricultural and domestic water functions on the river. Rice fields are unproductive and trigger the land conversion process. As a result, local communities are facing vulnerabilities in developing livelihood systems. Local people require a strategy to adapt to the new character of the local livelihood system. The impacts of the mining industry's socio-economic and ecological factors are shown in Figure 3.

Figure 3. Socio-economic and ecological impacts of andesite mining industries in Batujajar Village

4.2 Land transfer processs

The mechanism of power relations and its influence on asymmetric power should be identified from the process of how land has been transferred to investors. Figure 4 shows that the ability of each actor to access resources is quite different. According to Scoones [25] there are four livelihood capitals/assets that should be activated to support livelihood strategies. First, natural capital is in the form of natural resources (e.g., soil and water). Second, economic or financial capital is needed to mobilize other resources. Third, human capital is in the form of education, skills, or knowledge. Fourth, social capital is in the form of social relations and social networks. The latter is particularly important for people living in egalitarian societies [26].

In this study, we apply this approach to analyze the presence of asymmetric power. Figure 4. shows that there are different qualities of relationships among actors. Village and supra-village governments create a strong relationship to access land owned by local people. These actors have almost all types of capital required to access land. By utilizing their livelihood resources, such as human, social, financial, and physical capital, these actors have a great opportunity to access natural capital. In contrast, local people only have human (low quality) capital and natural capital for survival. Limited knowledge of the value of these resources causes the land transfer process to be effortless. The combination of the power of corporations and the government leads to the marginalization of local people.

IOP Conf. Series: Earth and Environmental Science **1503** (2025) 012002 doi:10.1088/1755-1315/1503/1/012002

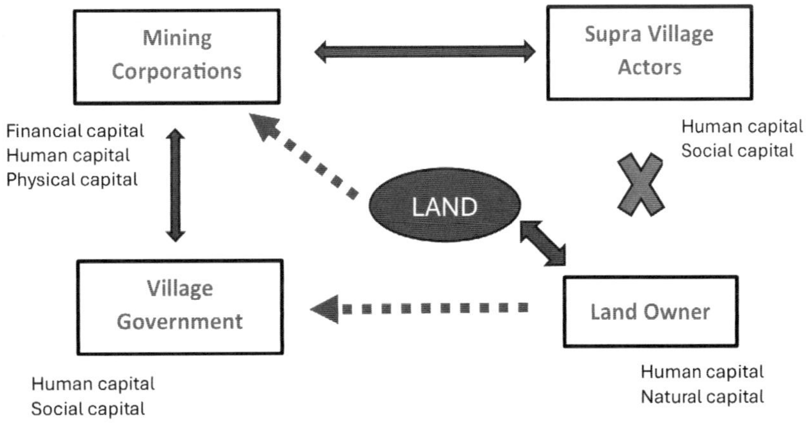

Figure 4. Utilization of livelihood capital among actors and power relations of land transfer process in Batujajar Mining Village

4.3 Mining extraction process

Basically, there are quite similar actors involved in the mining extraction process. The crucial actors that play an important role in this stage are the presence of "bogeg" and "berakan". Bogeg plays a liaison role in connecting companies and consumers. The benefit that "bogeg" gains is price reduction from the company, and this would be their profit margin. On the other hand, "berakan" is the reseller that supplies material tambang to direct consumers or building material shops. "Berakan" will buy stone from the company and convert it into small pieces for consumer purposes. They employ people who are excluded from their land as unskilled labor. In the situation of limited land ownership and the decreasing quality of agricultural land, job opportunities in the mining industry sector are the exit strategy for the farmers. However, this opportunity is very restricted, especially for those who intend to work as professional employees. Local people require more effort to meet their living needs. Selling unproductive agricultural land as the need for housing increases is a choice for some households.

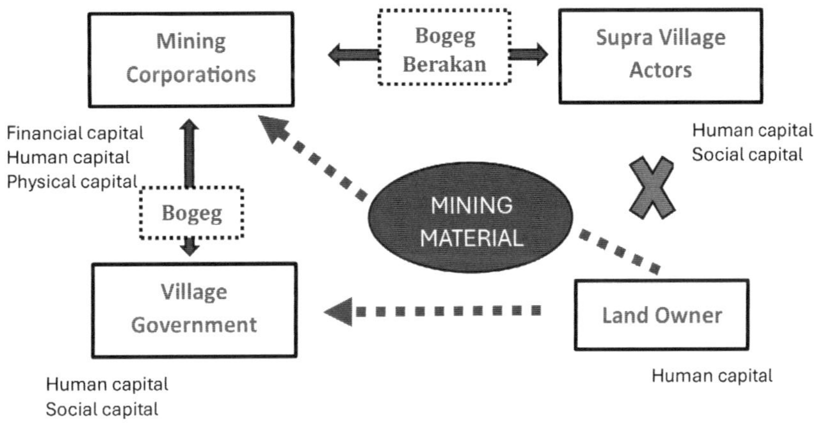

Figure 5. Utilization of livelihood capital among actors and power relations of mining extraction process in Batujajar Mining Village

It is obvious that control of specific livelihood capital (social, financial, physical, human, and natural capital) leads to the dominance of specific actors (as shown in Figure 6). Only those who seize opportunities can adapt to the socio-economic and ecological vulnerabilities caused by the presence of mining industries. This is because a limited number of people are involved in the mining industry, and the majority are excluded from their livelihood system. It is clear that in this resource contestation that leads to landscape transformations, corporations gain more access, local government and middlemen gain more but less than corporations, the community gains less access, while farmers are excluded from socio-economy and environmental resources.

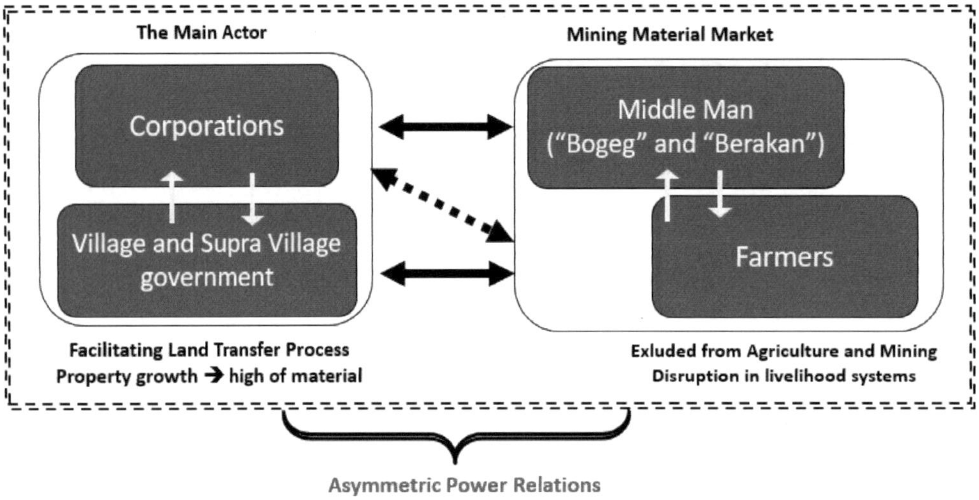

Figure 6. Asymmetric power among actors in Batujajar Mining Village

5. Conclusions

The land transfer process initiates landscape transformation. However, land-use changes are not immediately visible when the land transfer process occurs. The land-clearing process in the hilly area takes at least five years before the company starts the intensive extraction of andesite mining. Land transformation reaches its peak when 14 quarries began operating. This condition triggers the transformation of agricultural land for other purposes. Therefore, previously fertile rice fields are experiencing land conversion, especially for buildings and housing. Environmental changes encourage the decline of ecosystem services as a direct impact of transformations. The transformation trajectory for both dryland and wetland shows a food to non-food transformation trajectory type.

Each transformation stage involves different actors. In the land transfer process, village and supra-village governments collaborate to provide a mechanism for mining investors to access local resources. Meanwhile, landowners face the reality of continuous land purchase offers that are captured by the owner. Even though this village is not located in a very remote area, when someone wants to buy a large area, the amount of money obtained is very large for the local people. The second stage of the transformations shows that there are additional actors in the mining extraction process. When the land is operated intensively, the actors' orientation is to take maximum profit from the material. The involvement of "bogeg" and "berakan" as middlemen in the marketing process is important for connecting producers and consumers.

Applying the livelihood capital approach as a tool to identify unequal power relations among actors is the core of this research. Each actor utilizes their capital and exercises power to access and use resources for their own interests. Each actor maximizes their "capital" to gain more benefits. This process creates landscape transformation, which reflects the unequal power relations among actors. Ultimately, this unequal power leads to the simultaneous occurrence of exclusion and inclusion in different groups.

Acknowledgements

This study is part of the Doctoral Degree by Research Program of Rural Sociology Program Study, Bogor Agricultural University (IPB University). Funded by Bogor Agricultural University (IPB University) through Young Lecturer Research Scheme, Contract Number: 11423/IT3/PT.01.03/P/B/2023. The authors thank to IPB University.

References

[1] H. Nguyen *et al.*, "Transformation of Rural Landscapes in the Vietnamese Mekong Delta from 1990 to 2019: a Spatio-Temporal Analysis," *Geocarto Int.*, vol. 37, no. 26, pp. 13881–13903, 2022, doi: 10.1080/10106049.2022.2086623.

[2] S. Zukin, "Changing landscapes of power: Opulence and the urge for authenticity," *Int. J. Urban Reg. Res.*, vol. 33, no. 2, pp. 543–553, 2009, doi: 10.1111/j.1468-2427.2009.00867.x.

[3] H. Yu and I. Zahidi, "Assessing Environmental Degradation and Restoration: A Case Study of the Penjom Mine in Malaysia," *12th IGRSM Int. Conf. Exhib. Geospatial Remote Sens.*, vol. 1412, no. Earth and Environmental Science, pp. 1–11, 2024, doi: 10.1088/1755-1315/1412/1/012001.

[4] M. Antrop and V. Van Eetvelde, "Mechanisms in recent landscape transformation," in *WIT Transactions on the Built Environment*, 2008, pp. 183–192. doi: 10.2495/GEO080181.

[5] R. R. Chowdhury *et al.*, "A Multi-Scalar Approach to Theorizing Socio-Ecological Dynamics of Urban Residential Landscapes," *Cities Environ.*, vol. 4, no. 1, pp. 1–21, 2011, doi: 10.15365/cate.4162011.

[6] C. Paludan-Müller, "3. Actors and orders: the shaping of landscapes and identities," in *The Cultural Landscape and Heritage Paradox*, 2019. doi: 10.1515/9789048510962-005.

[7] G. Bächler, "Why environmental transformation causes violence: A synthesis," *Environ. Change Secur. Proj. Rep.*, vol. 4, pp. 24–44, 1998.

[8] D. Hall, P. Hirsch, and T. M. Li, *Introduction to Powers of Exclusion: Land Dilemmas in Southeast Asia*. 2011.

[9] U. Kasymov and D. Zikos, "Understanding Human Actions and Institutional Change: What Are the Impacts of Power Asymmetries on Efficiency in Pasture Use?," *Resources*, vol. 6, no. 4, p. 71, 2017, doi: 10.3390/resources6040071.

[10] T. M. Li, *Transforming the Indonesian uplands: marginality, power and production*. London: Taylor & Francis e-Library, 2000. doi: 10.5860/choice.37-2860.

[11] S. M. Borras and J. C. Franco, "Global land grabbing and trajectories of Agrarian change: A preliminary analysis," *J. Agrar. Chang.*, vol. 12, pp. 34–59, 2012, doi: 10.1111/j.1471-0366.2011.00339.x.

[12] M. D. Robards, M. L. Schoon, C. L. Meek, and N. L. Engle, "The importance of social drivers in the resilient provision of ecosystem services," *Glob. Environ. Chang.*, vol. 21, no. 2, pp. 522–529, 2011, doi: 10.1016/j.gloenvcha.2010.12.004.

[13] A. K. Jara-Guerrero, D. Maldonado-Riofrío, C. I. Espinosa, and D. H. Duncan, "Beyond the blame game: A restoration pathway reconciles ecologists' and local leaders' divergent models of seasonally dry tropical forest degradation," *Ecol. Soc.*, 2019, doi: 10.5751/ES-11142-240422.

[14] M. R. Felipe-Lucia, B. Martín-López, S. Lavorel, L. Berraquero-Díaz, J. Escalera-Reyes, and F. A. Comín, "Ecosystem services flows: Why stakeholders' power relationships matter," *PLoS One*, 2015, doi: 10.1371/journal.pone.0132232.

[15] K. Benessaiah and R. Sengupta, "How is shrimp aquaculture transforming coastal livelihoods and lagoons in Estero Real, Nicaragua?: The need to integrate social-ecological research and ecosystem-based approaches," *Environ. Manage.*, 2014, doi: 10.1007/s00267-014-0295-x.

[16] B. Martín-López *et al.*, "Uncovering ecosystem service bundles through social preferences," *PLoS One*, 2012, doi: 10.1371/journal.pone.0038970.

[17] B. A. Bryan, C. M. Raymond, N. D. Crossman, and D. H. Macdonald, "Targeting the management of ecosystem services based on social values: Where, what, and how?," *Landsc. Urban Plan.*, 2010, doi: 10.1016/j.landurbplan.2010.05.002.

[18] A. Nadaï and O. Labussière, "Wind power planning in France (Aveyron), from state regulation to local planning," *Land use policy*, vol. 26, no. 3, pp. 744–754, 2009, doi: 10.1016/j.landusepol.2008.10.018.

[19] A. Nadaï and O. Labussière, "Landscape commons, following wind power fault lines. The case of Seine-et-Marne (France)," *Energy Policy*, vol. 109, pp. 807–816, 2017, doi: 10.1016/j.enpol.2017.06.049.

[20] C. Zografos and J. Martinez-Alier, "The politics of landscape value: A case study of wind farm conflict in rural Catalonia," *Environ. Plan. A*, vol. 41, no. 7, pp. 1726–1744, 2009, doi: 10.1068/a41208.

[21] A. Church, P. Gilchrist, and N. Ravenscroft, "Negotiating recreational access under asymmetrical power relations: The case of inland waterways in England," *Soc. Nat. Resour.*, vol. 20, no. 3, pp. 213–227, 2007, doi: 10.1080/08941920601117298.

[22] T. Sikor and C. Lund, "Access and property: A Question of Power and Authority," in *The Politics of Possession: Property, Authority and Access to Natural Resources*, 2010. doi: 10.1002/9781444322903.ch1.

[23] J. C. Ribot and N. L. Peluso, "A theory of access," *Rural Sociol.*, vol. 68, no. 2, pp. 153–181, 2003, doi: 10.1111/j.1549-0831.2003.tb00133.x.

[24] D. Hall, P. Hirsch, and T. M. Li, "Powers of Exclusion: Land Dilemmas in Southeast Asia. Challenges of the Agrarian Transition in Southeast Asia," *Univ. Hawai'i Press. Honolulu*, vol. 34, no. 1, pp. 140–142, 2011.

[25] I. Scoones, "Sustainable Rural Livelihoods: a Framework for Analysis," *IDS Work. Pap.*, vol. 72, no. 22, 1998.

[26] A. H. Dharmawan and W. Manig, "Livelihood Strategies and Rural Changes in Indonesia : Studies on Small Farm Communities," *Dtsch. Tropentag*, no. Assessment of Poverty and Livelihood Strategies, pp. 1–9, 2000.

[27] A. H. Dharmawan, "Sistem Penghidupan dan Nafkah Pedesaan: Pandangan Sosiologi Nafkah (Livelihood Sociology) Mazhab Barat dan Mazhab Bogor," *Sodality J. Transdisiplin Sosiologi, Komunikasi, dan Ekol. Mns.*, vol. 01, no. 02, pp. 169–192, 2007, doi: 10.22500/sodality.v1i2.5932.

Household air pollution, birth weight, and children's respiratory health: trends and associations in Indonesia

Made Ayu Hitapretiwi Suryadhi

Department of Public Health and Preventive Medicine, Faculty of Medicine, Udayana University, Denpasar, Indonesia

*E-mail: hitapretiwi_suryadhi@unud.ac.id

Abstract. Exposure to household air pollution (HAP) during pregnancy and early childhood is associated with a range of adverse health outcomes. This study assessed the trend and association of HAP exposure with health outcomes based on residency in Indonesia. This study used the three latest demographic and health surveys for Indonesia (IDHS) to obtain information on mothers and children aged under five. The outcome of this study included low birth weight (LBW) and acute lower respiratory tract infection (ALRTI). The main HAP exposure included polluting fuel use and secondhand smoke (SHS). Trends, descriptive statistics, and odds ratios (OR) with a 95% confidence interval (CI) presented the results of this study. The logistic regression assessed the effect of exposure to the combined effects of polluting fuel, SHS, and kitchen location on LBW and ALRTI. The trend of polluting fuel use and SHS dwindled from 2007 to 2017, with the highest decrease observed for polluting fuel use (by more than 60%). Even so, polluting fuel use remained above 30%, whereas SHS exposure remained above 60% among rural dwellers. The latest IDHS (2017) showed that the exposure to SHS or polluting fuel and, in combination, significantly associated with LBW among rural mothers (OR 1.61, 95% CI 1.15 - 2.24; OR 1.77, 95% CI 1.19 - 2.64; and OR 1.60, 95% CI 1.13 - 2.27, respectively). Exposure to SHS or combined with polluting fuel was significantly associated with ALRTI among rural children aged under five (OR 1.51, 95% CI 1.13 - 2.02 and OR 1.61, 95% CI 1.18 - 2.19, respectively). The findings of this study for LBW and ALRTI, mainly among rural mothers and children aged under five, show an urgent need for targeted awareness and stricter enforcement of clean energy adoption regulations to mitigate health risks.

1. Introduction

Activities mostly spent indoors, such as sleeping, cooking, cleaning, heating, tobacco smoking, human occupancy, and building and furniture materials, may increase household air pollution (HAP) exposure (1, 2). Exposure to HAP during pregnancy and early childhood has been associated with several adverse health effects, such as miscarriage, stillbirth, and low birth weight (LBW), as well as deaths due to lower respiratory infection (3).

IOP Conf. Series: Earth and Environmental Science **1503** (2025) 012003 doi:10.1088/1755-1315/1503/1/012003

Regional reports indicate that the annual rates of LBW are highest in the Asian region at 20.2%, followed by the Central American region at 15.6%. In contrast, the African regions exhibit the lowest rates at 3.9% (4). Studies have shown the association between LBW and several adverse health effects in later life, such as stunting, lower IQs, diabetes, and cardiovascular diseases (5-7). As of 2023, Indonesia is recognized as one of the most densely populated countries globally, with a population surpassing 280 million individuals (8). There is growing emphasis on addressing stunting in Indonesia, as the country has reached a significant threshold indicating a very high prevalence (9). LBW is one prominent risk factor for other morbidities and mortalities in later life.

As for tobacco smoking, the latest Indonesian Global Adult Tobacco Survey (GATS) reported that 70.2 million adults, or 34.5%, used tobacco in 2021. Also, the Indonesian GATS reported that 121.6 million adults, or 59.3%, were exposed to indoor secondhand smoke (SHS) in 2021 (10). SHS is a prominent contributor to the global burden of disability-adjusted life years (DALYs) among young children (11).

For this reason, this study sought to analyze the trend and association of exposure to HAP and LBW and acute lower respiratory tract infection (ALRTI) in three Indonesian Demographic Health Survey (IDHS) waves.

2. Methods

2.1 Data and sample

This study used datasets from the 2007, 2012, and 2017 IDHS. The initial survey was conducted in 1987 and was originally titled the National Indonesia Contraceptive Prevalence Survey (NICPS); subsequent surveys have since been referred to as the IDHS. The IDHS collected extensive data on fertility, family planning, and maternal and child health. Participants were chosen using a two-stage stratified sampling technique. The survey was prepared rigorously in each wave to ensure reliable indicators and estimates with acceptable precision. This study did not include survey waves before the 2007 IDHS due to the absence of information on kitchen location and SHS. For the statistical analysis, KR and HR files were merged in each survey wave, then restricted to singleton births and dejure residences, which resulted in a total number of 13,367; 14,527 and 15,553 births and 4,662; 4,970; 5,639 children under five years old included from the 2007; 2012 and 2017 IDHS, respectively. The KR and HR files provided information on children and households. The dataset is available at the DHS Program after registering and granted approval (website: https://dhsprogram.com/).

2.2 Variable description

Dependent variables in this study included LBW and ALRTI. The IDHS collected information on the weight of live-born children five years preceding each wave, assessed by written records or a mother's report. For the statistical analysis, LBW was categorized dichotomously (Yes or No) and defined as a weight of less than 2.5 kilograms at birth. Similarly, ALRTI was categorized dichotomously (Yes or No) and described as children aged under five who experienced a cough with fast, short, rapid breaths or felt difficulty breathing in the last 2 weeks.

Cooking fuel use and SHS were the two main predictors in this study, which were obtained in the IDHS by surveying the type of fuel mainly used for cooking in a household. For the analysis, cooking fuel was defined as a dichotomous variable: "polluting fuel (kerosene, coal/lignite, charcoal, wood, straw/shrubs/grass, agricultural crop, and animal dung)" and "clean fuel (electricity, liquified petroleum gas (LPG), natural gas and biogas)." This category has been used in previous studies (12, 13). Another predictor in this study was SHS, which was sourced from the

IDHS based on the frequency of individuals smoking within the household. To account for the sample sizes associated with the weekly, monthly, and less than monthly categories, SHS was classified as a dichotomous variable consisting of "yes (daily)" and "no (weekly, monthly, less than monthly, and never)."

2.3 Statistical analysis

This study assessed residence differences in polluting fuel use, SHS, LBW, and ALRTI. The trend was presented using a bar graph for each survey wave (Figures 1 and 2). Descriptive statistics were used to show sociodemographic and kitchen locations in each survey wave (Table 1). Logistic regression was performed to evaluate the association between exposure and health outcomes, which are reported as odds ratios (ORs) with a 95% confidence interval (CI). The no-exposure category served as the reference group in the analysis (Table 2 and Table 3). First, the combined effect of exposure was assessed by categorizing the exposure into four categories: exposure to none, exposure to SHS, exposure to polluting fuel, and exposure to both (SHS and polluting fuel). Second, the combined effect assessed in this study included exposure to polluting fuels and the location of the kitchen, specifically whether the residence has a separate room designed for kitchen use. Previous studies have considered kitchen location as a significant factor when analyzing the impact of cooking fuel usage on health outcomes, including LBW (14, 15). The corresponding four categories were: has a separate kitchen and no exposure to polluting fuel, no separate kitchen, exposure to polluting fuel, and has no separate kitchen and has exposure to polluting fuel. The regression analysis for LBW was adjusted for the gender of the child, age at the time of birth, and education of the mother. Whereas the regression analysis for ALRTI adjusted for the age and gender of the child and the education of the mother. Finally, the sensitivity analysis was performed by adjusting for wealth index (i.e., poor, middle, and rich) in place of the mother's education level as a socioeconomic indicator (Supplemental Tables). The complex survey design in the IDHS was considered by applying weights to the analysis. The results were generated using Microsoft Excel software to produce the figures and STATA Software 15.1 for the statistical analyses.

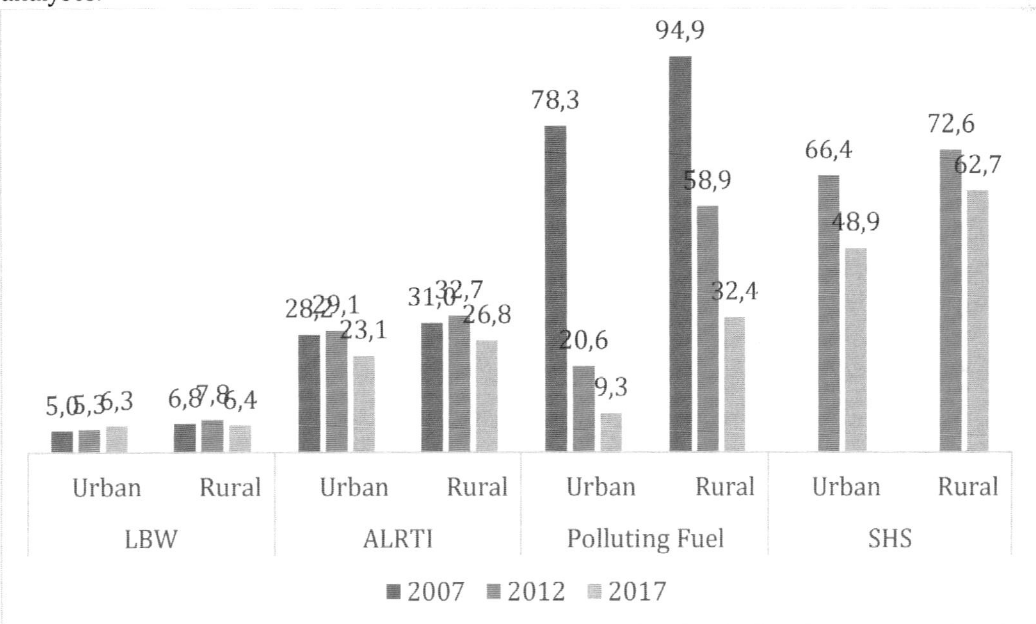

Figure 1. LBW, ALRTI, polluting fuel use, and SHS trend in Indonesia (2007 to 2017).

3. Results

The prevalence of ALRTI slightly increased from 2007 to 2012 and decreased from 2012 to 2017 among children aged under five living in the urban and rural. Concerning HAP, despite the prevailing results for polluting fuel use among rural dwellers at above 30% in 2017, there was a sharp decrease in polluting fuel use among urban and rural dwellers from 2007 to 2017. Also, SHS exposure was reduced among the urban and rural areas but remained above 60% among the rural areas (Figure 1).

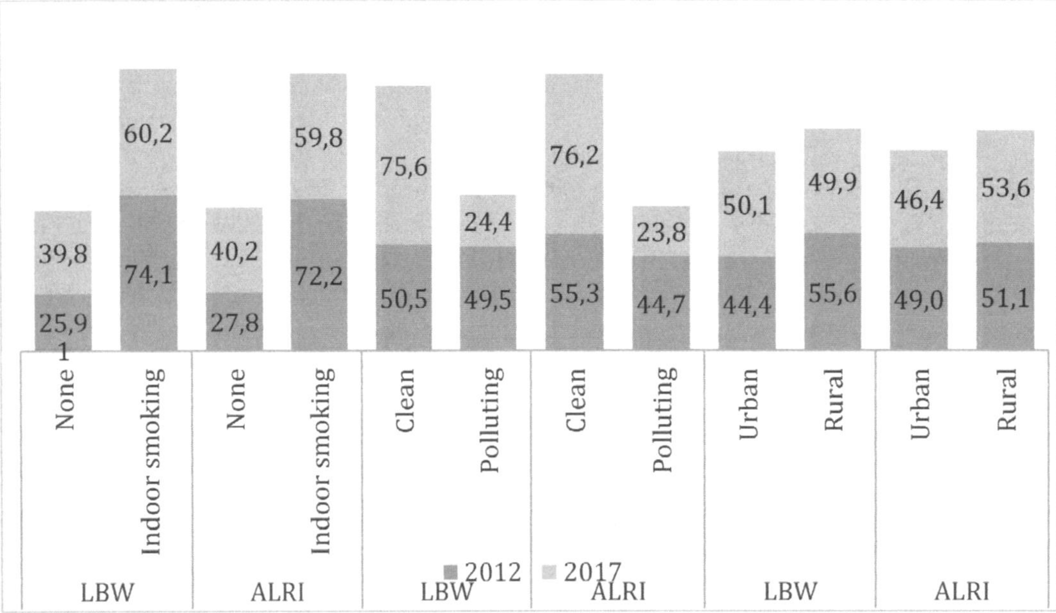

Figure 2. The prevalence of LBW and ALRTI according to residency and exposure to HAP in Indonesia (2012 and 2017).

Figure 2 presents the prevalence of LBW and ALRTI according to residency and exposure to HAP in Indonesia from 2012 to 2017. Here, results for 2007 were not included due to missing information on SHS. The figure shows that the cumulative sum prevalence of LBW was higher due to HAP from SHS (74,1% in 2012 and 60.2% in 2017) and polluting fuel use (49.5% in 2012 and 24.4% in 2017) compared to ALRTI (72.2% in 2012 and 59.8% in 2017 for SHS and 44.7% in 2012 and 23.8% in 2017 for ALRTI) (Figure 2). Rural mothers and children were also shown to have higher prevalences of LBW (55.6% in 2012 and 49.9% in 2017) and ALRTI (51.1% in 2012 and 53.6% in 2017) compared to the urban (44.4% in 2012 and 50.1% in 2017 for LBW and 49.0% in 2012 and 46.4% in 2017 for ALRTI).

Table 1. Characteristics of subjects and kitchen location according to residency in Indonesia (2007 to 2017).

Variable	2007			2012		
	U[a]	R[b]	T[c]	U[a]	R[b]	T[c]
Child's:						
Gender, n (%)						
Male	3294/6310 (52.6)	3795/7224 (51.8)	7089/13534 (52.2)	3863/7503 (50.0)	3730/7159 (52.1)	7593/14662 (51.0)

Female	3016/6310 (47.4)	3429/7224 (48.2)	6445/13534 (47.8)	3640/7503 (50.1)	3429/7159 (47.9)	7069/14662 (49.0)
Age, n (%)						
4	1275/6145 (20.6)	1386/6976 (19.3)	2661/13121 (19.9)	1422/7326 (19.5)	1384/6929 (18.7)	2806/14255 (19.1)
3	1184/6145 (19.0)	1340/6976 (19.1)	2524/13121 (19.0)	1469/7326 (19.9)	1286/6929 (18.4)	2755/14255 (19.2)
2	1170/6145 (19.2)	1380/6976 (20.0)	2550/13121 (19.6)	1451/7326 (19.7)	1319/6929 (19.8)	2770/14255 (19.7)
1	1217/6145 (19.3)	1409/6976 (20.1)	2626/13121 (19.7)	1466/7326 (20.0)	1459/6929 (21.7)	2925/14255 (20.8)
0	1299/6145 (21.9)	1461/6976 (21.5)	2760/13121 (21.7)	1518/7326 (21.0)	1481/6929 (21.5)	2999/14255 (21.2)
Mother's:						
Age, years (n (%))						
>=40	179/6310 (3.1)	272/7224 (3.5)	451/13534 (3.3)	247/7503 (3.7)	230/7159 (3.4)	477/14662 (3.6)
30-39	2183/6310 (34.4)	2214/7224 (31.6)	4397/13534 (32.9)	2642/7503 (35.4)	2220/7159 (30.7)	4862/14662 (33.2)
20-29	3559/6310 (56.5)	3895/7224 (52.5)	7454/13534 (54.4)	4110/7503 (54.3)	3915/7159 (55.1)	8025/14662 (54.7)
<20	389/6310 (6.0)	843/7224 (12.5)	1232/13534 (9.4)	504/7503 (6.7)	794/7159 (10.7)	1298/14662 (8.5)
Education Level, n (%)						
Higher school	937/6310 (13.8)	373/7224 (4.5)	1310/13534 (9.0)	1572/7503 (19.0)	661/7159 (7.6)	2233/14662 (13.7)
Secondary school	3932/6310 (59.7)	3459/7224 (45.3)	7391/13534 (52.2)	4599/7503 (60.0)	3670/7159 (51.2)	8269/14662 (55.9)
Primary school	1374/6310 (25.6)	3183/7224 (47.5)	4557/13534 (37.0)	1293/7503 (20.5)	2679/7159 (39.8)	3972/14662 (29.4)
No education	67/6310 (1.0)	209/7224 (2.7)	276/13534 (1.9)	39/7503 (0.5)	149/7159 (1.4)	188/14662 (0.9)
Kitchen located in a separate room, n (%)						
Yes	3702/5511 (72.5)	3702/5915 (69.6)	7404/11426 (71.0)	5157/6763 (79.8)	4524/6012 (77.0)	9681/12775 (78.5)
No	1809/5511 (27.5)	2213/5915 (30.4)	4022/11426 (29.0)	1606/6763 (20.2)	1488/6012 (23.0)	3094/12775 (21.5)

[a] Urban, [b] Rural, [c] Total

Table 1. *Continued.*

Variable	**2017**		
	U[a]	**R**[b]	**T**[c]
Child's:			
Gender, n (%)			
Male	4172/8126 (50.8)	3841/7512 (50.6)	8013/15638 (50.7)
Female	3954/8126 (49.2)	3671/7512 (49.5)	7625/15638 (49.3)

Age, n (%)			
4	1616/7940 (20.4)	1486/7338 (20.3)	3102/15278 (20.4)
3	1613/7940 (20.2)	1433/7338 (19.7)	3046/15278 (19.9)
2	1558/7940 (19.5)	1457/7338 (20.1)	3015/15278 (19.8)
1	1626/7940 (20.7)	1509/7338 (20.6)	3135/15278 (20.6)
0	1527/7940 (19.3)	1453/7338 (19.3)	2980/15278 (19.3)
Mother's:			
Age, years (n (%))			
>=40	364/8126 (4.3)	336/7512 (4.2)	700/15638 (4.2)
30-39	3331/8126 (41.5)	2762/7512 (35.6)	6093/15638 (38.6)
20-29	3934/8126 (48.5)	3716/7512 (49.9)	7650/15638 (49.2)
<20	497/8126 (5.8)	698/7512 (10.3)	1195/15638 (8.0)
Education Level, n (%)			
Higher school	1863/8126 (20.4)	1088/7512 (11.4)	2951/15638 (16.0)
Secondary school	4950/8126 (62.1)	4026/7512 (56.3)	8976/15638 (59.2)
Primary school	1282/8126 (17.1)	2304/7512 (31.5)	3586/15638 (24.2)
No education	31/8126 (0.4)	94/7512 (0.8)	125/15638 (0.6)
Kitchen located in a separate room, n (%)			
Yes	6341/7606 (83.9)	5377/6554 (84.5)	11718/14160 (84.2)
No	1265/7606 (16.1)	1177/6554 (15.6)	2442/14160 (15.9)

[a] Urban, [b] Rural, [c] Total

Table 1 provides information regarding the characteristics of subjects and the kitchen location based on residency. The mother's age at birth was mainly between 20 and 29 years in all IDHS waves, whereas urban mothers had higher numbers of higher school graduates than their rural counterparts (Table 1).

Table 2. Odds ratio for the association between exposure to SHS, polluting fuel, and kitchen location with LBW (OR, 95%CI).

Variables	2007			2012		
	U[a]	R[b]	T[c]	U[a]	R[b]	T[c]
Cooking fuel and secondhand smoke:						
None	-	-	-	Ref.	Ref.	Ref.
Secondhand smoke and no polluting fuel	-	-	-	1.23 (0.83 - 1.81)	0.89 (0.54 - 1.47)	1.09 (0.80 - 1.48)
Polluting fuel and no secondhand smoke	-	-	-	1.52 (0.84 - 2.73)	1.02 (0.63 - 1.66)	1.19 (0.82 - 1.71)

Secondhand smoke and polluting fuel	-	-	-	1.36 (0.86 - 2.13)	1.42 (0.91 - 2.23)	1.49 (1.10 - 2.03)
Cooking fuel and kitchen location:						
Separate kitchen and no polluting fuel	Ref.	Ref.	Ref.	Ref.	Ref.	Ref.
No separate kitchen and no polluting fuel	0.73 (0.28 - 1.87)	0.75 (0.22 - 2.61)	0.72 (0.33 - 1.54)	0.82 (0.52 - 1.29)	0.84 (0.44 - 1.63)	0.82 (0.56 - 1.19)
Separate kitchen and polluting fuel	1.12 (0.64 - 1.97)	1.17 (0.52 - 2.63)	1.14 (0.73 - 1.79)	1.15 (0.74 - 1.78)	1.26 (0.90 - 1.77)	1.18 (0.91 - 1.54)
No separate kitchen and polluting fuel	0.82 (0.45 - 1.51)	0.73 (0.32 - 1.65)	0.76 (0.47 - 1.23)	1.05 (0.53 - 2.08)	1.78 (1.09 - 2.92)	1.51 (1.02 - 2.24)

[a] Urban, [b] Rural, [c] Total

Table 2. *Continued.*

Variables	2017		
	U[a]	**R**[b]	**T**[c]
Cooking fuel and secondhand smoke:			
None	Ref.	Ref.	Ref.
Secondhand smoke and no polluting fuel	1.05 (0.82 - 1.35)	1.61 (1.15 - 2.24)	1.25 (1.02 - 1.51)
Polluting fuel and no secondhand smoke	1.27 (0.77 - 2.09)	1.77 (1.19 - 2.64)	1.44 (1.07 - 1.94)
Secondhand smoke and polluting fuel	1.18 (0.75 - 1.85)	1.60 (1.13 - 2.27)	1.32 (1.03 - 1.70)
Cooking fuel and kitchen location:			
Separate kitchen and no polluting fuel	Ref.	Ref.	Ref.
No separate kitchen and no polluting fuel	0.75 (0.52 - 1.07)	0.78 (0.49 - 1.24)	0.76 (0.57 - 1.02)
Separate kitchen and polluting fuel	1.06 (0.71 - 1.58)	1.05 (0.79 - 1.39)	1.04 (0.82 - 1.31)
No separate kitchen and polluting fuel	1.08 (0.54 - 2.15)	1.30 (0.79 - 2.16)	1.23 (0.82 - 1.84)

[a] Urban, [b] Rural, [c] Total

In 2007, no significant findings were found. In 2012, exposure to polluting fuels, in conjunction with the absence of a separate kitchen, was significantly associated with LBW among rural mothers (OR 1.78, 95% CI 1.09 – 2.92) compared to rural mothers without exposure to polluting fuel and had a separate room for the kitchen. On the other hand, an interesting finding is that exposure to polluting fuel and owning a separate room for the kitchen significantly associated with ALRTI among urban and rural children aged under five (OR 1.39, 95% CI 1.00 – 1.95 and OR 1.50, 95% CI 1.11 – 2.04, respectively). Nonetheless, the findings associated with kitchen location and LBW or ALRTI did not persist in 2017. In the 2017 IDHS, the single and combined effect of exposure to polluting fuel and SHS was significantly associated with LBW among rural mothers (OR 1.61, 95% CI 1.15 – 2.24; OR 1.77, 95% CI 1.19 – 2.64; and OR 1.60, 95% CI 1.13 – 2.27 for SHS, polluting fuel, and combined SHS and polluting fuel, respectively). In addition, exposure to SHS alone and combined with polluting fuel was significantly associated with ALRTI among rural children aged under five (OR 1.51, 95% CI 1.13 – 2.02 and OR 1.61, 95% CI 1.18 – 2.19 for SHS and combined SHS and polluting fuel, respectively). The results show no significant findings associated with kitchen location (Tables 2 and 3). The main findings of this study remained elevated in the sensitivity analysis (Supplemental Tables 1 and 2).

Table 3. Odds ratio for the association between exposure to SHS, polluting fuel and kitchen location with ALRTI (OR, 95%CI).

Variables	2007			2012		
	U[a]	R[b]	T[c]	U[a]	R[b]	T[c]
Cooking fuel and secondhand smoke:						
None	-	-	-	Ref.	Ref.	Ref.
Secondhand smoke and no polluting fuel	-	-	-	1.10 (0.77 - 1.56)	0.91 (0.60 - 1.39)	1.03 (0.78 - 1.35)
Polluting fuel and no secondhand smoke	-	-	-	1.26 (0.72 - 2.18)	0.90 (0.55 - 1.47)	1.03 (0.72 - 1.48)
Secondhand smoke and polluting fuel	-	-	-	1.37 (0.93 - 2.03)	1.42 (0.94 - 2.15)	1.45 (1.10 - 1.92)
Cooking fuel and kitchen location:						
Separate kitchen and no polluting fuel	Ref.	Ref.	Ref.	Ref.	Ref.	Ref.
No separate kitchen and no polluting fuel	0.73 (0.36 - 1.47)	3.45 (0.96 - 12.31)	1.02 (0.54 - 1.91)	1.10 (0.78 - 1.57)	1.09 (0.63 - 1.89)	1.11 (0.82 - 1.49)
Separate kitchen and polluting fuel	0.99 (0.55 - 1.76)	1.56 (0.8 - 3.04)	1.08 (0.68 - 1.74)	1.39 (1.00 - 1.95)	1.50 (1.11 - 2.04)	1.46 (1.16 - 1.83)
No separate kitchen and polluting fuel	1.11 (0.64 - 1.93)	1.86 (0.95 - 3.66)	1.25 (0.78 - 1.98)	1.48 (0.87 - 2.5)	1.39 (0.93 - 2.08)	1.39 (1.01 - 1.91)

[a] Urban, [b] Rural, [c] Total

Table 3. *Continued.*

Variables	2017		
	U[a]	R[b]	T[c]
Cooking fuel and secondhand smoke:			
None	Ref.	Ref.	Ref.
Secondhand smoke and no polluting fuel	0.93 (0.75 - 1.17)	1.51 (1.13 - 2.02)	1.14 (0.96 - 1.36)
Polluting fuel and no secondhand smoke	1.48 (0.87 - 2.51)	1.41 (0.93 - 2.13)	1.28 (0.93 - 1.76)
Secondhand smoke and polluting fuel	1.03 (0.66 - 1.61)	1.61 (1.18 - 2.19)	1.30 (1.03 - 1.63)
Cooking fuel and kitchen location:			
Separate kitchen and no polluting fuel	Ref.	Ref.	Ref.
No separate kitchen and no polluting fuel	1.15 (0.83 - 1.59)	0.89 (0.63 - 1.24)	1.02 (0.80 - 1.28)
Separate kitchen and polluting fuel	1.27 (0.85 - 1.92)	1.10 (0.84 - 1.44)	1.16 (0.93 - 1.45)
No separate kitchen and polluting fuel	1.06 (0.45 - 2.51)	1.38 (0.79 - 2.42)	1.32 (0.84 - 2.08)

[a] Urban, [b] Rural, [c] Total

4. Discussion and conclusion

The use of polluting fuel in Indonesia decreased considerably from 2007 to 2017 by more than 60%. However, the use of polluting fuel remained at over 30% among rural dwellers. Rural

IOP Conf. Series: Earth and Environmental Science **1503** (2025) 012003 doi:10.1088/1755-1315/1503/1/012003

mothers and children aged under five had higher prevalences of LBW, ALRTI, polluting fuel use, and exposure to SHS compared to the urban. Furthermore, LBW and ALRTI were higher among mothers and children exposed to SHS and living in the rural. Results from the latest survey suggest that rural mothers exposed to SHS or polluting fuel and, in combination, significantly associated with LBW compared to rural mothers with no exposure to SHS and/or polluting fuel. Also, rural children exposed to SHS or when combined with polluting fuel are significantly associated with ALRTI compared to rural children with no exposure to SHS and/or polluting fuel.

Indonesia began its policy to reduce the use of polluting fuels in 2007. The policy has been shown beneficial in substantially reducing the use of polluting fuel between 2007 and 2017, especially in urban areas. According to the findings from the 2012 IDHS, urban and rural mothers exposed to polluting fuel and had a separate room for the kitchen had a significantly higher risk for LBW than rural mothers not exposed to polluting fuel and had a separate room for the kitchen. However, this association was not apparent in the latest IDHS (2017). As a form of clean fuel, the government provides subsidized prices for the 3 kg liquid petroleum gas (LPG) cylinder, which eases families living in middle and lower economies in Indonesia. The findings from the latest survey in this study suggest that the existing policy is beneficial in reducing the use of polluting fuel (accounting for less than 10%), subsequently improving associated adverse health effects experienced by exposure during pregnancy and young-aged children, especially among urban dwellers. Nevertheless, polluting fuel use among rural mothers remained above 30%, which may affect the significant findings for LBW from the single exposure to polluting fuel use. To date, Indonesia is still facing challenges concerning LPG subsidies regarding its targeted distribution and policy reforms. Nonetheless, the government should explore the prospect of other environmentally friendly energy sources that are more sustainable.

The findings of this study align with a systematic review assessing the effects of exposure to polluting fuel among women and LBW, which reported statistically significant findings from four studies. The review reported 13 studies showing a significant association between cooking fuel and LBW, although the studies used varied approaches to define the study outcome and exposure (16). This study's findings from the most recent IDHS indicate that exposure to SHS is significantly associated with LBW and ALRTI, supporting previous research that documented adverse health effects linked to SHS exposure among women and young children (17-19). SHS is defined as the smoke inhaled from burning cigarettes or other tobacco products, as well as the exhaled smoke from smokers. The results of this study show a higher prevalence of LBW and ALRTI among mothers and children aged under five who were exposed to SHS, in comparison to those exposed to polluting fuels, based on data from the 2012 and 2017 IDHS. Furthermore, exposure to SHS was higher in the rural compared to the urban (63.7% and 48.9%, respectively). Indonesia still faces a big challenge as one of the countries with the highest cigarette consumption in the world (20). Indonesia's current standing in relation to the WHO Framework Convention on Tobacco Control may necessitate further examination, particularly in light of the adverse health effects associated with exposure to SHS, especially among pregnant women and young children.

With respect to countries with low and middle income, the findings of this study regarding LBW and ALRTI are consistent with results from other research conducted in similar contexts (21-23). Socioeconomic status within this area may influence the health outcomes from HAP exposure, especially SHS. For example, a study investigating exposure to SHS across 30 countries with low and middle income revealed that higher levels of education and greater wealth are linked to reduced exposure to SHS (24). In Indonesia, smoking households account for more than half of families in Indonesia, which in turn burdens the expenditure for other essential needs (25). This condition may influence outcomes such as LBW and ALRTI. However, the sensitivity analysis results in this study, using the wealth index, show no significant changes to the main findings.

This study presents several limitations and offers future recommendations. First, a limitation may arise from the reliance on mothers' reports of health outcomes; however, it is important to note that this reporting is nondifferential between the exposed and unexposed groups. Second, the datasets used in this research lack information regarding ventilation in the household or kitchen, which could significantly influence the level of exposure. While the study accounted for kitchen location, future research should consider incorporating ventilation as a critical variable. Third, future studies should examine the potential advantages of using continuous data to assess correlations between exposure and health outcomes when such data are available for a more comprehensive analysis. Additionally, it is recommended that future research employ a longitudinal approach to evaluate the long-term health effects associated with reduced HAP.

To conclude, Indonesia has had a declining trend of polluting fuel use and exposure to SHS from 2007 to 2017. However, polluting fuel use and exposure to SHS remained above 30% and 60%, respectively, among rural dwellers. Rural mothers exposed to SHS or polluting fuel and the combined effect of SHS and polluting fuel exposure are significantly associated with LBW. Also, exposure to SHS alone and combined with polluting fuel is significantly associated with ALRTI. The findings were mainly significant among rural mothers and children aged under five. There is an urgent need for targeted awareness in rural areas and stricter enforcement of clean energy adoption regulations to mitigate associated health risks in pregnant women and young children. In addition, factors such as energy access, cultural practices, and local policy barriers may influence the findings of this study. Collaboration with government or international organizations is crucial to integrate richer health and environmental data.

Acknowledgment
I would like to sincerely thank the Demographic and Health Surveys (DHS) Program for their valuable support.

IOP Conf. Series: Earth and Environmental Science **1503** (2025) 012003 doi:10.1088/1755-1315/1503/1/012003

References

1. Boor BE, Spilak MP, Laverge J, Novoselac A, Xu Y. Human exposure to indoor air pollutants in sleep microenvironments: A literature review. Building and Environment. 2017;125:528-55.

2. Baeza_Romero MT, Dudzinska MR, Amouei Torkmahalleh M, Barros N, Coggins AM, Ruzgar DG, et al. A review of critical residential buildings parameters and activities when investigating indoor air quality and pollutants. Indoor Air. 2022;32(11).

3. Lee KK, Bing R, Kiang J, Bashir S, Spath N, Stelzle D, et al. Adverse health effects associated with household air pollution: a systematic review, meta-analysis, and burden estimation study. Lancet Glob Health. 2020;8(11):e1427-e34.

4. Marete I, Ekhaguere O, Bann CM, Bucher SL, Nyongesa P, Patel AB, et al. Regional trends in birth weight in low- and middle-income countries 2013–2018. Reproductive Health. 2020;17(S3):176.

5. Aryastami NK, Shankar A, Kusumawardani N, Besral B, Jahari AB, Achadi E. Low birth weight was the most dominant predictor associated with stunting among children aged 12–23 months in Indonesia. BMC Nutrition. 2017;3(1):16.

6. Gu H, Wang L, Liu L, Luo X, Wang J, Hou F, et al. A gradient relationship between low birth weight and IQ: A meta-analysis. Scientific Reports. 2017;7(1):18035.

7. Jornayvaz FR, Vollenweider P, Bochud M, Mooser V, Waeber G, Marques-Vidal P. Low birth weight leads to obesity, diabetes and increased leptin levels in adults: the CoLaus study. Cardiovascular Diabetology. 2016;15(1):73.

8. U.S. Census Bureau. Population Clock: World [Internet]. [cited 2024 Jul 3]. Available from: https://www.census.gov/popclock/world/id#

9. United Nations Children's Fund WHOaTWB. Levels and trends in child malnutrition: UNICEF/WHO/World Bank Group Joint Child Malnutrition Estimates: Key findings of the 2023 edition. New York: UNICEF and WHO. Report No.: CC BY-NC-SA 3.0 IGO.

10. World Health Organization. 2021 GATS fact sheet Indonesia [Internet]. 2021 [cited 2025 Feb 7].

11. Xiang S, Chen Z, Dai Z, Wang F. Global burden of lower respiratory infections attributable to secondhand smoke among children under 5 years of age, 2010–2019: a systematic analysis of the global burden of disease study 2019. BMC Public Health. 2023;23(1):1920.

12. Woolley KE, Bagambe T, Singh A, Avis WR, Kabera T, Weldetinsae A, et al. Investigating the Association between Wood and Charcoal Domestic Cooking, Respiratory Symptoms and Acute Respiratory Infections among Children Aged Under 5 Years in Uganda: A Cross-Sectional Analysis of the 2016 Demographic and Health Survey. International Journal of Environmental Research and Public Health. 2020;17(11):3974.

13. Mondal D, Paul P. Effects of indoor pollution on acute respiratory infections among under-five children in India: Evidence from a nationally representative population-based study. PLOS ONE. 2020;15(8):e0237611.

14. Kanno GG, Anbesse AT, Shaka MF, Legesse MT, Andarge SD. Effect of biomass fuel use and kitchen location on maternal report of birth size: Cross-sectional analysis of 2016 Ethiopian Demographic Health Survey data. Public Health in Practice. 2021;2:100211.

15. Jabin N, Salam MT, Rahman MM, Sharna TI, Franklin M, Ahmed A, et al. Social inequality influences the impact of household air pollution on birth outcomes. Science of The Total Environment. 2022;822:153405.

16. Younger A, Alkon A, Harknett K, Jean Louis R, Thompson LM. Adverse birth outcomes associated with household air pollution from unclean cooking fuels in low- and middle-income countries: A systematic review. Environmental Research. 2022;204:112274.

17. Robinson O, Martínez D, Aurrekoetxea JJ, Estarlich M, Somoano AF, Íñiguez C, et al. The association between passive and active tobacco smoke exposure and child weight status among Spanish children. Obesity. 2016;24(8):1767-77.

18. Chen MM, Chiu C-H, Yuan C-P, Liao Y-C, Guo S-E. Influence of Environmental Tobacco Smoke and Air Pollution on Fetal Growth: A Prospective Study. International Journal of Environmental Research and Public Health. 2020;17(15):5319.

19. Sarfo JO, Amoadu M, Gyan TB, Osman A-G, Kordorwu PY, Adams AK, et al. Acute lower respiratory infections among children under five in Sub-Saharan Africa: a scoping review of prevalence and risk factors. BMC Pediatrics. 2023;23(1):225.

20. World Bank. The economics of tobacco taxation and employment in Indonesia: health, population, and nutrition global practice. WBG Global Tobacco Control Program. Washington, DC: World Bank Group; 2018 2018 [cited 2024 Jul 11].

21. Pathirathna ML, Abeywickrama HM, Sekijima K, Sadakata M, Fujiwara N, Muramatsu Y, et al. Effects of Prenatal Tobacco and Wood-Fuel Smoke Exposure on Birth Weight in Sri Lanka. Healthcare. 2017;5(4):64.

22. Hasan M, Tasfina S, Haque SMR, Saif-Ur-Rahman KM, Khalequzzaman M, Bari W, et al. Association of biomass fuel smoke with respiratory symptoms among children under 5 years of age in urban areas: results from Bangladesh Urban Health Survey, 2013. Environmental Health and Preventive Medicine. 2019;24(1):65.

23. Islam S, Mohanty SK. Maternal exposure to cooking smoke and risk of low birth weight in India. Science of The Total Environment. 2021;774:145717.

24. Reece S, Morgan C, Parascandola M, Siddiqi K. Secondhand smoke exposure during pregnancy: a cross-sectional analysis of data from Demographic and Health Survey from 30 low-income and middle-income countries. Tobacco Control. 2019;28(4):420-6.

25. Swarnata A, Kamilah FZ, Wisana IDGK, Meilissa Y, Kusnadi G. Crowding-out effect of tobacco consumption in Indonesia. Tobacco Control. 2024;33(Suppl 2):s81-s7.

IOP Conf. Series: Earth and Environmental Science **1503** (2025) 012003 doi:10.1088/1755-1315/1503/1/012003

Supplemental Table 1. Odds ratio for the association between exposure to SHS, polluting fuel, and kitchen location with LBW (OR, 95%CI).

Variables	2007			2012		
	U[a]	**R**[b]	**T**[c]	**U**[a]	**R**[b]	**T**[c]
Cooking fuel and secondhand smoke:						
None	-	-	-	Ref.	Ref.	Ref.
Secondhand smoke and no polluting fuel	-	-	-	1.22 (0.82 - 1.8)	0.90 (0.55 - 1.47)	1.12 (0.82 - 1.52)
Polluting fuel and no secondhand smoke	-	-	-	1.41 (0.78 - 2.57)	1.07 (0.66 - 1.71)	1.22 (0.84 - 1.76)
Secondhand smoke and polluting fuel	-	-	-	1.31 (0.81 - 2.12)	1.47 (0.94 - 2.30)	1.53 (1.11 - 2.10)
Cooking fuel and kitchen location:						
Separate kitchen and no polluting fuel	Ref.	Ref.	Ref.	Ref.	Ref.	Ref.
No separate kitchen and no polluting fuel	0.73 (0.29 - 1.85)	0.71 (0.20 - 2.55)	0.72 (0.34 - 1.54)	0.81 (0.51 - 1.27)	0.82 (0.43 - 1.59)	0.81 (0.55 - 1.18)
Separate kitchen and polluting fuel	1.09 (0.64 - 1.85)	1.13 (0.43 - 2.97)	1.13 (0.70 - 1.80)	1.11 (0.68 - 1.80)	1.26 (0.85 - 1.87)	1.18 (0.88 - 1.59)
No separate kitchen and polluting fuel	0.81 (0.45 - 1.46)	0.69 (0.25 - 1.86)	0.74 (0.45 - 1.23)	1.04 (0.52 - 2.06)	1.77 (1.03 - 3.04)	1.51 (0.99 - 2.28)

[a] Urban, [b] Rural, [c] Total

IOP Conf. Series: Earth and Environmental Science **1503** (2025) 012003 doi:10.1088/1755-1315/1503/1/012003

Supplemental Table 1. *Continued.*

Variables	2017		
	U[a]	**R**[b]	**T**[c]
Cooking fuel and secondhand smoke:			
None	Ref.	Ref.	Ref.
Secondhand smoke and no polluting fuel	1.10 (0.86 - 1.41)	1.58 (1.13 - 2.20)	1.26 (1.03 - 1.53)
Polluting fuel and no secondhand smoke	1.23 (0.74 - 2.04)	1.68 (1.11 - 2.55)	1.39 (1.02 - 1.89)
Secondhand smoke and polluting fuel	1.15 (0.72 - 1.85)	1.50 (1.05 - 2.13)	1.26 (0.97 - 1.64)
Cooking fuel and kitchen location:			
Separate kitchen and no polluting fuel	Ref.	Ref.	Ref.
No separate kitchen and no polluting fuel	0.74 (0.52 - 1.07)	0.77 (0.48 - 1.24)	0.76 (0.57 - 1.01)
Separate kitchen and polluting fuel	0.99 (0.66 - 1.48)	0.98 (0.72 - 1.33)	0.97 (0.76 - 1.24)
No separate kitchen and polluting fuel	1.03 (0.53 - 2.02)	1.23 (0.73 - 2.07)	1.17 (0.77 - 1.76)

[a] Urban, [b] Rural, [c] Total

IOP Conf. Series: Earth and Environmental Science **1503** (2025) 012003 doi:10.1088/1755-1315/1503/1/012003

Supplemental Table 2. Odds ratio for the association between exposure to SHS, polluting fuel and kitchen location with ALRTI (OR, 95%CI).

Variables	2007			2012		
	U[a]	R[b]	T[c]	U[a]	R[b]	T[c]
Cooking fuel and secondhand smoke:						
None	-	-	-	Ref.	Ref.	Ref.
Secondhand smoke and no polluting fuel	-	-	-	1.12 (0.79 - 1.59)	0.89 (0.57 - 1.37)	1.04 (0.79 - 1.36)
Polluting fuel and no secondhand smoke	-	-	-	1.23 (0.69 - 2.20)	0.73 (0.44 - 1.19)	0.94 (0.65 - 1.37)
Secondhand smoke and polluting fuel	-	-	-	1.38 (0.90 - 2.13)	1.11 (0.72 - 1.71)	1.29 (0.95 - 1.74)
Cooking fuel and kitchen location:						
Separate kitchen and no polluting fuel	Ref.	Ref.	Ref.	Ref.	Ref.	Ref.
No separate kitchen and no polluting fuel	0.73 (0.36 - 1.48)	3.28 (0.91 - 11.76)	1.01 (0.54 - 1.90)	1.11 (0.78 - 1.58)	1.06 (0.61 - 1.87)	1.09 (0.81 - 1.47)
Separate kitchen and polluting fuel	1.05 (0.63 - 1.77)	1.24 (0.66 - 2.32)	1.06 (0.69 - 1.64)	1.41 (0.98 - 2.02)	1.20 (0.88 - 1.63)	1.31 (1.03 - 1.65)
No separate kitchen and polluting fuel	1.20 (0.71 - 2.05)	1.42 (0.73 - 2.77)	1.21 (0.78 - 1.89)	1.58 (0.86 - 2.87)	1.10 (0.72 - 1.67)	1.25 (0.89 - 1.76)

[a] Urban, [b] Rural, [c] Total

IOP Conf. Series: Earth and Environmental Science **1503** (2025) 012003 doi:10.1088/1755-1315/1503/1/012003

Supplemental Table 2. *Continued.*

Variables	2017		
	U[a]	**R**[b]	**T**[c]
Cooking fuel and secondhand smoke:			
None	Ref.	Ref.	Ref.
Secondhand smoke and no polluting fuel	0.89 (0.72 - 1.12)	1.49 (1.11 - 1.99)	1.11 (0.94 - 1.32)
Polluting fuel and no secondhand smoke	1.25 (0.73 - 2.14)	1.33 (0.87 - 2.04)	1.16 (0.83 - 1.61)
Secondhand smoke and polluting fuel	0.86 (0.54 - 1.38)	1.48 (1.06 - 2.06)	1.16 (0.91 - 1.48)
Cooking fuel and kitchen location:			
Separate kitchen and no polluting fuel	Ref.	Ref.	Ref.
No separate kitchen and no polluting fuel	1.09 (0.79 - 1.51)	0.90 (0.65 - 1.26)	1.00 (0.79 - 1.26)
Separate kitchen and polluting fuel	1.12 (0.74 - 1.70)	1.00 (0.76 - 1.33)	1.04 (0.83 - 1.31)
No separate kitchen and polluting fuel	0.92 (0.38 - 2.23)	1.28 (0.73 - 2.23)	1.19 (0.75 - 1.88)

[a] Urban, [b] Rural, [c] Total

IOP Conf. Series: Earth and Environmental Science **1503** (2025) 012004 doi:10.1088/1755-1315/1503/1/012004

Land cover mapping and identification of local wisdom in spring

Trida Ridho Fariz[1*], Harun Syamsudin Nur Hidayah[1], Amnan Haris[1], Abdul Jabbar[1], Uswatun Rina Pamungkas[1], Unca Alia[1], Abi Manshurin[1], Wahid Darmawan[1], Astri Arum[1]

[1] Environmental science, Universitas Negeri Semarang, Indonesia

*E-mail: trida.ridho.fariz@mail.unnes.ac.id

Abstract. Springs are water resources that have long been used by the community since ancient times. Gunungpati District is one of the areas in Indonesia with quite a lot of spring potential, but has potential threats due to changes in land cover. Therefore, this study will map land cover and identify local wisdom in one of the areas of Gunungpati District, namely Pakintelan Village, the key area of this study is Jurang Springs in Pakintelan Village. Land cover mapping uses satellite imagery from the NICFI Basemap with visual interpretation. To identify local wisdom, this study uses literature studies, observations and in-depth interviews. The mapping results show that Pakintelan Village, which is the location of Jurang Springs, has land cover change. The change in land cover is dominated by mixed garden to settlement, but the change in land cover does not occur in the spring buffer zone. For local wisdom in Jurang Springs, it includes the configuration of the five springs associated with vegetation and traditions carried out at the spring location such as rasulan, sedekah bumi, and penjamasan. All forms of local wisdom aim to maintain and preserve the environment around Jurang Springs. To further preserve water resources and springs, the government needs preventive efforts such as spatial planning policies.

1. Introduction

As societies evolve, they often draw upon their cultural heritage to address contemporary challenges. Local wisdom, a rich tapestry of traditional knowledge and practices, provides valuable insights that can be applied to modern issues. Indonesia is one of the countries with a lot of cultural diversity and traditions that grow and develop in society as a legacy from ancestors. The existence of local wisdom, in addition to having a function as a characteristic of a group or region, also has other functions, one of which is as a form of environmental conservation efforts and climate change mitigation and adaptation [1].

Environmental conservation, especially water resource conservation, is a top priority for humans [2,3]. Water resources in particular have even become one of the goals in the SDG's (Sustainable Development Goals), namely clean water and sanitation. The dependence of human activities on water resources since ancient times has given rise to many local wisdoms that are closely related to water resources [4]. Forms of local wisdom in water resources management include traditional ceremonies and offerings at springs [5]. The objects of local wisdom in water resources are mostly directed at springs and rivers because they are the main water sources for Indonesian people since ancient times.

Gunungpati District in Semarang City is one of an area with many springs. This area is located at the foot of Mount Ungaran and has been identified with around 114 springs [6]. However, the springs in this area are threatened by massive land cover change and urbanization [6,7]. Pakintelan Village in Gunungpati District has many springs that are utilized by the community, one of which is Jurang Spring [7,8]. Considering changes in land cover such as increased settlement around springs can cause displacement and degradation, therefore this study will discuss land cover changes in the area [7]. In addition, this study will also identify forms of local wisdom in the management of Jurang Spring. Local wisdom should be examined as it holds value in shaping environmental policies. This is due to the fact that local wisdom represents a community's adaptive approach to addressing environmental challenges, making it a suitable foundation for guiding environmental policy development [9].

2. Methodology

2.1. Study location and data

The approach used in this study is quantitative and qualitative. This study is located at the Jurang spring located in Pakintelan Village, Gunungpati District, Semarang City, Central Java (Figure 1). Based on its function in spatial planning, Gunungpati District is a "lungs of the city" and educational designated area [10]. The stages of data collecting and data analysis consist of land cover mapping and identification of local wisdom.

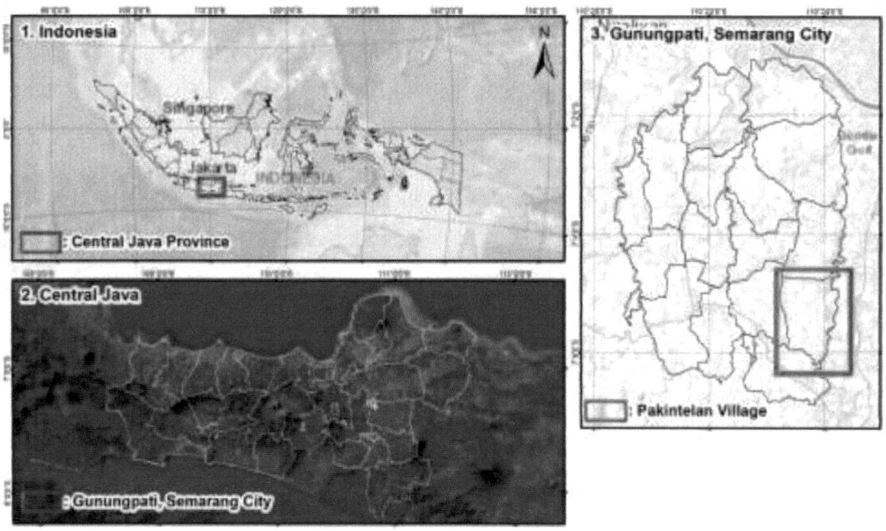

Figure 1. Study area

2.2. Land cover mapping and accuracy assesment

Satellite remote sensing is data and tools for land cover mapping [11]. Land cover mapping in this study applies the NICFI Satellite Data Program's Basemaps for Tropical Forest Monitoring—Asia dataset on the Google Earth Engine (GEE). The basemaps are created using multitemporal composites of PlanetScope Surface Reflectance images, which have been atmospherically corrected and provided by Planet Labs PBC [12]. Unlike individual PlanetScope images, the NICFI basemaps employ statistical methods to address challenges that arise when high-quality images are unavailable during specific time periods. Similar to standard PlanetScope imagery, the NICFI basemaps maintain a spatial resolution of 4.77 meters. However, they offer improved temporal resolution, as they are designed to provide a high likelihood of cloud-free, standard-quality imagery over five-day intervals in many regions globally [13].

The land cover mapping technique uses visual interpretation that involves several key interpretations. These keys are tone or color, shape, size, pattern, texture, shadow, association, and location. The visual interpretation process is carried out with a maximum magnification of 2.5x the map scale output [14]. Visual interpretation is chosen because it is more accurate in identifying land cover. In this study, land cover mapping in 2021 and 2024, due to the availability of image data from the NICFI basemap. Furthermore, in the results of land cover mapping, this study also looks at the land cover conditions of the buffer zone area. The determination of the buffer zone for spring is adopted from Indonesia's Ministry of Public Works and Housing, where the 200m area of the spring is the spring buffer zone [15].

2.3. Local wisdom identification

Local wisdom encompasses a wide range of tangible elements, such as rituals, tourism practices, traditional games, attire, folklore, artistic expressions, infrastructure, institutional frameworks, and systems for managing and cultivating land resources [9,16]. Local wisdom identification requires several data collection techniques, namely literature studies, observations and in-depth interviews. Data triangulation and methods to ensure the reliability criteria for the data collected [17]. Literature studies are collecting secondary data about Jurang Springs and their local wisdom. Observations to determine the socio-cultural environmental conditions of the community around the springs. In-depth interviews are direct interviews that are used in depth with individuals to obtain detailed data and information [17]. The techniques used in selecting informants are purposive sampling and snowball sampling. Purposive sampling is used to determine key informants, namely the managers of Jurang springs and snowball sampling is used to determine informants from village officials around the spring location. Secondary data comes from books, journals, and other sources to obtain all information related to the matter in question. All data obtained were then analyzed using qualitative descriptive analysis techniques so that conclusions could be drawn.

3. Results and discussion

3.1. Land cover in Pakintelan Village

Semarang City is known to have 139 springs, of which 114 are in Gunungpati District, one of which is in Pakintelan Village with 16 springs. The wealth of springs in Semarang City is because it is located at the foot of the Volcanic Mountain (Mt. Ungaran), and its area is located on an active fault line, namely the Garang fault, Gribik fault, Kreo fault and Karanganyar Gunung fault which have the potential to cause rock shifts that build springs [18]. Of the 16 springs in Pakintelan Village, there are 5 springs in the same area with the names Jurang Spring 1, Jurang Spring 2, Jurang Spring 3, Jurang Spring 4 and Jurang Spring 5 (Figure 2). These five Jurang springs are located on steep slopes and surrounded by cliffs.

IOP Conf. Series: Earth and Environmental Science **1503** (2025) 012004 doi:10.1088/1755-1315/1503/1/012004

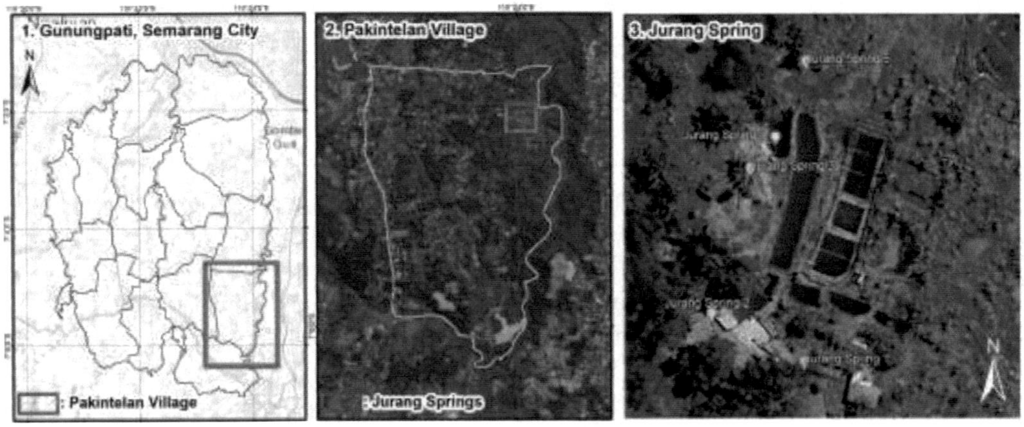

Figure 2. Jurang Springs area

The results of land cover mapping show that Pakintelan Village is dominated by mixed gardens (Figure 3). In 2021, mixed gardens covered an area of 208 Ha and decreased to 195.2 Ha in 2024 (Figure 4). Settlement and built-up areas are land cover classes that are increasing, in 2021 covering an area of 110.5 Ha and becoming 120.1 Ha in 2024. The growth of settlements and built-up areas is mostly due to the conversion of mixed gardens. The phase of change starts from mixed gardens being converted into bare soil, then from bare soil changing to settlement. Mixed gardens do not immediately change into settlements but first change into bare soil because it is a land clearing process.land clearing.

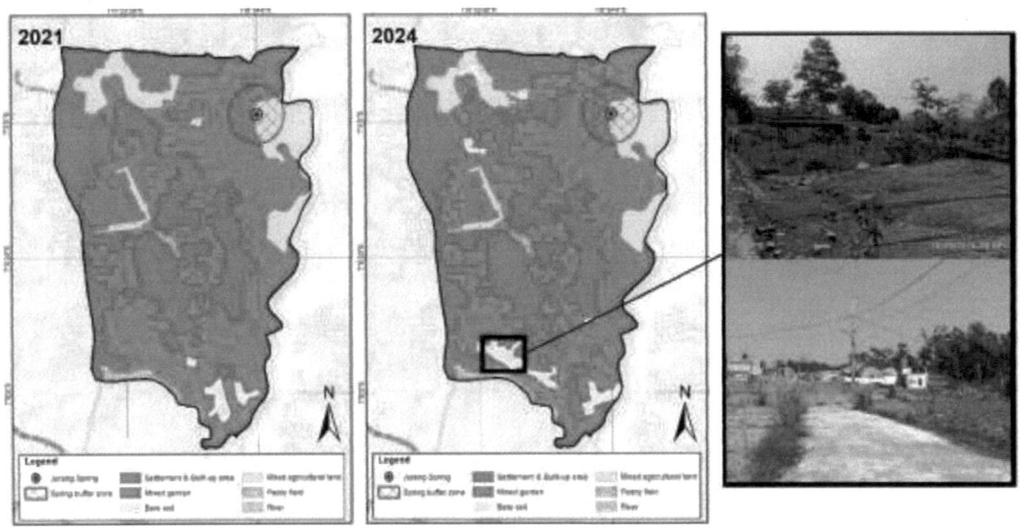

Figure 3. Land cover map in Pakintelan Village

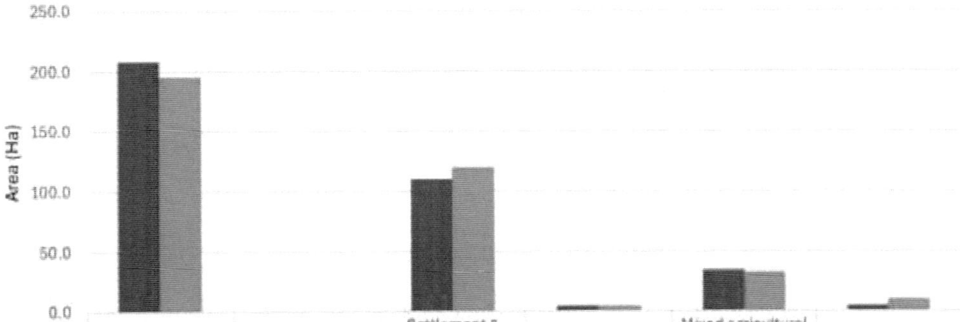

	Mixed garden	River	Settlement & Built-up area	Paddy field	Mixed agricultural land	Bare soil
2021	208.0	0.2	110.5	4.7	34.7	4.6
2024	195.2	0.2	120.1	4.7	32.6	10.0

Figure 4. Land cover changes area in Pakintelan Village

Land cover changes do not occur in the Jurang Springs area, even in the spring buffer zone area. However, land cover change control needs to be carried out for the conservation of Jurang Springs. This is because Jurang Springs is a source of water for the surrounding community for domestic and agricultural purposes. The use of water from Jurang Springs is through community-based water supply (PAMSIMAS) with a coverage area around Watusari Hamlet in Pakintelan Village and Patemon Village.

3.2. Local wisdom in Jurang Springs

Local wisdom in water resource management and conservation is not only a good strategy for water resource management, but also for climate change adaptation. The results of observations and interviews with key informants, there are 5 water sources in the Jurang area, but currently only 4 are still active. Jurang Spring 1 is located under a durian tree (*Durio zibethinus*) which is then surrounded by a wall as a protector (Figure 5). Jurang Spring 1 also has a name, *Bahurekso*, this is because there is a story about the early discovery of the spring through a ritual. After the ritual was completed, one of the people dreamed of meeting an old grandfather named *Bahurekso* who is also said to be considered the guardian of the spring.

Figure 5. Jurang Spring area

Jurang Spring 2 is located behind the Spring Manager's house, this spring has the strongest flow with the largest discharge (Figure 5). According to key informants, water from Jurang Spring 2 has a safe pH as a criterion for primary class water so it can be drunk directly, and is even believed to have health benefits. Jurang Spring 2 is named *Panguripan* which means source of life, where

IOP Conf. Series: Earth and Environmental Science **1503** (2025) 012004 doi:10.1088/1755-1315/1503/1/012004

this spring has the largest discharge so that it can support Watusari Hamlet. For Jurang Spring 3, it appears from under a mound of earth, and this spring has a fairly strong discharge. Jurang Spring 3 is named *Cakrabuana* which is said to be the name of the guardian of the spring.

Figure 6. Jurang Spring area

Jurang Spring 4 is located right under a large rock with clear water and a calm flow (Figure 6). Jurang Spring 4 also flows into the fishpond around it. Jurang Spring 4 is named Syafaat, because it is believed to provide benefits for people who are in trouble. The Jurang Spring 5 is located under a rock buried in the ground with a flow that is deliberately closed so that it becomes an inactive spring. According to key informants, Jurang Springs never dries up during the dry season even though it has been used as a source of domestic water for the surrounding community. According to key informants, the water quality of Jurang Springs is also good, this is also supported by observations that the water from Jurang Springs is not colorless, tasteless, and odorless. These results are inseparable from the efforts of the managers and local communities who have managed and conserved Jurang Springs.

One of the conservation efforts of Jurang Springs is the *Rasula*n tradition which is a form of expression of the Javanese people in maintaining traditions that are closely related to religious beliefs. Rasulan is a traditional ceremony that is still preserved by the Javanese people as an event to express gratitude to God Almighty for all the abundant sustenance and pleasures. The location used as a place for the Rasulan ritual is identical to the existence of large, lush trees that shelter a spring. Generally, large trees that are adjacent to springs will be considered sacred by local residents, a term that is usually called *sendang* by the local community [19].

Another tradition found in Jurang Springs is offerings or *sedekah bumi*. *Sedekah bumi* is one of the traditions carried out during the harvest month in Javanese society as a sign of gratitude for the abundance that has been given [19]. Based on key respondents, the sedekah bumi carried out by the community every Saturday night aims to appreciate the earth that has given life. Jurang Springs is used as a symbol that represents attitudes and values that are rich in wisdom. *Sedekah bumi* is used as a hope that the land that is lived on will remain fertile and free from drought. This ritual is intended for the ancestors, especially prophets, holy persons and elders to ask for blessings or *pangestu*. In the ritual, objects are also known in the form of objects that are used as the core of the *sedekah bumi* ritual [20]. The ritual at the Jurang spring uses a *keris* as an object of ritual objects that will be purified or *penjamasan*.

Penjamasan is bathing heirlooms such as *keris* inherited from elders or previous figures so that the value of the object is maintained [20]. The meaning of the *penjamasan keris* itself is to cleanse the soul and body, the ritual is carried out by washing the keris using water containing white roses, kenanga flowers, kantil flowers and jasmine flowers. Where each component has a meaning or symbolizes something, such as water which symbolizes life, kantil flowers symbolize attachment, jasmine flowers symbolize blessing marang gusti, white roses symbolize purity, and kenanga flowers symbolize protection. The series of penjamasan rituals then ends with a joint prayer across religions according to the beliefs of each community that follows the ritual.

Right above Jurang Springs there is a shack which is often used for prayer, meditation and prayer for people who have desires. *Tawasulan* is an activity of requesting prayers and praying for ancestors as an effort to get closer to the creator [19]. There is also a shack as a building used for wiridan or meditation right in front of the Jurang Spring 4, facing a large rock (Figure 7). As the name suggests, the Jurang Spring 4 is Syafaat, so it is believed to be able to provide benefits or requests for people who have this interest.

Figure 7. Tawasulan and Wiridan Place

The uniqueness of local wisdom is in line with the conservation treatment of water sources which can be in the form of maintaining the nature, condition and function so that the quality and quantity are adequate to meet the needs of the community now and in the future [21]. The presence of vegetation around the springs with strong roots and large canopies can facilitate the process of trees in absorbing water and the evapotranspiration process [22]. Around Jurang Springs there are also bamboo trees (*Bambusoideae*) which have very strong roots so that they can function to bind soil and water which is good for facilitating the flow of springs [22].

The form of local wisdom in Jurang Springs is mostly the sanctification of water sources to maintain sustainability. Moreover, this spring provides benefits for domestic water sources for the surrounding community. However, changes in land cover have the potential to threaten the sustainability of this spring. For this reason, the government needs to make preventive efforts through spatial planning policies. In detailed-scale spatial planning policies (RDTR), the Pakintelan Village area or even Gunungpati District should also consider the existence of springs. In detailed-scale spatial planning, low-density settlements and cultural heritage areas can be implemented for springs that have strong local wisdom. Preventive efforts, especially related to local wisdom, can be through ethnoscience reconstruction for the learning process in environmental and science education. It is because many studies have stated that the integration of culture or local wisdom into learning science is able to develop students' character as cultural conservationists [23]. This effort is certainly in line with the function of the study location according to Spatial Planning, where Gunungpati District is not only a buffer zone but also an educational area in Semarang City.

4. Conclusssion

Jurang Springs is located in Pakintelan Village, Gunungpati District, where there is a change in land cover. The change in land cover is dominated by mixed garden to settlement, but the change in land cover does not occur in the spring buffer zone. For local wisdom in Jurang Springs, it includes the configuration of the five springs associated with vegetation and traditions carried out at the spring location such as rasulan, sedekah bumi, and penjamasan. All forms of local wisdom aim to maintain and preserve the environment around Jurang Springs. To further preserve water resources and springs, the government needs preventive efforts such as spatial planning policies. In detailed scale spatial planning, low-density settlements and cultural heritage areas can be implemented for springs that have strong local wisdom in Gunungpati District.

Acknowledgement

Thank you to Nana Kariada Tri Martuti for providing the spring distribution data.

References

[1] Rahmah M and Sulistyono A 2024 The Integration of Traditional Knowledge and Local Wisdom in Mitigating and Adapting Climate Change: Different Perspectives of Indigenous Peoples from Java and Bali Island *Traditional Knowledge and Climate Change: An Environmental Impact on Landscape and Communities* (Springer) pp 61–80

[2] Alexandratos S D, Barak N, Bauer D, Davidson F T, Gibney B R, Hubbard S S, Taft H L and Westerhof P 2019 Sustaining water resources: Environmental and economic impact *ACS Sustain. Chem. Eng.* **7** 2879–88

[3] Trihadianta R, Gunawan M, Shabrina D, Mustikawati L H, Purwadi C E, Kholil P A, Fariz T R, Haris A and Jabbar A 2024 Water Availability Based on Seasons in Tembalang District, Semarang City *Indones. J. Earth Hum.* **1** 84–93

[4] Mawardi M 2012 Rekayasa Konservasi Tanah dan Air *Yogyakarta Bursa Ilmu*

[5] Pratiwi A S and Heriyanti A P 2024 Community Participation in Environmental Conservation Based on Dawuhan Tradition in Tegalwaton Village Spring, Semarang Regency *Indones. J. Earth Hum.* **1** 17–26

[6] Martuti N K T, Setyowati D L, Rahayuningsih M, Dewi N K, Sidiq W A B N and Mutiatari D P 2023 Springs Distribution and Biophysical Mapping as a Strategy to Preserve Water Sources in Semarang City, Indonesia *J. Environ. Manag. Tour.* **14** 113–25

[7] Sidiq W A B N, Sanjoto T B and Martuti N K T 2022 Land Use Change Analysis to Springs Conditions in Gunungpati Sub-District, Semarang City *Geosfera Indones.* **7** 150–64

[8] Purwadi C E, Shabrina D, Gunawan M, Trihadianta M R, Mustikawati L H, Jabbar A, Fariz T R and Haris A 2023 Sustainability of Community-Based Management of Lerak Spring in Semarang City *J. Environ. Sci. Educ.* **3** 127–34

[9] Fariz T R, Permana P I, Jabbar A, Saoki R N and Purwadi C E 2024 Local Wisdom on Mangrove Management in Nipah Panjang Village, Indonesia

[10] Pemerintah Kota Semarang 2021 *Rencana Tata Ruang Wilayah (Spatial Planning) Kota Semarang 2011-2031* (Indonesia)

[11] Fariz T R, Suhardono S and Fadhilla S 2024 Rapid mapping of land cover changes in tropical watershed in Indonesia using Google Earth Engine *IOP Conference Series: Earth and Environmental Science* vol 1414 (IOP Publishing) p 12077

[12] Vizzari M 2022 PlanetScope, Sentinel-2, and Sentinel-1 data integration for object-based land cover classification in Google Earth Engine *Remote Sens.* **14** 2628

[13] Lee C B, Martin L, Traganos D, Antat S, Baez S K, Cupidon A, Faure A, Harlay J, Morgan M and Mortimer J A 2023 Mapping the national seagrass extent in Seychelles using PlanetScope NICFI data *Remote Sens.* **15** 4500

[14] Fariz T R, Jatmiko R H and Mei E T W 2023 Pemanfaatan Foto Udata UAV untuk Pemetaan Kerentanan Fisik Rumah Terhadap Longsor di Sub-DAS Bompon *J. Ilmu Lingkung.* **21** 819–29

[15] Kementrian Pekerjaan Umum dan Perumahan Rakyat 2015 *Peraturan Menteri Pekerjaan Umum dan Perumahan Rakyat Republik Indonesia Nomor 28/Prt/M/2015 Tentang Penetapan Garis Sempadan Sungai dan Garis Sempadan Danau* (Indonesia)

[16] Wagiran W 2012 Pengembangan Karakter Berbasis Kearifan Lokal Hamemayu Hayuning Bawana (Identifikasi Nilai-Nilai Karakter Berbasis Budaya) *J. Pendidik. Karakter* **3**

[17] Khusniati M, Heriyanti A P, Aryani N P, Fariz T R and Harjunowibowo D 2023 Indigenous science constructs based on Troso woven fabric local wisdom: a study in ethnoscience and ethnoecology *J. Turkish Sci. Educ.* **20** 549–66

[18] Martuti N K T, Rahayuningsih M and Sidiq W A B N 2021 Kajian pemetaan potensi mata air di Kota Semarang *J. Riptek* **15** 1–7

[19] Barizi A and Kosbandhono E 2021 Ijtihad ekologis pesantren dalam upaya konservasi hutan dan sumber

daya air (studi pada Pesantren Nurussalam Singgahan, Tuban)
[20] Akhmad P 2017 *Membongkar Kesesatan perilaku syirik* (Quranic Healing Indonesia (Didukung: Adamssein Media))
[21] Sulistyorini I S, Edwin M and Arung A S 2016 Analisis kualitas air pada sumber mata air di kecamatan Karangan dan Kaliorang kabupaten Kutai Timur *J. hutan Trop.* **4** 64–76
[22] Triwanto J 2024 *Konservasi Lahan Hutan dan Pengelolaan Daerah Aliran Sungai* (UMMPress)
[23] Suja I W, Jayadiningrat I M G and Ardiyasa I N S 2023 Differences in the Triplechem Learning Model with Balinese Local Wisdom and the Discovery Learning Model in Influencing Students' Visual Literacy and Mental Models about Acid-Base Solutions *J. Pendidik. IPA Indones.* **12** 187–98

Spatial Pattern Analysis of the Distribution and Accessibility Level of Tourism Objects in Sragen Regency.

Amanda Oktavia Kurniawati[1], Kuswaji Dwi Priyono[2*], Umrotun[3]

[1]Faculty of Geography, Universitas Muhammadiyah Surakarta, Surakarta 57162, Indonesia

*E-mail: kdp130@ums.ac.id

Abstract. Sragen Regency is in Central Java Province. Its location adjacent to East Java Province has resulted in the tourism sector in Sragen Regency continuing to grow. Sragen Regency's regional income from the tourism sector has increased every year. In Sragen Regency, there are many tourist attractions with various characteristics and distribution patterns that spread in each region. Apart from that, the level of accessibility to tourist attractions in Sragen Regency, such as existing facilities and infrastructure, is quite good. However, in some tourist attractions there are still roads that are difficult to pass. In 2023, the number of tourist visiting Sragen Regency was 233,671 visitors, where the most visited tourist attraction was Sangiran Museum. To find out the distribution pattern and the influence of the level of accessibility, this research uses quantitative methods with descriptive analysis and surveys. As a result of this research, the spatial pattern of distribution and the level of accessibility of tourism objects in Sragen Regency can be known. The research results show that the distribution pattern of tourist attractions in Sragen Regency is dispersive.

Keywords : Tourism Objects, Sragen Regency, Accessibility Level

1. Introduction

Sragen Regency is in Central Java Province. Its location that borders the province of East Java results in the tourism sector in Sragen Regency to continue to grow. Sragen Regency's income from the tourism sector always increases every year [1]. In Sragen Regency there are many tourist attractions with various characteristics and distribution patterns that spread in every region, ranging from natural tourist attractions, ancient human fossils to man made tourist attractions.

Besides that, tourism activities is a travel process that is only temporary and carried out by a person or group of people with the purpose of a certain place where the place is different from the place of origin [2]. Tourism is an important supporting sector for economic progress. The existence of a growth rate and increasing development, infrastructure development, high foreign exchange flow of the country, and also the implementation of good governance will have a positive impact on various economic sectors and also tourism. But, the development of tourism will cause both positive and negative impacts. From a geographical point of view, tourism demand is the total number of people who want to travel or have already traveled to enjoy tourism services and facilities in places far from work and residence[3].

In addition, accessibility can be interpreted as a measure of ease to reach a location from another location through the transportation system. Factors that affect the level of accessibility are the infrastructure and facilities that support. Good quality infrastructure is also supported by several factors, one of which is supporting facilities. An area will become a tourist attraction area if it has many supporting elements. The level of affordability of a place is determined by its level of accessibility [4]. Apart from that, infrastructure is also a very important factor influencing public activities.

In Sragen Regency, it is not yet known whether there is a relationship between the level of accessibility and the number of visitors, so further studies are needed. In addition to the accessibility of supporting infrastructure, it is very important to support the development of tourism in the Sragen Regency. According to Jayadinata, infrastructure is a very potential and important factor to determine the future and direction of development in a region, this is because development will not run well without adequate infrastructure support[5]. In addition to the availability of infrastructure and accessibility, anoher determining factor in the progress of tourist attractions is the facilities provided. According to Suwantoro, tourist facilities are divided into 3 types, namely tourist attractions, facilities to tourist attractions, and restaurants [6]. Therefore, limited facilities and infrastructure are a problem that must be addressed.

In addition, network analysis is a method to find the shortest distance between line segments based on line geometry length [7]. The advantage of network analysis is that researchers do not need a field survey. Therefore, one of the requirements of the geographic information system is to be able to support and accommodate these needs [8]. Geographic information system can also present, analyze, and store both spatial and non-spatial data to determine a route [9]. The aim of this research is to analyze the spatial pattern of distribution of tourist attractions in Sragen Regency and determine the level of accessibility of tourist attractions in Sragen Regency.

2. Method

2.1 Research Location

The research location is Sragen District. Sragen Regency is in Central Java Province with the administrative center of Sragen District. This regency borders with Grobogan Regency in the north, Karanganyar Regency in the south, Boyolali Regency in the west, adnd Ngawi Regency in East Java Province in the east. The north side od Sragen Regency is in the form of hills which are part of the Kendeng Mountains. While a small part of the southern region is in the form of hills at the foot of Mount Lawu.

Geographically, Sragen Regency is located at 7°15' – 7°30' South Latitude and 110°45' – 111°10' East Longitude. The area is in the valley of the Bengawan Solo River which flows eastward. Most of it is lowland with a height of between 70-480 meters above sea level.

Sragen has a tropical climate with low temperatures throughout the day ranges from 19 - 31ºC. Average rainfall is below 3000mm per year with rainy days under 150 days per year. The hydrological situation in Sragen district is marked with the existence of rivers that flow within the watershed area Bengawan Solo and the Jratunseluna watershed (Jragung, Tuntang, Serang, Lusi and Juana). Rivers that flow in a single watershed Solo rivers are the Mungkung, Kenatan, Jenar, Kedungaren, Tanggan, Teseh and Kedung Dowo, while the rivers included in the Jratunseluna watershed unit, namely the

Serang.

IOP Conf. Series: Earth and Environmental Science **1503** (2025) 012005 doi:10.1088/1755-1315/1503/1/012005

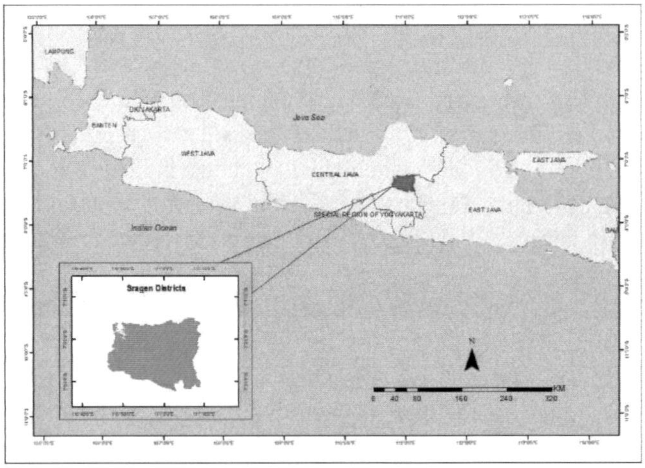

Figure 1. Administrative map of the location of Sragen district

Morphologically, Sragen Regency consists of lowlands and limestone mountains. This situation has quite an influence on tourism conditions in Sragen Regency. Apart from being fertile, Sragen Regency also has good tourism potential. Natural tourism, historical, religious, and so on can be found there. Location of Sragen Regency Based on the administrative map, Sragen Regency is located in the east of Central Java Province and directly borders the Ngawi region of East Java, which can actually be seen in Figure 1.

2.2 Data Collection and Processing

Data collection in this research used a quantitative approach. The quantitative approach used in this research is to use descriptive analysis of research data. Secondary data (survey) was obtained from search results on official agency websites, such as BPS and the Sragen Regency government website; related journals; and conduct field observations. The spatial pattern of the distribution of tourist attractions in Sragen Regency and the level of accessibility of tourist attractions in Sragen Regency can be determined using network analysis methods which are supported by creating maps of the distribution of tourist attractions, administrative maps of the Sragen Regency area, and route maps to get to tourist attractions in Sragen Regency. Table 1 presents information about the data used in this study.

Table 1. Data used in research

Data	Source	Function
Shapefile data for administrative boundaries (roads, rivers, etc.) of Central Java Province	Geospatial Information Agency (BIG) 2019	Create an administrative map of Sragen Regency
Point coordinates and tourist attraction addresses	Google Earth Pro	Determine the location of the tourist attraction along with the best route
Number of visitors to tourist attractions	Central Statistics Agency (BPS) Sragen	Knowing the number of tourists who visit tourist attractions in Sragen Regency

The data processing stages in this research were carried out by a) adding all administrative shapefile data into ArcGIS to create an administrative map of Sragen Regency and b) plotting the coordinate data of tourist attractions in Sragen Regency that had been obtained into ArcGIS (which had been created previously) to produce a map of the distribution of tourist attractions and the best routes to tourist attractions in Sragen Regency. Determining distribution routes and the best routes to tourist attractions using network analysis techniques.

3. Result and Discussions

3.1 Results

Sragen Regency has areas with high altitudes which varies between 71 m to 500 m above sea level. The plateau is in the north and southeast, while in The middle part is a relatively flat and watered valley Bengawan Solo River. This results in Sragen Regency having fertile land.

Administratively, Sragen Regency has 20 sub-districts, 208 villages/sub-districts, 2,519 hamlets and 5,228 RTs and has an area of 94,155 Ha. The subdistricts with the largest number of villages are Plupuh and Tanon, each consisting of 16 villages. Meanwhile, the fewest are Gesi, Tangen and Jenar sub-districts, each with 7 villages.

The tourism potential in Sragen Regency is influenced by the location of Sragen Regency which is close to mainstay tourist attractions in Central Java, including: Surakarta palace cultural tourism, Borobudur Temple tourism in Magelang, and volcano tourism in Selo, Boyolali Regency. Apart from that, the discovery of many ancient human fossils in Sragen Regency makes it a special attraction for visitors. Therefore, the ancient human museum is the leading tourist attraction in Sragen Regency with the most visitors and provides the highest distribution of regional income. Apart from the ancient human museum, in Sragen Regency there are also natural and man-made tourist attractions spread across various regions. Table 2 shows the distribution of tourist attractions in Sragen Regency.

Table 2. Coordinates and Addresses of Tourist Attractions in Sragen Regency

Name of Tourist Attraction	X Coordinate	Y Coordinate	Address of Tourist Attraction
Sangiran ancient human museum	481706	9175878	Kebayanan I, Bukuran, Kalijambe, Sragen Regency, Central Java 57275
Kemukus mountain tour	481189	9188099	Jl.Gn.Kemukus, Kebayanan II, Pendem, Sumberlawang, Sragen Regency, Central Java 57272
Bukuran Cluster ancient man museum tour	484810	9176382	Dusun 3, Bukuran, Kalijambe, Sragen Regency, Central Java 57275

Table 2. *(continued)*

Name of Tourist Attraction	X Coordinate	Y Coordinate	Address of Tourist Attraction
Ndayu park sragen	499090	9175007	Gembong RT08/RW04, Saradan, Karangmalang, Sragen Regency, Central Java 57222
Bayanan hot spring bath	513526	9169808	Gamping, Jambeyan, Sambirejo, Sragen Regency, Central Java 57293
Sragen Kridoanggo Park	503101	9179161	Jl. Sukowati No.488, Krajoyok, Sragen Wetan, Sragen, Sragen Regency, Central Java 57211
Doeng Cuo Swimming Pool	503181	9175283	Jl Sragen – Batujamus KM4, Bolorejo, Puro, Karangmalang Puro, Karangmalang, Sragen Regency, Central Java 57291
Sendang Kun Gerit Tour	483919	9180298	Jl. Gemolong – Plupuh Sidorejo RT02, Kukungerit Dusun I, Jatibatur, Gemolong, Sragen Regency, Central Java 57274
Bahulak Market Tour	489905	9180307	Sawahan, Karungan, Plupuh, Sragen Regency, Central Java 57283
waterboom jambangan permai	509954	9177600	Gondang – Kaliwedi, Dimoro, Kaliwedi, Gondang, Sragen Regency, Central Java 57254
Bandut Lake Tour	516332	9168351	Dk. Jetis, RT11, Pondok, Sukorejo, Sambirejo, Sragen Regency, Central Java 57293

Table 2. *(continued)*

Name of Tourist Attraction	X Coordinate	Y Coordinate	Address of Tourist Attraction
Kartika Swimming Pool	502692	9178295	Kroyo, Karangmalang, Sragen Regency, Central Java 57211
gemolong edupark	481438	9182652	Dusun 2, Gemolong, Sragen Regency, Central Java 57274
Kedung ombo reservoir tour	478600	9194225	Gilirejo Baru, Miri, Sragen Regency, Central Java 57276
Joko Tingkir's Grave	488971	9175494	Dusun II, Gedongan, Plupuh, Sragen Regency, Central Java 57283

After getting the coordinates of each tourist attraction in Sragen Regency, the coordinates were input into ArcGIS to create a map of the distribution of tourist attractions. Figure 2 below is a map of the distribution of tourist attractions in Sragen Regency.

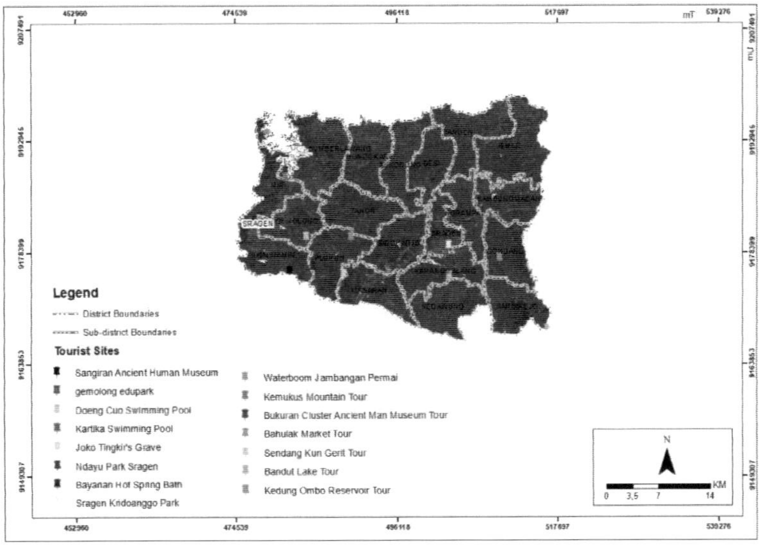

Figure 2. Map of the Distribution of Tourist Attractions in Sragen Regency

Once the spatial pattern of the distribution of tourist attractions in Sragen Regency is known, network analysis can be carried out. The results of the network analysis of tourist attractions in Sragen Regency can be seen from Figure 3. Network analysis produces a distribution map of the location routes of existing tourist attractions.

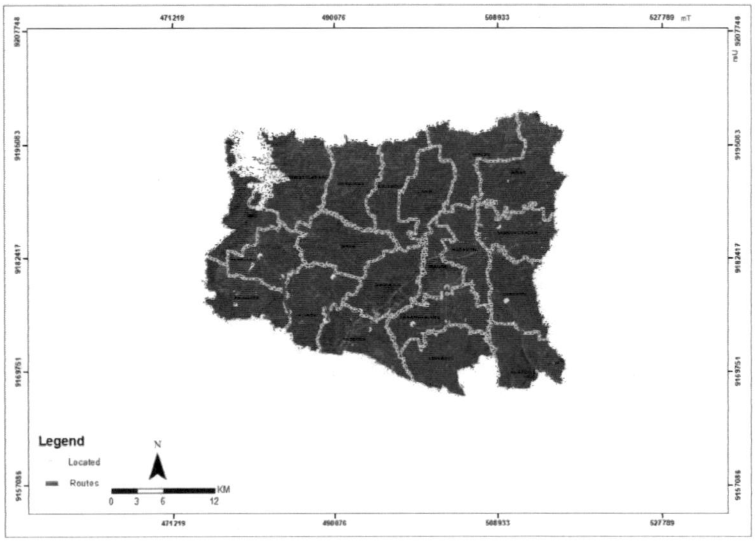

Figure 3. Map of Route Distribution of Tourist Attractions in Sragen Regency (Results from Network Analysis)

3.2 Discussion

Based on the map image (figure 2), tourist attractions in Sragen Regency are spread from the east side, namely Sambirejo District, to the west side, namely Miri District. Each existing tourist attraction has its own charm. The most famous attraction in Sragen Regency is the ancient human site at the Sangiran Museum. The UNESCO award for the Sangiran Museum has made the museum better known to many people.

The research results obtained show that most of the tourist attractions in Sragen Regency are in lowland areas and are border areas with other regencies/cities, such as Karanganyar Regency, Boyolali Regency and Surakarta City. Apart from its strategic location, this tourist attraction area is also always adding tourism support facilities. With the addition of tourism supporting facilities, it will increase tourist attraction.

Tourist attractions in Sragen Regency are very diverse, there are historical tourist attractions about ancient humans, natural tourist attractions such as Mount Kemukus and the Kedung Ombo reservoir, religious tourist attractions such as the Joko Tingkir tomb, to man-made tourist attractions such as the Doeng Cuo swimming pool. In urban areas, the majority of tourist attractions are man-made tourist attractions. An example is the Kridoanggo Park tourist attraction in Sragen. This is because in urban areas there are more changes in land use for settlements and industry. Kridoanggo Park, Sragen, is equipped with a theater performance venue.

However, the existence and development of tourist attractions has an impact on society and the environment, this impact can be negative or positive. The positive impact for society is to advance the economy and improve the welfare of people around tourist attractions, but it will have a negative impact on the environment, namely environmental damage due to exploitation of natural resources and land conversion. In terms of negative impacts due to the development of tourist attractions, in Sragen Regency it is still very small.

Based on Figure 3 regarding the route distribution map for tourist attractions in Sragen Regency, tourist attractions dominate in the east, which are border areas. Meanwhile, on the west side it is still not dominant, there are even several sub-districts that do not yet have tourist attractions, such as Mondokan, Tanon, Jenar, Tangen and Gesi sub-districts. The main reason is that these sub-districts are in the middle of the district and do not border other districts so their location is not strategic. This results in the development of facilities being slower and the level of accessibility still lacking. However, there are still several tourist attractions that are difficult to access, such as Joko Tingkir's grave, Bandut lake, and Bayanan hot springs.

On the other hand, tourist attractions in border areas have a greater number of visitors than tourist attractions in the city. This may happen because in the city the tourist attractions are only monotonous, while the tourist attractions in border areas have various types and the supporting facilities are better.

The level of accessibility of tourist attractions in Sragen Regency is assessed in terms of infrastructure and facilities in the field as quite good for each tourist attraction. The condition of the road used to get to the tourist attraction location is quite good and even. The existing roads have been paved or cast. However, the problem is that the roads are uphill and narrow, such as the road to the tourist attraction Joko Tingkir's grave. The road is narrow and cannot be passed by buses.

According to data published by BPS Sragen Regency on February 28 2024, the largest number of visitors to tourist attractions in Sragen Regency is the Sangiran Early Man Museum, namely 143,476 visitors. This happens because the Sangiran Museum has very easy road access and is bypassed by the main route from Jalan Solo-Purwodadi. The number of visitors to a tourist attraction really depends on the level of accessibility. The better the level of accessibility of a tourist attraction, the greater the number of visitors.

4. Conclusion

The research results show that most of the tourist attractions in Sragen Regency are located in the district/city border areas. In fact, the leading tourist attraction in Sragen Regency, namely the Sangiran early human museum, is directly adjacent to Karanganyar Regency. This tourist attraction, which is synonymous with ancient human fossils, continues to grow every year and has resulted in the conversion of land from agriculture to museum buildings. Thus, the development of a tourist attraction will definitely have positive and negative impacts.

Apart from that, the level of accessibility in Sragen Regency for each tourist attraction is quite good. Almost all the roads have been paved or cast, which makes it easier for visitors to tourist attractions. However, there are still narrow roads that cannot be passed by buses, such as the road to the Joko Tingkir Tomb and Telaga Bandut tourist attractions.

The spatial pattern of the distribution of tourist attractions in Sragen Regency has a dispersive nature. Apart from that, the existing level of accessibility apparently influences the number of visitors.According to data published by BPS Sragen Regency on February 28 2024, the largest number of visitors were in border areas, specifically at the Sangiran Museum which directly borders Karanganyar Regency. There were 143,476 visitors out of 233,671 visitors visiting the Sangiran Museum. This happens because the Sangiran Museum has a very strategic location and has very easy access by vehicles, even large buses.

References

[1] Badan Pusat Statistik Kabupaten Sragen, 2024, "Statistik Daerah Kabupaten Sragen," Sragen : Badan Pusat Statistik.

[2] Wicaksono, A., Farda, N. M., Khakhim, N., & Wibowo, T. W, 2021, "Flickr Photos Analysis for BeachTourismManagementinBantulRegency,Indonesia:PopularityandTouristAttractions." Forum Geografi, 35(1), 85–101. https://doi.org/10.23917/forgeo.v35i1.13007

[3] Wahab, 2003, "Manajemen Pariwisata," Pradya Paramitha : Jakarta.

[4] Nugraheni, I., Yudana, G., & Rini, E. F, 2022, "Kesesuaian aksesibilitas kawasan wisata budaya di Kota Surakarta ditinjau berdasarkan jalur pedestrian sebagai NMT," Region: Jurnal Pembangunan Wilayah dan Perencanaan Partisipatif, 17(2), 254 268.

[5] Jayadinata, 2015, "Tata Guna Tanah Dalam Perencanaan Pedesaan Perkotaan & Wilayah," Bandung : ITB Bandung.

[6] Suwantoro, G, 1997, "Dasar-dasar tentang Pariwisata," Yogyakarta: Andi.

[7] Guo Bo, Poling A.D & Poppe M.J, 2000, "GIS/GPS in Transportation," Real World Experiences.

[8] Jumadi,J.,&Priyana,Y, 2016, "DevelopmentandEvaluationofWebGISApplicationforGround water Management of Karanganyar Regency," Forum Geografi, 29(2), 165. https://doi.org/10. 23917/forgeo.v29i2.1482

[9] Jumadi&Sigit Widiadi, 2009, "Manajemen Pemanfaatan Air Tanah Berbasis Web Menggunakan PHP, JAVADANMYSQLSPATIAL (Studi Kasus di Kabupaten Banyumas)," Forum Geografi,23, 123-138. https://doi.org/10.23917/forgeo.v23i2.5006

IOP Conf. Series: Earth and Environmental Science **1503** (2025) 012006 doi:10.1088/1755-1315/1503/1/012006

Analysis of the Spatial Transformation of Poverty and Education in Central Java, Indonesia Using Bivariate Moran's I in 2013 and 2023

I R Dani[1], H Z Hadibasyir[1*], N L Anggani[1]

[1] Faculty of Geography, Universitas Muhammadiyah Surakarta, Surakarta 57162, Central Java, Indonesia

*E-mail: hamim.zaky.h@ums.ac.id

Abstract. Poverty remains a fundamental issue that requires serious attention from the government, especially in provinces with large populations like Central Java. Education plays a crucial role in helping lower-middle-income countries absorb modern technology and achieve sustainable growth. A region's social welfare and human growth are greatly impacted by poverty and education, two interconnected aspects. Investigating the spatial correlation between poverty and education indicators more especially, the Average Length of Education in Central Java Province from 2013 to 2023 is the aim of this study. This research uses descriptive quantitative methods by utilizing secondary data. Data analysis employs the Moran Index and LISA to analyze the spatial correlation between the percentage of poverty and the Average Length of Education. The research results show that the spatial pattern of the Moran's Global test for the percentage of poverty is I = 0.327825 (p-value = 0.048) in 2013 and I = 0.267285 (p-value = 0.10) in 2023, indicating a change in pattern from clustering to dispersing. For the Average Length of Education, the Moran's Global test is I = 0.084529 (p-value = 0.052) in 2013 and I = 0.073046 (p-value = 0.56) in 2023, indicating a pattern that tends to disperse. The spatial correlation of the Bivariate Moran's test shows that areas with a high percentage of poverty tend to be adjacent to areas with low education levels. The Bivariate LISA test in 2013 and 2023 indicates a significant increase in the number of regions from 3 to 5 districts/cities. It is recommended that local governments prioritize equitable access to education and region-based poverty alleviation programs, especially in impoverished areas included in quadrant IV.

1. Introduction

Poverty is a worldwide concern as it impacts numerous countries and poses a major challenge by hindering the fulfillment of essential life needs [1]. As in Indonesia, poverty is still a fundamental problem that requires serious attention from the government, especially in provinces with large populations such as Central Java. In densely populated areas, the challenge of overcoming poverty is more complex due to the variety of factors that influence the socio-economic conditions of the community. Central Java Province's poverty rate is consistently higher than the national average, according to data from the Central Statistics Agency in 2013 it had 14.44% of the 33.26 million population and in 2023 it will have 10.77% of the total population of 37.5 million people [2].

Table 1. Table of the Percentage of Poverty

Percentage of poverty	2013	2015	2017	2019	2021	2023
Central Java	14,44%	13,32%	12,23%	10,80%	11,25%	10,77%
Indonesia	11,47%	11,13%	10,12%	9,22%	9,71%	9,36%

People who experience poverty are categorized as those with monthly per capita expenditure below the poverty line [3]. The poverty line is used to measure the condition of the population by showing the minimum amount of expenditure needed for basic food and non-food needs. The poverty rate is broken down by district in the Central Java Provincial Statistics Agency (BPS) data in the form of a percentage, where the figure is obtained from the number of people experiencing poverty compared to the population in Central Java Province.

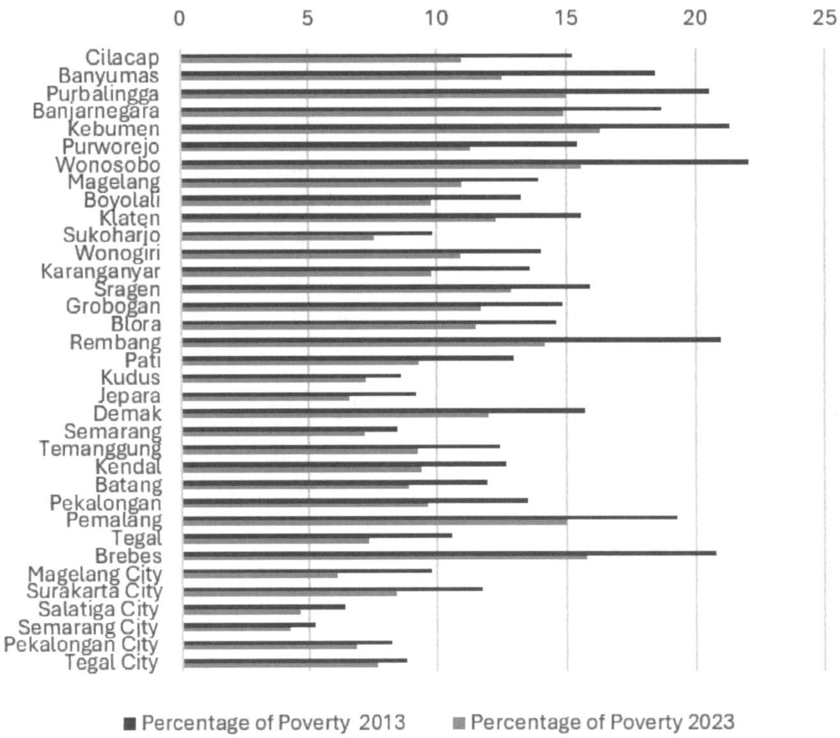

Figure 1. Diagram of the Percentage of Poverty in Central Java

Education plays an important role in helping lower-middle-income countries absorb modern technology and achieve sustainable growth [4]. Highly educated individuals can generate new ideas on how to produce more goods and services. If these ideas are accepted by society, then the benefits will be felt by many people, not just by the individuals themselves. The vicious circle theory of poverty explains that underdevelopment must be broken to overcome poverty, and education is one effective way to do so. The higher a person's education, the higher their productivity and income, which in turn can help them escape poverty [5]. This study adapts from previous research that analyzed poverty and its relationship with stunting cases [6]. Besides that, it has also been done to analyze the link between covid cases that occur and the poverty of an area [7]. Unlike previous studies, the originality of this research lies in its focus on poverty and education variables

Poverty and education are two factors that influence each other and have a major impact on human development and the social welfare of a region [16]. In Indonesia, poverty is still one of the main issues that needs to be addressed seriously, especially in provinces with large populations such as Central Java. The Indonesian government mandates that pupils finish their education up to the high school level through the 12-year compulsory education program. Reducing school dropout rates and increasing educational access and quality are the goals. The program also aims to create educational equality in all parts of Indonesia, including remote areas. One of the key metrics frequently used to assess a region's educational quality is the average number of years of schooling. It displays the average number of years spent in formal education by those who are 15 years of age and older. Average Length of Education is the average number of years that people have completed at all levels of formal education; this is calculated when they are 25 years of age or older [8]. When Average Length of Education is low, it often reflects limited access to education, which in turn can exacerbate poverty. Many regions in Central Java have not yet reached the target of the government's 12-year compulsory education program.

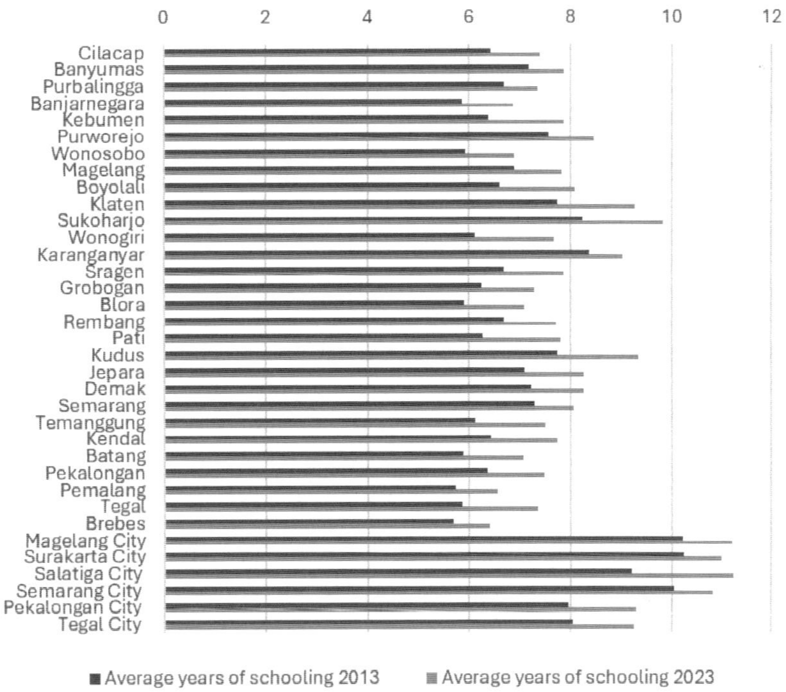

Figure 2. Diagram of Average Length of Education in Central Java

Central Java Province, which has a large population, continues to overcome the challenges of poverty and improve access to education. The local and central governments have implemented various policies to improve these conditions. However, despite these measures, disparities in poverty and education across regency/cities in Central Java are still evident. The low level of education as measured by Average Length of Education is of particular concern, as education is key to reducing poverty and improving overall community welfare. Referring to previous research, poverty and the Average Length of Education have a significant impact on poverty variables. This study produces a regression coefficient consistent with Human Capital Theory, which assumes that investment in education can improve societal productivity. If the quality of education is high, the overall quality of society also improves. Education is essential for assisting people in escaping the cycle of poverty .

This study aims to analyze the spatial patterns of poverty and Average Length of Education, and to examine the spatial correlation between the two in Central Java Province in 2013 and 2023. Using a spatial analysis approach, this study will explore the relationship between poverty and education, to understand how poverty conditions can affect the quality of education in the region. The results of this analysis are expected to provide greater insight into the challenges faced by Central Java Province in its efforts to improve education and reduce poverty.

2. Methodology

This research uses a quantitative approach with a descriptive-analytical method. The quantitative approach is used because this research focuses on statistical measurements and spatial patterns. Descriptive methods are used to describe the distribution of poverty, while correlation analysis is conducted to understand the relationship between poverty percentage and Average Length of Education. The population of this study is the poverty percentage and Average Length of Education per district in Central Java Province in 2013 and 2023. The sampling method is Non Probability Sampling, namely saturated sampling which involves selecting all members of the population as samples, so that every individual in the population has the opportunity to be included without considering probability.

The secondary data collection method in this research is carried out by relying on information from government agencies, especially documents issued by the Central Java Province Statistics Agency (BPS) and will also retrieve spatial data such as Earth Shape Maps and SHP (Shapefile) files needed through the LapakGIS website. Secondary data processing techniques that have been extracted per district are then processed to determine the spatial pattern and spatial correlation of the two variables tested for districts/cities in Central Java Province.

Data analysis was conducted using Moran's Index and Local Indicators of Spatial Association (LISA). Moran's Index is used to determine the pattern formed from two variables globally, while LISA is used to understand the relationship between the two variables locally.

Univariate Moran's I serves to determine the spatial pattern of distribution of each variable studied based on the I value obtained. This value will classify the distribution pattern of research objects into clustered, spread, or random categories. Univariate Moran's I was also used to identify local spatial autocorrelation patterns, which aided the interpretation of these patterns and made the analysis easier to understand [9]. On the other hand, the Bivariate Moran's I Test is used to analyze the spatial relationship between the variables of poverty and Average Length of Education, which can result in a positive or negative correlation. Bivariate Moran's I can be used in various fields, including environmental, social, and urban planning sciences, to understand the spatial relationship between two highly critical variables [10].

IOP Conf. Series: Earth and Environmental Science **1503** (2025) 012006 doi:10.1088/1755-1315/1503/1/012006

The computation results will be visualized using a Moran's Scatterplot, which is separated into four distinct quadrants. namely, Quadrant I, II,III and IV. This scatterplot serves to illustrate the spatial distribution of observation areas based on high and low values. Quadrant I (High-High) represents areas with high observed values surrounded by other high-value areas. Quadrant II (Low-High) indicates areas with low observed values that are adjacent to high-value areas. Quadrant III (Low-Low) consists of areas with low observed values surrounded by similarly low-value areas. Meanwhile, Quadrant IV (High-Low) includes areas with high observed values but is surrounded by low-value areas[11]. In addition, Bivariate LISA will be used to analyze spatial relationships locally, with results displayed in the form of a spatial clustering map (Cluster Map) and a significance map of each region (Significance Map).

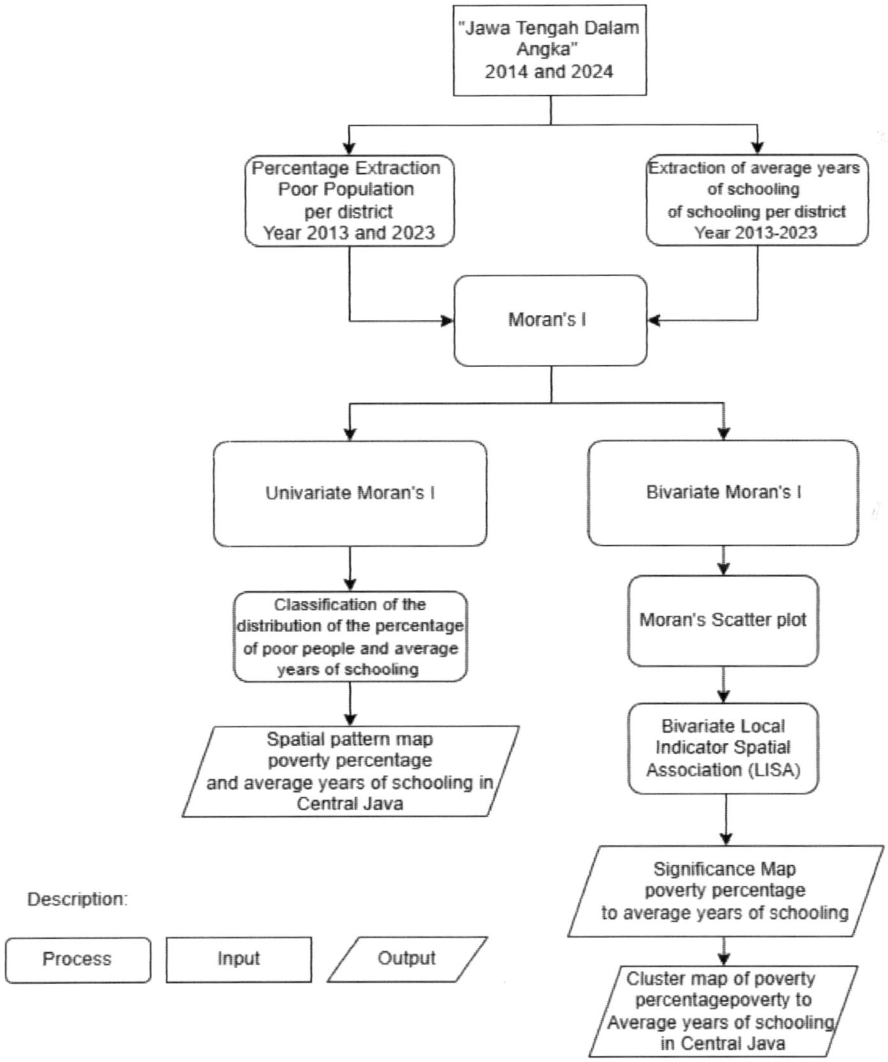

Figure 3. Research Flow Chart

IOP Conf. Series: Earth and Environmental Science **1503** (2025) 012006
doi:10.1088/1755-1315/1503/1/012006

3. Results and Discussion

3.1 Spatial Pattern of Poverty Percentage in Central Java

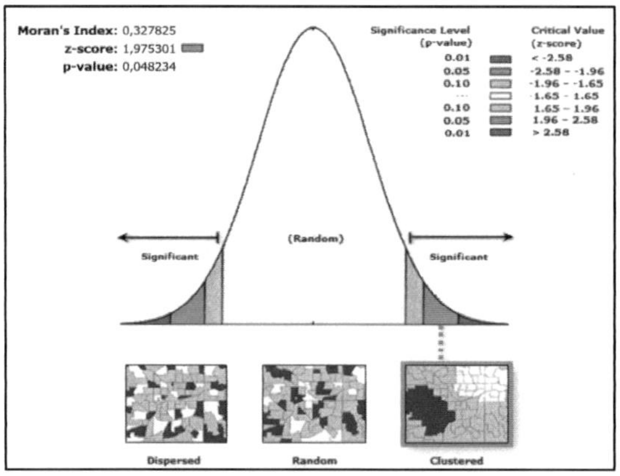

Figure 4. Poverty Percentage in 2013

According to the results of the univariate Moran's test on the poverty percentage variable in 2013, the distribution of the percentage of poverty in Central Java shows a positive correlation, meaning that areas with high percentages of poverty are typically close to areas with high values. The Moran's Index value for this variable is 0.327825, or has a positive value. Conversely, regions with low rates of poverty also frequently have low rates of poverty in close proximity to one another. A p-value of 0.048234 suggests that the observed spatial pattern is clustered.

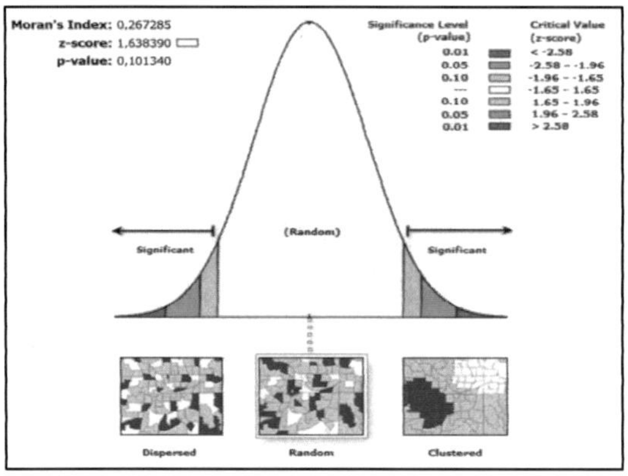

Figure 5. Poverty Percentage in 2023

The poverty percentage in 2023 shows a Moran's Index value of 0.267285 or has a positive value, indicating a positive correlation or clustering tendency. Although this value shows a clustering tendency, the p-value obtained of 0.101340 is higher than 0.05. This shows that the spatial pattern that occurs in 2023 is not significant enough to be categorized as a real clustering pattern and tends to be in the random category.

These findings demonstrate that the regional structure of the poverty percentage changed between 2013 and 2023. In 2013, the distribution of poverty in Central Java shows a clustered pattern, while in 2023 this pattern changes to become more dispersed or random. This can be observed through the distribution map of the poverty percentage in 2013 and 2023 in **Figure 6.**

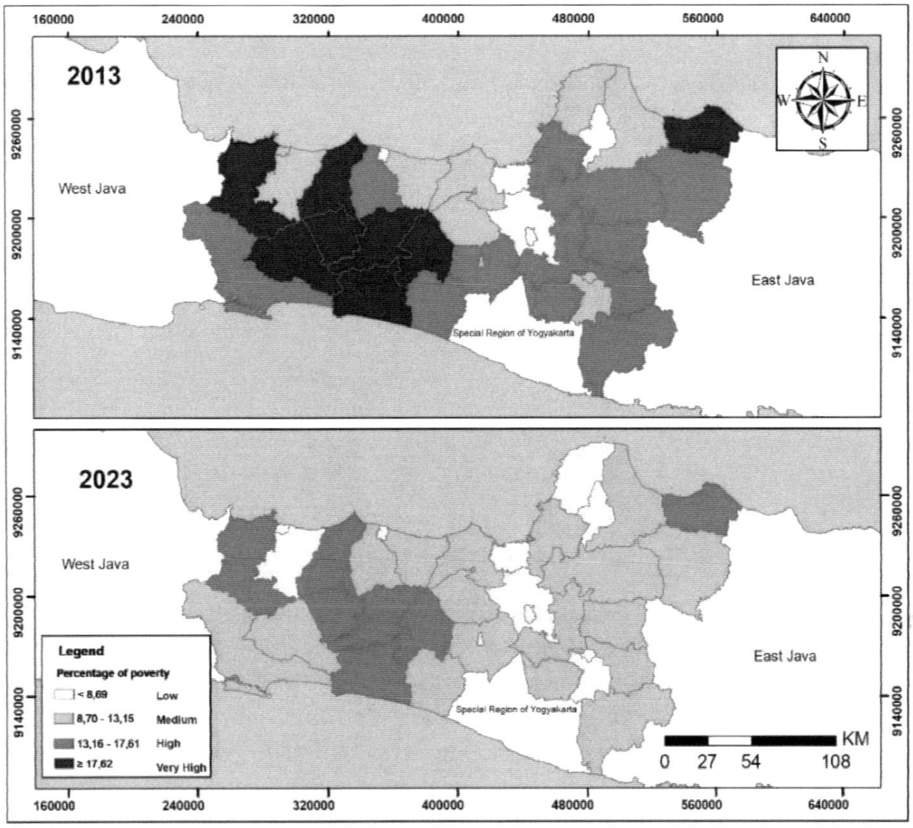

Figure 6. Poverty Distribution Map in Central Java

The percentage of poverty in Central Java in 2013 and 2013 is visualized in the form of a map that is divided into 4 classes, for 2013 the results show, Low class includes Kudus, Semarang, Salatiga City, Pekalongan City, and Semarang City. Medium class includes Pati, Jepara, Temanggung, Kendal, Sukoharjo, Batang, Tegal, Magelang City, Surakarta City and Tegal City. High class includes Wonogiri, Karanganyar, Sragen, Grobogan, Blora, Demak, Boyolali, Klaten, Magelang, Purworejo, Pekalongan and Cilacap. Very high class includes Rembang, Wonosobo, Banjarnegara, Kebumen, Purbalingga, Pemalang, Banyumas and Brebes. Referring to the distribution map of the poverty percentage in Central Java in 2013, the classes with very high values are clustered around Banjarnegara and Purbalingga.

While the poverty percentage map in Central Java in 2023 shows very different results, in the map there are only 3 classes, namely low class includes Kudus, Semarang, Tegal, Jepara, Sukoharjo, Salatiga City, Tegal City, Semarang City, Pekalongan City, Surakarta City and Magelang City. Medium class includes Wonogiri, Karanganyar, Blora, Sragen, Grobogan, Demak, Pati, Boyolali, Klaten, Magelang, Purworejo, Pekalongan, Temanggung, Kendal, Batang, Banyumas, and Cilacap. High class includes Rembang, Kebumen, Banjarnegara, Purbalingga, Wonosobo, Pemalang, and Brebes. In 2023, there were no longer any regencies/cities classified as very high. This can be interpreted as a sign that there is no significant disparity between regions in Central Java, and the fact that many regencies experienced a downgrade in classification reflects the increasing economic development equity in the province.

3.2 Spatial Pattern of Average Length of Education in Central Java

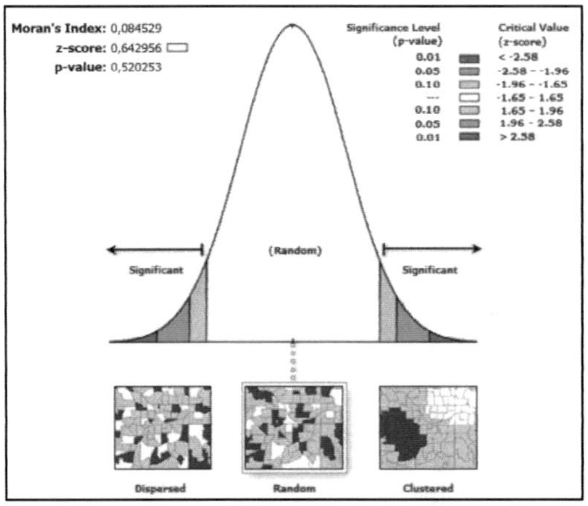

Figure 7. Average Length of Education 2013

Moran's univariate testing on the Average Length of Education variable in Central Java in 2013 the results of the Moran's Index value of 0.084529 indicate a positive correlation in the distribution of Average Length of Education in Central Java which means that areas with high Average Length of Education tend to be adjacent to areas that have high values. Vice versa, areas that have low Average Length of Education tend to be close to areas that have low values. At 0.520253, the p-value is greater than 0.05. This demonstrates that the geographical pattern that emerges in 2023 is more likely to fall into the random group and is not significant enough to be classified as a true clustering pattern [17].

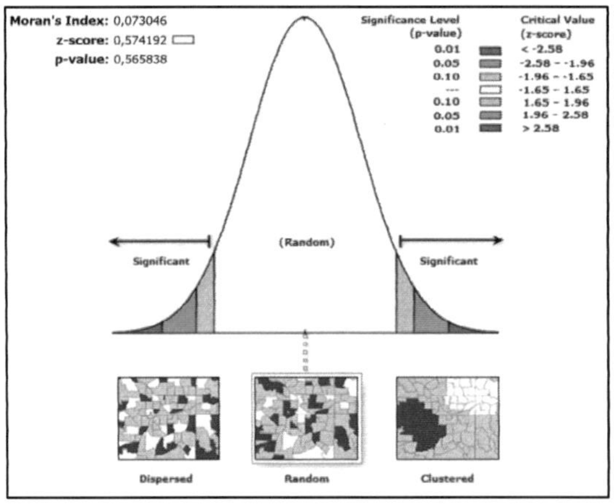

Figure 8. Average Length of Education in 2013

The results of the univariate Moran's test of the Average Length of Education in Central Java in 2023 show a Moran's Index value of 0.073046 or has a positive value, indicating a positive correlation or clustering tendency. Although this value indicates a clustering tendency, the p-value obtained is 0.565838 or higher than 0.05. This indicates that the spatial pattern in 2023 is not significant enough to be considered a clear clustering pattern and tends to fall into the random category as in 2013.

IOP Conf. Series: Earth and Environmental Science **1503** (2025) 012006 doi:10.1088/1755-1315/1503/1/012006

Based on these results, the spatial pattern of Average Length of Education in Central Java in 2013 and 2023 shows no change. This indicates that the distribution or spread of the Average Length of Education in each region tends to be stable from year to year, without any striking changes in the pattern of grouping or distribution between 2013 and 2023. However, when observed through the spatio-temporal map of the Average Length of Education in Central Java in 2013 and 2023, there are many changes. Changes in color gradations that were originally dominated by bright colors in 2013 became dominantly dark in 2023 as illustrated in **Figure 9**.

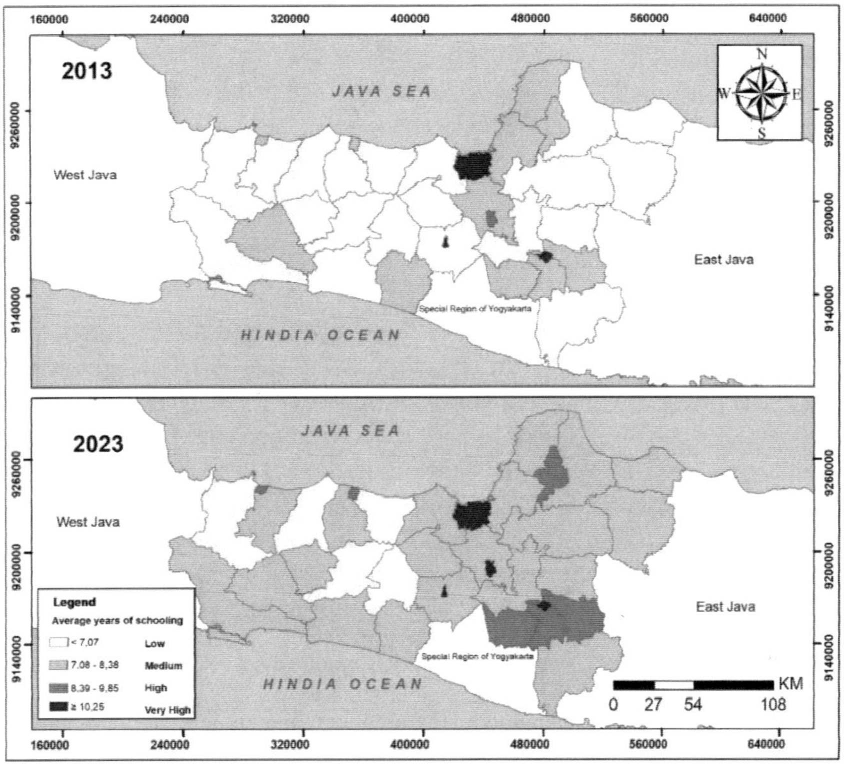

Figure 9. Distribution Map of Average Length of Education in Central Java

The distribution pattern of the Average Length of Education in 2021 and 2023 is the result of processing which is visualized in the form of a map of the distribution pattern of the Average Length of Education presented in Figure 8. Based on the test results, the regions in Central Java are divided into 4 classes, namely low, medium, high and very high. The average length of education in Central Java Province for the very high class both in 2013 and 2023 is always in large cities such as Surakarta City, Salatiga City, Magelang City and Semarang City. In this average duration of schooling variable, practically all districts in Central Java have grown. It is evident that there is a change in color that becomes darker in all districts in Central Java Province. When associated with the results of Moran's Index, it is in line because from 2013 and 2023 all experienced simultaneous increases in almost all parts of Central Java Province. However, from all districts/cities in Central Java, there are several districts that did not experience an increase in the average number of years of schooling and are still classified in the low class, such as Brebes, Pemalang, Banjarnegara and Wonosobo. This shows that there are still gaps in educational attainment in some areas, which may be caused by socio-economic factors, limited access to education, or differences in the level of educational infrastructure development.

3.3 Spatial Correlation of Poverty Percentage and Average Length of Education in Central Java

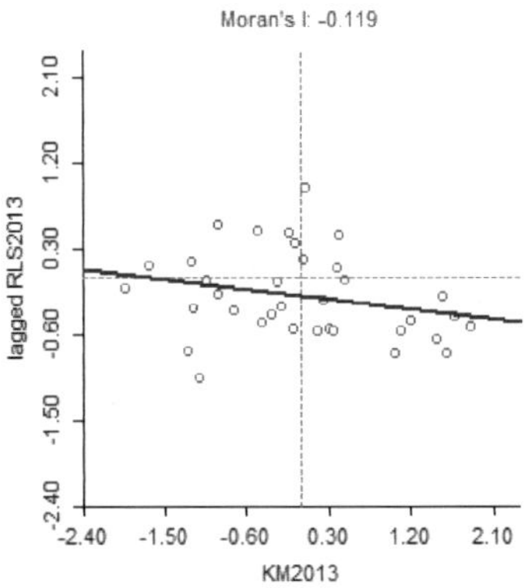

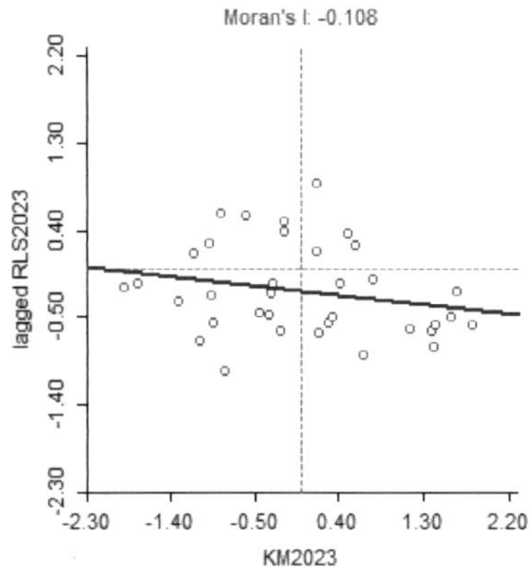

Figure 10. Bivariate moran's scatterplotin 2013 **Figure 11.** Bivariate moran's scatterplot in 2023

The Bivariate Moran's test in 2013 shows a negative spatial correlation (I = -0.119) between the percentage of poverty and the Average Length of Education in Central Java. The same thing happened in 2023, where the Bivariate Moran's Test also showed a negative spatial correlation (I = -0.108) between the percentage of poverty and the Average Length of Education in Central Java. This means that areas with a high poverty percentage tend to be surrounded by nearby areas with low Average Length of Education. Conversely, areas with a low poverty percentage are usually surrounded by areas with a high Average Length of Education. Thus, this result shows that in 2013 and 2023 there is an inverse relationship locally or in specific areas between the poverty percentage and the Average Length of Education in the surrounding areas.

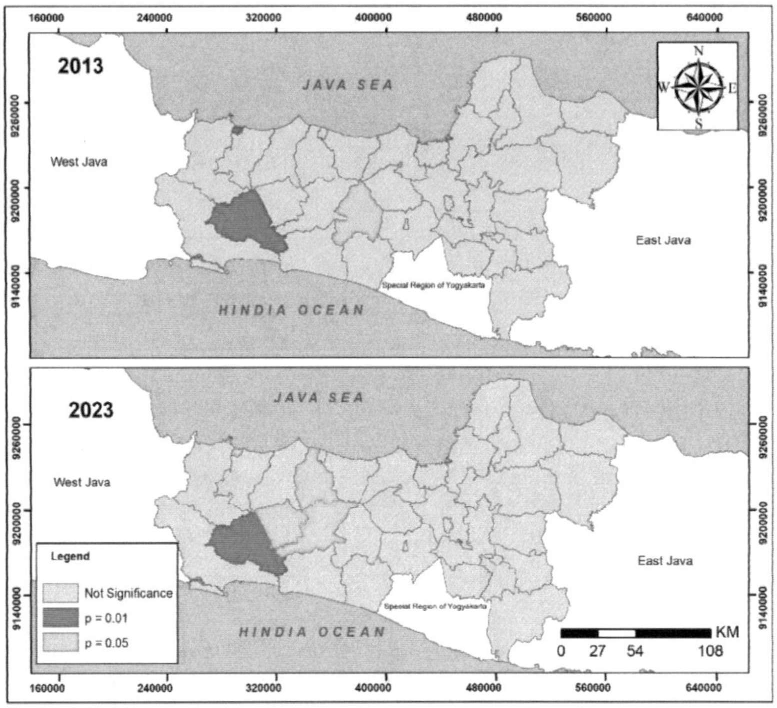

Figure 12. LISA Bivariate Significance Map

IOP Conf. Series: Earth and Environmental Science **1503** (2025) 012006 doi:10.1088/1755-1315/1503/1/012006

Figure 12 shows that there are several regions that are significant or have spatial correlation. In 2013 there were 3 regencies/cities, namely the city of tegal, Banyumas Regency and Wonosobo Regency. Meanwhile, in 2023 the number of regencies/cities increased to 5, namely Banyumas Regency, Purbalingga Regency, Banjarnegara Regency, Pekalongan Regency, and Tegal City. The increase in the number of regencies/cities with significant spatial correlation and the change in Wonosobo Regency from significant to insignificant indicates a change or shift in spatial patterns.

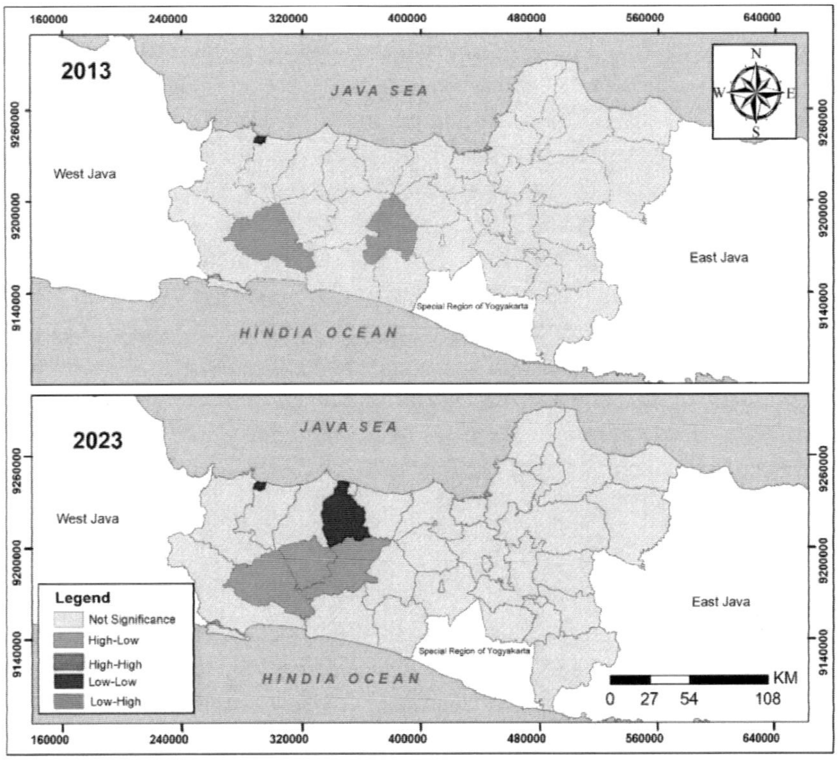

Figure 13. LISA bivariate Cluster Map

Based on Figure 13, in 2013, Tegal City was included in Quadrant III (Low-low) or positively correlated where there was a low poverty percentage value surrounded by neighboring areas with low Average Length of Education values. Meanwhile, Banyumas Regency and Wonosobo Regency are included in Quadrant IV (High-Low) or negatively correlated, which means that both regencies have a high percentage value of poverty and are surrounded by neighboring areas that have a low Average Length of Education.

The 2023 cluster map shows that Tegal City and Pekalongan Regency have a positive spatial correlation and are included in quadrant III (Low-low), which means that there is a low percentage of poverty surrounded by neighboring regions with low Average Length of Education. Kabupaten Banyumas, Kabupaten Purbalingga and Kabupaten Banjarnegara are included in quadrant IV (High-Low) or have a negative spatial correlation, which means that these three kabupaten have a high percentage of poverty and are surrounded by kabupaten with low Average Length of Education.

This is in line with data published by the Central Java Statistics Agency, which shows that the three districts classified in Quadrant IV (H-L) have a high poverty severity index value (0.50+)

[12]. The increasing number of districts in Quadrant IV (High-Low) in 2023 indicates the need for special attention to high-poverty areas surrounded by low Average Length of Education. This indicates that inequality in access to education may be contributing to poverty. The renewal in this study lies in identifying the location of the relationship between the two variables. Unlike previous studies, which relied solely on simple correlation analysis to determine the relationship between data, this study incorporates spatial correlation analysis. By considering spatial factors, this study offers deeper insights into how the relationship between variables varies across different regions, revealing spatial patterns that may have been missed in previous studies such as in the study by Faritz in 2020 [13].

Banyumas, Purbalingga and Banjarnegara districts should be the top priority for development interventions. The negative spatial correlation in these areas indicates high poverty which is likely related to low access to education in the surrounding areas. Investing in education programs in these and surrounding districts can open up new economic opportunities and help break the cycle of poverty in a sustainable way. The structural poverty that shackles communities is not easy to overcome. The difficulty in improving community welfare is not only influenced by a number of internal weaknesses, but also by external factors. Limited education, lack of opportunities to access and master more modern technology, often complicate efforts to improve people's lives [14]. The government is encouraged to enhance education quality by offering scholarships to disadvantaged communities and constructing schools in underserved and remote areas. These efforts aim to improve the overall quality of population [15]. By improving access and quality of education, communities will be more empowered, have relevant skills, and be ready to face the dynamics of the labor market, thus having a positive impact on the welfare of the local economy and future generations.

4. Conclusions

The distribution of poverty in Central Java changed from a pattern that was predominantly clustered in 2013 to a more evenly spread pattern in 2023, indicating progress in economic equality in several areas. Meanwhile, the Average Length of Education also increased in almost all regencies, although the distribution pattern remained relatively stable, signaling improved access to education. The spatial relationship between poverty percentage and education shows a negative correlation, where areas with high poverty tend to be located near areas with low education levels, both in 2013 and 2023. In 2023, more areas fell into the High-Low category, such as Banyumas, Purbalingga, and Banjarnegara, highlighting the need for efforts to improve education in these regions to help reduce poverty levels.

Acknowledgments

Appreciation was expressed to the Geography faculty, Muhammadiyah University of Surakarta (UMS) for organizing the 2024 International Conference of Geography and Disaster Management (ICGDM). This conference is part of the 2024 International Summit on Science, Technology and Humanity (ISETH), which aims to hold collaboration between beginners, practitioners and policy makers in the field of geography and disaster management. This event is a forum for sharing research, understanding the challenges of natural disasters, and the importance of sustainable development.

References

[1] Y. Yacoub, 2012, "Pengaruh Tingkat Pengangguran terhadap Tingkat Kemiskinan Kabupaten/Kota di Provinsi Kalimantan Barat," *Jurnal EKSOS*, vol. 8, no. 3, pp. 176–185.

[2] BPS Jawa Tengah, 2024, "Badan Pusat Statistik Provinsi Jawa Tengah. Jumlah dan Persentase Penduduk Miskin Menurut Kabupaten/Kota di Provinsi Jawa Tengah, 2023. [Online]. Available https://jateng.bps.go.id/id/statisticstable/3/UkVkWGJVZFNWakl6VWxKVFQwWjVWeTlSZDNabVFUMDkjMw==/jumlah-dan-persentase-penduduk-miskin-menurut-kabupaten-kota-di-provinsi-jawa-tengah--2023.html?year=2023."

[3] R. Rahmawati, D. Safitri, and O. U. Fairuzdhiya, 2015, "Analisis Spasial Pengaruh Tingkat Pengangguran terhadap Kemiskinan di Indonesia (Studi Kasus Provinsi Jawa Tengah)," *Media Statistika*, vol. 8, no. 1, pp. 23–30.

[4] T. Michael P and S. Stephen C, 2006, "Pembangunan Ekonomi, 9th" *Jakarta: Erlangga,* vol. 1.

[5] P. Mosley and A. Verschoor, 2005, "Risk Attitudes and the 'Vicious Circle of Poverty,'" *Eur J Dev Res*, vol. 17, no. 1, pp. 59–88, 2005, doi: 10.1080/09578810500066548.

[6] S. A. Puspitasari and C. Amin, 2024, "Analisis Korelasi Spasial Kemiskinan Terhadap Kasus Stunting di Provinsi Jawa Timur Tahun 2021 dan 2022," Skripsi thesis, Universitas Muhammadiyah Surakarta, Surakarta, 2024.

[7] C. Amin, B. Hidayah, D. N. Sari, K. D. Priyono, and C. A. T. Noermartanto, 2023, "Investigating Spatial Correlation Between COVID-19 Incidence and Poverty in Klaten Regency, Central Java," 2023, pp. 775–790. doi: 10.2991/978-94-6463-284-2_83.

[8] A. Hadi, 2019, "Pengaruh Rata-rata Lama Sekolah Kabupaten/Kota terhadap Persentase Penduduk Miskin Kabupaten/Kota di Provinsi Jawa Timur Tahun 2017.," *Media Trend*, vol. 14, no. 2, pp. 148–153, Dec. 2019, doi: 10.21107/mediatrend.v14i2.4504.

[9] L. Mason, B. Hicks, and J. Almeida, 2024, "Demystifying Spatial Dependence: Interactive Visualizations for Interpreting Local Spatial Autocorrelation," *in 2024 IEEE Visualization and Visual Analytics (VIS), St. Pete Beach, FL, USA: IEEE*, pp. 71–75. doi: 10.1109/VIS55277.2024.00022.

[10] D. Pathmanathan, I.-M. Dabo, T. H. Khoo, A. Ali-Hassan, and S. Dabo-Niang, 2024, "Spatial Principal Component Analysis and Moran Statistics for Multivariate Functional Areal Data," Cornel University.

[11] N. L. Anggani, H. M. Amrullah, and D. S. A. Gemilang, 2023, "Moran I Autocorrelation Study For Level Spatial Pattern Analysis," *Jurnal Indonesia Sosial Teknologi*, vol. 4, no. 9, pp. 1285–1291. doi: 10.59141/jist.v4i9.686.

[12] Badan Pusat Statistik Provinsi Jawa Tengah, 2024, "Indeks Keparahan Kemiskinan (P2)," [Online]. Available https://jateng.bps.go.id/id/statistics-table/2/NzgjMg%3D%3D/indeks-keparahan-kemiskinan--p2-.html."

[13] M. N. Faritz and A. Soejoto, 2020, "Pengaruh Pertumbuhan Ekonomi dan Rata-rata Lama Sekolah Terhadap Kemiskinan di Provinsi Jawa Tengah," *Jurnal Pendidikan Ekonomi (JUPE),* vol. 8, no. 1, pp. 15–21, Mar. 2020, doi: 10.26740/jupe.v8n1.p15-21.

[14] E. Y. Pinem, S. W. Widiono, and I. Irnad, 2019, "Kemiskinan Struktural Komunitas Nelayan di Kelurahan Sumber Jaya, Kecamatan Kampung Melayu, Kota Bengkulu," *Jurnal Sosiologi Nusantara*, vol. 5, no. 2, pp. 91–112, Dec. 2019, doi: 10.33369/jsn.5.2.91-112.

[15] D. A. Kusumaningrum, 2022 "Kemiskinan di Provinsi Jawa Tengah dan Faktor-Faktor yang Memengaruhinya," Skripsi Thesis, Universitas Muhammadiyah Surakarta, Surakarta, 2022.

[16] R. Mulyana, I. W. K. Suryawan, I. Y. Septiariva, S. Suhardono, M. M. Sari, N. Ulhasanah, N. N. Arifianningsih, W. Prayogo, & D. M. A. Buana, 2024, "Localising Global Standards in Toba, In-donesia Through a Tourist-Centric Per-spective of UNESCO Guidelines" *Forum Geografi*, vol. 38, no. 1, pp. 28-39. doi: 10.23917/forgeo.v38i1.3022

[17] L. Somantri, 2024, "Spatial Modeling Analysis of Potential Expansion of Ci-panas City Candidate Area as an Autono-mous City," *Forum Geografi,* vol. 38, no, 1, pp. 57-73. doi: 10.23917/forgeo.v38i1.3579

ANALYSIS OF MINIMARKET DISTRIBUTION TOWARDS TRADITIONAL MARKETS IN BLORA DISTRICT

Elzaharani Hermawan[1], Choirul Amin[2], M. Iqbal Taufiqurrahman[3]*, Umar El Izzudin Kiat[4], [5]
Faculty of Geography, Universitas Muhammadiyah Surakarta, 57162 Surakarta City, Indonesia

*) Corresponding Author (email: mit375@ums.ac.id)

ABSTRACT

Business competition between traditional markets and modern minimarkets is crucial as both types of markets play a role in the economy, especially in low-income countries such as Indonesia. Traditional markets are often the center of the local economy that provides livelihood for many small traders, while modern minimarkets offer convenience and competitive prices, this study aims to determine the distribution of minimarkets towards traditional markets using coordinate point data for each location. The results show that the distribution of minimarkets is clustered in two sub-districts.

Keywords: Distribution, Minimarket, Blora, Traditional Market

1. Introduction

Research on business competition between traditional markets and modern minimarkets is important as both types of markets play a role in the economy, especially in low-income countries such as Indonesia. Traditional markets are often the center of the local economy providing livelihoods for many small traders, while modern minimarkets offer convenience and competitive prices, often backed by efficient distribution networks and strong capital [1]. The presence of retail as a form of public service is an essential part of the infrastructure that supports urban development. Service itself is defined as an effort to provide for the needs of others [2] . A minimarket is a small store that offers a variety of goods, similar to those found in traditional markets. Essentially, a minimarket operates like a grocery store selling a wide range of items and food. However, the difference is that minimarkets implement a self-service system, where customers can pick the items they need from the shelves and pay for them at the cashier. [3]Competition between the two not only affects the welfare of small traders, but also affects people's consumption patterns, the price of goods, and the sustainability of the local economy. Unconsciously, economic, social, and spatial disparities have continued to widen [4].

This research can provide insights into how these two types of markets can adapt and compete healthily, as well as offer policy recommendations to support inclusive and sustainable economic development. Marketing performance is a concept used to evaluate the achievements of a business's marketing activities. This concept is essential for understanding how well the business competes in the market as an indicator of its success. Marketing performance is multidimensional, encompassing various objectives and types of organizations. The most prominent activities in the desakota area are the growth and development of industrial, commercial, and residential zones [5].

IOP Conf. Series: Earth and Environmental Science **1503** (2025) 012007 doi:10.1088/1755-1315/1503/1/012007

The development of minimarkets every year is increasing and uncontrolled but unevenly distributed in several regions resulting in business owners of grocery stores and traditional markets becoming deserted and going out of business. In November 2022, the number of Indomaret outlets spread across Java, Bali, West Nusa Tenggara, East Nusa Tenggara, Sumatra, Batam, Kalimantan, Sulawesi, Maluku, Sorong and Papua amounted to 21,026 outlets. While as of December 2023, [6]

Alfamart has 22,310 retail outlets and 360 *stock point* outlets with a percentage of 27% Jabodetabek, 40% Java (Non-Jabodetabek), and 33% Outside Java.[7] Regulations related to "Modern Stores" are regulated in Presidential Regulation (Perpres) No. 112 of 2007 concerning the arrangement and development of traditional markets, shopping centers, and modern stores, which was signed by President Susilo Bambang Yudhoyono on December 27, 2007 [8]. According to this regulation, modern stores are defined as stores with a self-service system that sells various types of goods at retail, including minimarkets, supermarkets, department stores, hypermarkets, and wholesalers. Modern stores are required to consider the socio-economic conditions of the surrounding community as well as the distance from existing traditional markets [9]. Article 3 of PR 112/2007 states that the building area of a minimarket must be less than 400m^2. In addition, the establishment of modern stores must follow the provisions in the local Regency / City Spatial Detail Plan (RTRW).

Previous studies conducted by Tiara Toyyibatul Arofah. M.Awaluddin, Arief Laila Nugraha (2017) on "GIS-based modern minimarket Distribution Analysis (Georafical Information System) in Kudus Regency" which has results in the form of conventional maps, tables of the number of minimarkets in each sub-district, minimarket growth maps, minimarket distance maps that are less than 500m2 terharap traditional markets [10]. Blora Regency is one of the regencies in Central Java Province. It is located between Grobogan, Rembang, Ngawi, and Tuban. Blora Regency has potential in the field of petroleum, especially in Cepu and Sambong Subdistricts, while Blora Subdistrict and Cepu Sub district itself are the economic centers in Blora Regency.[11] Therefore, an information system is needed to manage minimarket data in Blora Regency. Geographic information system technology can play an important role in analyzing business competition between traditional markets and modern minimarkets by providing in-depth spatial visualization of the location, distribution, and attractiveness of each type of market.[12] With GIS, researchers can map the location of traditional markets and modern minimarkets. It also enables distance analysis between traditional markets and modern minimarkets. Identify areas of high competition, and evaluate the potential impact of new minimarkets on existing traditional markets. Thus, GIS becomes a very useful tool to understand the dynamics of competition and assist in strategic decision-making by both the government and businesses to ensure the balance and sustainability of the local economy. [13] The use of collected data can be optimized through map visualization and spatial statistical analysis. In the context of urban spatial planning, the development of healthy residential areas must be balanced by providing environmental buffer spaces such as parks, green belts, or urban forest [14].. Map visualization will facilitate understanding of minimarket distribution patterns, while spatial statistical analysis can qualitatively explain distribution patterns, influencing factors, and projections of future developments. Therefore, the Geographic Information System (GIS) method is indispensable, and that is the reason researchers are interested in conducting research on "Minimarket Distribution Analysis of Traditional Markets in Blora Regency".

IOP Conf. Series: Earth and Environmental Science **1503** (2025) 012007 doi:10.1088/1755-1315/1503/1/012007

2. METHODS

2.1. Study location

Figure 1 is an illustration Blora Regency. The object of this research is mimarkets and traditional markets in Blora Regency. Meanwhile, the population of this research is the people of Blora.

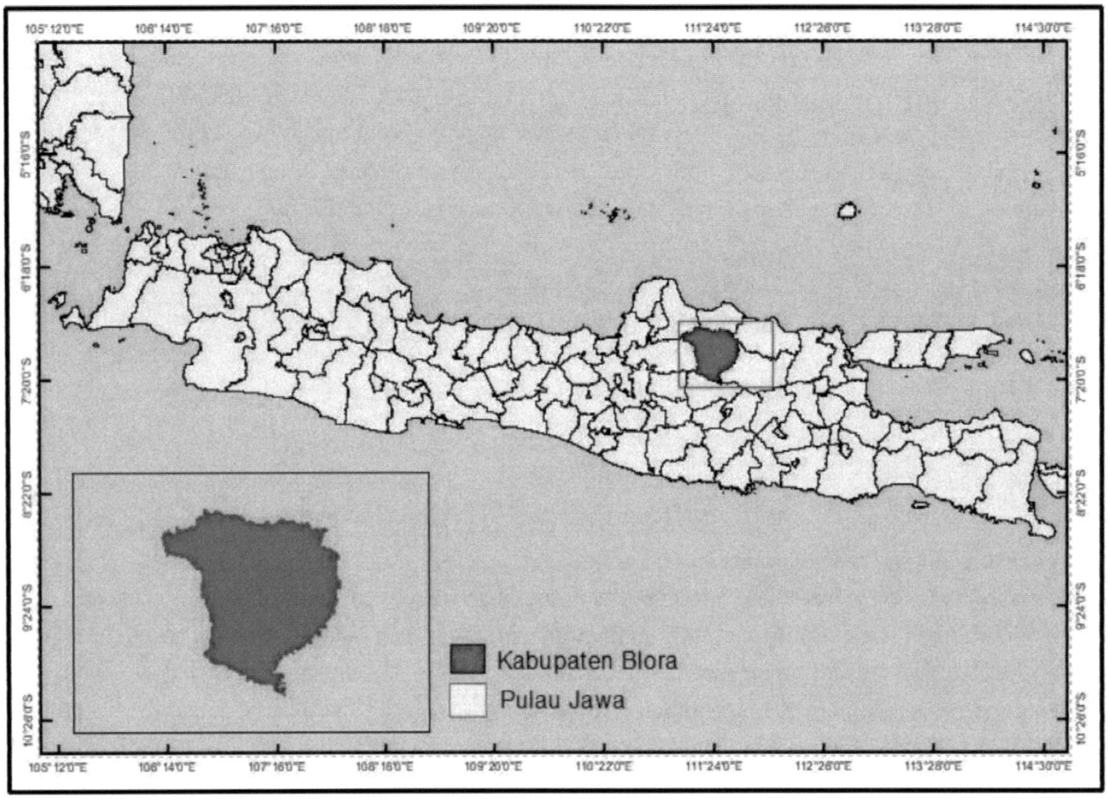

Figure 1: Location of Blora Regency

2.2. Tools and materials

Initial data was searched through Google Earth to determine the locations of minimarkets and traditional markets, which were then saved in .csv format in Excel. For further explanation, see Table 1 and Table 2 below.

Table 1. Tools used in the study

Tool name	Function
ArcMap 10.6	Spatial data processing and creation of distribution maps
Table 2	Data to be used

IOP Conf. Series: Earth and Environmental Science **1503** (2025) 012007 doi:10.1088/1755-1315/1503/1/012007

Table 2. -Data used in the study

Data name	Function	Source
Minimarket Coordinates	To find out the distribution of minimarkets in Blora Regency	Google Earth Pro
Traditional Market Coordinates	To find out the market location point traditional in Blora Regency	Google Earth Pro
District Administration Shapefile Blora	Basic materials in map making	Bappeda Blora Regency

2.3. Stages of data processing and analysis

In this research, data processing is carried out using qualitative data obtained from related institutions and websites. The researcher will analyze the data in detail and thoroughly through several stages[15]. First, we will collect information on the number of minimarkets and traditional markets operating in Blora Regency. Next, we will determine the location coordinates of each minimarket to create a distribution map using ArcGIS software [16]. After that, Average Nearest Neighbor analysis will be conducted to identify the distribution pattern of minimarkets in the area. A Hot Spot map will also be created using the Hot Spot Analysis tool to measure the density of the distance between minimarkets [17]. In addition, researchers will conduct a Buffer analysis to determine the extent to which minimarkets affect traditional markets. Finally, the findings of this entire analysis process will be described descriptively and concluded in writing, as well as further analyzed based on the results of the research that has been conducted.

2.3.1 Nearest Neighbor Analysis

In this study, to determine the distance between minimarkets, researchers first collected secondary data on the location of minimarkets. The data is then entered into Google Earth to find the location coordinates of each minimarket. After that, the location points found will be processed using ArcGIS software to map the distribution of minimarkets in Blora Regency. To measure the distance between minimarkets, researchers use the Euclidean distance tool available in ArcGIS. This tool will calculate the closest distance between each raster cell and the nearest source in Blora Regency, which will then be converted into kilometers.

2.3.2 Hotspot Analysis

Hotspot analysis is a method used to identify and analyze locations or areas that have a high level of occurrence or activity compared to other areas. The term "hotspot" itself refers to a location or area that shows a high density or intensity of an activity. [18]

certain phenomena. The result of this analysis is Heat Maps, which are data visualizations that show density or intensity in the form of color maps. Areas with high intensity will be marked with more striking colors.

3. RESEARCH RESULTS

total number of minimarkets in Blora Regency is 8 outlets, 38 Alfamart outlets and 43 Indomaret outlets.

X	Y	NAMA
565791.48	9209778.54	ALFAMART SURABAYA CEPU
565926.33	9210633.43	ALFAMART NGARENG
566633.89	9211232.23	ALFAMART GIYANTI CEPU
566207.51	9210799.17	ALFAMART SOROGO
561498.92	9205726.45	ALFAMART KAPUAN
564813.11	9209492.18	ALFAMART RONGGOLAWE
564858.14	9211127.32	ALFAMART WONOTEJO KARANGBOYO
563285.92	9209007.12	ALFAMART TAMBAKROMO
562798.22	9213739.01	ALFAMART SAMBONG
552211.71	9229156.12	ALFAMAART RAYA CEPU BLORA
556094.46	9226248.26	ALFAMART JIKEN
526253.82	9220603.00	ALFAMART JL RAYA KUNDURAN
526937.31	9220672.62	ALFAMART JL RAYA WONOSARI
520891.60	9232695.95	ALFAMART TODANAN
553731.03	9228927.12	ALFAMART JEPON
554850.54	9228444.58	ALFAMART SPBU JEPON
542542.03	9203767.73	ALFAMART RANDUBLATUNG
545317.35	9204882.29	ALFAMART JL RAYA CEPU-RANDUBLATUNG
557686.21	9232777.84	INDOMARET BOGOREJO BLORA
532062.23	9232683.90	ALFAMART JAPAH
550532.25	9202001.94	ALFAMART SUMBER BLORA
463050.42	9146641.29	ALFAMART JATISARI
524878.37	9220703.80	ALFAMART GAGAAN BLORA
534407.41	9225405.78	ALFAMART NGAWEN BLORA
549767.39	9229364.51	ALFAMART STAI BLORA

545070.69	9229509.79	ALFAMART GATOT SUBROTO
546390.19	9229501.67	ALFAMART JL PEMUDA
547105.70	9229982.54	ALFAMART JL BHAYANGKARI
545877.42	9229358.93	ALFAMART ARUMDALU BLORA
546132.60	9229883.26	ALFAMART DR SUTOMO
547286.80	9229432.72	ALFAMART JENSUD BLORA
545402.04	9229024.66	ALFAMART H345 MR ISKANDAR BLORA
545542.92	9230485.14	ALFAMART AGIL KUSUMADYO
547095.78	9231223.37	ALFAMART AHMAD YANI
555134.37	9207581.73	ALFAMART KEDUNGTUBAN
542548.80	9229847.14	ALFAMART TUNJUNGAN BLORA
543251.58	9229718.38	ALFAMART SUKORAME BLORA
548427.60	9235462.90	ALFAMART MEDANG

Table. 3 Coordinate point data of Alfamart minimarkets in Blora Regency

IOP Conf. Series: Earth and Environmental Science **1503** (2025) 012007 doi:10.1088/1755-1315/1503/1/012007

X	Y	NAMA
565422.48	9209931.47	INDOMARET DIPONEGORO 48
565539.82	9210634.47	INDOMARET PEMUDA 57
565375.83	9209450.10	INDOMARET DIPONEGORO
565903.19	9210504.80	INDOMARET NGARENG CEPU
565711.59	9209920.96	INDOMARET KEBUN KELAPA
565109.82	9208516.12	INDOMARET BALUN
564743.07	9209459.72	INDOMARET RONGGOLAWE
564291.97	9209276.30	INDOMARET RONGGOLAWE 2
563296.80	9208993.84	INDOMARET TAMBAKROMO
556126.45	9226252.31	INDOMARET JIKEN
524910.40	9220654.59	INDOMARET SAMBITORO
527393.26	9220748.90	INDOMARET JAGONG
520491.81	9232830.35	INDOMARET TODANAN
532393.02	9205934.54	INDOMARET DOPLANG BLORA
528694.97	9205699.16	INDOMARET GABUSAN
546778.39	9230241.98	INDOMARET KARANGJATI
543216.62	9203840.38	INDOAMRET RANDUBLATUNG
552518.55	9229115.80	INDOMARET JEPON 3
553494.07	9228960.23	INDOMARET RAYA JEPON
552979.08	9229031.46	INDOMARET JEPON
549935.35	9229323.89	INDOMARET SESO
548111.76	9229441.25	INDOMARET BANGKLE
534587.61	9225646.55	INDOMARET NGAWEN
535131.28	9226371.28	INDOMARET RAYA NGAWEN
550447.96	9206223.17	INDOMARET PETING RANDUBLATUNG
542569.39	9203796.34	INDOMARET RANDUBLATUNG-DOPLANG
543216.47	9203840.29	INDOMARET RANDU BLATUNG
560860.95	9208797.96	INDOMARET MERNUNG
548729.11	9200000.53	INDOMARET MENDENREJO
463478.45	9146358.99	INDOMARET KRADENAN
545320.72	9226443.56	INDOMARET KAMOLAN
544277.94	9229578.15	INDOMARET GATOT SUBROTO
546307.07	9229519.45	INDOMARET PEMUDA

545547.97	9229635.97	INDOMARET ALUN-ALUN BLORA
546002.17	9228563.18	INDOMARET CENDANA
546207.58	9230432.65	INDOMARET GUNUNG LAWU
545596.27	9230034.57	INDOAMRET RA KARTINI-KUNDEN
546435.95	9229816.88	INDOMARET RSU BLORA
548306.18	9234599.55	INDOMARET BLORA REMBANG 76
555080.53	9207618.66	INDOMARET KEDUNGTUBAN
555235.74	9205011.94	INDOMARET PULO KEDUNGTUBAN
562318.39	9214298.00	INDOMARET SAMBONG
542823.81	9229824.58	INDOMARET TUNJUNGAN

Table. 4 Coordinate point data of Indomaret minimarkets in Blora Regency

IOP Conf. Series: Earth and Environmental Science **1503** (2025) 012007 doi:10.1088/1755-1315/1503/1/012007

X	Y	NAMA
545226.21	9227415.88	PASAR RAKYAT BLORA SIDO MAKMUR
526151.86	9220565.69	PASAR TRADISIONAL KUNDURAN
532484.97	9232599.90	PASAR TRADISIONAL JAPAH
554527.48	9208013.42	PASAR TRADISIONAL KEDUNGTUBAN
546917.00	9229505.12	PASAR RAKYAT RAJAWALI
535863.35	9230687.34	PASAR TRADISIONAL PENGKOLREJO
534730.21	9225842.60	PASAR NGAWEN
527407.24	9214828.22	PASAR TRADISIONAL NGRONGGAH
553034.67	9229055.54	PASAR JEPON
519782.35	9233098.73	PASAR TRADISIONAL JEPON
546507.49	9230926.18	PASAR PON BLORA
543209.33	9203668.93	PASAR RAKYAT RANDUBLATUNG
548552.08	9235861.83	PASAR RAKYAT MEDANG
531854.54	9205967.42	PASAR DAERAH DOPLANG
519782.19	9233065.83	PASAR TRADISIONAL TODANAN
555493.91	9204485.36	PASAR TRADISIONAL WADO
555207.90	9207079.41	PASAR TRADISIONAL BAJO
516650.86	9231273.02	PASAR NGUMBUL

Table 5. Coordinate point data of Traditional Markets in Blora Regency

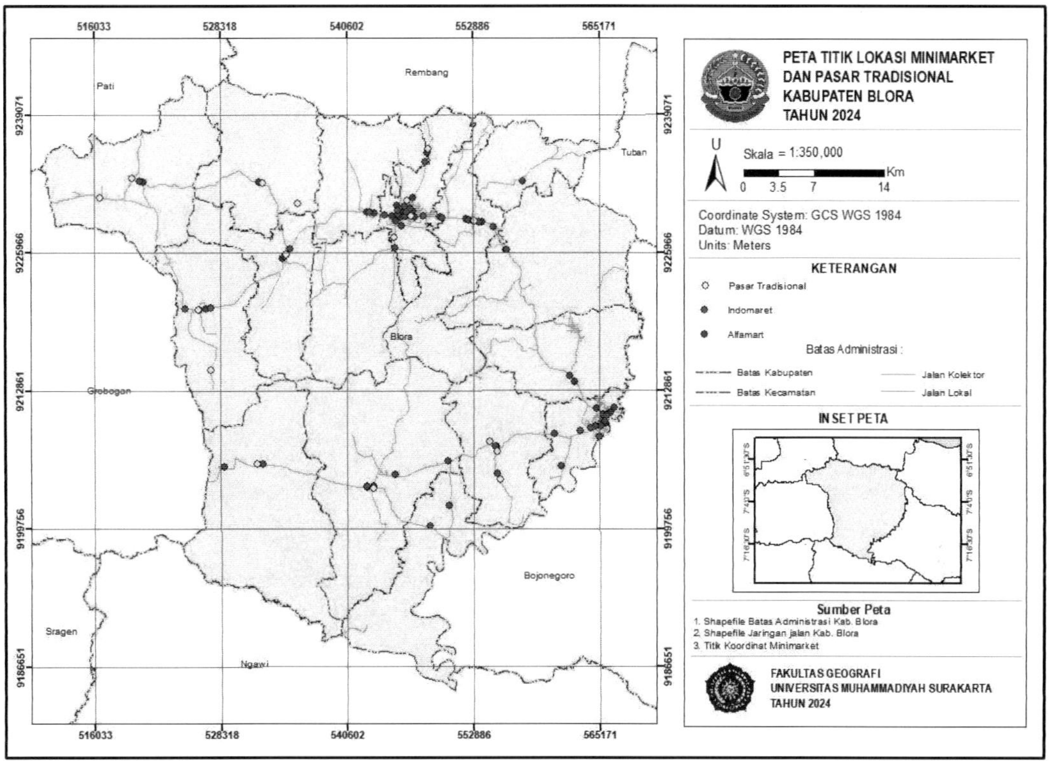

Figure 2. Map of minimarket location points and traditional markets in Blora district.

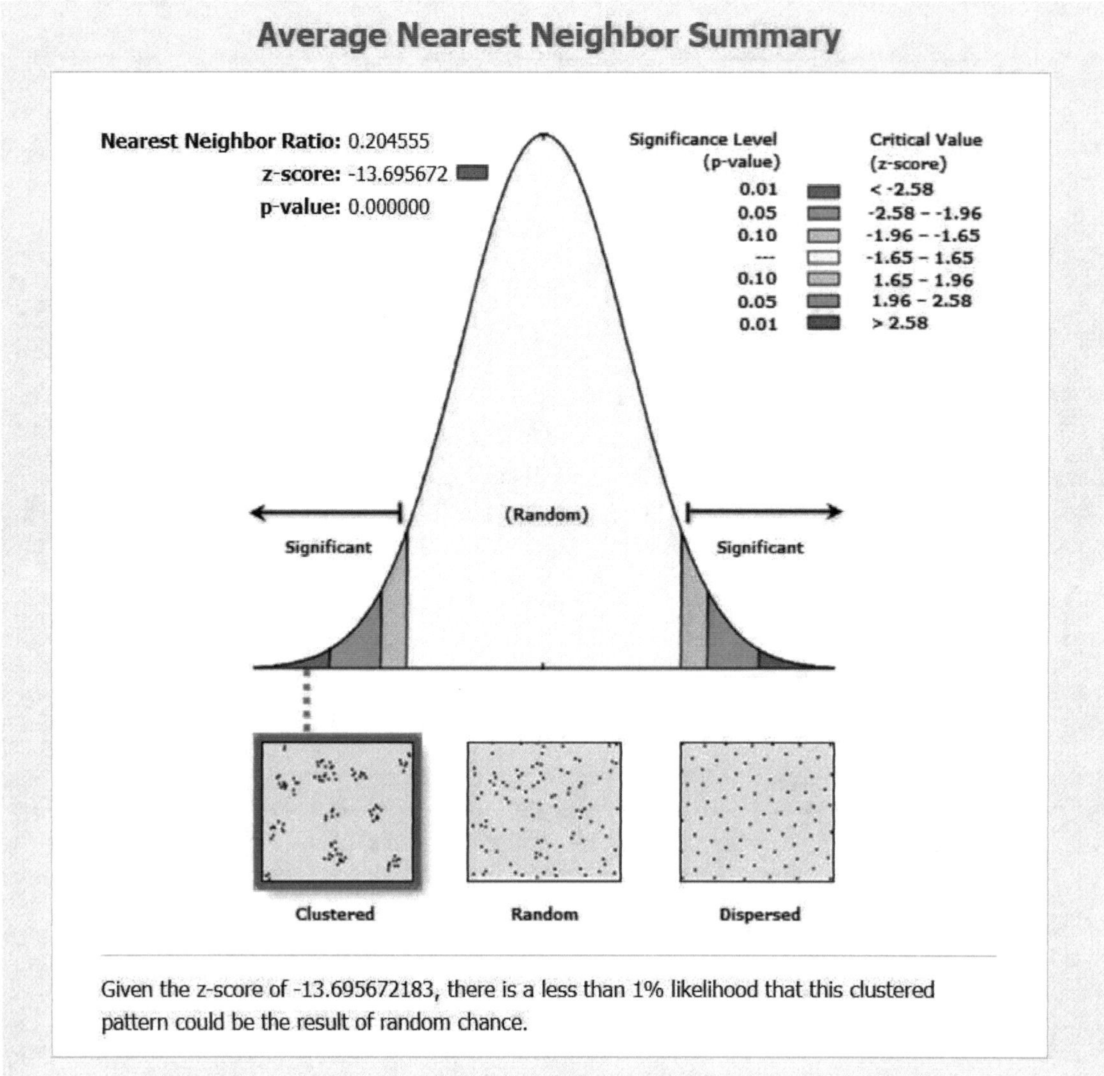

Figure 3. Average Nearest Neighbor Results

In the figure, the result of the Nearest Neighbor Ratio is 0.204555 with a Z-score of -13.69 which means it falls into the significant level 0.01 category and gets a clustered result.

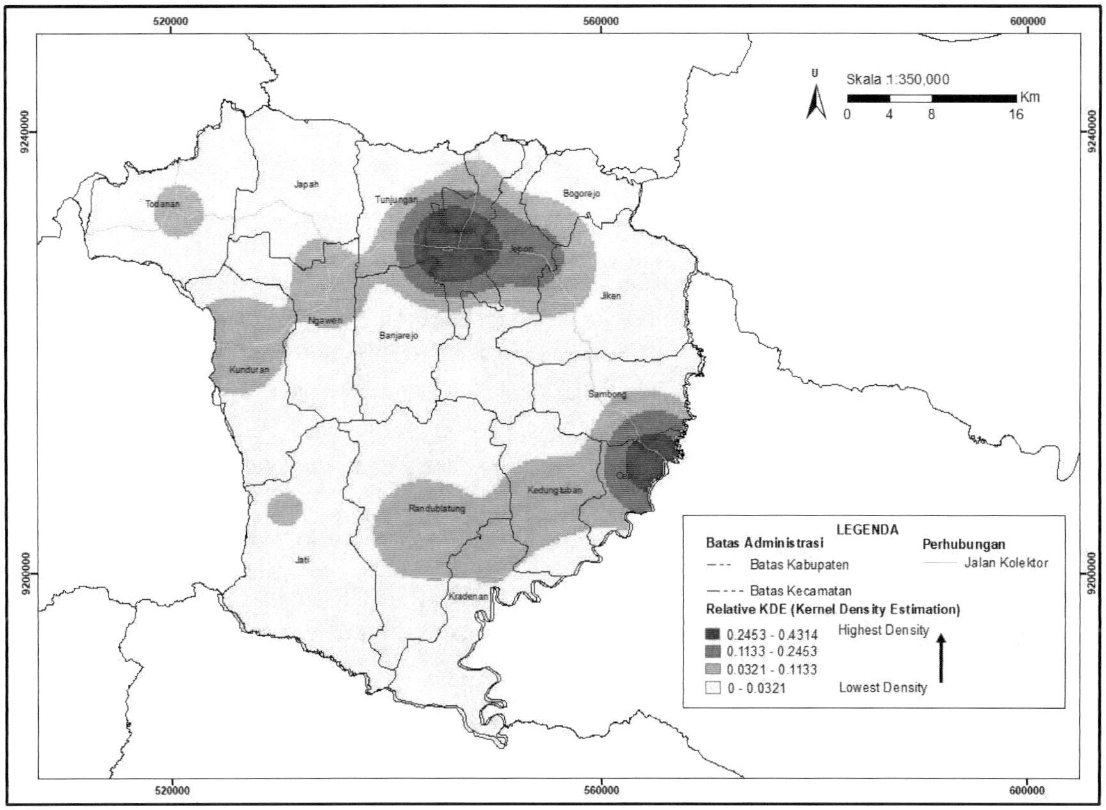

Figure 4.Hotspot Map of Blora Regency

From the hotspot map of minimarket distribution in Blora Regency using the Kernel Density Estimation tool, the red color indicates the highest density of minimarkets.

4. DISCUSSION

Land suitability for specific purposes is adjusted according to the objectives of the analysis [19]. Strategic site selection plays an important role in the economic success of a retail project. In choosing a location, it is necessary to consider the optimal combination of various characteristics, one of which is accessibility and visibility [20]. Based on the results of nearest neighbor analysis, it shows that the distribution of minimarkets in Blora Regency is mostly in two sub-districts, namely Blora and Cepu sub-districts. This is because Blora Sub-district is the center of government and administration so that it provides many public services while Cepu Sub-district has an important role in the economy and is strategically located close to the oil and gas industry. In addition, these two kecamatan have a higher population density than other kecamatan. As a result, there is inequality between regions and consequently inequality in access to services. This is why there are few minimarkets in other sub-districts and many residents still shop at traditional markets.

With a comparison of 18 traditional markets and 81 minimarkets, of course, there are positive and negative impacts. The impact of the existence of a more dominant minimarket facilitates consumers in accessing daily necessities and flexibility to consumers because many minimarkets are already operating 24 hours.

5. CONCLUSIONS

This study shows the distribution of minimarkets towards traditional markets in Blora Regency. The results show that the distribution of minimarkets is clustered in Blora District and Cepu District. Recommendations for the government are the strengthening of traditional markets in improving facilities such as cleanliness and parking areas. Then minimarket regulation to regulate the even distribution of minimarkets so that they are not too close to traditional markets. Then empowerment of small traders such as training to be more competitive and creative.

6. REFERENCES

[1] Susilowati, K. D. S. (2019). The impacts of modern market to traditional traders (A case in Malang City - Indonesia). International Journal of Technical Research and Applications, 2(Special Issue 8), 38–44
[2] Rachmawati, R. (2008). Pengembangan pusat pelayanan ekonomi di pinggiran kota sebagai alternatif penanganan problematik ruang di Kota Yogyakarta. Majalah Geografi Indonesia, 22(1), 73–90.
[3] Erlin Elisa, 2018, "Market Basket Analysis at Minimarket Ayu with Apriori Algorithm," Resti Journal, vol. 2, no. 2, pp. 472 – 478
[4] W. A. Astuti, & M. Musiyam, 2009, "Kemiskinan dan perkembangan wilayah di Kabupaten Boyolali," Fakultas Geografi, Universitas Muhammadiyah Surakarta.
[5] M. Dilahur, & C. Amin, 2001, "Orientasi penggunaan rumah di kawasan desakota: Studi kasus di bagian timur Kecamatan Kartasura, Kabupaten Sukoharjo," Fakultas Geografi, Universitas Muhammadiyah Surakarta.
[6] Gerai Indomaret, 2022, Indomaret.co.id
[7] Annual Report and Sustainability Report, PT Sumber Alfaria Trijaya Tbk, 2023, Alfamart
[8] Presidential Regulation of the Republic of Indonesia Number 112 of 2007 concerning the Compliance and Development of Traditional Markets, Shopping Centers and Modern Stores.
[9] Arief Laila Nugraha, Tiara Toyyibatul Arofah, M. Awaluddin, 2017, "GIS-based Modern Minimarket Distribution Analysis (Geographic Information System) in Kudus Regency," Undip Geodesy JournalAqli Wafirul. 2010. Buffer Analysis in Geographic Information System for Space Are Planning. INERSIA, vol. VI, no. 2
[10] Tiara Toyyibatul Arofah, M. Awaluddin, Arief Laila Nugraha, 2017, "GIS-based Modern Minimarket Distribution Analysis (Geographic Information System) in Kudus Regency," Undip Geodesy JournalAqli Wafirul. 2010. Buffer Analysis in Geographic Information System for Space Are Planning. INERSIA, vol. VI, no. 2
[11] Blora Regency Statistics Bureau, 2024, "Blora in Figures 2024", Blora: Central Bureau of Statistics.
[12] Effandi, & Faris, 2010, "Minimarket Distribution Patterns with Traditional Retailers Business Performance in Small Towns (Case Study: Serang City, Tanjungsari, and Lembang)," Journal of Urban and Regional Planning, vol 21, no.3, pp.183-196.
[13] Nugraha, Sigid Aditya, 2013, "Analysis of Distribution Pattern of Traditional Market and Modern Market in Surakarta City with Geographic Information System (GIS) Application," Thesis, University of Muhammadiyah Surakarta.
[14] Dahroni, 2005, "Analisis keruangan terhadap perubahan dan persebaran permukiman kumuh di Kecamatan Banjarsari Kota Surakarta tahun 2001–2005," Fakultas Geografi, Universitas Muhammadiyah Surakarta.
[15] F. Wibowo, A. U. Khasanah, & F. I. F. S. Putra, 2022, "Analisis dampak kehadiran pasar modern terhadap kinerja pemasaran pasar tradisional berbasis perspektif pedagang dan konsumen di Kabupaten Wonogiri," Benefit: Jurnal Manajemen dan Bisnis, vol. 7, no. 1, pp. 53-65.
[16] B. K. K. Ahammed, I. W. G. A. Karang, I. W. Nuarsa, I. C. G. S. Indrawan, N. M. B. S. Dewi, & A. C. Pandey, 2024, "Mapping of the Spatio Spectral Dynamics of Mangrove Chlorophyll Con-centrations via Sentinel-2 Satellite Imagery" Forum Geografi, vol. 38, no. 2, pp. 244-246. doi: 10.23917/forgeo.v38i2.4518
[17] A. Ahmad, T. Masron, S. N. Junaini, M. H. Barawi, M. S. Redzuan, Y. Kimura, N. Jubit, L. H. Bismelah, & A. S. M. Ali, 2024, "Criminological Insights: A Comprehensive Spatial Analysis Crime Hot Spots of Property Offenses in Malaysia's Urban Centers," Forum Geografi, vol. 38, no. 1, pp. 94-109. doi: 10.23917/forgeo.v38i1.4306
[18] Valgunadi, A. N. ., Zidanarta, M. B., Rahmalia, A., & arrasyid, R. (2023). Analisis Hotspot (Getis Ord Gi*) dan Average Nearest Neighbour (ANN) pada Sebaran Pariwisata di Kabupaten Wonosobo. Jurnal Pendidikan Geografi Undiksha,

11(2), 204–214

[19] A. A. Sigit, 2007, "Analisis Proses Degradasi Lahan dan Dampaknya Terhadap Produktivitas Lahan Pertanian di Kabupaten Klaten", Fakultas Geografi Unibersitas Muhammadiyah Surakarta

[20] H. Agustin, S. Astutik, M. A. Mujib, E. I. Pangastuti, & F. A. Kurnianto, 2024, "Impact Analysis Distribution of Minimarkets on the Income of Traditional Retailers in Yosowilangun Village, Lumajang Regency Using Moran's I Analysis," Geography Learning Magazine, vol. 7, no. 1, pp. 83-92.

IOP Conf. Series: Earth and Environmental Science **1503** (2025) 012008 doi:10.1088/1755-1315/1503/1/012008

Sustainable Water Management in the ITDC Nusa Dua Area, Bali: Strategies, Impacts, and Hotel Awareness Levels

Meitolo Hulu[1*], Nathalie Syaloom Angelina Gandi[2], Patricia Dafhanie Mulyana[3]

[123]Faculty of Hospitality and Tourism, Universitas Pelita Harapan, Indonesia

*E-mail: meitolo.hulu@uph.edu

Abstract. Water scarcity has become a critical issue in Bali, exacerbated by mass tourism, which has led to water pollution, freshwater shortages, and environmental degradation. In response, the Nusa Dua area, managed by the Indonesia Tourism Development Corporation (ITDC), has implemented several water management strategies aimed at mitigating these challenges and promoting sustainable tourism. The aim of this study is to examine the water management efforts implemented by hotels in the Nusa Dua ITDC area, Bali, in response to water scarcity and wastewater pollution. Specifically, it seeks to explore the strategies hotels use to conserve water, manage wastewater, and mitigate environmental impacts. This study employs a qualitative research approach with a case study design to explore wastewater management practices and their impact on environmental sustainability in the Nusa Dua ITDC tourism area, Bali. The research involves several data collection methods, including in-depth interviews, Focus Group Discussions (FGD), participatory observation, and document analysis, to gain a comprehensive understanding of water management efforts in the area. This study investigates the efforts made by hotels in the Nusa Dua ITDC area to address water scarcity and wastewater management. The research explores the role of wastewater recycling systems, the use of Sea Water Reverse Osmosis (SWRO) technology, and the implementation of water-saving equipment. It highlights the positive impacts, such as cost reductions and environmental benefits, while also addressing challenges like high investment costs and the potential environmental risks associated with advanced water treatment technologies. The findings underscore the importance of integrated water management and collaboration among stakeholders, including the government, tourism managers, and local communities, to ensure long-term sustainability. The study concludes that while significant progress has been made, further innovation and policy support are essential to maintaining Bali's status as an eco-friendly tourism destination amidst increasing water demand.

1. Introduction

Water scarcity has emerged as a significant barrier to socio-economic progress and a growing threat to livelihoods in many regions around the world [1]. Bali has experienced significant impacts from the development of mass tourism, including water pollution, freshwater shortages,

IOP Conf. Series: Earth and Environmental Science **1503** (2025) 012008 doi:10.1088/1755-1315/1503/1/012008

environmental degradation, damage to aquatic vegetation, and sanitation issues [2] and water scarcity problem [3]. A study on water scarcity in Bali shows that the demand for water in the tourism sector increased by 20.8 million m^3 (295%) between 1988 and 2013 [4]. Excessive groundwater extraction for tourism needs in Bali has caused a significant decline in water levels on the island, resulting in more than 260 rivers out of over 400 rivers in Bali drying up [5]. As a result, seawater has begun to infiltrate Bali's groundwater reserves. In addition, efforts to reduce water consumption must also pay great attention to wastewater management, as Bali faces water pollution. The Indonesia Tourism Development Corporation (ITDC), established under the state-owned corporation in 1974, took over unproductive coastal land to develop a tourism destination in the Nusa Dua area, Bali. To promote sustainable tourism, the Nusa Dua ITDC developed a wastewater recycling system called Lagoon Nusa Dua, which treats wastewater before it is released into water bodies and used for non-potable purposes. They also implemented a Water Stabilization Pond technology system.

Curiosity arose about how this integrated tourism area, consisting of 19 star-rated hotels, manages water both individually and collectively. This raised several questions about the efforts of the hotels in the Nusa Dua ITDC to manage water to minimize water shortages, the positive and negative impacts faced by these hotels in water management, and their awareness level of water management in relation to water scarcity. Furthermore, the water management efforts in the Nusa Dua area also involve the implementation of environmentally friendly practices aimed at minimizing negative impacts on the local ecosystem. The implementation of advanced technological solutions and practices that enhance water use efficiency, along with the utilization of alternative (non-traditional) water resources, is emphasized and explored as a strategy to tackle increasing environmental and health concerns, while also minimizing emerging conflicts among water users [6]. Hotels in the area are increasingly aware of the importance of water conservation, with many starting to implement water-saving technologies, such as real-time water consumption monitoring systems and the use of water-efficient equipment. Good water management not only reduces the impact on the availability of clean water but also enhances the image of the tourism area as an environmentally friendly destination. Therefore, sustainable water management is a key factor in maintaining the appeal of the tourism area while also preserving the sustainability of natural resources.

However, despite various efforts to address water scarcity issues, significant challenges remain in terms of collaboration among relevant stakeholders. The government, tourism area managers, and local communities need to work together to create policies that support efficient and sustainable water management. One important step is strengthening monitoring and regulation systems to ensure that every hotel and other tourism facilities comply with established water management standards. Additionally, investment in more innovative and environmentally friendly water management technologies is crucial to address the projected growth in the number of tourists in Bali. This study aims to explore the extent to which the hotels in Nusa Dua ITDC recognize their efforts in maintaining wastewater management for the sustainability of the surrounding environment, as well as the strategies they have implemented to address water scarcity and water pollution issues around ITDC Nusa Dua, Bali. This study also examines the positive impacts or potential threats that may arise from these water management processes.

2. Theoritical Perspectives

The principle of sustainability encompasses three main components: environment, economy, and socio-culture in tourism development. The implementation of 12 aspects of sustainable tourism can provide a sense of safety and comfort for both domestic and foreign tourists at The Nusa Dua ITDC [7].Wastewater management includes collection, treatment, and reuse of wastewater, which is an essential strategy for preserving water resources [8]. Previous study identified three main sources of wastewater production: domestic, industrial, and agricultural [9]. Domestic wastewater comes from household activities, industrial wastewater comes from sectors such as food, chemicals, paper and pulp, nuclear, thermal power, laundry, pharmaceuticals, mining, and steel, while agricultural wastewater comes from livestock management, irrigation, and the use of pesticides and fertilizers.

In general, there are two approaches to wastewater treatment. A centralized system, often managed by the public sector, requires large pipes, significant excavation, and access holes to collect and treat large volumes of wastewater, especially in large cities [10]. In contrast, decentralized systems handle wastewater from individual households, household communities, subdivisions, and isolated commercial establishments at or near the point of wastewater generation [11]. Andini and Arida (2019) stated that wastewater management in the Nusa Dua Lagoon is considered effective because it can process 4,000-7,000 m^3 per day and is utilized for irrigation [12].

Implementation of integrated water management is one of the pillars of robust sustainable development [13]. Water management strategies must include the integration of wastewater quality and its vulnerability. According to the Global Water Partnership, Integrated Water Resources Management (IWRM) is defined as a coordinated process in managing water, land, and related resources aimed at maximizing economic and social benefits without compromising the sustainability of vital ecosystems [14]. Good water management, by minimizing water consumption for more efficient use, can help reduce costs. The Dublin Statement document states that the benefits of water recycling can reduce water consumption in industrial areas by 50% or more, which will also have a positive impact in limiting pollution [15].

In the context of the hospitality industry, efficient water management is not only important for environmental sustainability but also for reducing significant operational costs. As demand increases for environmentally friendly hospitality facilities, hotels worldwide are beginning to implement more efficient water management strategies, such as the use of water-saving technologies in bathrooms, kitchens, and public areas. According to Lee et al. (2017), the application of water-saving technologies in hotels can reduce water consumption by up to 30%, which directly impacts cost savings and reduces negative environmental impacts [16]. With growing awareness of water scarcity issues, sustainable water management has become an important part of responsible hotel operational strategies.

Furthermore, hotels are increasingly interested in internal wastewater treatment systems to reduce their reliance on external water sources. Effective wastewater treatment systems enable hotels to recycle used water, reduce the burden on public water management systems, and mitigate pollution impacts on the surrounding environment. The application of technologies such as biofilter systems, membrane-based wastewater treatment, and the use of treated wastewater for non-potable purposes (e.g., for irrigation or air cooling) has proven effective in reducing the consumption of clean water. Implementation of wastewater recycling systems in the hospitality sector not only reduces water consumption but also contributes to reducing carbon emissions and the impacts of climate change [17]. On the other hand, good water management can also strengthen a hotel's image in the eyes of travellers who are increasingly concerned with environmental issues. Hotels that implement eco-friendly practices, including efficient water management, often receive higher appreciation from guests who choose accommodations based on sustainability criteria. Previous study found that 60% of travellers are more likely to choose a hotel with an environmentally friendly policy, including good water management [18]. This indicates that efficient water management is not only an operational aspect but also a marketing strategy that can enhance a hotel's attractiveness.

The importance of efficient water management is also aligned with global efforts to reduce the ecological footprint of the hospitality industry. International organizations, such as the World Tourism Organization (UNWTO), have emphasized the importance of water management in the tourism sector to support long-term sustainability. UNWTO in 2017 reminded that the tourism sector, especially hospitality, must take responsibility in managing natural resources, including water, to reduce negative impacts on local and global ecosystems. Therefore, it is crucial for hotels in major tourist areas like Nusa Dua ITDC to adopt sustainable water management principles in their operations to support sustainable development goals and preserve environmental quality for future generations.

3. Research Method

This research uses a qualitative approach with a case study to delve deeper into wastewater management and its impact on environmental sustainability in the Nusa Dua ITDC tourism area, Bali. The methods used include in-depth interviews with hotel managers and relevant staff, aiming to gain an understanding of their awareness and practices of water management, as well as the positive and negative impacts of these efforts [19]. Additionally, a Focus Group Discussion (FGD) was conducted to gather collective views from hotel managers on the challenges faced and the solutions implemented [13]. Participatory observation was conducted to directly monitor the wastewater treatment systems applied in these hotels and to evaluate the effectiveness of the technologies used, such as those implemented in the Nusa Dua Lagoon, Bali [20]. Furthermore, analysis of related documents, such as water management policy reports and hotel case studies, will be used to examine existing water management policies and practices [21]. The case study approach allows for an in-depth analysis of various hotels in the ITDC area that implement different water management policies and technologies, to understand the successes and challenges in implementing sustainable tourism [22]. The collected data will be analyzed using thematic analysis to identify patterns and themes related to water management and its impact on the environment and tourism sustainability in Bali [11].

The key informants in this study include 9 hotel managers by selecting four 5-star hotels and four 4-star hotels. The selection of these hotels is based on star classification and their location around Nusa Dua, Bali. The key informants also from ITDC Nusa Dua Management, laundry industry and local residents around Nusa Dua. The interview process consists of a series of questions aimed at exploring the efforts currently being implemented by stakeholders to reduce water usage and minimize wastewater impacts. Additionally, the interviews aim to identify the positive and negative impacts of these efforts and the level of awareness regarding water

IOP Conf. Series: Earth and Environmental Science **1503** (2025) 012008 doi:10.1088/1755-1315/1503/1/012008

management in the Nusa Dua ITDC area. The following Informants, including nine-star hotels and two four-star hotels, ITDC Nusa Dua Management, Industry, The questions and data sources in table 1.

Table 1. Data Sources

Nu.	Data	Sources
1	How do the hotels in the Nusa Dua ITDC area manage water resources individually and collectively?	Hotel
2	What efforts have been made by the hotels in the Nusa Dua ITDC area to reduce water consumption in response to water scarcity?	Hotel
3	To what extent are the hotels in the Nusa Dua ITDC area aware of the importance of water management for the sustainability of the surrounding environment?	Hotel
4	What positive impacts have the hotels in the Nusa Dua ITDC area experienced related to their wastewater management?	Hotel, ITDC Nusa Dua Management, Industries around Nusa Dua, Local citizens
5	What negative impacts have arisen from the water management practices of the hotels in the Nusa Dua ITDC area on the surrounding environment?	Hotel, ITDC Nusa Dua Management, Industries around Nusa Dua, Local citizens
6	What strategies have been implemented by the hotels in the Nusa Dua ITDC area to address water scarcity issues in the area?	Hotel ITDC Nusa Dua Management
7	How does the wastewater recycling system in the Nusa Dua ITDC area contribute to reducing water pollution and addressing water scarcity?	Hotel ITDC Nusa Dua Management
8	What challenges do the hotels in the Nusa Dua ITDC area face in their efforts to maintain sustainable water management?	Hotel ITDC Nusa Dua Management

[a] Source: Researcher, 2023

IOP Conf. Series: Earth and Environmental Science **1503** (2025) 012008 doi:10.1088/1755-1315/1503/1/012008

4. Research Findings

4.1. Efforts Implemented by Hotels in Water Management

The following, in Table 2,3, and 4 are the results of the interview with the informant regarding the efforts made to manage water for use in operational activities at ITDC Nusa Dua, Bali.

Table 2. Data Sources

Nu.	Informant	Efforts
1	ITDC Nusa Dua Management	Integrated Wastewater Treatment Plant Installation (Lagoon), Irrigation from Recycled Water, Socialization, Regulation Creation, Investor Selection, Recycled Water Quality Monitoring, Routine Maintenance, Sustainability Certification, Technology Development Division, Water Transfer to Mangrove.
2	Four Stars Hotel	Replacement of Deep Wells as Water Source, Installation of Water Saving Equipment, Compliance with Regulations, Signage, Socialization, Initial Wastewater Treatment, Irrigation from Recycled Water, Outsourcing Laundry.
3	Five Stars Hotel	Socialization - Training, Replacement of Deep Wells as Water Source (Local Water Company & SWRO), Irrigation from Recycled Water, Water Usage Monitoring, Installation of Water Saving Equipment, Signage, Sustainability Certification, Initial Wastewater Treatment, Monitoring of Chemicals in Water, Rainwater Utilization, Compliance with Regulations, Seawater Lagoon.
4	Industries around Nusa Dua	Compliance with Regulations, Use of Sustainable Laundry Chemicals, Investment in Sustainable Machines, Wastewater Treatment, Water Recycling.

[a] Source: Processed Data, 2023

The water management efforts implemented by hotels in the The Nusa Dua ITDC area aim to reduce water scarcity and support environmental sustainability. One of the main measures applied is the operation of the Integrated Wastewater Treatment Plant (Lagoon ITDC Nusa Dua), which has been in operation since 1974. In this system, all actors in the area, including hotels, restaurants, and other businesses, are required to channel their wastewater to this plant for treatment. The treated wastewater can then be used for irrigation or discharged into water bodies, in accordance with the Minister of Environment Regulation No. 5/2022 [23]. The hotels in this area are also equipped with accumulation tanks to store wastewater before it is channeled to the Lagoon ITDC. The cost of wastewater treatment varies depending on the hotel's occupancy rate. Some five-star hotels even utilize treated wastewater for garden irrigation [24].

Additionally, hotels in the area replace their main water sources with deep wells or use local water supplies as backups. Hotels such as the St. Regis Bali Resort have even invested in Sea Water Reverse Osmosis (SWRO) technology, which allows them to produce clean water from seawater and process it into drinking water [23]. Other hotels also treat wastewater independently before channeling it to Lagoon ITDC [25]. ITDC also enforces strict regulations regarding environmental management, including a ban on deep well drilling by hotels in the area. One of the sustainable practices supporting eco-friendly water management is the policy applied by laundry service providers, such as Texkleenindo, who comply with wastewater treatment regulations [11]. To raise environmental awareness among hotel staff, ITDC holds various socialization and training programs. Some hotels even organize monthly training sessions and establish internal committees called 'Green Committees.' Guest awareness is raised in ways that do not disrupt their comfort, such as through information provided in rooms or common areas [20]. Hotels have also installed water-saving equipment, such as low-flow showerheads and toilets, to reduce water consumption. Water quality is routinely monitored, including the control of chemical dosages used in water treatment, to ensure that treated wastewater can be safely reused without harming the environment [8].

4.2. Positive and Negative Impacts of Water Management at ITDC Nusa Dua

The following, in Table 3, are the results of the interview with the informant regarding the awareness of hotels in The Nusa Dua ITDC complex regarding water management.

Table 3. The positive impacts and downsides that may result from managing water in The Nusa Dua ITDC management

Nu.	Informant	Negative Impacts	Positive Impacts
1	ITDC Nusa Dua Management	Long-term effect of SWRO, Odor emission, Local water company's water demand increases due to new settlements.	The bird sanctuary boosts tourism marketing, e competition, while providing a healthy nutrien Additionally, it ensures the local water quality ar preserved
2	Four Stars Hotel	Additional cost for the Lagoon	Cost efficiency in integrated water management (f labor), accurate data monitoring from the lagoon, reduction
3	Five Stars Hotel	Groundwater absorption by SWRO, high investment cost for SWRO, occasional distribution delays, odor emissions.	Cost efficiency in integrated water management (fre and technology investment), recognition as a sustai hotel marketing), improved water quality and quan savings with SWRO, healthy nutrients for plant: impact, well-preserved local water quality and quan scarcity, and reduced water pollution
4	Industries around Nusa Dua	Detergent waste can damage the environment if not properly treated, and the high cost of well water	Cost efficiency for hotels, minimal environmental ir water scarcity
	Local citizens	Odor emission	Healthy nutrients for plants, indirect long-tern preserved local water quality and quantity

[a] Source: Processed Data, 2023

IOP Conf. Series: Earth and Environmental Science **1503** (2025) 012008 doi:10.1088/1755-1315/1503/1/012008

Integrated water management in the ITDC Nusa Dua area has had several positive impacts, particularly in terms of cost efficiency for hotels, especially four-star hotels. These hotels have been able to reduce the costs of clean water usage and labor. Some hotels even outsource their laundry services to providers like TexKleenindo to save on operational costs (Massoud et al., 2008). The Sea Water Reverse Osmosis (SWRO) system used by some five-star hotels helps reduce operational costs, as the cost per cubic meter of produced water is much cheaper compared to water supplied by the local water utility (PDAM) [8]. Furthermore, effective wastewater treatment helps reduce pollution while also meeting legal obligations related to waste processing. The treated wastewater has also been shown to be beneficial for irrigation, even supporting plant growth, as demonstrated by Kayumanis Nusa Dua Villa [20]. The sustainability efforts implemented in this area also enhance the image of ITDC Nusa Dua as an environmentally friendly destination. This not only attracts international tourists but also helps ITDC Nusa Dua receive prestigious awards and green certifications, further increasing the appeal of the area in the international market [7].

However, water management in this area is not without its negative impacts. Some hotels, such as The Grand Nusa Dua, opt to use groundwater, which is cheaper, even though the cost of distributing water from Lagoon ITDC is quite high. Groundwater use also requires substantial treatment costs, particularly due to the hardness of the groundwater, which can damage laundry equipment [5] The Sea Water Reverse Osmosis (SWRO) technology, while effective in producing clean water, requires a significant initial investment and high operational costs, especially concerning the energy needed to pump seawater [26]. Additionally, wastewater treatment at Lagoon ITDC can generate unpleasant odors, especially during the rainy season, which may reduce the comfort of residents in the area [11]. The application of SWRO technology also carries the risk of damaging the marine ecosystem, as this process requires large amounts of seawater, which could disrupt the balance of the surrounding ecosystem. Increasing demand for water also places pressure on the existing water resources, necessitating more sustainable management to ensure water availability for all parties in the area [14].

4.3. Environmental Awareness

Environmental awareness in water management plays a crucial role in achieving sustainable development, especially in regions like Bali, where tourism places significant pressure on natural resources. Water conservation efforts must begin early, particularly to address the over-extraction of groundwater driven by tourism activities [27]. In response, several hotels, including the Sofitel, have implemented sustainability standards that align with the United Nations' Sustainable Development Goals [7]. Furthermore, wastewater management has emerged as an important issue, with properties such as ITDC Nusa Dua and Grand Hyatt recognizing the need for sustainable wastewater disposal methods to protect the environment and improve public health. These efforts are supported by Bali's cultural values, particularly the "Tri Hita Karana" philosophy, which advocates for a harmonious relationship between humans, nature, and the divine. This philosophy is reflected in Bali Provincial Regulation No. 5 of 2020, which mandates that tourism development adhere to these principles to preserve Bali's natural and cultural heritage [4]. In the context of ITDC Nusa Dua, these water management efforts not only promote environmental sustainability but also support the development of eco-friendly tourism [28]. Moreover, Bali's integrated water management and incorporation of traditional wisdom offer valuable lessons for sustainable tourism practices globally ([22]; [29]; [30]; [31].

The following, in Table 1, are the results of the interview with the informant regarding the awareness of hotels in The Nusa Dua ITDC complex regarding water management.

Table 4. Awareness of hotels in The Nusa Dua ITDC complex regarding water management

Nu.	Informant	Efforts
1	ITDC Nusa Dua Management	Corporations have an environmental responsibility. Cultural values play a role in preserving the environment, and environmental damage should be eliminated. Sustainable tourism should be supported, and discharge must be environmentally friendly
2	Four Stars Hotel	Groundwater capacity is limited, and natural resources are a valuable inheritance
3	Five Stars Hotel	Cultural values play a key role in preserving the environment. Global warming is a pressing issue, and sustainability standards must be followed. Groundwater capacity is limited, so discharge must be environmentally friendly. An integrated WWTP is beneficial, and management should be both sustainable and advantageous. Water is a precious gift
4	Industries around Nusa Dua	Compliance with Regulations, Use of Sustainable Laundry Chemicals, Investment in Sustainable Machines, Wastewater Treatment, Water Recycling.

[a] Source: Researcher, 2023

The coastal area of Nusa Dua, Bali, further underscores the urgency of sustainable water management, as the region's beaches and marine ecosystems are highly susceptible to pollution and erosion caused by unregulated tourism activities. The rapid development of beachfront resorts and infrastructure has contributed to coastal degradation, including shoreline retreat and the loss of natural habitats for marine life [32]. To address these challenges, integrated coastal zone management (ICZM) initiatives have been introduced, focusing on beach nourishment, mangrove restoration, and the regulation of wastewater discharge to protect the fragile marine environment. These efforts align with Bali's broader sustainability goals, ensuring that tourism growth does not come at the expense of the region's ecological integrity while maintaining Nusa Dua's reputation as a premier eco-friendly tourism destination [27]. The following, in Table 1, are the results of the interview with the informant regarding the awareness of hotels in The Nusa Dua ITDC complex regarding water management.

5. Conclusions

Water management in the ITDC Nusa Dua area has become a strategic initiative to address water scarcity and mitigate the adverse effects of mass tourism in Bali. Various efforts, such as the operation of the Integrated Wastewater Treatment Plant (Lagoon ITDC Nusa Dua), the implementation of Sea Water Reverse Osmosis (SWRO) technology, and the installation of water-saving equipment, highlight the commitment of hotels to preserving water resources. These measures significantly reduce clean water consumption while enabling wastewater treatment for non-potable uses, such as irrigation, thereby minimizing water pollution in the area. Additionally, the success of these initiatives enhances Nusa Dua's reputation as an eco-friendly and sustainable tourist destination, attracting international visitors and earning prestigious sustainability awards. However, several challenges persist, including the high investment costs of advanced technologies, potential risks to marine ecosystems due to SWRO operations, and the unpleasant odor from wastewater treatment, which may affect visitor comfort. Moreover, the rising demand for water, coupled with increasing pressure on local resources, necessitates a more sustainable approach to water management to maintain ecological balance.

Water management in the ITDC Nusa Dua area plays a pivotal role in environmental conservation and the advancement of sustainable tourism. The implementation of wastewater treatment facilities, water-saving technologies, and desalination through SWRO has successfully reduced clean water consumption and minimized pollution, demonstrating a proactive approach to resource preservation. However, challenges such as high investment costs, potential environmental risks, and increasing water demand require continuous improvements and adaptive strategies.While existing measures have yielded positive results, further collaboration among the government, area managers, and local communities is crucial. A more integrated, innovative, and eco-friendly approach is needed to ensure long-term efficiency and sustainability, particularly in anticipation of the projected rise in tourism to Bali. By addressing these challenges and expanding sustainable initiatives, Nusa Dua can strengthen its reputation as a global model for environmentally responsible tourism.

Acknowledgement

The author would like to express gratitude to the Research and Community Service of Universitas Pelita Harapan for providing research funds and full support from the Faculty of Hospitality and Tourism (P-024-M/FPar/X/2023).

References

[1] J. Liu *et al.*, "Water scarcity assessments in the past, present, and future," *Earth's Futur.*, vol. 5, no. 6, pp. 545–559, 2017, doi: 10.1002/2016EF000518.
[2] N. Sudini and G. Sujana, "Impacts of mass tourism development on Bali's environment: Water pollution, freshwater shortages, and ecological degradation," vol. 2507, no. February, pp. 1–9, 2020.
[3] D. Adhariani, *Unravelling Stakeholders' Perceptions for Sustainable Tourism: The Case of Water Scarcity in Bali*, vol. 306, no. 1. 2019.
[4] K. S. K. Yamamoto, S. Watanabe, "Tri Hita Karana- A cultural approach to sustainable tourism in Bali.pdf." .
[5] D. . Abdullah, G. Sujana, and I. N. Sudini,Porter, "The impact of excessive groundwater extraction on Bali's water resources," vol. 13, no. 5, p. 2010, 2010.
[6] V. A. Tzanakakis, N. V. Paranychianakis, and A. N. Angelakis, "Water supply and water scarcity," *Water (Switzerland)*, vol. 12, no. 9, pp. 1–16, 2020, doi: 10.3390/w12092347.
[7] R. B. A. Apliant, R. Suharto, "Sustainable hotel practices and the United Nations SDGs: A study of water conservation in Bali," *Nucl. Phys.*, vol. 13, no. 1, pp. 104–116, 2023, doi: 10.1016/j.ijhm.2023.102518.
[8] N. Roozbahani, "Wastewater management systems and their role in sustainable water conservation," vol. 19, no. 4, p. 6, 2021.
[9] V. Manasa and K. Mehta, "Sources of wastewater and their impacts on environmental sustainability," vol. 2507, no. February, pp. 1–9, 2020, doi: 10.1021/es1012345.
[10] A. A. O. M. A. Massoud and M. R. Nasr, "Wastewater treatment in large urban areas: Centralized versus decentralized systems," *Environ. Eng. Sci*, vol. 25, no. 6, pp. 61–64, 2008, doi: 10.1089/ees.2007.0162.
[11] H. D. S. G. Tchobanoglous, F. L. Burton, "Wastewater Engineering: Treatment and Reuse," p. 2004, 2004.
[12] M. A. Andini and M. AridaRodríguez, Velastequí, "Evaluation of wastewater management system in Nusa Dua Lagoon, Bali," vol. 15, no. 2, pp. 1–23, 2019.
[13] M. K. N. Sdiri and N. Gharbi, "Integrated water resources management: A key element for sustainable

development," *Nucleic Acids Research*, 2018. http://dx.doi.org/10.1016/j.gde.2016.09.008%0Ahttp://dx.doi.org/10.1007/s00412-015-0543-8%0Ahttp://dx.doi.org/10.1038/nature08473%0Ahttp://dx.doi.org/10.1016/j.jmb.2009.01.007%0Ahttp://dx.doi.org/10.1016/j.jmb.2012.10.008%0Ahttp://dx.doi.org/10.1038/s41598-018-2212.

[14] L. Xie, *Integrated water resources management: A framework for managing water resources sustainably*. 2006.

[15] UNEP, "The Dublin Statement on Water and Sustainable Development," United Nations Environment Programme," 1992..

[16] S. L. M. Lee, Y. Kim, "Water-saving practices in hotels: A case study in Korea," *J. Akunt.*, vol. 11, no. 5, p. 9669582, 2017, doi: 10.1080/09669582.2016.1279134.

[17] Z. C. X. Tang, S. Lee, "Adoption of wastewater treatment technologies in the hospitality industry: Implications for sustainability," vol. 2507, no. February, pp. 1–9, 2020, doi: 10.1177/1096348020916985.

[18] L. N. P. Bohdanowicz, P. Zientara, "Sustainable tourism in hospitality: An analysis of the industry's trends and practices," *Int. J. Hosp. Manag.*, vol. 31, no. 2, pp. 329–337, 2012, doi: 10.1016/j.ijhm.2011.04.013.

[19] F. O. M. A. Massoud and M. Al-Ali, "Wastewater management in urban areas: A review," vol. 7, no. 2, pp. 61–64, 2008.

[20] N. Andini and P. Arida, "Wastewater management practices in Nusa Dua Lagoon: Effectiveness and sustainability," vol. 14, no. 3, pp. 1–23, 2019.

[21] H. Roozbahani, "Wastewater management strategies for sustainable development: A case study of wastewater treatment in tourism zones," *Pharmacogn. Mag.*, vol. 75, no. 17, pp. 399–405, 2021, doi: 10.1016/j.jenvman.2020.105621.

[22] M. A. et Al., "The role of sustainable tourism practices in enhancing visitor satisfaction at Nusa Dua ITDC, Bali," *Int. J. Tour. Res.*, vol. 21, no. 6, pp. 790–800, 2019, doi: 10.1002/jtr.2305.

[23] G. S. F. Apliant, I. N. Sujana, "Sustainable tourism practices at The Nusa Dua ITDC: A case study of Bali's tourism development," *AT-TAWASSUTH J. Ekon. Islam*, vol. VIII, no. I, pp. 1–19, 2023.

[24] L. U. S. Apliant, H. Prasetyo, "Sustainable tourism and its impact on visitor satisfaction in The Nusa Dua ITDC, Bali," *AT-TAWASSUTH J. Ekon. Islam*, vol. VIII, no. I, pp. 1–19, 2023, doi: 10.1234/jthm.2023.014.

[25] M. S. A. and M. A. Ummah, "Evaluasi sistem pengelolaan air limbah di Lagoon Nusa Dua, Bali," *Sustain.*, vol. 11, no. 1, pp. 1–14, 2019, [Online]. Available: http://scioteca.caf.com/bitstream/handle/123456789/1091/RED2017-Eng-8ene.pdf?sequence=12&isAllowed=y%0Ahttp://dx.doi.org/10.1016/j.regsciurbeco.2008.06.005%0Ahttps://www.researchgate.net/publication/305320484_SISTEM_PEMBETUNGAN_TERPUSAT_STRATEGI_MELESTARI.

[26] R. Manasa and P. Mehta, "Sources and treatment of wastewater in industrial sectors," *Kaos GL Derg.*, vol. 8, no. 75, pp. 147–154, 2020, [Online]. Available: https://doi.org/10.1016/j.jnc.2020.125798%0Ahttps://doi.org/10.1016/j.smr.2020.02.002%0Ahttp://www.ncbi.nlm.nih.gov/pubmed/810049%0Ahttp://doi.wiley.com/10.1002/anie.197505391%0Ahttp://www.sciencedirect.com/science/article/pii/B9780857090409500205%0Ahttp:

[27] I. K. A. I. M. Sudiajeng, N. G. Putra, "Water conservation efforts in Bali: Addressing over-extraction and groundwater sustainability," vol. 11, no. 1, pp. 92–105, 2017.

[28] N. H. S. Sdiri, H. Boughanmi, "Integrated water management as a key for sustainable development: The role of IWRM in the tourism sector," *Water Resour. Manag.*, vol. 32, no. 13, pp. 4359–4374, 2018, doi: 10.1007/s11269-018-2062-x.

[29] A. W. N. Anggraeni, D. Kurniawati, "Wastewater management in sustainable tourism: The case of Bali," *Braz Dent J.*, vol. 33, no. 1, pp. 1–12, 2022, doi: 10.1016/j.jenvman.2022.113539.

[30] P. N.] R. Prabowo, H. Wijaya, "Integrated water management in Bali: Balancing tourism and sustainability," *Sustain. Dev.*, vol. 31, no. 1, pp. 492–509, 2023, doi: 10.1002/sd.2405.

[31] I. Kurniawan and A. Setiawan, "Tri Hita Karana and sustainable tourism development: A case study of Bali," vol. 9, no. 2, pp. 356–363, 2022, doi: 10.1080/15313220.2022.2040924.

[32] D. S. A. Santoso, I. M. Wijaya, "Coastal erosion and water pollution in Nusa Dua: Impacts of tourism and mitigation strategies," vol. 214, p. 105901, 2021.

IOP Conf. Series: Earth and Environmental Science **1503** (2025) 012009 doi:10.1088/1755-1315/1503/1/012009

Identification of Geothermal Manifestation by Land Surface Temperature Method in Dieng Plateau, Banjarnegara Regency, Central Java

Annisa Trisnia Sasmi[1*], Annisa Rizky Tiara Dita[1], Hian Risnandar[1]

[1] Faculty of Geography, Universitas Muhammadiyah Surakarta, Surakarta, Indonesia
*email: ats678@ums.ac.id

Abstract

This study aims to examine the geothermal manifestations in Dieng Plateau using remote sensing methods and provide recommendations for their potential utilization as geothermal resources. The main research questions addressed are the condition of the geothermal manifestations in this region based on remote sensing methods and the recommendations for their utilization based on temperature classification from data processing results. The objective of this study is to analyse the condition of geothermal manifestations in the Dieng Plateau and to provide recommendations for their potential use as geothermal resources based on temperature classifications. The results showed that the maximum surface temperature in Sileri Crater and Industrial Area reached 31.03°C, while the minimum temperature in the Forest Area is 13.31°C. Temperature classification at various research locations showed low to very low temperature variations, with the highest temperature in Sileri Crater and the lowest in the Forest Area was 13.31°C. low, with the highest temperature at Sileri Crater and the lowest at Sipandu Spring. This temperature variation is influenced by area, where manifestations with smaller areas are utilized as water baths. Manifestations that are utilized as hot springs tend to have lower temperatures than manifestations with larger areas. lower temperatures compared to larger manifestations that are not utilized utilized. The results of the data processing show that the surface temperature in the research object has not met the requirements as a geothermal energy source, which requires a minimum temperature of 180°C-250°C. Correlation analysis between LST and NDVI shows a downward trend, where the higher the surface temperature, the lower the vegetation density lower the vegetation density. This is evidenced by the conditions in the geothermal manifestations that lack vegetation due to soil and rock structures that do not support plant growth. Do not support plant growth. This study concludes that although there are geothermal manifestations in the Dieng Plateau, the surface temperatures do not meet the criteria for geothermal energy sources. The correlation between surface temperature and vegetation density provides additional insights into the environmental conditions around the geothermal manifestations.

1. Introduction

Indonesia has geothermal resources in several regions Indonesia. A total of 276 geothermal locations are scattered in the formation path of the volcanoes from Sumatra, Java, Nusa Tenggara, Sulawesi and Maluku. Some of these geothermal resources are located in volcanic landforms, however of the 276 geothermal locations scattered in several regions of Indonesia, only 31% of the locations have been surveyed in detail and used as a source of power generation energy [3]. Improved research and development of geothermal energy as a national energy source is required in order to employ geothermal energy as a power plant that can help reduce greenhouse gas emissions [2].

Seeing from the emergence of geothermal heat spread in various regions of Indonesia, Indonesia ranks second as the largest country in the world. Indonesia, then Indonesia ranks second as the largest producer of geothermal energy. Producer of energy from geothermal, with an increase in electricity supply to 1,924 MWe (MegaWatt) [10]. Whereas the first place was occupied by the United States, and in third place was occupied by the Philippines. Outside the island of Java, it is estimated that the potential for geothermal energy is potential when compared to the island of Java. The total potential of geothermal energy Indonesia is estimated to reach 23.7 GW. Of this potential, the government has established Geothermal Working Areas (WKP) & Preliminary Survey and Exploration Assignment Areas (WPSPE) that are ready to be Assignment of Preliminary Survey and Exploration (WPSPE) that are ready to be developed by [12].

Geothermal WKP is the area stipulated in the IUP (Mining Business License). The creation and establishment of WKP is the authority of the central government, in this case the Ministry of Energy and Mineral Resources (Law No. 27 Year 2003). The Dieng plateau is included in one of the WKP managed by PT Geodipa.

The Dieng plateau has a WKP area of 113,400 Ha in the District of Banjarnegara, Central Java Province. With the total resources from the distribution of hotspots on the Dieng plateau with a total summation in the Dieng area, Candradimuka, and Mangunan - Wanayasa in the class of proven reserves of 352 Mwe [11]. Proven reserve class is a classification of reserves with an estimate of their energy potential based on the results of detailed investigations, tested with exploration wells, delineation and development as well as feasibility studies development and feasibility studies [9].

The geothermal activity in Dieng Plateu and Banjarnegara Regency is linked to the region's geological history, including volcanic activity. The presence of volcanic rocks and faults contributes to the formation of geothermal reservoirs. The Sunda Arc, a tectonically active region that is a part of the Pacific Ring of Fire [7], is what defines Banjarnegara, Indonesia's geological history. Due to the Indo-Australian Plate subducting beneath the Eurasian Plate, which resulted in extensive volcanic activity and the construction of geothermal systems [7], this area has a rich volcanic history. Situated in Banjarnegara, the Dieng Plateau is a notable structure sculpted by volcanic processes, exhibiting traces of previous eruptions and geothermal expressions like fumaroles and hot springs. The majority of the volcanic rocks in this area are andesite and basalt, which have undergone changes due to hydrothermal activities. Research shows that Banjarnegara has a significant geothermal potential [1]. Several geothermal reservoirs with different temperatures and heat loss potentials have been found [13][14]. Numerous geothermal manifestations are supported by a complex hydrothermal system that was generated by the interaction of meteoric water and heated volcanic rocks [15]. In addition to highlighting the region's geothermal potential, this geological background also emphasizes how vulnerable it is to volcanic dangers, making continuous monitoring and research necessary to comprehend its dynamics and ramifications for nearby communities.

Geothermal in the Dieng plateau has the potential to be further developed as a power plant for local economic growth, national, and international economic growth. Utilization of geothermal potential in the Dieng plateau can be through direct utilization in geothermal manifestations such as hot water for lodging in tourist areas, making hot springs, geothermal power plants, and others. Although the potential 2 utilization of geothermal manifestations in the Dieng plateau is high, however research and studies on these geothermal manifestations are still lacking, and it is hoped that they can be further developed. less, and with the hope that it can be further developed.

In order to analyze objects without physically visiting the target region, remote sensing is essential. It makes things easier and lowers the danger associated with the object under study [17]. Earth observation, including Land Surface Temperature (LST) measurements, can now be carried out using satellite observations in multiple temporal methods at different spatial and temporal resolutions due to the quick development of remote sensing technologies [6]. Landsat 8 image data processing is used to generate surface temperature by processing the thermal sensor, band 10. Surface temperature by processing the thermal sensor, namely band 10 with a spatial resolution of 100 meters and wavelength for band 10 of 10.60 meters. Spatial resolution of 100 meters and a wavelength for band 10 of 10.60 -11.19 μm and band 11 of 11.50 -12.51 μm [14].

One of the basic methods to assess the geothermal potential is by identifying the geothermal manifestations surround an area. Natural phenomena known as "geothermal manifestations" are signs of the existence of geothermal energy beneath the surface of the Earth, which is frequently the result of tectonic or volcanic action. Hot springs, fumaroles, mud pots, and geysers are a few examples of these manifestations; they happen when subsurface water gets heated by geothermal heat sources like magma. Utilizing remote sensing methods to locate and examine thermal anomalies connected to geothermal manifestations is part of the process of using Landsat 8 for LST estimation in geothermal zones. Landsat 8 is an invaluable resource for geothermal study since it offers critical information on variations in surface temperature that suggest the presence of geothermal potential. From this reason, this study aims to utilize LST as a tool to understand the potential of geothermal manifestations surround Dieng Plateau.

2. Method

Remote sensing methods can be used to detect geothermal manifestations. For example, the distribution of LST linked to surface manifestations of geothermal activity can be mapped using Landsat 8 imagery. This aids in locating regions with strong thermal anomalies that are suggestive of geothermal activity. This research examines the identification of geothermal manifestations by using the using Landsat 8 satellite image data processing and field survey methods. The image data processing method is carried out to obtain the results of LST temperature range classification data. While the survey method aims to accuracy and validation of the image processing results that have been carried out.

The object of this research is in the form of geothermal manifestations, where the research was conducted at the was conducted at geothermal hotspots in the Dieng Plateau region, Banjarnegara Regency, Central Java. Banjarnegara, Central Java. Namely, Candradimuka Crater, Sileri Crater, Sikidang Crater, Kalianget Mangunan, and Sipandu Spring.

2.1 Data Collection Methods

The data collection method is a method used by a researcher to obtain the necessary data. With the right data collection method in a study, it will be possible to solve problems validly and objectively. In this study, there are 2 methods, namely:

2.2.1 Primary Data

Primary data is data that is taken directly to adjust research instruments. Primary data includes data taken from the field validation process. Field validation in remote sensing is a critical process that ensures the accuracy and reliability of remote sensing products by comparing them with ground-based observations. This validation is essential to verify the performance of remote sensing systems and to correct any discrepancies that may arise from the inherent limitations of satellite data. In this occasion, the field validation was implemented on February 2024 by capturing documentations, plotting the local coordinates, and recording the area descriptions from the selected geothermal manifestations.

2.2.2 Secondary Data

Secondary data is supporting data that has been provided by a government and private agency or from another body, closely related to the use of data in the study. We utilized Landsat 8 TIRS/OLI

IOP Conf. Series: Earth and Environmental Science **1503** (2025) 012009 doi:10.1088/1755-1315/1503/1/012009

Path 120 Row 65 Image of August 2022, and Digital administrative data of the Dieng plateau area, Central Java which was provided by BIG (Badan Informasi Geospasial). Landsat 8 is a satellite for earth observation which was developed by the U.S. Geological Survey (USGS) and NASA. It was launched on February 11, 2013, and its two main scientific instruments are the Thermal Infrared Sensor (TIRS) and the Operational Land Imager (OLI). With a spatial resolution of 30 meters, the OLI records data in nine spectral bands, including the visible, near-infrared, and short-wave infrared bands. It also has a panchromatic band with a 15-meter resolution, which improves image detail. With a spatial resolution of 100 meters, the TIRS provides surface temperature data in two thermal infrared bands. Every sixteen days, Landsat 8 can cover the entire planet and take hundreds of pictures every day.

2.2 Data Processing Technique

To accurately recover LST data, there are numerous crucial steps in the Radiative Transfer Equation (RTE)-based remote sensing data processing procedure. First, thermal infrared data must be obtained from satellites such as Landsat 8, whose Thermal Infrared Sensor (TIRS) can detect heat radiation in two bands (Band 10 and Band 11). To ensure that the data are in a consistent unit and free from biases peculiar to a particular sensor, the raw data are then radiometrically calibrated to transform digital numbers (DN) into radiance values. Subsequently, the RTE method is utilized to apply atmospheric correction, which estimates and subtracts the atmospheric influence on the surface thermal radiation. This yields the intensity of surface thermal radiation. Next, utilizing the relationship function between thermal radiation intensity and surface temperature, this adjusted radiance is translated into LST. When it comes to separating the temperature and emissivity components, the RTE approach works better than other algorithms like the split-window and single-window methods, yielding more precise LST estimations. In order to guarantee accuracy—a crucial step for geothermal exploration and other environmental applications—the extracted LST values are lastly verified using data from ground-based observations. This all-encompassing strategy guarantees accurate processing of remote sensing data, yielding trustworthy information for a range of practical and scientific uses. The following steps below are the stages of LST data processing using the Radiative Transfer Equation (RTE) [8], which was conducted in this study:

2.2.1 Cropping or image cutting

Cropping is an image processing process that aims to minimize the research area, because the downloaded Landsat 8 image still adjusts based on path & row. Image cropping in this study was carried out in accordance with the research area, namely the Dieng plateau, the administrative boundaries of Banjarnegara Regency.

2.2.2 Radiometric Correction

Radiometric correction is a crucial procedure in remote sensing that improves the quality and accuracy of satellite and aerial data through the correction of numerous distortions induced by atmospheric conditions, sensor characteristics, and illumination factors. The raw digital numbers (DN) that sensors capture are not absolute measurements; rather, they are subject to variations between different images and times due to air scattering, absorption, and sensor noise. Through radiometric correction, these DN values are transformed into calibrated reflectance or radiance values using precise algorithms, providing dependable data for uses like as land cover classification and climate monitoring. Radiometric correction is a process to improve the visual quality of the image, in terms of correcting pixel values that do not correspond to the actual spectral reflectance or radiance spectral object. This conversion is also a radiometric correction/sensor calibration process (conversion of digital number to TOA radiance). By using formula (1):

IOP Conf. Series: Earth and Environmental Science **1503** (2025) 012009 doi:10.1088/1755-1315/1503/1/012009

Radiometric Correction *(DN to Radiance) = (Ml*Qcal)+Al*...(1)
Ml : Multi Band Radiance (metadata)
Qcal : Band
Al : Radiance Add Band (metadata)

2.2.3 Radiance to Brightness Temperature
The process of turning radiance, which is a measurement of electromagnetic energy intensity, into a temperature value is known as radiation to temperature correction. In several disciplines, including materials science, climatology, and remote sensing, this procedure is essential. This is a temperature extraction process that begins with the conversion from radiance to effective satellite temperature. By using formula (2):

$$\text{Radiance to Celcius} = \frac{k2}{ln\, ln\left[\left(\frac{k1}{L\lambda}\right)+1\right]} - 273.15 \quad (2)$$

K1, K2 : Constant Band (metadata)
Lλ : Radiance value on thermal band

2.2.4 NDVI
One of the most widely used vegetation indices is the NDVI, which has values ranging from -1 to 1 (Sari et al., 2017). It is a method that can be used to calculate vegetation density where the higher the NDVI value, the higher the vegetation density. By using formula (3):

$$NDVI = \frac{NIR-RED}{NIR+RED} \quad (3)$$

NIR : Near Infrared Band (Band 5 on Landsat 8)
RED : Band Red (Band 4 on Landsat 8)

2.2.5 PV (Proportion of Vegetation)
The percentage of an area covered by vegetation is referred to as the proportion of vegetation (PV) in the context of remote sensing. This metric is essential for many applications, such as forestry, agriculture, urban planning, and environmental monitoring. PV is determined by the ratio of the vertical projection of vegetation on the ground to the total vegetation area. It is particularly important for estimating emissivity parameters, which are derived from previously processed NDVI values. By using formula (4):

$$PV = \left(\frac{(NDVI-NDVIs}{(NDVIv-NDVIs)}\right)^2 \quad (4)$$

NDVI : NDVI results
NDVIs : Soil value or the lowest value of the NDVI results
NDVIv : Vegetation value or the highest value of the NDVI results

2.2.6 LSE (Land Surface Emissivity)
A crucial metric in remote sensing, Land Surface Emissivity (LSE) measures how well a surface emits heat radiation in comparison to a blackbody at the same temperature. It is crucial for comprehending energy balance and surface processes as well as for the precise recovery of LST. LSE is a significant indication for various land cover types, such as plant, soil, and urban areas. It varies depending on material composition, moisture content, and surface roughness. It is a proportional factor that measures the degree of blackness to calculate the radiation emission capable of carrying thermal energy across the surface in the atmosphere. Using formula (5):

$$\varepsilon = 0.004*PV + 0986 \quad (5)$$

ε : Emissivity
PV : Proportion of Vegetation

IOP Conf. Series: Earth and Environmental Science **1503** (2025) 012009 doi:10.1088/1755-1315/1503/1/012009

2.2.7 LST

The radiative skin temperature of the Earth's surface is measured by LST, a crucial environmental parameter that is derived from solar radiation and is commonly measured with thermal infrared sensors in remote sensing. Different from air temperature, LST is crucial for assessing processes of land surface and land-atmosphere exchange, limiting surface energy budgets, and giving global and important regional temperature changes. Albedo, vegetation cover, and soil moisture are some of the elements that affect LST. For example, surfaces with a high albedo reflect more solar radiation and hence have lower temperatures, whereas thick vegetation generally has lower surface temperatures because of its insulating qualities. LST can be derived using formula (6):

$$LST = \frac{BT}{\left\{1+\left[\frac{\lambda BT}{C2}\right] ln\, ln\, \varepsilon\right]\right\}} \quad (6)$$

BT : Brightness Temperature

λ : Central wavelength of emitted radiance (Landsat 8 band 10 is 10.8 μm and band 11 is 12 μm)

C2 : c2 : h*c/s = 1.4388*10-2mK = 14388 μm K

lnln ε : LSE

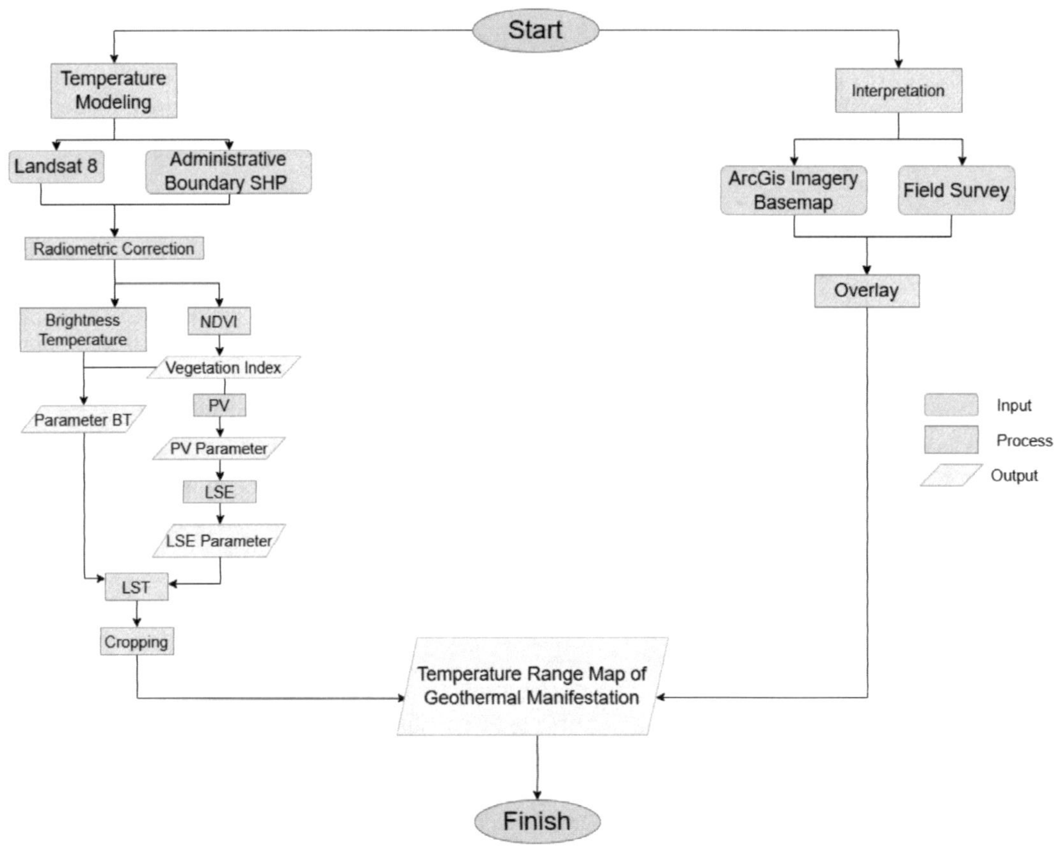

Figure 1. Research Flow Chart

3. Results and Discussion

3.1 Field Validation Results

The Dieng plateau in Banjarnegara Regency has a manifestation of geothermal manifestations, namely Kalianget Mangunan which is located in Kalibening District, then Sikidang Crater, Sileri Crater, Candradimuka Crater, and Sipandu Spring in Batur Sub-district. Sipandu Springs located in Batur District. The geothermal manifestation object, partly used as a tourist attraction, and partly already utilized by the PT Geodipa Energi, which is located in Dieng itself.

1. Kalianget Mangunan

   Kalianget Mangunan has a coordinate location point at -7.23379079012279, 109.63990117553087. Is a hot spring tourism hot springs in Kalibening District, Banjarnegara Regency, the location of the Kalianget Mangunan which is in the middle of the forest, and is approximately 1km away from the nearest settlement, can make Kalianget Mangunan as a tourist spot that is worth visiting because it can be accessed by two-wheeled and four-wheeled vehicles. also by two wheeled and four-wheeled vehicles. Kalianget Mangunan as a hot spring tourism, not only provides hot springs, but also provides camping only, but also provides campsites and restaurants, very suitable for vacation activities together. For vacation activities together. Manifestations in Kalianget Mangunan are in the form of hot springs where a pool has been built for swimming and bathing.

Figure 2. Kalianget Mangunan

2. Sikidang Crater

   Sikidang Crater has a coordinate location point at -7.22022811939088, 109.903950235382. It is a geothermal tourism in Batur District, Banjarnegara Regency, the location of Sikidang Crater is in the middle of the forest, but access to the location is equipped with a road that can be accessed by two-wheeled and four-wheeled vehicles, the road is accessed by other tours, namely the Temple Area is access to other tours, namely the Arjuna Temple Area. Crater Sikidang as a geothermal tourist spot, has several craters in one area. The main crater is equipped with a bridge around it that can be passed by visitors, and there are other craters that are above but there is no access to the crater. In the crater area, the surface conditions have a dry soil condition, yellow, and warm. Then for the condition of the rocks around the crater is characterized by medium size, hard and there is sulfur around the crater area.

Figure 3. Sikidang Crater

3. Sipandu Spring

Sipandu spring has a coordinate location point at -7.183758461324608, 109.88955694095041. It is a tourist hot spring in Batur Subdistrict, Banjarnegara Regency, the location of Sipandu Spring is in a residential area. Sipandu spring is located in a residential area. Sipandu spring is worth to visited because it can also be accessed by two-wheeled and four-wheeled vehicles. Sipandu spring only provides hot springs, in contrast to Kalianget Mangunan which also provides hot springs. with Kalianget Mangunan which also provides other facilities for tourism traveling. The manifestation at Sipandu spring is in the form of hot springs which has been built in the form of a pool for swimming and soaking.

Figure 4. Sipandu Spring

4. Sileri Crater

Sileri Crater has a coordinate location point at -7.193126114493816, 109.88354422883214. It is the largest crater in the Dieng area, located in Batur District, Banjarnegara Regency. Then, there are pipes from PT Geodipa around the crater, and there is no access for tourists due to the eruption. for tourist visitors due to the eruption of Sileri Crater in 2018, causing damage to the crater. Eruption in 2018, causing damage around the crater and making it dangerous for tourists to enter area and dangerous for tourist visitors to enter. For access to Sileri Crater, is not equipped with adequate access because it is not a tourist attraction, to get to Sileri Crater can not be by 4-wheeled vehicles, can only be passed by 4-wheeled vehicles, can only be passed by 2-wheeled vehicles or pedestrians and with steep road conditions, the and with steep road conditions, so that it can only be observed with a radius of approximately 1km. a radius of approximately 1km. In the crater area, surrounded by with the state of barren land.

Figure 5. Sileri Crater

5. Candradimuka Crater

Candradimuka Crater, has a coordinate location point at -7.188763156362128, 109.85330459226097. It is a geothermal tourism in Batur District, Banjarnegara Regency, the location of Candradimuka Crater is in the middle of the forest but the access to the location of the Candradimuka crater is in the middle of the forest but access to the location is equipped with a road that can be accessed by vehicles equipped with a road that can be accessed by two-wheeled vehicles and four-wheeled vehicles, the road is an access with the the road is accessed by two-wheeled and four-wheeled vehicles, the road is accessed by another tour, namely Telaga Dringgo. Towards the Candradimuka crater, which is located below, can be accessed via stairs of about 20 meters, then the Candradimuka crater itself has 2 craters, separated by a small bridge that can be passed. In the crater area, the surface conditions have a soil condition wet, ash-slightly yellow in color, and warm. Then for the condition of rocks around the crater are characterized by medium-large size, hard and there is sulfur around the crater area.

Figure 6. Candradimuka Crater

3.2 Data Processing Results

The results of the data processing stage using the Landsat 8 TIRS/OLI Path 120 Row 65 image in July 2022.

1. Radiometric Correction
   The result of this radiometric correction aims to eliminate disturbances caused by optical system errors in the sensor, atmospheric disturbances, and the influence of the sun's elevation angle. By changing the data on the image stored in the form of a digital number.

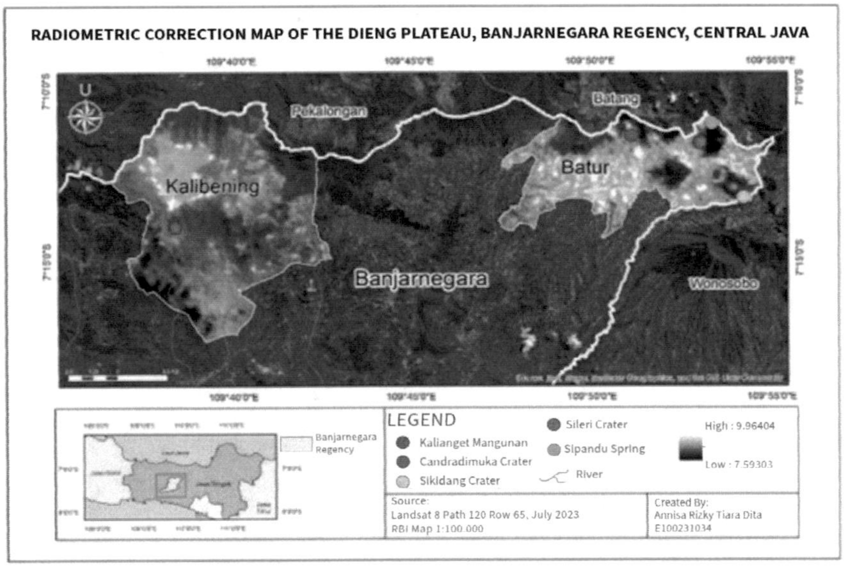

Figure 7. Radiometric Correction Map

2. Radiance to Brightness Temperature
   The temperature extraction process is done by converting radiance to effective satellite effective satellite temperature, the results of this process are needed in the final The final step is for LST processing.

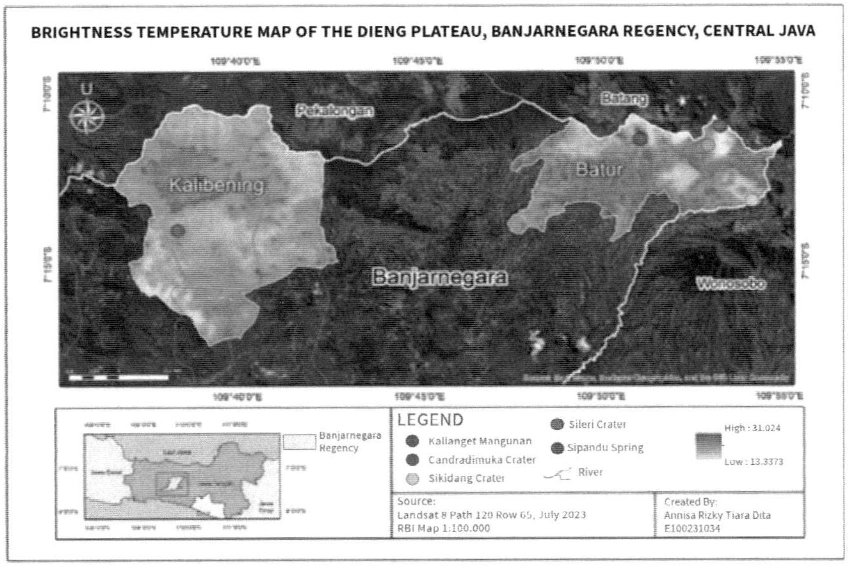

Figure 8. Brightness Temperature Map

3. NDVI

NDVI is one of the vegetation indices, needed to calculate the vegetation density of an area. The result of this process are needed in the next step of PV processing.

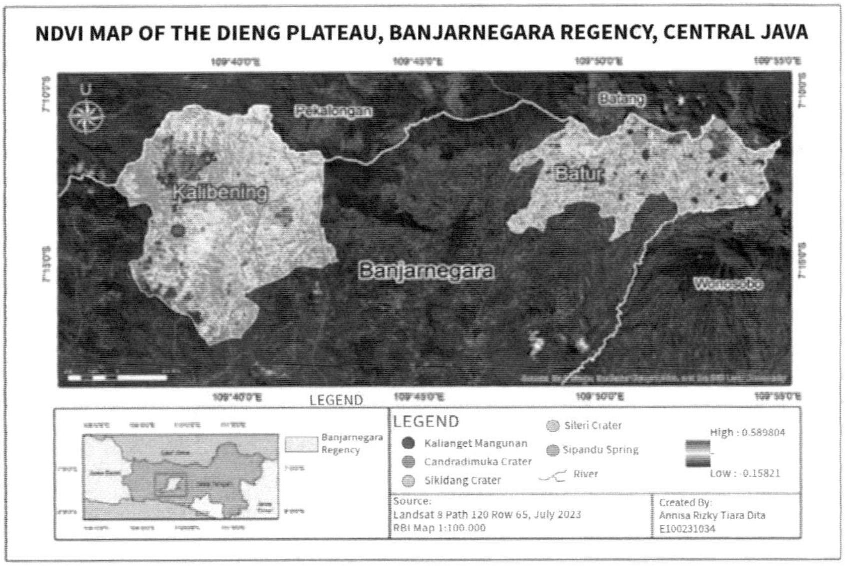

Figure 9. NDVI Map

4. PV

PV is the ratio of the vertical projection of vegetation area at ground level to the total vegetation area. The result of this process is needed in the next step, which is LSE processing.

 IOP Publishing
doi:10.1088/1755-1315/1503/1/012009

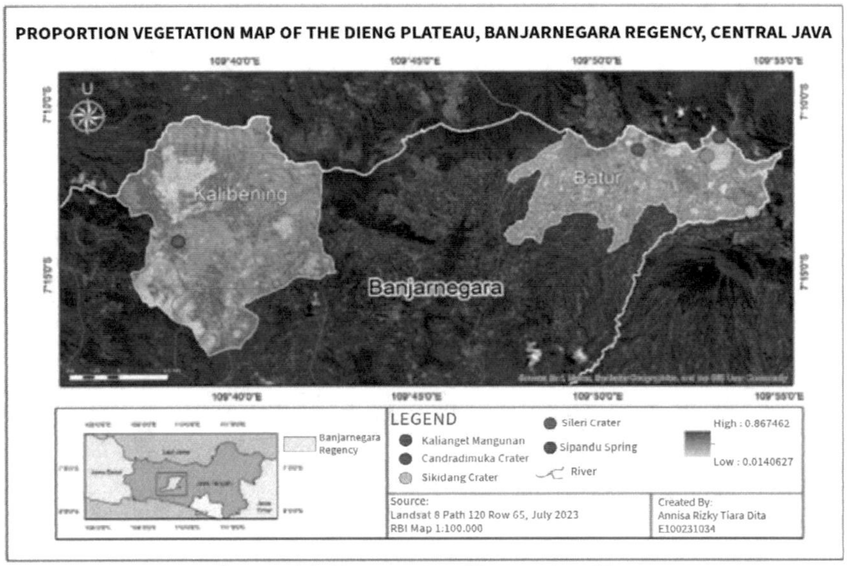

Figure 10. Proportion Vegetation Map

5. LSE
   LSE is a pro-social factor that measures the degree of blackness to calculate radiant emissions, so as to carry thermal energy across surfaces in the atmosphere. This LSE process requires the results from PV, and the results from LSE are needed for the final process, namely LST processing.

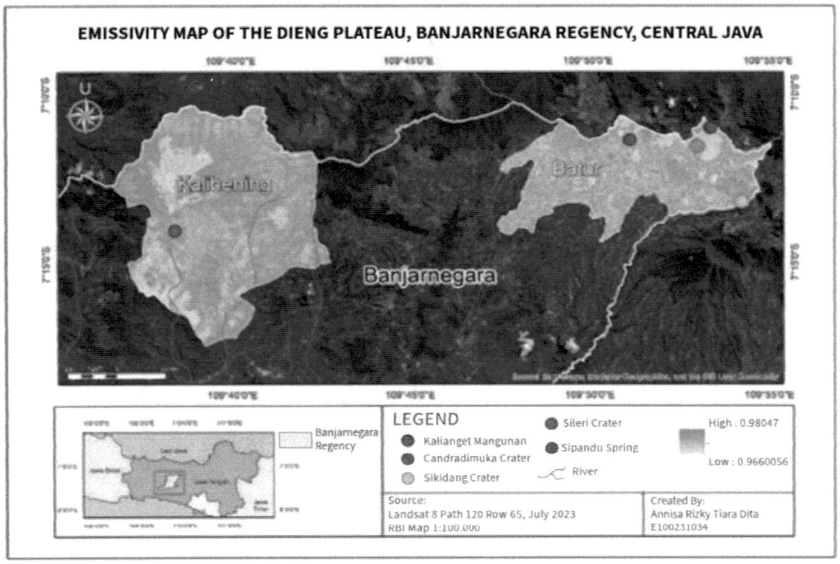

Figure 11. Emisivitas Map

6. LST
   Is the final result in this study, which is to determine surface temperature, this final result requires the results of several data processing stages that have been carrie out, namely from

the BT results, and stages of data processing that have been carried out, namely from the results of BT, and LSE.

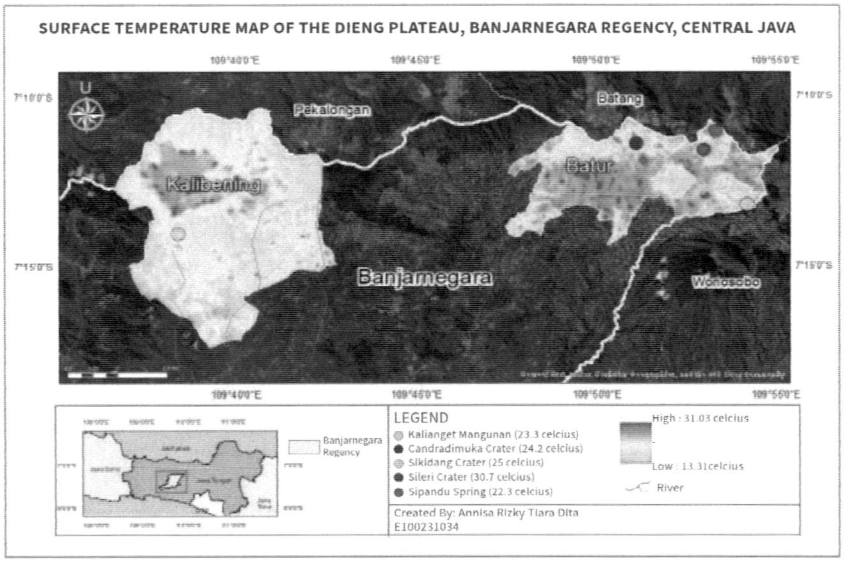

Figure 12. Surface Temperature Map

3.3 Surface Temperature Analysis

The results of surface temperature processing can be seen in Figure 17. Surface Temperature, which has gone through several surface temperature processing processes. Namely from the radiometric correction process, radiane to brightness

temperature, NDVI, PV, LSE, to the final process of LST. Surface temperature processing processing uses Landsat 8 Path 120 Row 65 Image data for the month of July 2023.

On the surface temperature map, shows the results of the surface temperature processing surface that the temperature class with the final result of the maximum temperature is at a temperature of 31.03°C and the minimum temperature is 13.31°C. Based on interpretation, the maximum temperature is at the location of the research object, namely Sileri Crater and Industrial Area, then for the minimum temperature is in the Forest Area [19].

It can be seen that the Kalianget Mangunan location is at a temperature of 23.3 °C (very low), Candradimuka Crater is at 24.2 °C (very low), Sikidang Crater is at 25 °C (very low), Sileri Crater is at 30.7 °C (very low), and Sipandu Spring is 22.3 °C (very low), all locations of the object of research are in qualitative classification very low, the classification class results are in table 1. With the highest temperature. The highest temperature of all research objects is at the Sileri Crater location and the lowest temperature is at the Sipandu Spring location.

Table 1. Qualitative Classification of LST of Research Object Locations

Research Object Location	Surface Temperature	Qualitative Classification
Kalianget Mangunan	23.3 °C	Very Low
Candradimuka Crater	24.2 °C	Very Low

Sikidang Crater	25 °C	Very Low
Sileri Crater	30.7 °C	Very Low
Sipandu Spring	22.3 °C	Very Low

Based on the results of surface temperature analysis, there are several factors that affect the acquisition of surface temperature data using the remote sensing method. One of them is the limitation of obtaining images that are [4]. Cloud-free images are usually obtained from 46 recording during the dry season. However, limited access to rainfall data in 2023 in Banjarnegara Regency causes a lack of data sources to determine the dry season in that year.

Based on Table 2, the lowest rainfall in Banjarnegara District in 2022 occurs in May. Thus, the dry season in 2022 is expected to take place in May and June, with rainfall amounts < 100 mm. From the data for 2022, it is assumed that in 2023 there will be no significant changes in the amount of rainfall and the range of the dry season. Then the image data used is the recording time in July 2023 and with the least comparison of cloud cover when compared to cloud cover comparison is the least when compared to other months in 2023. The extent or magnitude of the area affected by geothermal manifestations will determine how significant the effect on temperature is will determine how significant its influence on surface temperature is. The area of the geothermal manifestation research object location can be seen in table 2.

Table 2. Area of Research Object Location

Research Object Location	Broad (m^2)
Kalianget Mangunan	892
Candradimuka Crater	1214
Sikidang Crater	53614
Sileri Crater	22871
Sipandu Spring	48

One of the factors that can affect the high and low temperatures at the location of the research object is in the area. Based on table 2, the location of the Sikidang Crater research object has the largest area. Table 2, the Sikidang Crater research object location has the largest area which is 53614 m^2 and the location of the Sipandu Spring research object has the smallest area of 48 m^2.

After identifying the geothermal temperature with remote sensing method, namely by processing Landsat 8 images, then based on the results of the data processing, the research object cannot be identified. Data processing, the research object cannot yet be categorized as geothermal energy, because it has not met the geothermal energy, because it has not fulfilled the existing requirements, one of which is that the minimum temperature of heat must be around 180°C-250°C [14]. Where the temperature of the object of research based on data processing is not up to the temperature to be categorized as geothermal energy.

IOP Conf. Series: Earth and Environmental Science **1503** (2025) 012009
doi:10.1088/1755-1315/1503/1/012009

3.4 Image Interpretation Analysis and Field Validation

Based on the results of image interpretation analysis using images from Arcgis Imagery Basemap, it is seen from the elements of image interpretation for the research objects of Sikidang Crater, Sileri Crater, Candradimuka Crater, Kaliange Mangunan, and Sipandu Spring. With the first element, namely hue and color, the manifestation research objects in the form of craters, namely Sikidang Crater, Sileri Crater, and Candradimuka Crater have a predominantly white area on the surface, the white surface layer of soil indicates that the soil contains lime (calcium carbonate), has the least nutrients, and white hot steam.

Then the size element, namely the size of the research object if compared to surrounding objects, has size and area larger than the size of the housing area, but has the size of the area is smaller compared to the size of the forest area. Furthermore, the element of texture, all research objects have a smooth texture at the research location of craters and hot springs. Indicates that there is water inside the research object, while around the crater it shows. The rough texture shows the crater is surrounded by rock. Then, for element of association that all research objects are related to geothermal activity.

For field validation analysis, that there is conformity between image interpretation with field validation of the research object, Sikidang Crater, Sileri Crater, Candradimuka Crater, Kalianget Mangunan, and Air Springs Sipandu, based on the match between the coordinates of the location of the research object, image interpretation results, then results from field validation [18].

3.5 Correlation Analysis of LST and NDVI Results

Based on the results of LST and NDVI data processing, the relationship between the two can be seen the relationship between the two, where the relationship is visualized in the Scatter Plot in Figure 5.1. Scatter plot processing plot processing is done using ArcGis, namely by inputting LST and NDVI data that is already in the form of a fishnet. NDVI data that is already in the form of a fishnet.

The scatter plot uses LST as the Y-axis and NDVI as the X-axis, where each data processing is done using the LST data X axis, where each data processing has its own range of values range. Namely LST has a value range of 13.3145 to 31.0337 and NDVI has a value range of -0.15821 to 0.589804. The value in the LST results for the highest results are in the industrial area (31.0337) and the lowest results are in the vegetation cover area (13.3145). As for the value of in NDVI results for the highest results are in the vegetation cover area (0.589804) and the lowest result is in the water body area (-0.15821).

The scatter plot illustrates a negative correlation between LST and NDVI in the Dieng Plateau, Banjarnegara Regency, Central Java. This means that as LST increases, NDVI decreases, suggesting that higher surface temperatures are associated with lower vegetation density. To examine their linear relationship, a correlation test was conducted, confirming that areas with high NDVI tend to have lower LST. This negative correlation aligns with previous findings by [5].

The correlation coefficient between LST and NDVI is also influenced by the type of land cover and land use. cover type and land use. Variations in cover type and land use land use have a significant influence on the strength of the relationship between LST and NDVI. The correlation between these two variables is very strong in areas covered by green vegetation, while covered by green vegetation, while the relationship is moderate in built-up areas and bare land. and bare land. This correlates with validation in the field, that the conditions in the research object area of the point of geothermal manifestation points, no lush vegetation was found, because the structure of the soil and rock structure cannot be overgrown by plants.

4. Conclusion

The condition of geothermal manifestations in the Dieng Plateau is obtained from the results of field validation results, that Kalianget Mangunan, located in the middle of the forest in Kalibening sub-district about 1 km from the nearest settlement, provides hot springs, camping areas, and restaurants, and can be accessed by two-wheeled vehicles. hot springs, camping areas, and restaurants, and can be accessed by two-wheeled and four-wheeled vehicles. and four-wheeled vehicles. Sikidang Crater in Batur Sub-district, accessed via a road connected to the connected to the Arjuna Temple area, has several craters in one area with a bridge around the crate. Mata Air Sipandu, which is located in the residential area of Batur Sub-district, only provides hot springs without additional facilities but is still easily accessible to the by two and four wheeled vehicles. four-wheeled vehicles. Sileri Crater, the largest crater in Dieng located in Batur Sub-district, has been inaccessible to visitors since the 2018 eruption due to dangerous conditions, and can only be reached by vehicles. Dangerous conditions, and can only be reached by two-wheeled vehicles or on foot through a steep through a steep road. Candradimuka Crater, also in Batur Sub-district, is in the middle of the forest with easy access via a road. the middle of the forest with easy access via a road that connects to Telaga Dringgo. The crater has two craters separated by a small bridge, and the ground is wet and steep bridge, and the ground is wet and warm.

This research has analyzed the condition of geothermal manifestations in the Dieng Plateau using remote sensing methods. Dieng Plateau using remote sensing methods. Temperature map shows significant temperature variations in the study area. Further analysis shows that the surface temperature in various geothermal manifestations, such as Kalianget Mangunan, Candradimuka Crater, Sikidang Crater, Sileri Crater, and Sipandu Spring, are in the low temperature range, with the highest temperature in Sikidang Crater, Sileri Crater, and Sipandu Spring. With the highest temperature in Sileri Crater (30.7 °C) and the lowest in Sipandu Spring (22.3 °C). Sipandu Spring (22.3 °C). The main factor influencing this temperature difference is the size of the manifestation area. The results of data processing show that the research object has low potential because it has not met the requirements to be categorized as geothermal energy, where geothermal energy requires a minimum temperature of about 180°C-250°C [16].

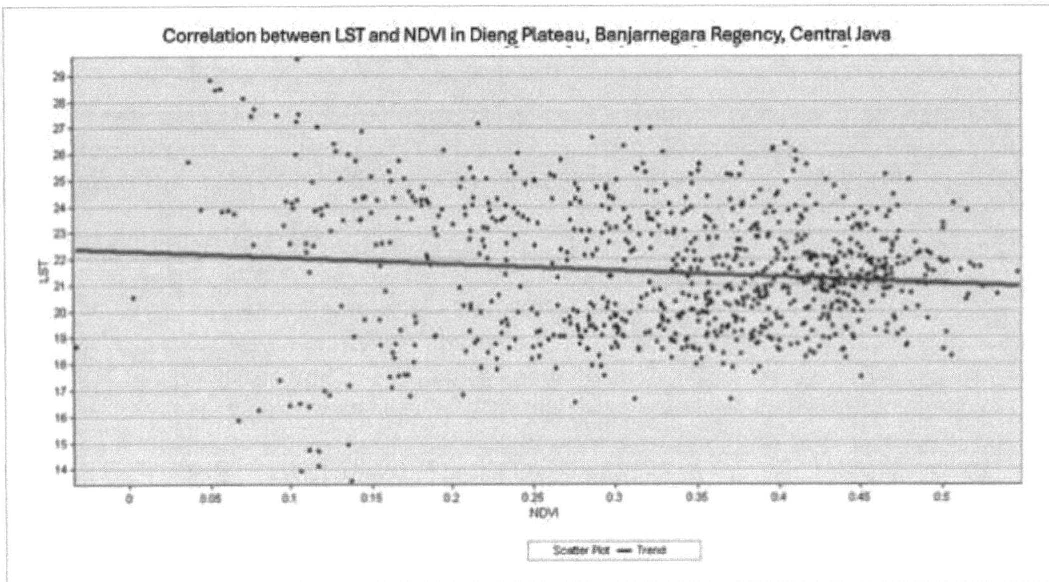

Figure 13. Scatter Plot between LST and NDVI

References

[1] C. Ansori, & F. A. Ramdhani, 2015, "Penentuan Tipe Fluida, Geotermometer Reservoir Dan Hilang Panas Alamiah Berdasarkan Analisis Data Geokimia Panas Bumi Di Kabupaten Banjarnegara, Jawa Tengah," *Buletin Sumber Daya Geologi,* vol. 10, no. 3.

[2] E. N. Dachlan, V. Putiksari, & L. B. Prasetyo, 2015, "Geothermal Energy Utilization in the Kamojang Nature Reserve, West Java, Indonesia," *Forum Geografi*, vol 29, no. 2, pp. 128-142. doi: 10.23917/forgeo.v29i2.999

[3] Farras, Naufal, Abdi Sukmono, & Nurhadi Bashit, 2017, "Analisis Estimasi Energi Panas Bumi Menggunakan Citra Landsat 8 (Studi Kasus: Kawasan Gunung Telomoyo),"*Jurnal Geodesi UNDIP*, vol. 6, no. 4.

[4] N. I. Fawzi & M. Y. Iswari, 2019, "Hubungan kategori tutupan lahan dan suhu permukaan menggunakan Landsat-8" Jurnal Spasial STKIP PGRI Sumatera Barat Indonesia, vol. 6, no. 1, pp. 27-35.

[5] S. A. Gorgani, M. Panahi, & F. Rezaie, 2013, "The Relationship between NDVI and LST in the urban area of Mashhad, Iran," *International Conference on Civil Engineering Architecture & Urban Sustainable Development*, vol 27 & 28, pp. 1-7.

[6] H. Z. Hadibasyir, S. S. Rijal, & D. R. Sari, 2020, "Comparison of Land Surface Temperature During and Before The Emergence of Covid-19 Using Modis Imagery in Wuhan City, China," *Forum Geografi,* vol. 34, no. 1, pp. 1-15. doi: 10.23917/forgeo.v34i1.10862

[7] R. Hall, 2012, "Late Jurassic-Cenozoic reconstructions of the Indonesian region and the Indian Ocean," *Tectonophysics*, vol. 570–571, pp. 1–41 . doi: 10.1016/j.tecto.2012.04.021.

[8] Hayati, R. N. Ati, 2019, "Pemanfaatan Citra Landsat 8 Untuk Mengetahui Perubahan Suhu Permukaan Tanah (Land Surface Temperature) Di Kabupaten Ngawi Tahun 2015, 2017, dan 2019. Skripsi," Fakultas Teknik Sipil dan Perencanaan. Malang: Institut Teknologi Nasional.

[9] Indonesian Ministry of Energy and Mineral Resource (Kementerian ESDM), 2017, "Potensi Panas Bumi," Jakarta: Direktorat Panas Bumi Kementrian ESDM.

[10] Indonesian Ministry of Energy and Mineral Resource (Kementerian ESDM), 2017, "Indonesia Peringkat 2 Produsen Listrik Panas Bumi Lampaui Filipina" (https://ebtke.esdm.go.id/post/2018/04/28/1948/indonesia.peringkat.2.produsen.listrik.panas.bumi) In Indonesian. *Accessed on September 2024.*

[11] Kementrian ESDM. 2020. Potensi Pengembangan Energi Panas Bumi di Indonesia. [Online]. Available https://ebtke.esdm.go.id/lintas/id/investasi-ebtke/sektor-panas- bumi/potensi

[12] Lapan, 2018, "Landsat-8 Citra Satelit Resolusi Menengah," [Online]. Available https://inderaja-catalog.lapan.go.id/application_data/default/pages/about_Landsat-8.html, diakses pada 25 September 2023

[13] Y. Ramadhan, K. Channel, & N. R. Herdianita, 2013, "Hotwater Geochemistry for Interpreting The Condition of Geothermal Reservoir, Dieng Plateau Case, Banjarnegara-Wonosobo Regency, Central Java" *Indonesian Journal of Geology,* vol. 8, no. 2, pp. 89-96.

[14] M. S. Rosid, and C. Sibarani, 2021, "Reservoir identification at Dieng geothermal field using 3D inversion modeling of gravity data" *Journal of Physics : Conference Series*, vol. 1816, no. 2021, pp. 012083.

[15] A. T. Sasmi, A. D. Nugraha, M. Muzli, S. Widiyantoro, S. Syuhada, M. Muttaqy, Z. Zulfakriza, S. Wei, A. Priyono, H. Afif, P. Supendi, Y. M. Husni, B. S. Prabowo, & A. F. N. Sarjan, 2023, "Shear wave splitting of the 2018 Lombok earthquake aftershock area, Indonesia," *Geoscience Letters,* vol. 10, no. 7. doi: 10.1186/s40562-022-00258-3

[16] C. Setiawan, M. Muzani, W. Warnadi, F. R. A'Rachman, Q. Qismaraga, & E. Ermalia, E, 2020, "Remote Sensing Imagery and GIS for Monitoring the Pyroclastic Material of Mount Sinabung" *Forum Geografi*, vol. 33, no. 2, pp. 184-195. doi: 10.23917/forgeo.v33i2.9223

[17] N. Suhartono, 2012, "Pola Sistim Panas dan Jenis Geothermal Dalam Estimasi Cadangan Daerah Kamojang," *Jurnal Ilmiah MTG*, vol. 5, no.2.

[18] K. M. Islam, A. I. M. Al-Muaybid, M. F. A. Al-Saqer, M. S. R. Al-Nagada, K. S. A. Al-Newaihel, R. Akter, M. A. Gul, M. M. Rahman, & Z. Shatnawi, Z., 2024, "Assessing Survey Data to Study Traffic Flow Characteristics: An in-depth analysis of King Fahad Road, Al-Ahsa, Saudi Arabia," *Forum Geografi,* vol. 38 no.2, pp. 167-177. doi: 10.23917/forgeo.v38i2.4629

[19] F. Hematang, A. Murdjoko, F. Kesaulija, J. Marwa, & A. Ungirwalu, 2024, "Grouping Land Cover Using Orthophoto in Small Islands: An Application of Low-Cost UAV on Mansinam Island," *Forum Geografi*, vol. 38, no. 2, pp. 189-202. doi: 10.23917/forgeo.v38i2.4870

Effect of Temperature Variation and Reaction Time on Transesterification Efficiency of Nyamplung Oil Using Heterogeneous Catalysts

Lilla Puji Lestari[1], Andiyan Andiyan[2*], Iin Arianti[3], Nyimas Yanqoritha[4] and Wetri Febrina[5]

[1] Department of Mechanical Engineering, Universitas Maarif Hasyim Latif, Sidoarjo, Indonesia
[2] Department of Architecture, Universitas Faletehan, Bandung, Indonesia
[3] Department of Civil Engineering, Politeknik Negeri Pontianak, Pontianak, Indonesia
[4] Department of Civil Engineering, Universitas Prima Indonesia, Medan, Indonesia
[5] Department of Industrial Engineering, Sekolah Tinggi Teknologi Dumai, Riau, Indonesia

*E-mail: andiyanarch@gmail.com

Abstract. The increasing reliance on fossil fuels has raised environmental and economic challenges, prompting the need for sustainable energy alternatives. Biodiesel, derived from renewable sources, presents a viable solution. Nyamplung (*Calophyllum inophyllum*) oil is a promising biodiesel feedstock due to its high oil content and adaptability to marginal lands. This study explores the impact of temperature and reaction duration on the transesterification efficiency of Nyamplung oil using heterogeneous calcium oxide (CaO) catalysts. Oil extraction was performed using n-hexane, followed by esterification and transesterification at 60°C, 70°C, and 80°C for 30–90 minutes. The maximum biodiesel yield of 85% was obtained at 70°C with a reaction time of 60 minutes. Longer reaction durations and higher temperatures resulted in reduced yields due to potential methyl ester degradation. Biodiesel quality was assessed through ASTM GC-MS and physicochemical analyses, confirming its compliance with international standards. The study provides valuable insights into optimizing biodiesel production, contributing to the development of sustainable biofuels. These findings emphasize the potential of Nyamplung oil as a biodiesel source and highlight the importance of precise reaction conditions for enhancing transesterification efficiency.

1. Introduction

The search for alternative energy sources has intensified in response to the growing ecological and financial concerns associated with the global use of fossil fuels. Because it may lessen our dependency on nonrenewable resources and reduce emissions of greenhouse gases, biodiesel has attracted significant attention as a possible substitute[1,2].Biodiesel, derived from various biological materials, offers a sustainable and environmentally friendly solution to the energy crisis

Because it can be made from sustainable sources like vegetable oils, biodiesel is a potential substitute[3]. As a potential biodiesel source, the nyamplung plant (*Calophyllum inophyllum*)

shows great promise[4].This plant has a high oil content and is able to thrive in barren soil, making it a promising source of biodiesel[5].Research provided evidence that the transesterification of Nyamplung oil with a Lewis base catalyst efficiently generates biodiesel. Findings from the research highlight the significance of reaction time and temperature in optimizing biodiesel output[6].Furthermore, processed Nyamplung oil through esterification using a ZnO-MCM-48-CaO catalyst. The results showed that the proper catalyst selection can increase the selectivity and efficiency of biodiesel production[3].Research expedited the transesterification process by using microwave heating[7].Results with high biodiesel efficiency were best achieved at 70°C with a reaction time of 60 minutes, according to the findings. This study confirms the importance of proper operating parameters in enhancing biodiesel production from Nyamplung oil. The high oil yield and adaptability to various environmental conditions make this plant a good choice for sustainable energy production[8].

Research into using nyamplung oil as a biodiesel feedstock has been extensive because of the plant's favorable fatty acid composition and high oil content (approximately 71.4% by seed weight)[9,10]. The transesterification process of Nyamplung oil using heterogeneous catalysts has shown significant efficiency in producing biodiesel that meets international standards[11].

The benefits of heterogeneous catalysts in speeding up the transesterification process have made their use a topic of interest. Microwaves were used for the transesterification of Nyamplung oil with a heterogeneous catalyst based on natural zeolites, and it was discovered that the production efficiency of biodiesel could be enhanced[12].Similar results were reported in which the use of a zeolite catalyst in transesterification produced high-quality biodiesel[13].One important and well-known area of the global agricultural business is the palm oil sector[14].

Research investigated the use of eco-friendly and sustainable catalysts derived from Moringa oleifera leaf ash in the transesterification of Nyamplung oil[15]. The research results indicate that green catalysts, such as Moringa oleifera leaf ash, have become the focus of various studies due to their potential in reducing dependence on harmful chemicals. These natural materials offer a more environmentally friendly alternative in biodiesel production, helping to improve the sustainability of the transesterification process. Several studies support their use to improve reaction efficiency and accelerate the transition toward more sustainable biodiesel production methods. This suggests that the drilling method negatively affects the environment, especially soil conditions, vegetation, and the depletion of fossil fuels[16].

Examining how temperature variations and reaction times affect the efficiency of the transesterification of Nyamplung oil using heterogeneous catalysts is the primary goal of this study[17].The results are expected to provide ways to optimize the transesterification process and increase biodiesel production from Nyamplung oil. This study identifies ideal operating parameters to provide a scientific foundation for enhancing efficient and sustainable biodiesel production, thus aiding worldwide initiatives to diminish reliance on fossil fuels and alleviate climate change[18].

The results of this research are anticipated to enhance the efficiency and sustainability of biodiesel production processes, aid in worldwide initiatives to decrease reliance on fossil fuels and combat climate change, and bolster efforts to transition to renewable energy sources. Nyamplung oil can be obtained through several stages of the process, namely [19].

1. Stripping the seeds from the tough skins.
2. Shredding into thin slices.
3. Sun drying for 2 days.
4. After two hours of pounding, steaming is performed.

5. Using a hydraulic press or extruder to manually extract black oil that contains skin contaminants and chemical compounds such as chlorophyll, alkaloids, phosphatides, and carotenoids is effective[20].The oil from 2.5 kg of dried, skinless nyamplung seeds may be extracted into 1 liter of oil
6. Using a 1% phosphoric acid solution, the gum and impurities are extracted from the oil as part of the gum removal process.

Vegetable and animal oils may be used to make biodiesel[21].Biodiesel and vegetable oils are both chemical substances that fall within the category of fatty acid esters. Biodiesel is a monoester of fatty acids linked to methanol, in contrast to vegetable oils, which are triglycerides of fatty acids linked to glycerol[22].Considerable attention goes into determining its potential as a diesel engine fuel due to the major effects of fluctuations in molecular composition[23].

1. Biodiesel, specifically methyl esters, has a molecular weight that is much lower than that of triglycerides, which are oils derived from vegetables. This property makes triglycerides very susceptible to hydrolysis when heated in an oxygen-free environment[2].
2. The viscosity of vegetable oil is much higher than that of biodiesel or diesel. Because of this, when diesel engines are fed vegetable oil, the fuel injection pump cannot effectively mist the fuel for combustion[2].
3. The lower cetane number for vegetable oils compared to methyl esters is a result of the higher degree of branching in vegetable oil molecules[2].
4. Diesel fuel's combustibility is quantified by the cetane number. The process of making esters from free fatty acids is called esterification[2].

The process of alcoholysis, or transesterification, involves reacting triglycerides (present in vegetable oils) with alcohols to produce alkyl esters.The process yields glycerol as an additional product.Methanolysis refers to the process of obtaining alkyl groups from monohydric alcohols, and methanol is the most common choice because of its low cost and high reactivity[2].Therefore, biodiesel may be considered equivalent to fatty acid methyl esters[24].In reality, UF is an environmentally harmful substance that comes at a premium cost[25].

Research conducted by investigators examined whether Nyamplung oil might be transesterified using heterogeneous catalysts[12,26].Biodiesel production efficiency may be enhanced by using heterogeneous base catalysts like KOH and CH3COONa in conjunction with microwave assistance, as shown in this work.The best result obtained was a 96% biodiesel yield with KOH as the catalyst under optimal operating conditions. The escalation of the global population has resulted in heightened energy demands[27].

The next research was conducted using electromagnetic induction to study how temperature and reaction time affect biodiesel synthesis efficiency[28,29].Optimal conditions were determined to be 70°C with a reaction period of 60 minutes, according to this research, which also indicated that reaction length significantly affected biodiesel production[30–32]. Additionally, the research demonstrated that the transesterification efficiency of Nyamplung oil could be enhanced by using the electromagnetic induction approach

Using a heterogeneous catalyst composed of calcium oxide doped with potassium, the experiment to produce biodiesel from Nyamplung oil was also carried out by[33].The findings demonstrated that the use of a CaO-K catalyst greatly enhanced the biodiesel conversion of Nyamplung oil, resulting in a yield of 93%.The research also found that 65°C and 50 minutes of reaction time were the best working conditions[2].

The biodiesel's acid number, viscosity, density, and methyl ester concentration were among the several characteristics tested in a research study by[32].The findings proved that Nyamplung oil might be a sustainable source of biodiesel since the biodiesel it produced satisfied all of the international criteria. Green technology is becoming increasingly popular[2,34].

This research will be one of the first studies to systematically explore the effect of temperature variation (60°C, 70°C, and 80°C) and reaction duration (30 min, 60 min, and 90 min) on the transesterification efficiency of Nyamplung oil using a CaO heterogeneous catalyst[35,36]. Thus, this study can fill the gap in the existing literature and provide more comprehensive empirical data.The use of a CaO (calcium oxide) heterogeneous catalyst is well known; this study will combine it with a detailed analysis of variations in reaction temperature and time for the first time.This is anticipated to yield novel insights into the optimization of the transesterification process and the efficacy of biodiesel generation[37].

Research on biodiesel derived from Nyamplung oil corroborates the attainment of many Sustainable Development Goals (SDGs). First, in accordance with SDG 7, biodiesel from Nyamplung provides a clean and affordable renewable energy alternative that can be grown on marginal lands without disrupting food production. In addition, this research also supports SDG 12 by utilizing green catalysts, which increase production efficiency while reducing chemical use and waste. Lastly, this biodiesel helps achieve SDG 13 by decreasing emissions of greenhouse gases, facilitating the shift to renewable energy sources, and lowering global warming temperatures in accordance with the goals set forth in the Paris Agreement. An impending energy disaster might be on the horizon as a result of the depletion of fossil fuel reserves, and our dependence on nonrenewable energy sources reduces our ability to sustain life on Earth due to carbon dioxide emissions[2,27].

Although there have been a number of research studies on the optimization of biodiesel synthesis from Nyamplung oil using CaO heterogeneous catalysts, this study adds to the literature by investigating variations in temperature (60°C, 70°C, and 80°C) and reaction duration (30 to 90 minutes).Biodiesel yield is the primary focus of this research, which also thoroughly analyzes the physicochemical properties of the biodiesel produced using GC-MS methods in compliance with ASTM requirements.Therefore, this research contributes to our understanding of how to optimize the transesterification of Nyamplung oil, which should lead to a sustained improvement in biodiesel production efficiency[2].

2. Materials and Methods

2.1 Materials

This study primarily utilizes Nyamplung seeds obtained from Baluran National Park, Situbondo, Indonesia.The seeds were air-dried for three days, manually dehulled, and finely pulverized into powder.Using the Soxhlet procedure with n-hexane for six hours, the oil extraction was carried out. To ensure the oil's stability for future research and replication, it was then stored in an amber glass container at 4°C to prevent oxidation and degradation[2].

The study also utilized various reagents, including n-hexane, heterogeneous CaO catalysts, sodium hydroxide solution, hydrochloric acid (33% purity), and Merck methanol (99.9% purity), all obtained from PT Kurniajaya in Surabaya.

2.2 Experimental Procedure

This research was executed in many phases: catalyst preparation, oil extraction from Nyamplung seeds utilizing n-hexane solvent, characterization of the recovered oil, esterification to diminish free fatty acid content, and transesterification to generate biodiesel.Using ASTM procedures and the Indonesian National Standard (SNI 04-7182-2006), the biodiesel yield was ascertained, and its properties, such as density, viscosity, cetane number, flash point, and methyl ester content, were assessed.Calcium oxide was used as a heterogeneous catalyst in the transesterification process. A previous study indicated that the ideal conditions for methyl ester conversion were a reaction duration of 30, 60, or 90 minutes and temperatures of 60, 70, or 80 degrees Celsius[2].Lower temperatures may lead to incomplete conversion, whereas excessive heat can trigger side reactions and degrade the final product. To evaluate the effects of extreme conditions, control experiments were performed at temperatures exceeding 80°C and reaction durations longer than 90 minutes.These conditions were assessed for potential degradation effects, such as excessive methanol evaporation and reduced biodiesel yield.The control parameters ensured the identification of the most efficient conditions for transesterification.

2.3 Heterogeneous Catalyst Preparation

The heterogeneous catalyst selected was calcium oxide (CaO) due to its high catalytic activity, low cost, and low environmental impact.One of the heterogeneous catalysts often used in the transesterification process for biodiesel synthesis is calcium oxide (CaO).In this investigation, the following procedures were used to create a CaO heterogeneous catalyst: eggshells or other shells that are rich in calcium carbonate (CaCO3) were used as the raw materials. After that, the calcination step follows.Calcination is the process that transforms calcium carbonate (CaCO3) from its natural or synthetic origins into calcium oxide (CaO).Heat is applied to CaCO3 as part of the calcination process.The CaCO3 material is heated in a furnace for two to three hours at temperatures between 800 and 900 degrees Celsius.Calcium oxide (CaO) and carbon dioxide (CO2) gas are byproducts of this process. After that, storing and cooling occurs.The final CaO is allowed to cool to room temperature when the calcination process is finished. Since CaO is hygroscopic and easily reacts with water to form calcium hydroxide (Ca(OH)2), it is necessary to store the cooled catalyst in a desiccator to prevent atmospheric moisture absorption.The high basicity of CaO makes it an ideal catalyst for the transesterification reaction, which speeds up the process of converting triglycerides into biodiesel.In addition to being a sustainable and cost-effective option for biodiesel production, CaO is highly recyclable[2].

2.4 Extraction of nyamplung oil

The seeds of the nyamplung were dehydrated and then pulverized into a powder. After six hours of Soxhlet extraction in an n-hexane solvent, the powdered nyamplung seeds were extracted.The solvent was removed from the extracted oil by drying it at 60°C after isolating it using a rotary evaporator[2,3].Pure nyamplung oil was obtained by distilling the extract from the solvent mixture.

2.5 Free fatty acid (FFA) content analysis

The assessment of free fatty acid levels in Nyamplung oil is crucial for evaluating the oil's quality and its suitability for transesterification[10]. Free fatty acids in the oil may interact with the base catalyst during transesterification, leading to soap formation and diminishing the efficiency of biodiesel production. Therefore, it is essential to know the FFA content before transesterification. The following steps outline the process of analyzing FFA content:

1. Prepare the Nyamplung oil samples to be analyzed. If the oil is in solid form, heat it to room temperature. Ensure that the oil is homogeneous before taking the sample.
2. Take 50 milliliters (mL) of a solvent combination and dissolve about 1 to 2 grams of Nyamplung oil. Mixed with a few drops of phenolphthalein indicator, the most typical solvent combination consists of one part ethanol and one part toluene[2].

3. Combine the oil and solvent, and then titrate the mixture with a 0.1 N NaOH solution to complete the titration procedure. When the titration reaches the endpoint, slowly add NaOH while stirring. A steady pink hue in the solution that lasts for about 30 seconds indicates the endpoint[2].

2.6 Esterification

A strong acid catalyst, namely concentrated HCl, is used to react nyamplung oil with methanol at a mole ratio of 1:20 during the esterification process, which lowers free fatty acid levels. A three-neck flask is used to mix 370 mL of methanol with nyamplung oil. A concentrated HCl solution, comprising 10% of the oil content, is added as a catalyst. Two hours are then spent reacting the mixture at 60 °C. After the reaction is finished, the methyl esters are separated from the mixture using a funnel, and their free fatty acid content is measured[2].

2.7 Transesterification

Specifically, zeolite and bentonite were used as modified catalysts in the transesterification process. With the catalyst making up 10% of the oil's weight, the nyamplung oil-to-methanol mole ratio is 1:6. In a three-neck flask, 100 mL of esterified nyamplung oil was added, followed by 24.3 mL of methanol and a 10% by weight catalyst of modified zeolite or bentonite to conduct the experiment. After that, the reaction was stirred for 5 hours at various temperatures (50, 60, 65, and 70 °C) using a magnetic stirrer. When the reaction was finished, the biodiesel and glycerol mixture was filtered through a Buchner funnel to remove the catalyst. The next step was to use a separating funnel to separate the glycerol and biodiesel. To further remove any remaining water, the biodiesel was subjected to additional drying in a rotary evaporator. Next, the production and characteristics of the biodiesel produced were evaluated[2].

3. Results and Discussion

3.1 Analysis of Nyamplung Oil

Nyamplung oil (*Calophyllum inophyllum*) is extracted as a raw material for biodiesel production. It has several important characteristics that determine its quality and efficiency in the transesterification process. The following is the analysis data for the extracted Nyamplung oil[2]:

Table 1. Characteristic Analysis of Nyamplung Oil.

Characteristics	Analysis Result	Description
Free Fatty Acid Content (FFA)	7,06 %	High, esterification is required before transesterification.
Fatty Acid Composition		
- Oleic Acid (C18:1)	46 %	Dominant in Nyamplung oil
- Linoleic Acid (C18:2)	24%	
- Palmitic Acid (C16:0)	18%	
- Stearic Acid (C18;0)	12%	
Viscosity	35.6 mm^2/s (at 40°C)	Must be reduced during the transesterification process.
Density	0.920 g/cm^3 (at 15°C)	Higher compared to biodiesel.
Acid Number	14.1 mg KOH/g	Indicates the amount of free fatty acids present.

Water Content	0,25 %	Should be minimized for optimal transesterification efficiency.
Impurity Content	0,02 %	Affects catalyst efficiency.
Calorific Value	39,5 MJ/kg	Indicates the energy that can be produced from biodiesel.

Nyamplung oil has to undergo esterification before it can be utilized in transesterification due to its high FFA level of 7.06%. To avoid soap formation, which may disrupt the biodiesel production process, esterification can assist in lowering the FFA concentration. The primary component of nyamplung oil is oleic acid, while linoleic acid, palmitic acid, and stearic acid make up the remainder. Because it affects the physical and chemical properties of the produced biodiesel, including its oxidative stability and viscosity, the composition is crucial[38]. In order to create biodiesel that satisfies international standards, the transesterification process must lower the crude oil's viscosity if it is high. An elevated acid value indicates that the oil has a significant proportion of free fatty acids. The value must be reduced by esterification to avert soap production during the transesterification process. High water content can interfere with the transesterification process and reduce catalyst efficiency. Therefore, the water content should be minimized after the esterification process to ensure optimal process efficiency. The calorific value indicates the energy that can be produced from biodiesel. Nyamplung oil has a fairly high calorific value, making it a potential source of energy after being processed into biodiesel.

Table 2. Characteristics of Nyamplung oil after the esterification process.

Characteristics	Analysis Result	Description
Kandungan Asam Lemak Bebas (FFA)	1.5 %	Decreases after esterification ready for transesterification.
Fatty Acid Composition		
- Oleic Acid (C18:1)	45 %	Slightly Reduced
- Linoleic Acid (C18:2)	23 %	Slightly Reduced
- Palmitic Acid (C16:0)	17 %	Slightly Reduced
- Stearic Acid (C18;0)	11 %	Slightly Reduced
Viscosity	5.2 mm^2/s (at 40°C)	Reduced, approaching biodiesel standards.
Density	0.88 g/cm^3 (at 15°C)	Slightly reduced.
Acid Number	1,0 mg KOH/g	Significantly reduced.
Water Content	0,1 %	Decreased.
Impurity Content	0,01 %	Decreased.
Calorific Value	40,0 MJ/kg	Slight increase due to FFA reduction.

3.2 Effect of reaction temperature and reaction time on biodiesel yields

Over the course of 60 minutes at 60°C, the biodiesel output increased, peaking at 75%. The yield did drop slightly, however, when the response time was increased to 90 minutes[39]. The optimal conditions for biodiesel production of 85% over a 60-minute reaction period were at a temperature of 70°C. As the response time was increased to 90 minutes, the biodiesel output dropped to 80%. After 60 minutes of reaction time at 80°C, the biodiesel production dropped to 78%, which was lower than at 70°C. Extending the reaction time to 90 minutes decreased the yield to 73%. Temperatures in transesterification reactions can cause methyl ester degradation, leading to lower biodiesel yields. In addition, reaction times may result in unwanted side reactions that negatively affect biodiesel quality. The data highlight the necessity of controlling temperature and reaction time to prevent such issues and optimize production efficiency. The use of CaO as a catalyst proved effective in achieving high biodiesel yields under controlled conditions.

3.3 Analysis Results of Biodiesel Characteristics at the Highest Gain

Using GC-MS procedures in accordance with ASTM standards and physicochemical characteristic tests, the properties of the biodiesel created under ideal circumstances (70°C and 60 minutes of reaction time) were examined. The results showed the following: an 85% biodiesel yield, a 96% methyl ester content, a viscosity of 4.5 mm^2/s (meeting the requirements of the ASTM D445 standard), a density of 0.88 g/cm^3 (meeting the requirements of the ASTM D4052 standard), and an acid number of 0.35 mg KOH/g (meeting the requirements of the ASTM D664 standard). Under ideal circumstances, the analytical results demonstrated that the biodiesel made from Nyamplung oil satisfied all international biodiesel quality criteria[2]. Based on its physicochemical characteristics that meet ASTM requirements and its high methyl ester concentration, this biodiesel shows promise as a greener, more sustainable fuel option.

The results of this investigation proved that the amount and quality of biodiesel made from Nyamplung oil are greatly affected by temperature and reaction time. Results showing 85% biodiesel and physicochemical characteristics meeting ASTM standards were obtained from a transesterification process that ideally took place at 70°C for 60 minutes[40]. The fact that the yield drops at 80°C suggests that the transesterification process must be performed at a specific temperature to prevent the breakdown of methyl ester or the creation of undesirable by-products. Longer reaction times also do not help much and could lower biodiesel production[2].

These results make an important contribution to optimizing biodiesel production from Nyamplung oil using CaO heterogeneous catalysts and offer a more efficient and sustainable method for biodiesel production as a renewable energy source.

3.4 Sustainability Considerations

The use of CaO as a green catalyst reduces reliance on hazardous chemical catalysts, minimizing toxic waste generation and environmental pollution. Additionally, Nyamplung oil cultivation promotes carbon sequestration and utilizes non-arable lands, reducing competition with food crops and promoting ecosystem sustainability.

Environmental Impact: The use of CaO as an environmentally friendly catalyst minimizes reliance on hazardous chemical catalysts, thereby reducing toxic waste and environmental pollution. Additionally, cultivating Nyamplung oil contributes to carbon sequestration, prevents deforestation, and utilizes non-arable lands, promoting overall ecological sustainability.

The findings of this research are in line with several SDGs. Renewable and sustainable energy sources are being promoted through the transformation of Nyamplung oil into biodiesel, in accordance with SDG 7's requirements for affordable and clean energy. Sustainable Development Goal 12 (Responsible Consumption and Production): Reduced waste and increased resource efficiency are achieved through the use of CaO catalysts produced from waste. SDG 13 (Climate Action): Biodiesel made from Nyamplung oil is a greener alternative to fossil fuels that helps reduce emissions of greenhouse gases and lessen the effects of climate change.

4. Conclusion

This study concludes that oil from Nyamplung seeds has significant promise as a sustainable biodiesel source, particularly due to the plant's ability to thrive in arid conditions and its high oil yield. This study examines the influence of temperature and reaction time on the transesterification process utilizing heterogeneous catalysts. The findings indicate that the effectiveness of transesterification is markedly influenced by temperature and reaction duration. The ideal temperature for optimum biodiesel synthesis was 70°C, with a reaction time of 60 minutes, yielding a transesterification efficiency of 85%. This study establishes a scientific foundation for enhancing transesterification techniques for biodiesel synthesis from Nyamplung oil, which is anticipated to augment the efficiency and sustainability of global renewable energy resources.

References

[1] Qadariyah L, Bhuana D S, Selaksa R, As Shodiq J and Mahfud M 2018 Biodiesel production from Calophyllum inophyllum using base lewis catalyst *ASEAN J. Chem. Eng.* **18** 53–9

[2] Hindarso H, Ayucitra A, Dewi D R S and Asri N P 2024 Biodiesel Synthesis from Calophyllum Inophyllum Utilizing Modified Zeolite and Bentonite Catalysts. *Annales de Chimie Science des Matériaux* vol 48

[3] Kolo L, Firdaus F, Taba P, Zakir M and Soekamto N H 2022 Selectivity of the new catalyst ZnO-MCM-48-CaO in esterification of calophyllum inophyllum oil *Automot. Exp.* **5** 217–29

[4] Kurniati S, Galla W F and Syam S 2021 Optimizing production of biodiesel from Nyamplung oil with microwave power setting *IOP Conference Series: Materials Science and Engineering* vol 1034 (IOP Publishing) p 12086

[5] Kurniati S, Soeparman S, Yuwono S S and Hakim L 2018 Characteristics and Potential of Nyamplung (Calophyllum inophyllum L .) Seed Oil from Kebumen , Central Java , as a Biodiesel Feedstock *Int. Res. J. Adv. Eng. Sci.* **3** 148–52

[6] Alsaiari R A, Musa E M, Alqahtani H and Rizk M A 2023 Biodiesel production from date seed oil via CaO-derived catalyst from waste eggshell *Biofuels* **14** 743–50

[7] Kurniati A, Rahardjo A and Widiastuti S 2021 Optimasi kondisi operasi untuk produksi biodiesel dari minyak Nyamplung menggunakan reaksi transesterifikasi *J. Energi dan Lingkung.* **2** 45–53

[8] Budianto A and Sumari S 2018 Biofuel Production From Nyamplung Oil Using Catalytic Cracking Process With Zn-HZSM-5 / γ Alumina Catalyst

[9] Malabadi R B, Kolkar K P and Chalannavar R K 2023 Biodiesel production : An updated review of evidence

[10] Arvanitoyannis I S and Vlachos A 2007 Implementation of physicochemical and sensory analysis in conjunction with multivariate analysis towards assessing olive oil authentication/adulteration *Crit. Rev. Food Sci. Nutr.* **47** 441–98

[11] Ramdhani D A and Trisunaryanti W 2023 Study of green and sustainable heterogeneous catalyst produced from Javanese Moringa oleifera leaf ash for the transesterification of Calophyllum inophyllum seed oil **8** 124–33

[12] Juwono H, Triyono, Sutarno, Triwahyuni E, Ulfin I and Kurniawan F 2017 1 production of biodiesel from seed oil of nyamplung (Calophyllum inophyllum) by AL-MCM-41 and its performance in diesel engine *Indones. J. Chem.* **17** 316–21

[13] Hartono R, Iswanto A H, Sucipto T and Lubis K M 2018 Effect of particle pre-treatment on physical and mechanical properties of particleboard made from oil palm trunk *IOP Conference Series: Earth and Environmental Science* vol 166 (IOP Publishing) p 12006

[14] Febrina W, Saryono S, Awaluddin A, Helwani Z, Andiyan A, Lestari L P and Yanqoritha N 2025 Review of biodegradation of palm oil mill effluent (POME) in perspective of microbial involved on different type of bioreactor *IOP Conference Series: Earth and Environmental Science* vol 1441 (IOP Publishing) p 12005

[15] Ramdhani A and Trisunaryanti W 2023 Penggunaan katalis hijau dari abu daun Moringa oleifera dalam transesterifikasi minyak Nyamplung *J. Teknol. Hijau* **5** 123–34

[16] Rosyidah M, Andiyan A, Listyorini H, Prayitno P H, Yuswardi Y and Yuhanah Y 2022 LCA methodology for detecting environmental impacts on natural gas drilling process *IOP Conf. Ser. Earth Environ. Sci.* **1041** 12035

[17] Patel P, Garaniya V, Abbassi R, Baalisampang T and Aryai V 2022 Experimental analysis of pitting corrosion in offshore structures *Advances in Construction Safety: Proceedings of HSFEA 2020* (Springer) pp 65–87

[18] Andrifar M, Goembira F, Ulfah M, Putri R, Yuliarningsih R and Aziz R 2022 Optimization of sustainable biodiesel production from waste cooking oil using heterogeneous alkali catalyst **16** 66–71

[19] Gunawan S, Aparamarta H W, Taufany F, Prayogo A, Putri H A and Wijaya C J 2020 Separation and purification of triglyceride from nyamplung (Calophyllum inophyllum) seed oil as biodiesel feedstock by using continuous countercurrent extraction **16** 18–22

[20] Qadariyah L, Ansori A, Wibowo S A, Muchammad R S C, Bhuana D S and Mahfud M 2019 Biodiesel production from Calophyllum inophyllum L oil using Microwave with Calcium Carbonate catalyst *IOP Conference Series: Materials Science and Engineering* vol 543 (IOP Publishing) p 12072

[21] Uzun B B, Kılıç M, Özbay N, Pütün A E and Pütün E 2012 Biodiesel production from waste frying oils:

Optimization of reaction parameters and determination of fuel properties *Energy* **44** 347–51

[22] Fathallah A Z M and Pinto F 2022 The Influence of NaCl dissolved on biodiesel of used cooking oil on performance and its degradation of main components of diesel engine *IOP Conference Series: Earth and Environmental Science* vol 972 (IOP Publishing) p 12030

[23] Sisca V and Jamarun N 2020 Biodiesel Production from Waste Cooking Oil Using Catalyst Calcium Oxide Derived of Limestone Lintau Buo **11** 8–14

[24] Zheng D, Zhu Y, Sun X, Sun H, Yang P, Yu Z and Zhu J 2024 Equilibrium Moisture Mediated Esterification Reaction to Achieve Over 100 % Lignocellulosic Nanofibrils Yield **2402777** 1–12

[25] Arianti I, Rafani M, Fitriani N, Nurhidayati E, Febrina W and Andiyan A 2024 Environmentally Friendly Particle Board from Sawdust with the Addition of PP and PET Plastic Waste without Using UF Adhesive *Civ. Eng. Archit.* **12** 3172–8

[26] Hartono R, Wijanarko A and Hermansyah H 2018 Synthesis of biodiesel using local natural zeolite as heterogeneous anion exchange catalyst *IOP Conference Series: Materials Science and Engineering* vol 345 (IOP Publishing) p 12002

[27] Haslinah A, Thamrin T, Bandrang T N, Taryana T, Karyasa T B, Purwoko B and Andiyan A 2024 Green technology's function in the production of renewable energy and mineral extraction *Casp. J. Environ. Sci.* **22** 93–102

[28] Shahabi Mohammadabadi S, Goli M and Naji Tabasi S 2022 Optimization of Bioactive Compound Extraction from Eggplant Peel by Response Surface Methodology: Ultrasound-Assisted Solvent Qualitative and Quantitative Effect *Foods* **11**

[29] Al-Hamamre Z and Yamin J 2014 Parametric study of the alkali catalyzed transesterification of waste frying oil for Biodiesel production *Energy Convers. Manag.* **79** 246–54

[30] Kartika I A, Yani M, Ariono D, Evon P and Rigal L 2013 Biodiesel production from jatropha seeds: Solvent extraction and in situ transesterification in a single step *Fuel* **106** 111–7

[31] Basumatary S F, Brahma S, Hoque M, Das B K, Selvaraj M, Brahma S and Basumatary S 2023 Advances in CaO-based catalysts for sustainable biodiesel synthesis *Green Energy Resour.* 100032

[32] Kurniati S and Syam S 2024 Enhancing Biodiesel Production from Nyamplung Oil : Kinetic Analysis of Transesterification via Electromagnetic Induction Heating 21–8

[33] Gaurav K, Kumari S and Dutta J 2021 Utilization of Waste Chicken Eggshell as Heterogeneous CaO Nanoparticle for Biodiesel Production **12** 49–57

[34] Judijanto L, Liestyowati D, Andiyan A, Umam R and Indah D Y 2025 Harmonizing work behaviour, social equity, and green culture in urban living: the moderating role of sustainable cities green technology *IOP Conference Series: Earth and Environmental Science* vol 1441 (IOP Publishing) p 12004

[35] Khurram M S, Al-muhtaseb A H, Inayat A and Shah N S 2022 Enhancing the Catalytic Activity of Eggshell-Derived CaO Catalyst and Its Application in Biodiesel Production from Waste Chicken Fat

[36] Ansori A, Wibowo S A, Kusuma H S, Bhuana D S and Mahfud M 2019 Production of biodiesel from nyamplung (Calophyllum inophyllum L.) using microwave with CaO catalyst from eggshell waste: optimization of transesterification process parameters *Open Chem.* **17** 1185–97

[37] Ulakpa W C, Aseibichin C, Chimezie O A, Olaseinde A A, Odeworitse E, Onoriode E and Adaeze I M 2024 Nano-CaO as a heterogeneous catalyst for biodiesel synthesis by transesterification of Jatropha oil *J. Trace Elem. Miner.* 100183

[38] Rachmaditasari R, Darojat M I and Mahfud M 2024 Production of biodiesel (isopropyl ester) from coconut oil by microwave assisted transesterification : parametric study and optimization

[39] Esan A O, Olabemiwo O M, Smith S M and Ganesan S 2021 A concise review on alternative route of biodiesel production via interesterification of different feedstocks *Int. J. Energy Res.* **45** 12614–37

[40] Pulidindi I N and Gedanken A 2021 The catalytic production of biofuels (biodiesel and bioethanol) using sonochemical, microwave, and mechanical methods *Nontraditional Activation Methods in Green and Sustainable Applications* (Elsevier) pp 171–239

MNDWI AND BSI FOR MONITORING LAND DEGRADATION IN GRESIK REGENCY

F. N. S. Herlita, Jumadi*, A. A. Sigit, D. N. Sari[1]

[1] Geography, Universitas Muhammadiyah Surakarta, Surakarta, Indonesia

*E-mail: jumadi@ums.ac.id

Abstract. Land degradation or land productivity declined in Indonesia, reaching 25.1% in 2014. However, in East Java, land degradation showed a downward trend from 2006 to 2018. This land degradation process was expected to affect the condition of agriculture and plantations in the region. This research aimed to map land degradation in Gresik Regency in 2000, 2005, 2010, 2015, and 2023 using Google Earth Engine (GEE). This land degradation mapping was intended to help in policymaking for the local government to develop the region. Observations of the extent of land degradation in Gresik Regency focused on the years 2000, 2005, 2010, 2015, and 2023 to obtain significant changes and an overview of long-term changes in the area. The MNDWI and BSI methods were used to determine degraded land. The observation showed a decrease in the area of land degradation in Gresik in 2023. The enactment of Regional Regulation No. 7 of 2015, which governed land use compliance with its designated purpose, aimed to prevent irrigation channel damage and maintain soil fertility. Land degradation must be addressed properly to realize the spatial plan map for agricultural areas.

Keywords: Land, Land Degradation, Google Earth Engine, Economy, Agriculture

1 Introduction

Land degradation is a condition of land damage that results in a decrease in the level of productivity of land. Degradation can be caused by erosion, soil that contains toxins for plants reduced nutrients and organic matter, and soil saturation by water [1].In addition to these elements, land degradation can be caused directly or indirectly by geomorphological processes. According to the United Nations Convention to Combat Desertification (UNCCD), land degradation is defined by Feng et al. [2] as the reduction or loss of land's biological and economic complexity and productivity due to unsustainable land use, deforestation, overgrazing, and climate change, this term emphasizes how human actions hasten the process of degradation.

Based on various expert interpretations, it may be inferred that land degradation is a process of declining land function and quality, which can be either temporary or permanent. This process impacts productivity, the land's capacity to support life, and the ecosystem as a whole. Degradation is also accelerated by human activities, including deforestation, climate change, and irresponsible land use. Land degradation can be caused by natural and man-made forces, endangering food production, soil carrying capacity, and ecosystem sustainability.

In 2014, Indonesia had a degraded land level of 25.1% of the total area of Indonesia itself [3]. Due to degraded land, crops cannot grow optimally, which causes a decrease in land productivity. Land degradation can occur due to deforestation, which can trigger natural disasters [4].

Indonesia, as the most significant agricultural country, certainly has a lot of farmland and plantations. Gresik Regency is part of East Java Province, located in the southwest of Surabaya City. This makes Gresik an area that is the centre of economic growth and is expected to be able to boost the economy of the surrounding area. One of the leading sectors in the economy of Gresik Regency is the processing sector. The dominance of the processing sector in Gresik Regency makes a lot of land conversion from agriculture to factories[5]. Urbanization and industrialization are the main factors for increasing land degradation in Gresik Regency[6].

Technological developments in remote sensing systems allow humans to repeatedly observe the earth's shape with satellite imagery. This can help in the identification and observation of landform changes [7]. Mapping land degradation in a certain period will help determine the condition of land in Gresik Regency. Mapping to observe the extent of land degradation requires image processing, while image processing often requires the support of high-quality devices and extensive storage. Google Earth Engine provides a platform for processing images that is fast, cheap and easy to process images into maps [8]. Google Earth Engine is a platform that can perform cloud-based image processing. Google Earth Engine processes images with the help of JavaScript coding. Data processing with Google Earth Engine does not take up space on local computer storage as long as the computer has an internet network and is capable of running a modern web browser. Data analysis methods with BSI and MNDWI will help in identifying land degradation on dry land [9].

With all its convenience, Google Earth Engine is an alternative to observe changes in land degradation areas in the Gresik Regency. Land degradation mapping itself can help the government in making land use policies. In addition, the land degradation map can be used as an evaluation material for the government's efforts to reduce the level of land degradation every year or within a certain period.

In the research of Khoirurrizqi et al. 2023[10] which aims to monitor the conversion of paddy fields in Kulon Progo, using NDVI, NDBI, and NDWI indices to identify paddy field objects. MNDWI and BSI methods detect dry land related to land degradation. Google Earth engine platform is used as a tool in this research. Discussions on land degradation in Indonesia with MNDWI and BSI as indicators of bare land are still rarely used. This study uses these two indices to determine which areas are considered bare land from the process of changing forests or agriculture into settlements. Land degradation maps can help in the formulation of spatial planning policies in an area. The research conducted in Gunungpati sub-district shows that land use change from non-built to built-up causes a lack of water catchment areas and an increased risk of landslides. These results help in policy development by the Gunungpati local government [11]. In the case of land degradation, it is possible to identify areas where land quality has declined due to uncontrolled land conservation.

2. Research Method

2.1 Research Location

Gresik Regency, which is the location of this research, is located at 112° - 113° East Longitude and 7° - 8° South Latitude. Gresik Regency is a lowland at an altitude of 2-12 meters above sea level. Only the Pancang sub-district is at an altitude of up to 25 meters above sea level. Geographically, Gresik Regency is bordered by Sidoarjo and Mojokerto in the south and Madura Strait and Surabaya City in the east. In the west, it borders Lamongan Regency and in the north, it borders the Java Sea. (BPS, 2020). Analysis of changes in land degradation area in Gresik Regency based on conditions in 2000, 2005, 2010, 2015 and 2023. The range of years used is per five years, and the latest year data is 2023.

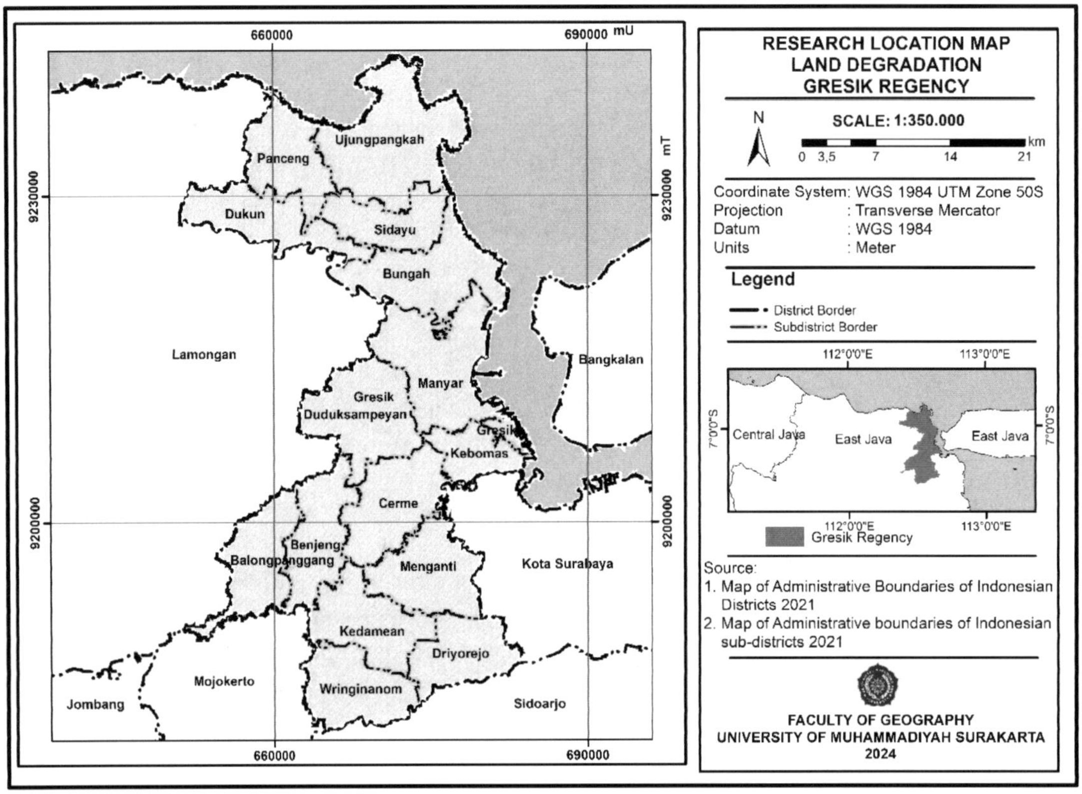

Figure 1. Research Location

2.2 Analysis Method

A clip was made for the Gresik Regency area to obtain its SHP and entered into the GEE script. The GEE added a script for calling Landsat 5, 7, 8 and 9 images, the use of Landsat imagery as a tool for observing land degradation in Gresik Regency because Landsat imagery can record the entire research area [12]. The image data is then processed with a cloud correction script to reduce the effect of cloud cover. Determining the image band that will be processed with the MNDWI method to determine the land with healthy vegetation. Then, an indication is also made for vacant land with BSI, where vacant land includes less productive land. All these processes rely on coding scripts in the GEE worksheet. Five year interval chosen to observe land degradation is an ideal interval to capture medium-term changes, as well as to avoid bias from annual fluctuations. In addition, the processes of land use change, erosion and deforestation usually develop gradually, and within 5 years the changes in land conditions will be more significant. Land management policies in the medium term use a 5-year cycle so this interval can be used to evaluate the impact of policies on land conditions periodically.

The first stage is that the SHP of Gresik Regency is entered into the script on Google Earth Engine. This aims to crop the research location [8]. The dataset used comes from Landsat 5, 7, 8, and 9 images with Level-2, Collection 2, and Tier 1 correction levels. Level-2 correction ensures that the images have undergone atmospheric and radiometric correction processes, minimizing image anomalies and producing more accurate surface reflectance data for spatial analysis [13]. Landsat imagery was used in this study due to its better data completeness than sentinel [14]. The resolution of Landsat imagery is 30 meters, which is sufficient for regional-scale monitoring such as identifying land cover change [15]. Landsat 7 dataset in 2010 suffered damage in the form of a cosomh area due to SLC damage, the gapfill correction process used Landsat 5 images of the same year[16].

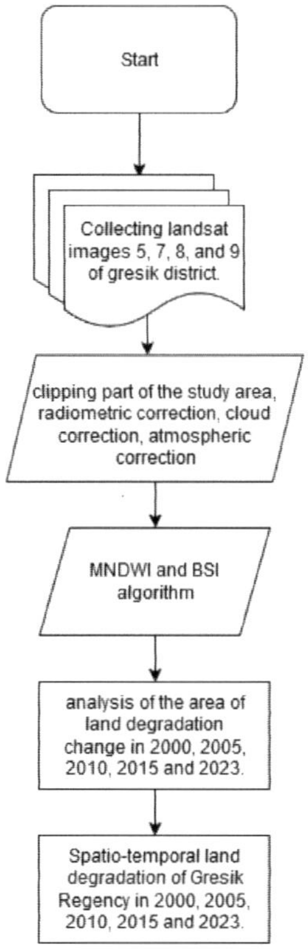

Figure 2. Research Flow Chart

The method used comes from secondary data processing in the form of map images processed with Google Earth Engine. The SHP of the Gresik District boundary is required to delimit the study area. This image data is then processed with MNDWI and BSI to determine degraded land areas. The Bare Soil Index itself is an indicator used to obtain soil variation. A combination of red, blue, NIR and SWIR bands is required to get the Bare Soil Index [17].

MNDWI, or Modified Normalized Difference Water Index combines the green band and SWIR band. It aims to detect water content in vegetation. The SWIR band is susceptible to the water content of a plant. Plants with good water content indicate healthy plants and can indicate productive land. Furthermore, the formula of MNDWI is also entered into the script so that it can be processed. The representation of soil condition and moisture is better with MNDWI and BSI than using NDVI. NDVI still captures the presence of low vegetation as a positive value, making it less accurate for identification of arid areas.

IOP Conf. Series: Earth and Environmental Science **1503** (2025) 012011 doi:10.1088/1755-1315/1503/1/012011

The Bare Soil Index is an indicator used to obtain soil variation. A combination of red, blue, NIR and SWIR bands is required to obtain the Bare Soil Index. The BSI calculation formula is entered into the script to facilitate processing.

BSI= (RED + SWIR) - (NIR + BLUE)((RED + SWIR) + (NIR + BLUE)

Description
RED = Red band spectral reflectance value
BLUE = Blue band spectral reflectance value
NIR = NIR spectral band reflectance value
SWIR = SWIR spectral band reflectance value

Higher BSI values indicate predominantly open land areas. If the BSI value increases over time, it indicates an increase in the area of bare soil, which could be an indicator of land degradation.

Table 1. BSI Value

BSI Value	Interpretation	Land Condition
0 to 1	Areas of open or degraded land	Highly degraded land, without vegetation cover
-0.1 to 0.1	Transition area between open land and vegetation	Land in transition
-1 to -0.1	Vegetated or more covered land	Land with vegetation cover or watered land.

Source: [18]

MNDWI or Modified Normalized Difference Water Index combines green band and SWIR band. It aims to detect the water content of vegetation. The SWIR band is very sensitive to the water content of a plant. Plants with good water content indicate healthy plants and can indicate productive land. Furthermore, the formula of MNDWI is also entered into the script so that it can be processed.

MNDWI= (GREEN-SWIR)(GREEN + SWIR)

Description
GREEN = Green band spectral reflectance value
SWIR = Reflectance value of SWIR spectral band
Higher MNDWI values indicate the presence of water or wet soil. A decrease in MNDWI values over time may indicate land degradation due to moisture loss.

Tabel 2. MNDWI Value

MNDWI Value	Interpretation	Land Condition
>0,5	Areas of open or degraded land	Highly degraded land, without vegetation cover
0 to 0,2	Transition area between open land and vegetation	Land in transition
-1 to -0.1	Vegetated or more covered land	Land with vegetation cover or watered land.

Source: [19]

Next, add a script to show empty land. Then, create a script to calculate the area of degraded land and other scripts that support data processing and analysis such as adding legends. After downloading, data processing continued with the help of ArcGIS to combine polygons of land degradation conditions in Gresik Regency. Afterwards, a map of the land degradation area was created with the respective layouts for each period. Finally, a script was inserted to create a graph showing changes in urban areas. After the land degradation map was processed with GEE, it was downloaded in .tif form to be processed later with ArcGIS. The results of the graph can be downloaded into png and also csv.

The graph of land degradation in Gresik Regency is used to analyze changes in land degradation area. This was done to determine the extent of changes in land degradation in 2000, 2005, 2010, 2015 and 2023 in Gresik Regency. Trend analysis is carried out in analyzing the results of the maps and graphs obtained. Analyzing the trend of changes in land degradation area based on graphs generated from GEE processing.

3. Results and Discussion

3.1 Result

Land Degradation Map in Gresik Regency.

The results of processing in Google Earth Engine are downloaded to be able to be layered with ArcGIS (Figure 1). Five maps of land degradation conditions match the desired year range, namely 2000, 2005, 2010, 2015 and 2023. Land degradation is symbology with red color, while land that is not degraded is symbology with yellow color. The result of calling the 2010 image with Google Earth Engine was that there were gaps in the image. This requires gap fill correction by processing the 2010 image using ENVI. This correction aims to obtain accurate data on the extent of land degradation in 2010.

Land Degradation in Gresik Regency Year 2000-2023

In addition, the land degradation area that occurred from 2000 to 2023 was also obtained in Figure 2). The graph shows changes in land degradation in Gresik Regency in units of hectares (Ha) in the range 2000-2023. The graph shows the increase and decrease in the vacant land area in Gresik Regency. The rise in land degradation area continues to occur from 2000 to 2015. Then, after 2015, land degradation in Gresik Regency experienced a fairly visible decline.

IOP Conf. Series: Earth and Environmental Science **1503** (2025) 012011 doi:10.1088/1755-1315/1503/1/012011

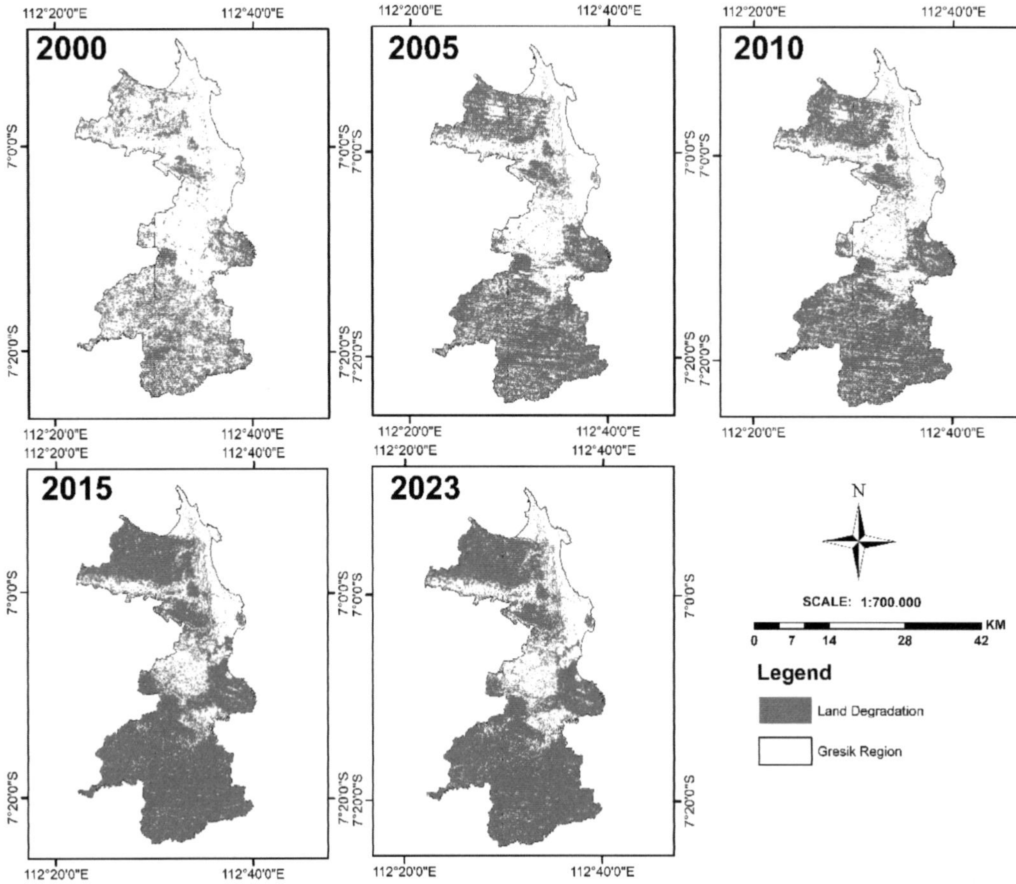

Figure 3. spatio temporal land degradation Gresik Regrency 2000, 2005, 2010, 2015 and 2023

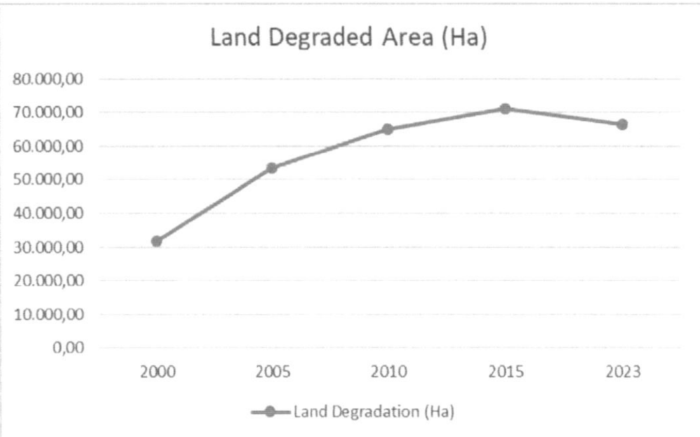

Figure 4. Land Degradation Area Changes in Gresik 2000, 2005, 2010, 2015, and 2020

This image data can be processed into tabular data, as shown in Table 3. In Table 3, the highest degraded land area value is in 2015. The lowest degraded land area value was in 2000. This shows that there was a rapid increase in the area of degraded land in Gresik Regency from 2000 to 2015. the graph shows a decrease in the area of degraded land in 2023.

3.2 Discussion

Monitoring land degradation change with the MNDWI and BSI methods will identify degraded land in the form of dry land. For MNDWI processing, areas that have negative spectral values indicate high land degradation because they are drylands [20]. The MNDWI method is very well used to detect areas with healthy vegetation or water bodies, so it can select built-up land well[21]. The BSI method is also applied to determine the location of degraded land, where land with a spectral value above 0 is included in degraded land[18]. After that, the overlay is done to determine the areas that are included in the critical land area.

Table 3. Land Degradation Area

Year	Land Degradation (Ha)
2000	31.848,12
2005	53.370,09
2010	65.044,80
2015	70.940,16
2023	66.456,54

Source: Author, 2024

The results of the identification of degraded land are processed into data on the extent of land degradation in 2000, 2005, 2010, 2015 and 2023. Changes in the area of degraded land in Gresik Regency showed a downward trend from 2015 to 2023. The highest land degradation area occurred in 2015. The decrease in land degradation area is supported by an increase in rice production in Gresik Regency in 2020 by 0.11% compared to 2019. In 2020, the area of agricultural land in Gresik Regency was 32 920.45 Ha, and the area of plantation land was 29 337.21 Ha. This agricultural and plantation land area will certainly be very beneficial for Gresik Regency in terms of agriculture. However, the process of land degradation is also not something that can be avoided. Mapping land degradation in a certain period of time will help find out the condition of the land in Gresik Regency. Gresik Regency is one of the regions in East Java, that showed an increase in agricultural productivity figures in 2019. This shows that the agricultural sector can be developed in Gresik Regency. In 2000, it was found that Gresik Regency had 31.848,12 ha of degraded land, which increased from 67.58% in 2005 to 53.370,09 ha.

The 68% increase in land degradation can be attributed to a decrease in nutrients in agricultural lands as well as land conversion from agriculture to settlements. Areas that tend to experience land degradation are the southern, northern and sub-districts of Gresik and Kebonmas. A significant decrease in green open space occurred in 2002-2008 in Gresik District [22]. Green open space land that decreased included rice fields, ponds and dry land [23]. The Gresik Regency government's policy to make Gresik an industrial area also helped reduce land degradation. Given that one of the causes of land degradation is the lack of utilization of land. However, the construction of factories or industrial land can also cause high levels of land degradation if the industrial waste is not managed properly [24]. Periurban areas in Gresik Regency are targeted for industrial development by the government to prevent urban sprawl and optimize land use.

This decrease in green open space area certainly has an impact on land productivity where in 2011 crop yields fell by 1.1% from 2007 [23]. From the observations on the Land Degradation

Map of Gresik Regency in 2000, 2005, 2010, 2015 and 2023, it was found that the sub-districts that experienced a lot of land degradation were those close to Surabaya City and Lamongan Regency. As it is known that Surabaya City is the capital of East Java Province, this affects the physical and social conditions of the surrounding areas. People tend to choose to live in areas close to the city center. For the Gresik Sub-district itself, land degradation occurs due to the high land conversion in the sub-district. This is supported by the Gresik Sub-district, which is the centre of government, economy and education in Gresik Regency. The high rate of industrial development in Gresik Regency can certainly be a factor in land degradation in the region [22].

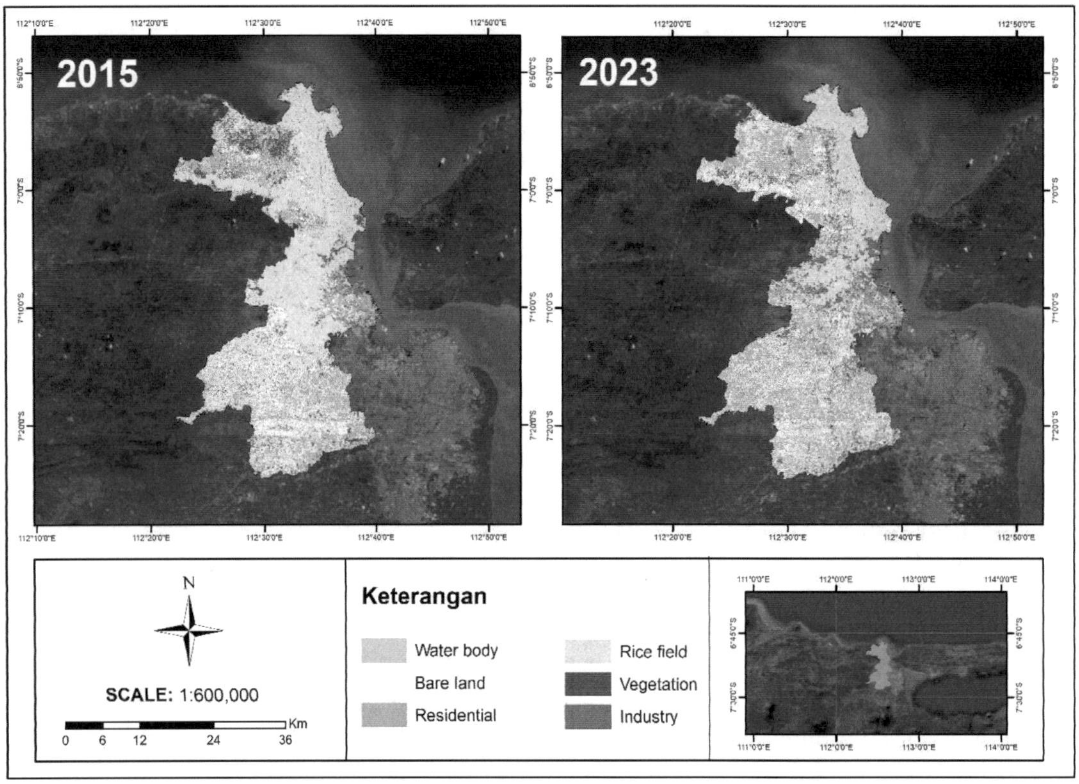

Fig 5. Gresik Landuse 2015 and 2023

The classification of land use in gresik district produces 6 classes with an accuracy level of kappa value of 82%. the six classes of land use identified are water body, reidental, bare land, rice field, industry and vegetation. land use changes in 2015 to 2023 occurred most in the industry. where the industrial land area in 2023 increased by 200%. this shows that the decrease in land degradation in 2023 is not influenced by the increase in industrial development in Gresik.

By 2015, land degradation increased by 32,94% from 2005 and decreased by 6,32% by 2023. This decrease in land degradation occurred because the vacant lands in 2015 had been utilized properly and became more productive. In addition, the efforts of the Gresik district government to reduce the high rate of land degradation can be said to be going well. Post-mining management is one of the efforts of the Gresik Regency government to reduce land degradation[6]. Agricultural conservation is also carried out to reduce the level of land degradation in Gresik district. With the reduction of land degradation, the negative impacts of land degradation can also be suppressed.

The negative impacts of land degradation have been experienced by several areas, especially areas in Gresik Regency that often experience flooding[25].

4. Conclusion

Land degradation mapping with GEE provides results that can be used as a reference for regional development analysis by the Gresik District government. In the last eight years and five years, the Gresik Regency government has made many efforts to reduce the level of land degradation. These efforts provide satisfactory results, where the area of land degradation in Gresik Regency has decreased in 2023. Areas experiencing land degradation are areas close to economic centres. Sub-districts directly adjacent to Surabaya City are areas experiencing land degradation. From 2015 to 2023, land degradation decreased slightly by 6,32% . This can occur because the vacant lands in 2015 have been utilized properly in 2023. Increased industrial development in Gresik shows appropriate land use so that land degradation in Gresik can decrease. To help optimize the recovery of degraded land, several policies can be developed by the government such as increasing forest protection and conservation areas. Community involvement in critical land rehabilitation with environmental education will help in optimizing the reduction of land degradation. Land degradation that occurs in Gresik Regency results in flood disasters that often hit the area[26]. The Gresik Regency government's policy to make Gresik an industrial area also helped reduce the level of land degradation. Given that one of the causes of land degradation is the lack of utilization of land. However, the construction of factories or industrial land can also cause high levels of land degradation if the industrial waste is not managed properly [24].

References

[1] S. Arsyad, 2010, "Konservasi Tanah & Air", 2nd ed. Bogor, 2010.

[2] S. Feng, W. Zhao, T. Zhan, Y. Yan, and P. Pereira, 2022, "Land degradation neutrality: A review of progress and perspectives", *Ecol. Indic.*, vol. 144, no. August, pp. 109530.

[3] Wahyunto and A. Dariah, 2014, "Degradasi Lahan di Indonesia: Kondisi Existing, Karakteristik, dan Penyeragaman Definisi Mendukung Gerakan Menuju Satu Peta," *J. Sumberd. Lahan*, vol. 8, no. 2, pp. 81–93, 2014.

[4] J. Jumadi, D. N. Sari, U. Umrotun, M. Musiyam, C. Nurmantyo, S. F. Muhammad and M. H. Ibrahim., 2024 "Remote Sensing And Gis Driven Model For Flood Susceptibility Assessment In The Upper Solo River Watershed," *Geogr. Tech.*, vol. 19, no. 2, pp. 39–50.

[5] Z. A. Afdal and U. Linarti, 2023 "Preventive Maintenance Analysis Using Monte Carlo Simulation and Failure Mode and Effect Analysis (FMEA)," *J. Ilm. Tek. Ind.*, vol. 22, no. 2, pp. 251–262.

[6] R. Nedd and A. Anandhi, 2024, "A synthesis on land degradation in the context of sustainable development goals," *L. Degrad. Dev.*, vol. 35, no. 13, pp. 3937–3964, 2024.

[7] V. N. Fikriyah, D. Danardono, M. I. T. Sunariya, M. Cholil, T. A. Hafid, and M. I. Ismail, 2022, "Spatio-temporal analysis of built-up area and land surface temperature in Surakarta using Landsat imageries," *Sustinere J. Environ. Sustain.*, vol. 6, no. 2, pp. 92–101, 2022.

[8] B. Semedi, 2023, "Pemanfaatan Google Earth Engine Untuk Memantau Perubahan Luasan Hutan Mangrove Di Probolinggo," *JFMR-Journal Fish. Mar. Res.*, vol. 7, no. 2, pp. 79–87, 2023.

[9] D.P. Costa, S. M. Herrmann, R. N. Vasconcelos, S. G. Duverger, W. J. S. Franca Rocha, E. C. B. Cambuí, J. S.B. Lobão, E. M.R. Santos, J. Ferreira-Ferreira., 2023, "Bibliometric Analysis of Land Degradation Studies in Drylands Using Remote Sensing Data", *A 40-Year Review*, vol. 12, no. 9.

[10] Y. Khoirurrizqi, R. Sasongko, N. L. E. Utami, A. Irbah, and S. Arjasakusuma, 2023 "Machine Learning-Based Rice Field Mapping in Kulon Progo using a Fusion of Multispectral and SAR Imageries," *Forum Geografi*, vol. 37, no. 2, pp. 134–148. doi: 10.23917/forgeo.v37i2.20304

[11] N. K. Dewi and I. Rudiarto, 2014, "Pengaruh Konversi Lahan terhadap Kondisi Lingkungan di Wilayah Peri-urban Kota Semarang (Studi Kasus: Area Berkembang Kecamatan Gunungpati)," *J. Pembang. Wil. Kota*, vol. 10, no. 2, p. 115.

[12] U. Rafsenja, L. Muh, G. Jaya, and S. Rahim, 2020, "Analisis Perbandingan Citra Landsat 8 dan Citra Sentinel 2-A untuk Mengidentifikasi Sebaran Mangrove," *Jurnal Geografi Aplikasi Dan Teknologi*, vol. 4, no. 1, pp. 63–70.

[13] R. Lukiawan, E. H. Purwanto, and M. Ayundyahrini, 2019, "Standar Koreksi Geometrik Citra Satelit Resolusi Menengah Dan Manfaat Bagi Pengguna," *Djambatan*, vol. 21, no. 1, pp. 45–54, 2019.

[14] M. V. da Silva , J. F. de Oliveira-Júnior, J. L. B. da Silva, Almeida, G. L. P. de Almeida, A. A. de Assunção Montenegro, M. Mesquita, M. B. Ferreira, T. C. Santana, G. T. B. Marinho, A. R. da Costa, P. R. Giongo, 2022, "Remote sensing techniques via Google Earth Engine for land degradation assessment in the Brazilian semiarid region, Brazil," *J. South Am. Earth Sci.*, vol. 120, no. October, 2022.

[15] R. D. Putra, H. S. Napitupulu, A. H. Nugraha, M. P. Suhana, A. R. Ritonga, and T. E. Y. Sari, 2022 "Pemetaan Luasan Hutan Mangrove Dengan Menggunakan Citra Satelit Di Pulau Mapur, Provinsi Kepulauan Riau," *J. Kelaut. Trop.*, vol. 25, no. 1, pp. 20–30.

[16] L. M. Irsan, N. Hasanah, R. Musyawarah, E. H. Garusu, and S. Aldiansyah, 2024, "Analisis Transformasi Lahan Menggunakan Citra Satelit Landsat Multi Temporal," *J. Penelit. Pendidik. Geogr.*, vol. 9, no. 1, pp. 34–43.

[17] A. Gasmi, C. Gomez, A. Chehbouni, D. Dhiba, and H. Elfil, 2022, "Satellite Multi-Sensor Data Fusion for Soil Clay Mapping Based on the Spectral Index and Spectral Bands Approaches," *Remote Sens.*, vol. 14, no. 5, pp. 1–22.

[18] C. Polykretis, M. G. Grillakis, and D. D. Alexakis, 2020, "Exploring the impact of various spectral indices on land cover change detection using change vector analysis: A case study of Crete Island, Greece," Remote Sens., vol. 12, no. 2.

[19] D. Eramudadi and C. A. Rokhmana, 2024, "Ekstraksi Permukiman dari Kombinasi Citra Sentinel-2 dan Sentinel-1 dengan Pendekatan Object-Based Image Analysis," *JGISE J. Geospatial Inf. Sci. Eng.*, vol. 7, no. 1, pp. 71.

[20] F. Oñate-Valdivieso, A. Oñate-Paladines, and R. Díaz, 2024, "Soil degradation in andean watersheds: a case study using remote sensing," *Front. Earth Sci.*, vol. 12, no. February, pp. 1–12.

[21] M. Z. Hasan, I. P. A. Citra, and A. S. A. Nugraha, 2019, "Monitoring Perubahan Garis Pantai Di Kabupaten Jembrana Tahun 1997 – 2018 Menggunakan Modified Difference Water Index (Mndwi) Dan Digital Shoreline Analysis System (DSAS)," *J. Pendidik. Geogr. Undiksha*, vol. 7, no. 3, pp. 93–102.

[22] K. M. Sari, E. Budiyanto, M. Muzayanah, and A. Kurniawati, 2021, "Ketersediaan Ruang Terbuka Hijau Menggunakan Metode Normalized Difference Vegetation Index Di Kabupaten Gresik Tahun 2021," *J. Geogr. Geogr. dan Pengajarannya*, vol. 19, no. 1, pp. 49.

[23] A. Ghozali and P. G. Ariastita, 2013, "Optimasi Penggunaan Lahan Melalui Pendekatan Telapak Ekologis di Kabupaten Gresik," *Jurnal Wilayah dan Lingkungan*, vol. 1, no. 1, pp. 67–78.

[24] M. B. El Fuadi, "Analisis Penataan Ruang Perkotaan Di Kabupaten Gresik Tahun 2019-2024," *AS-SIYASAH J. Ilmu Sos. Dan Ilmu Polit.*, vol. 7, no. 2, p. 78.

[25] A. E. Saputra, S. Nugraha, and R. Noviani, 2023, "Analysis of Urbanisation's Relationship with Clean Water Supply Ecosystem Services in Sukoharjo Regency, Indonesia," *Forum Geografi*, vol. 37, no. 2, pp. 178–190. doi: 10.23917/forgeo.v37i2.21218

[26] G. G. T. Dewandaru and U. Lasminto, 2014, "Studi Penanggulangan Banjir Kali Lamong Terhadap Genangan di Kabupaten Gresik," *Jurnal Teknik Pomits*, vol 3, no. 2, pp. C35-C40.

IOP Conf. Series: Earth and Environmental Science **1503** (2025) 012012 doi:10.1088/1755-1315/1503/1/012012

The clash of interests in Karimunjawa : Environment conservation versus economic value

Ignatius Agung Satyawan[1*]

[1] Department of International Relations and ASEAN Study Center, Universitas Sebelas Maret, Surakarta, Indonesia

*E-mail: agungsatyawan@staff.uns.ac.id

Abstract. Indonesia has a central role in the global chain of coral reef biodiversity. Laid in the north of Central Java, Karimunjawa Islands are one of the National Marine Parks in Indonesia, covering 111,625 hectares with 27 islands and a population of around 8,000 people. This marine park is home to 500 types of fish and 300 species of coral reefs. This park which was formed in 2001, has tasked in preserving the environment with collaboration of local residents and NGOs. The underwater beauty of this archipelago has made this place a tourism destination. However, recently environmental sustainability has been threatened due to the existence of shrimp ponds which have high export value. For several years, these shrimp ponds have polluted the environment, threatening the lives of various types of fish and coral reefs. Apart from that, as the number of electric power plants increases on the island of Java, the need for coal is also increasing, which is brought from Kalimantan and other places in eastern Indonesia. Barges carrying coal dropped anchor in the Karimunjawa marine park which resulted in damage to coral reefs. The economic interests shown by the existence of shrimp ponds and coal barges clash with the interests of environmental conservation, resulting in clashes between pond and barge owners and local residents who are assisted by NGOs. This article discusses the conflict of interest that occurred in Karimunjawa and looks for alternative solutions so that economic and environmental conservation interests can run harmoniously.

1. Introduction

Karimunjawa is a group of islands composed of 27 islands and situated in Java Sea 86 km northern direction from Jepara regency. The population of Karimunjawa is 10,609 people and they live on 4 islands, mostly inhabit in Karimunjawa Island, the biggest island in this archipelago [1]. The majority of the population's livelihoods are traditional fishermen and a small number are farmers. Therefore, their life depends on the sea.

Karimunjawa is rich of biodiversity, especially coral reefs which are found in almost all shallow waters on the Karimunjawa islands. These coral reefs are marine tourism objects so that Karimunjawa is the main tourist destination in Central Java. Year to year tourist visits to Karimunjawa increase. In 2022, domestic tourists were recorded at 59,203 people and grew in 2023 by 77,770 people. Meanwhile, foreign tourists numbered 3,649, rose to 7,850 in 2023. [2]. The large number of tourist visits enable to add people's livelihoods and open businesses in the tourism sector such as home stay, rental transportation and culinary.

Unfriendly fishing practices and uncontrolled tourist visits can cause damage to coral reefs clean environment. This condition causes fish catches to decrease and Karimunjawa is no longer a tourist destination, which will boomerang for local residents. Therefore, environmental protection and conservation are actively carried out by the government, for example through the establishment of a Marine Nature Reserve in the waters of Karimunjawa in 1986 and in 1988 it was changed to a Nature Conservation Area. Then in 1999 the government declared the Karimunjawa Islands with its water as the Karimunjawa National Park (KNP) [3].

Unfriendly activities to the environment have occurred recently, namely the increasing number of shrimp ponds and the traffic of coal barges that anchor in the waters of Karimunjawa. These two activities have caused damage to coral reefs and severe environmental pollution in Karimunjawa.

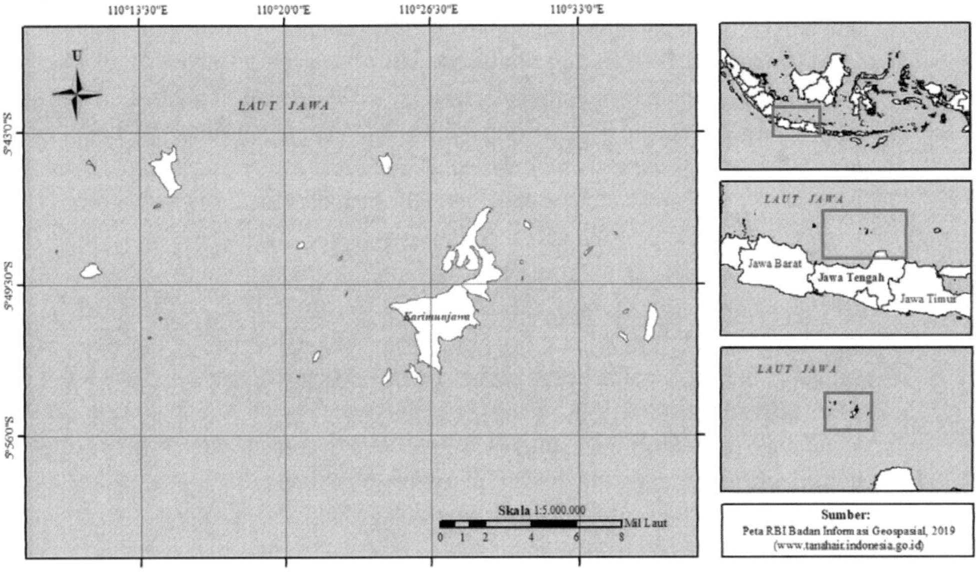

Figure 1. Karimunjawa Map [4]

This article analyses the clash of interests between environmental conservation interests and economic interests. Environmental conservation interests are supported by the KNP and local residents who depend on marine resources for their livelihoods. Meanwhile, the economic interests that play a role are the establishment of illegal shrimp farms and coal barge traffic. This article also analyses the solutions taken to align economic interests with environmental conservation interests.

2. Method

This research employs a descriptive qualitative method with a case study approach. This approach allows to describe the phenomena that occur in a certain case by answering the questions what, who, why and how [5, 6]. Through library research techniques, the data collected is secondary data consisting of various literature and documents. Then the data is analysed through interpretive procedures.

3. Result and discussion

3.1. Environment conservation interest

Prior to 1980s, Karimunjawa which has a wealth of biodiversity, especially coral reefs and mangrove forest areas, experienced damage due to environmentally unfriendly fishing practices such as the use of dynamite and trawl ships, clearing of mangrove forests and tropical forests for construction materials. In addition, environmentally unfriendly tourism practices are often found as well as limited knowledge of local peoples concerning environment have a negative impact on conservation efforts [7].

Therefore, since 1986, the government has designated Karimunjawa as a marine nature reserve that prohibits the exploitation of marine resources using unfriendly techniques and provides public awareness of the importance of environmental conservation. The government then designated the status of Karimunjawa as a National Park in 1999 which includes a sea area of 1.101 km^2, tropical forest area is 13 km^2 and 3 km^2 of mangrove forest area [8]. The mechanism used is the zoning system where the KNP is divided into 9 zones with different functions, but there are three important zones, namely the core zone, protection zone, land and marine utilization zones [9]. This KNP is operated by the Karimunjawa National Park Office (KNPO), an institution under the Ministry of Environment and Forestry. This institution is responsible for maintaining the sustainability of the environmental ecosystem in Karimunjawa together with other stakeholders, especially with the local community.

The existence of KNP can be said that the waters and island clusters in Karimunjawa are Marine Protected Areas (MPA), namely a marine area that is protected and managed in the long term for the conservation of marine resources, maintenance of ecosystems and also cultural heritage through a sustainable zoning system [10, 11]. Based on the status of Karimunjawa as a National Park and MPA, it is expected that the sustainability of coral reef ecosystems, mangrove forests and coastal biodiversity can be maintained so that fish stocks are sufficient and increase tourism attractions which in turn can improve community welfare.

3.2. Economic interest

3.2.1. Shrimp pond

Shrimp commodities have high economic value. Indonesian shrimp exports are able to reach US$ 1.7 billion, thus attracting investors to open shrimp ponds, including in Karimunjawa [12]. Shrimp production from Karimunjawa is estimated at 1,600 tons per year or worth Rp 131 billion and all of its product exported to foreign countries. However, there are side effects that need to be considered, namely that shrimp pond waste can increase the concentration of nitrate, phosphate, and ammonia which causes damage to coral reef ecosystems [13,14] and pollutes the environment so that fish stocks decrease and harm for the fishermen. The opening of shrimp ponds has an impact on the clearing of mangrove forests and the reduction of its area, resulting in seawater intrusion into the mainland, thus polluting the availability of clean water.

The existence of illegal shrimp ponds began to enter Karimunjawa since 2016 and continued to increase in the following years. In 2023, there were 39 shrimp ponds with an area of 31 hectares [15]. If illegal shrimp ponds expand further, it could result in horizontal conflict with its surrounding community. In addition, an expert from Diponegoro University warned that if shrimp ponds expanded without control, Karimunjawa could potentially sink [16]

3.2.2. Coal barge

The need for electricity on the island of Java has increased in the last decade. In 2023 there was a step-up in electricity consumption of 14% [17]. Since 2011 electricity consumption was climbing in every year [18]. To meet these needs, the government has built several power plants in several places, one of which is the Tanjung Jati Power Plant in northern part of city of Jepara. The power plant uses cheap energy, namely coal imported from outside Java, especially from Kalimantan.

Coal from Kalimantan has been transported using barges to Jepara and other power plants on the north coast of Java since 2012. While waiting in line to unload and shelter from storms and large waves, the barges drop anchor in the waters of Karimunjawa. As a result, the barge anchors damage thousands of square meters of coral reefs. In fact, in 2017 there were 30 coal barges anchored in the waters of NPK and some of them stranded in those waters [19]. Anchoring the barges has damaged 1,687 m^2 coral reef [20].

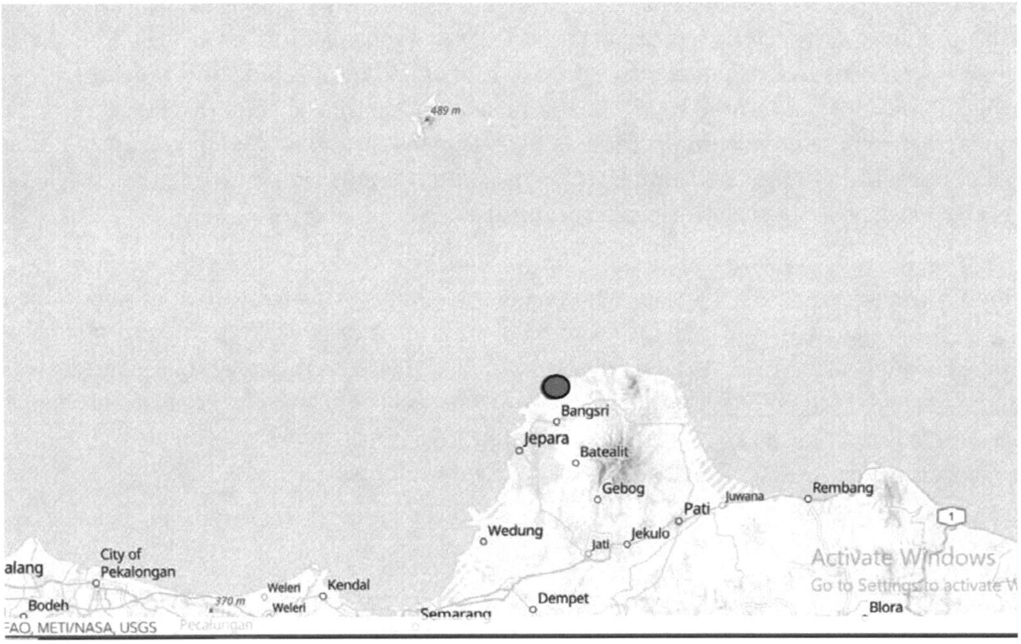

Figure 2. Karimunjawa-Tanjung Jati map [21]

3.3. Conflict resolution

The population of Karimunjawa consists of various ethnicities including Javanese, Bugis, Madurese and Mandar. Therefore, relations between these ethnicities are prone to conflict. However, so far, internal conflicts between ethnicities can be resolved through customary procedures [22]. While conflicts with residents outside Karimunjawa are resolved by national law.

Since 2017, the local people together with several non-governmental organizations (NGOs) have conveyed the problem of illegal shrimp farming practices that pollute the environment and harm to the community. However, this complaint did not receive a response from the government. To get a response from the wider community and from public opinion, several NGO activists uploaded the shrimp farming problem on social media with the title #SaveKarimunjawa. This upload received a response from the public and actually resulted in the activists being arrested by the authorities and accused of causing unrest in the community based on Law Number 19 2016 regarding the Amendments to Law No. 11/2008 on Information and Electronic Transactions. The activists were sentenced to 7 months in prison by the Jepara court [23]. However, when they appealed to a higher court, these activists were then released.

Regarding the issue of environmental damage caused by coal barges, local residents and NGOs protested. Even in May 2018, a Greenpeace ship named Rainbow Warrior blocked and expelled coal barges from NPK waters which are indeed prohibited for commercial ship traffic without permission. For three months no coal barges docked in Karimunjawa waters, but after that the barges returned to Karimunjawa waters [24].

The facts mentioned above show that law enforcement efforts are very weak, even KNPO and other authorities are powerless to take action and sanction lawbreakers. To overcome this and in an effort to attract the attention of domestic and international communities, the local people and assisted by NGO activists held a campaign by displaying posters in the waters of Karimunjawa containing the messages "#savekarimunjawa from shrimp ponds and coal barges" and "end coal go renewable" [15, 25]. This poster was then photographed from the air and uploaded to social media so that it went viral.

Figure 3. Poster campaign #savekarimunjawa [15]

Figure 4. Campaign anti coal barge [25]

Due to pressure from the community, the government, especially KNPO and the Jepara Regency Government, immediately implemented regulations and sanctions for violations. The Jepara Regency Regional People's Representative Council made a regional regulation on the Spatial Planning Plan of Jepara Region for 2023–2043 mentioning that shrimp ponds are prohibited in Karimunjawa and Coal barge also prohibited to pass KNP water.

4. Conclusion

Clashes between economic interests and environmental conservation interests, although both have good intentions, often occur everywhere. In fact, these two interests are similar to a zero-sum game. If economic interests are prioritized, conservation interests will be ignored. Likewise, if conservation interests are prioritized, economic interests will be reduced. It is not easy to align these two interests.

One way that can be done to align these interests is to enforce the rule of law firmly and impartially. However, sometimes the implementation of the rule of law will lose to the owners of capital and this often happens in developing countries where corruption rates are high.

The common way taken by Civil Society is to form public opinion through various means, for example through social media that raises collective awareness so that it can urge the government to implement the rule of law appropriately. The term that is currently popular for implementing the rule of law properly is "no viral, no justice".

References

[1] BPS Jepara 2024 *Kecamatan Karimunjawa dalam angka*. (Badan Pusat Statistik Kabupaten Jepara)

[2] infojateng.id 24 Oktober 2024 *Kunjungan Wisatawan Mancanegara di Karimunjawa Naik 115 Persen*, https://infojateng.id/read/57074/kunjungan-wisatawan-mancanegara-di-karimunjawa-naik-115-persen/

[3] Cioutra W, 21/02/2022 https://regional.kompas.com/read/2022/02/21/135635878/taman-nasional-karimunjawa-sejarah-flora-dan-fauna-hingga-tempat-wisata-di?page=all

[4] Magfiroh A N, Zairion, Fahrudin A 2020 *J. Il. dan Tek. Kelautan Tropis* **12** 2 http://doi.org/10.29244/jitkt.v12i2.29262

[5] Crowe S, *et.al* . *BMC Med Res Methodol* **11** 100. https://doi.org/10.1186/1471-2288-11-100

[6] Yin R K 2009 *Case study research, design and method*. (London: Sage Publications Ltd)

[7] Sutrisno A.D, Chen Y, Suryawan J.I.W.K.; Lee C.H 2023 *Water*, **15** 1784.

[8] Cambell S J, Kartawijaya T, Yulianto I, Prasetia R, Clifton J 2013 J. *Mar. Pol.* **41** 72–79 https://doi.org/10.1016/j.marpol.2012.12.022

[9] Yuliana E, Fahrudin A, Boer M, Kamal M M, Pardede S T 2016 *AACL Bioflux*, **9** 3 http://www.bioflux.com.ro/aacl

[10] Ocean Exploration 2019 *What is a marine protected area (MPA)?* https://oceanexplorer.noaa.gov/facts/mpas.html

[11] Green, A.L., Fajariyanto, Y., Lionata, H., Ramadyan, F., Tighe, S., White, A., Gunawan, T., Rudyanto 2019 *A Framework for Designing Marine Protected Areas and Marine Protected Area Networks in Indonesia* Report prepared by The Nature Conservancy for the USAID Sustainable Ecosystems Advanced Project

[12] Siagian R. A. S, Sabdono A, Sunaryo S, Trianto A, Dirgantara D 2022 *Ind. J. Mar. Sci.* **27** 3 267-278 https://doi.org/10.14710/ik.ijms.27.3.267-278

[13] Mustafa A, *et. al.* 2022 *Sustainability* **14** 2659 https://doi.org/10.3390/su14052659

[14] Herbeck L S, Unger D, Wu Y, Jennerjahn T C 2013 *Cont. Shelf Research* **57** 92-104 https://doi.org/10.1016/j.csr.2012.05.006

[15] KumparanNews, 20 September 2023 https://kumparan.com/kumparannews/warga-karimunjawa-gelar-aksi-tolak-tambang-udang-ilegal-yang-cemari-laut-21E3R6c2ro2/full

[16] KumparanNews, 24 September 2023 https://kumparan.com/kumparannews/ahli-prediksi-pulau-karimunjawa-akan-tenggelam-karena-tambak-udang-21FeoiaPvuC/full

[17] Firda, D M, 18 January 2024 https://www.cnbcindonesia.com/news/20240118150151-4-507002/konsumsi-listrik-ri-di-2023-melejit-14-ini-pemicunya

[18] Stefanus J, Ari M G, 2022 *Media Ekonomi*, 30 1 109–120 http://dx.doi.org/10.25105/me.v30i1.10662

[19] Apriando T, 6 May 2018 https://www.cnnindonesia.com/nasional/20180505124457-20-295911/mengusir-kapal-tongkang-pembunuh-terumbu-karang-karimunjawa

[20] Munasik, Muhammad H, Rikoh M S, Suharsono. 2020 *J. of Mar. Research*, **9** 3 343-354, https://doi.org/10.14710/jmr.v9i3.28239

[21]https://www.arcgis.com/apps/mapviewer/index.html?layers=ea6514e3d87d44848bf89112583caf54

[22] Anis W, 2015 *Yustisia*, 4 3, 602-616, https://doi.org/10.20961/yustisia.v4i3.8692

[23] Gokkon B, 4 April 2024 https://news.mongabay.com/2024/04/daniel-tangkilisan-karimunjawa-marin-national-park-indonesia-environmental-activist-jailed/

[24] Greenpeace Indonesia, 19 September 2023, , https://www.greenpeace.org/indonesia/siaran-pers/57120/selamatkan-karimunjawa-dari-limbah-tambak-dan-tongkang-batu-bara/

[25] Utami R R, 14 Okt 2020 https://www.its.ac.id/seatrans/id/pengalihan-alur-pelayaran-batu-bara-menyelamatkan-karimunjawa/

IOP Conf. Series: Earth and Environmental Science **1503** (2025) 012013 doi:10.1088/1755-1315/1503/1/012013

Understorey and Seedling Dynamics: Keys to Unlocking Biodiversity Exploration in Surakarta's Green Space

Almas Halawatal Imani[1*] and Santhyami[1,2]

[1] Department of Biology Education, Faculty of Teacher Training and Education, Universitas Muhammadiyah Surakarta, Jl. A. Yani Tromol Pos I, Pabelan, Kartasura, Surakarta 57162, Jawa Tengah, Indonesia
[2] Enviromental Study Centre, Universitas Muhammadiyah Surakarta, Jl. A. Yani Tromol Pos I, Pabelan, Kartasura, Surakarta 57162, Jawa Tengah, Indonesia

*Email: a420200109@student.ums.ac.id

Abstract. Surakarta City is a city that has urban planning with Green Open Space (GOS). The available green land certainly has a variety of plants and plants that exist, including understorey and seedlings. This study aimed to determine the structure and composition of understorey plants and seedlings that grow in natural green open spaces in Surakarta City. This research was conducted on the banks of the Bengawan Solo River, precisely in the Bengawan Solo Urban Forest Park. The research design used the square meter plot method with 1x1 meter quadratic plots. After data collection was carried out, the data was calculated by looking at the number of plants in each box. Knowing this amount, the data used was entered into the Shannon-Wiener diversity index formula. This index refers to the H '1-3 scale. The results of this research on the composition and structure of understorey plants and seedlings obtained the number of understorey plants and seedlings as many as 22 species from 13 different family groups, with the Fabaceae and Poaceae families dominating. The highest Important Value (IV) results from *Brachiaria mutica* (Forssk.) Stapf (buffalo grass) plants with an IV value of 126.95. The total diversity index in the study area amounted to 1.77, which indicates that the riverbanks of the Urban Forest Park have a moderate level of plant diversity. Understorey and seedlings found at the research site certainly also have sustainable potential benefits, such as utilization as antibacterial, antioxidant, anticholesterol, biopesticide, herbal/traditional medicine, animal feed, and use as food crops. With the enormous diversity of understory plants and seedlings, the potential benefits of each species can be further recognized and optimally managed.

1. Introduction

Urban areas have a variety of different points in their arrangement. Urban planning within the scope of the environment has principles. With these principles, urban and regional planning can provide a spatial framework to protect and manage the natural and built environment, including biodiversity, land, and natural resources, and ensure integrated and sustainable development [40]. One of the points contained inurban areas is a green zone, which can also be known as Green Open Space (GOS). Green open space is an area that is expanded and grouped, more open in use, where plants grow, both naturally planted and deliberately planted. This green area ideally covers 40% of the surface area. Green open space, with its ecological function, aims to support the physical sustainability of a city. The green open space is a form of green open space that is located, sized and has a definite shape within a city area [1]. GOS have biodiversity in them. Including the plant community. plants that spread in GOS have the potential of tree vegetation to reduce air temperature in the Urban Heat

Island (UHI) phenomenon, which is done by analyzing air temperature based on land cover type [2]. Among the plants scattered in GOS are understory plants and seedlings. In this study, the variety of vegetation focused only on understorey and seedlings. Understorey itself is a type of basic vegetation that grows at the bottom of the surface of a green space.

The type of understorey is annual, biennial (requires 2 seasons), perennial, and its distribution pattern can occur randomly, clumped or in groups, and evenly distributed. Lower plants have a fairly high diversity and have an average height of about ≤ 1 meter [3]. In addition to understorey plants, seedlings are scattered in the study area. Seedlings are tree saplings that are counted from sprouts to less than 1.5 m in height. In addition to seedlings, other tree phases are saplings, poles, and mature trees that have reached the peak of a tree's growth [4]. The quality of seedlings can be analyzed to illustrate the quality of the seedlings themselves. The higher the seedling, the more fertile the seedling growth and the better the photosynthesis process. Seedling diameter is an important morphological trait that can be used as a parameter to determine seedling quality. This is because the diameter of the seedlings affects their resistance to being transplanted to the field [5]. Understorey and seedlings that grow on the riverbanks of the research location tend to be fertile and spread along the banks. Human activities around riverbanks affect the life of wild plants around the river [6]. This also affects the ability of wild plants on riverbanks to absorb water and carry out biogeochemical cycles properly.

The spread of understorey and seedlings on riverbanks results in the rapid growth of wild plants. The diversity of species of understorey plants and scattered seedlings can be known for a variety of necessary uses. As a basis for preserving the diversity of understorey and seedling species on the riverbanks of the research location, vegetation data is needed. This data is used for scientific purposes, such as research, or used by the local government to regulate the management of riverbanks on understorey plants and seedlings. In addition, this area has a variety of vegetation ranging from trees to understorey plants. Since inventory research on understorey plants and seedlings is still limited, research is needed to collect data on the types of understorey plants and seedlings on the banks of the Bengawan Solo River. Identification of understorey and seedling species is also needed to determine the potential usefulness of each species. Therefore, this study aims to identify plant species, structure, composition, and potential for sustainable use of understorey plants and seedlings.

2. Materials and Methods

2.1 Sampling and Sampling Location

This research on understorey plants and seedlings was conducted on the banks of the Bengawan Solo River. The selected area is Bengawan Solo Urban Forest Park, Pucangsawit District, Jebres, Surakarta City, Central Java. The research site is located on the banks of the river as shown in Figure 1. The coordinate points recorded are -7. 5710885. 110. 8477297 with a land elevation of 62 meters above sea level (masl). This riverbank area has quite a lot of trees, poles, saplings, and wild plants scattered around. During the research, the temperature and humidity were recorded at 33°C with 90% humidity. According to the geographical data of Surakarta City in the Department of Investment and One-Stop Integrated Services, the average temperature of the city is around 26.55°C-29.10°Cwith humidity ranging from 68% to 86%. The recorded temperature and humidity at the research site may be due to the rainy season during the research, so the measured humidity tends to be high. The temperature at the study site is also influenced by the sunlight available and can be a factor in the growth of understorey plants and seedlings.

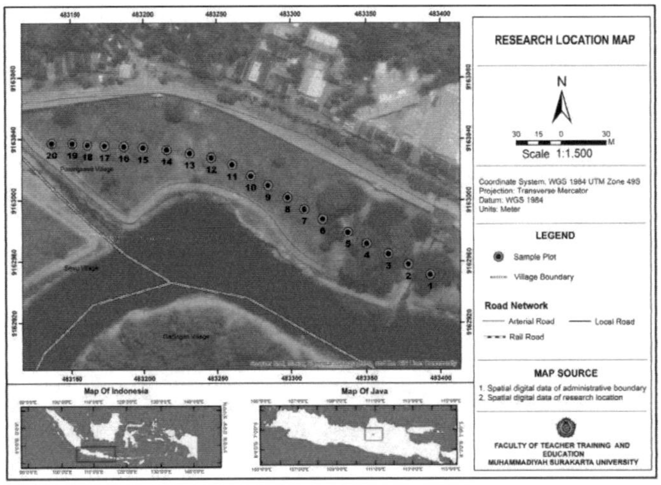

Figure 1. Research Location Map, Bengawan Solo Urban Forest Park

Understorey and seedlings to be identified will be sampled first. The sampling method is the plot method. The plot used is a 1x1 meter plot for sampling understorey plants and seedlings. The research plot is made into a 1x1 grid tool m^2. The data collection will use a plot with a total of 20 points. The grid tool is attached to a mattress rope so that it can form a square meter square with 100 square decimeters frames. This tool serves to measure and delimit 1x1 m^2 plots to perform calculations in the field. A rectangular plot shape will obtain a higher species richness or diversity value than a rectilinear plot because this shape will cover a longer area and cover a greater height difference than a square [3].

Figure 2. Sampling tools square meters (1x1 m)

2.2 Identification

Sampling of understorey plants and seedlings was continued into dry form. The understorey plants and seedlings taken were made into herbarium. Furthermore, the herbarium is used to identify the type of understorey plants and seedlings obtained. Identification was carried out at the University of Muhammadiyah Surakarta assisted by references from the literature. Identification and nomenclature is done through literature studies using the book Plant Taxonomy (Schizophyta, Thallophyta, Bryophyta, Pteridophyta), Flora, and the website Power of the Plants Online (POWO).

2.3 Data Analysis

The Index of Importance (IV) value is calculated based on the sum total of the three aspects. IV is calculated by summing the values of relative frequency (RF), relative density (RD), and relative dominance (RDo). The Shannon Wiener diversity index is calculated to determine the species diversity of an area. The Shannon-Wiener diversity index refers to the H' 1-3 scale. If H'≤1, the diversity in the area is low. If the value of H' is between 1 and 3 (1<H'<3), the diversity is moderate and if the value of H'≥3, the diversity is at a high level.

3. Results and Discussion

The riverbank ecosystem that coincides in Bengawan Solo Urban Forest Park has several species of understorey plants and seedlings scattered in the area. Understorey and seedlings can be distinguished from other plants that also grow in the riverbank area. Generally, lower plants and seedlings have characteristics that can be seen from their height. The maximum height owned by lower plants and seedlings is no more than 4.5 meters [7]. If distinguished further, lower plants have differences with seedlings. Understorey plants can be seen as a layer of plants that crowd an area and are generally located under tree stands. Understorey plants consist of herbs, vines, and shrubs [8]. Meanwhile, seedlings are tree saplings or seedlings that are seen from the start of sprouts to have a height of about 1.5 meters [9].

The understorey and seedlings sampled from 20 plots and found throughout the study site had a total of 22 species belonging to 13 different families. The composition of the understorey plants and seedling species and the calculation of the Index of Importance (IV) and Diversity Index are further described in Table 1 with each identified species.

Table 1. Composition, structure, Important Value (IV), and Diversity Index of understorey plants and riverbank seedlings of Bengawan Solo Urban Forest Park

No.	Name of Species	Family	Local Name	RD (%)	RF (%)	RDo (%)	IV (%)	H'
1.	*Ruellia tuberosa* L.	Acanthaceae	*Kencana ungu*	1.51	3.08	1.49	6.07	0.06
2.	*Alternanthera sessilis* (L.) DC.	Amaranthaceae	*Kremah*	5.09	4.62	5.20	14.90	0.15
3.	*Alstonia scholaris* (L.) R. Br.	Apocynaceae	*Pulai*	0.04	1.54	0.04	1.62	0.003

No.	Name of Species	Family	Local Name	RD (%)	RF (%)	RDo (%)	IV (%)	H'
4.	*Colocasia esculenta* (L.) Schott	Araceae	*Talas*	0.15	1.54	0.17	1.85	0.01
5.	*Bidens alba* L.	Asteraceae	*Jarum spanyol*	0.11	1.54	0.50	2.14	0.01
6.	*Bidens Pilosa* L.		*Ketul*	2.87	3.08	2.76	8.72	0.10
7.	*Calystegia sepium* (L.) R. Br.	Convolvulaceae	*Terompet surga*	0.77	3.08	3.47	7.32	0.04
8.	*Ipomoea obscural* (L.) Ker Gawl.		*Ki papesan*	0.59	1.54	2.76	4.89	0.03
9.	*Coccinia grandis* (L.) Voigt	Cucurbitaceae	*Kemarungan*	0.48	1.54	0.45	2.47	0.03
10.	*Cyperus rotundus* L.	Cyperaceae	*Rumput teki*	0.44	3.08	0.45	3.97	0.02
11.	*Euphorbia hirta* L.	Euphorbiaceae	*Patikan kebo*	1.99	3.08	2.02	7.09	0.08
12.	*Calopogonium mucunoides* Desv.	Fabaceae	*Kalopo*	13.30	15.39	12.62	41.31	0.27
13.	*Crotalaria pallida* Aiton		*Orok-orok*	0.11	1.54	1.77	3.42	0.01
14.	*Grona triflora* (L.) H.Ohashi & K.Ohashi		*Rumput jarem*	9.43	10.77	7.47	27.67	0.22
15.	*Indigofera tinctoria* L.		*Tarum*	0.07	1.54	0.41	2.03	0.01
16.	*Mimosa invisa* Mart. ex Colla	Mimosaceae	*Sikejut besar*	1.07	3.08	0.87	5.01	0.05
17.	*Mimosa pudica* L.		*Putri malu*	1.58	3.08	0.91	5.57	0.07
18.	*Passiflora foetida* L.	Passifloraceae	*Rembusa*	0.44	1.54	2.39	4.37	0.02
19.	*Brachiaria mutica* (Forssk.) Stapf	Poaceae	*Rumput kerbau*	52.43	27.69	46.82	126.95*	0.34
20.	*Eleusine indica* (L.) Gaertn.		*Rumput belulang*	0.15	1.54	0.17	1.85	0.01
21.	*Imperata cylindrica* (L.) Raeusch.		*Rumput ilalang*	3.43	4.62	3.14	11.18	0.12

No.	Name of Species	Family	Local Name	RD (%)	RF (%)	RDo (%)	IV (%)	H'
22.	*Pennisetum purpureum* Schumach.		*Rumput raja*	3.94	1.54	4.13	9.61	0.13
		TOTAL		**100**	**100**	**100**	**300**	**1.77**

Note: RD= Relative Density; RF= Relative Frequency; RDo= Relative Dominance; IV= Important Value; H'= Diversity Index; * = The Highest IV

The classification of understorey plants and seedlings in Table 1 shows that the families that are often found throughout the study site are from the Fabaceae and Poaceae family groups with 4 species each belonging to these groups. The presence of non-timber understorey plants is also influenced by temperature and humidity characteristics. Measurements of temperature and humidity parameters showed relatively low temperatures and high humidity, which led to the high number of understorey groups scattered throughout the study area [10]. The parameter needed to state the level of dominance of a species in a plant community is IV. Plant species that have the highest importance value are the most dominant species [11].

The plant that has the highest IV value at the research site is *Brachiaria mutica* (Forssk.) Stapf (*Rumput kerbau*) which comes from the Poaceae family. This plant has an important value of 126.95. The high value of IV is influenced by density, frequency, and dominance. Meanwhile, the relative density is 52.43%, with relative frequency and relative dominance values of 27.69% and 46.82%, respectively, which can be seen further in table 1. *Brachiaria mutica* (Forssk.) Stapf is most commonly found because it is a superior grass species that can be found easily in tropical areas. *Brachiaria mutica* (Forssk.) Stapf is quite adaptive in a tropical environment with a fairly high temperature and low humidity [12]. In addition, *Brachiaria mutica* (Forssk.) Stapf is included in C4 plants that can live in environmental conditions that tend to be hot. This adaptation functions as an osmoregulator and osmoprotectant in the process of plant physiology to maintain osmotic balance and cell turgor [13]. While the species that has the lowest IV is *Alstonia scholaris* (L.) R. Br. (*Pulai*) with an IV value of 1.62. Species organized by density, frequency, and small dominance have the potential to disappear from the ecosystem due to their low level of presence [14]. The total IV value is directly proportional to the relative frequency (RF) and relative density (RD) values. The higher the value of RF and RD, the higher the IV value of a plant species in the area [15].

Understorey and seedlings scattered in the study site have diverse species. Diversity itself is a description of the condition of the ecosystem in a place and can be caused by differences in shape, number, color, texture, appearance, and size [16]. The Shannon Wiener diversity index at the research site was recorded at 1.77 as in table 2.

Table 2. Shannon-Wiener Diversity Index in different studies and locations

No.	Location	H'	Description	Source
1.	Urban Forest Park, Bengawan Solo	1.77	Medium	This research
2.	Payeti Watershed, East Sumba	2.70	Medium	[17]
3.	Cakrabuana Mount, Majalengka	3.27	High	[18]
4.	Panjang watershed, Semarang	1.71	Medium	[19]

The Shannon-Wiener diversity index (H') calculated from the total of all understorey plants and seedlings is 1.77. The magnitude of the value of H' determines the high and low diversity of species in an area, where the definition of the value of species diversity as a reference is the Shannon-Wiener index. With the data obtained, the Shanon-Wiener index shows H' data more than 1 and less than 3, therefore the diversity at the research site is at a moderate level. Biotic factors in the location greatly affect the species that grow in that place [20].

The vegetation at the different sites also had medium and high Shannon-Wiener diversity index descriptions. The three different sites have different land elevations, temperatures, and sunlight. The temperature recorded from the three different locations in Table 2 ranges from 26°C33°C which is a temperature that can still be tolerated by plants to grow. On Mount Cakrabuana, a high diversity index was recorded. This is because the land position is higher than the other three locations. Sunlight also affects and optimizes temperature, which is one of the important roles for understorey plants to develop. The presence of non-timber understorey plants is also influenced by temperature and humidity characteristics. Temperature and humidity show that relatively low temperatures and high humidity will cause high groups of understorey to spread [21]. Soil conditions in different locations have a pH that tends to be neutral. The relationship between soil properties, especially pH, organic matter, and the content of available phosphor (P) and potassium (K) nutrients, is quite strong and favorable to plant productivity [22].

Understorey and seedlings that have been further identified can be known for their potential benefits for survival. As already described in table 3, understorey plants and seedlings have quitea lot of benefits.

Table 3. Potential benefits of understorey and seedlings on the riverbank of Bengawan Solo Urban Forest Park

No.	Benefits	Species	Local Name	Benefit Description
1.	Antibacterial	*Alstonia scholaris* (L.) R. Br.	*Pulai*	It contains butanol in its bark which is quite strong. In addition, the leaves contain 70 types of alkanoid

No.	Benefits	Species	Local Name	Benefit Description
				compounds that act as antibacterials [23].
		Eleusine indica (L.) Gaertn.	*Rumput belulang*	Bioactive compounds found in *Eleusine indica* (L.) Gaertn. positively contain anthraquinones. Utilization of bioactive compounds containing steroids, tannins, and flavonoids can be used as antibacterials [24].
2.	Biopesticides	*Mimosa invisa* Mart. ex Colla	*Sikejut besar*	This plant is used as a vegetable pesticide because of the chemical content in it. The chemical content which is a secondary metabolite such as mimosine, alkaloids, sterols, tannins, etc. can control plant pest organisms [25].
3.	Antioxidants	*Mimosa pudica* L.	*Putri malu*	*Mimosa pudica* L. is identified to contain 5 secondary metabolite compounds. Among them are alkaloids, flavonoids, saponins, coumarins, and phenolics that function as antidiabetics and antioxidants [26].
		Euphorbia hirta L.	*Patikan kebo*	Antioxidant activity in *Patikan Kebo* plant is powerful. This can be seen from the percentage of optimum antioxidant activity of 99.21% [27].
4.	Herbal Medicine	*Ruellia tuberosa* L.	*Kencana ungu*	Some Indonesians usually use this plant to cure diabetes. That is because this plant contains antidiabetic content such as flavonoids, triterpenoids, etc. [28].
		Ipomoea obscural (L.)	*Ki papesan*	The leaves of *Ipomoea obscural* (L.) have properties that can heal outside the outside [29].
		Grona trifolra (L.) H.Ohashi & K.Ohashi	*Rumput jarem*	The benefits of this grass include treating diarrhea, asthma, flatulence, and antidote. The content contained in this grass is alkanoid and flavonoids for health benefits [30].
5.	Animal Feed	*Pennisetum purpureum* Schumach.	*Rumput raja*	*Pennisetum purpureum* Schumach. can be formulated by increasing its nutrients to support the productivity of aquaculture farms. This grass is easy to find and

No.	Benefits	Species	Local Name	Benefit Description
				contains silage fiber, which is good for livestock [31].
		Calopogonium mucunoides Desv.	*Kalopo*	This group of legumes is used for animal feed, and it contains a source of crude protein and crude fiber in the leaves and stems [32].
		Brachiaria mutica (Forssk.)	*Rumput kerbau*	Until now, *rumput kerbau* has generally been used as animal feed. The part that is utilized is the young and soft leaves. The stem is separated because it is hard [33].
6.	Food Crops	*Coccinia grandis* (L.) Voigt	*Kemarungan*	The fruit of this plant is characterized as a small cucumber. Usually the ripe fruit with a reddish color is eaten directly or processed into dishes [34]
7.	Anticholesterol	*Passiflora foetida* L.	*Rembusa*	Secondary metabolites in *rembusa* found include tannin which can reduce cholesterol levels. Saponin content also showed a decrease in cholesterol levels when the leaves were extracted [35].

Information on the potential benefits of understorey plants and seedlings is important to consider when choosing control techniques. Based on the information in Table 3, the potential benefits of understorey plants and seedlings on the riverbanks of the Bengawan Solo Urban Forest Park can be used as antioxidants, antibacterials, anticholesterol, animal feed, biopesticides, and food crops. In addition to its role in nutrient cycling, reducing erosion, increasing infiltration, and serving as a source of germplasm and medicines, understory vegetation plays a very significant role in the ecosystem [36].The most common benefits found based on table 3 are lower plants as herbal medicines and animal feed. In plants, secondary metabolites such as alkaloids, phenolics, lipophilic, and terpenoids accumulate in the leaves, bark, and roots so that they can be an efficient alternative raw material for herbal medicines [37]. The utilization of plants in herbal medicines is due to theease with which plants are obtained and the fact that they contain many active compounds such as flavonoids [48]. In lower plants and seedlings, tannins are also found that can inhibit enzyme inhibitors so that they have the potential to be utilized in various aspects such as vegetable insecticides, herbal medicines, and animal feed in order to balance blood sugar in animals [39].

4. Conclusions

Observations and research conducted on the riverbanks of the Bengawan Solo Urban Forest Park resulted in data on the importance value index and diversity index of understorey plants and seedlings. Understorey and seedlings found in the research location amounted to 22 species from 13 different family groups. The species that has the highest IV value is *Brachiaria mutica* (Forssk.)

Stapf or rumput kerbau which has an IV value of 126.95 and a RD value of 52.43% and RF and RDo values of 27.69% and 46.82% respectively. The Shannon Wiener diversity index recorded at the research location has a value of 1.77, which means that the diversity of species on the riverbanks of the Bengawan Solo Urban Forest Park is at a moderate level. The value obtained is due to various factors, including land elevation, temperature, humidity, and soil acidity. With the diversity of species scattered, the understorey plants and seedlings found have a variety of potential benefits. The content contained in each understorey and seedling has a variety of potential benefits ranging from antioxidants, antibacterials, anticholesterol, biopesticides, food crops, herbal medicines, and animal feed.

5. Acknowledgments

The author would like to thank various parties who have contributed to and helped the research, especially the University of Muhammadiyah Surakarta, which has facilitated the research, and the Surakarta Environmental Agency, which has helped, directed, and taken care of permits to conduct research on the banks of the Bengawan Solo River.

References

[1] Mashur, D., & Rusli, Z. (2018). Efforts and implications of providing Green Open Space (GOS). *Jurnal Kebijakan Publik*, *9*(1), 45–52. [Bahasa]

[2] Rushayati, S. B., Shamila, A. D., & Prasetyo, L. B. (2018). The Role of Vegetation in Controlling Air Temperature Resulting from Urban Heat Island. *Forum Geografi*, *32*(1), 1–11. https://doi.org/10.23917/forgeo.v32i1.5289

[3] Destaranti, N., Sulistyani, S., & Yani, E. (2017). Structure and Understory Vegetation of Pine Stands in RPH Kalirajut and RPH Baturraden Banyumas. *Scripta Biologica*, *4*(3), 155–160. https://doi.org/10.20884/1.sb.2017.4.3.407. [Bahasa]

[4] Kasmadi, D. (2015). Composition and structure of tree species in Ake Oba-Tanjung Wayamli-Ake Oba limited production forest. *Journal Universitas Sam Ratulangi*. [Bahasa]

[5] Suhartati, T. (2013). Variation of Seedling Density on Growth. *Jurnal Wana Tropika*, *3*(2), 1–10. [Bahasa]

[6] Mosyaftiani, A., Kaswanto, R. L., & Arifin, H. S. (2018). Potential of Wild Plants in the Built-Up Area of Ciliwung River in Bogor City as a River Ecosystem Restoration Effort. *Jurnal Risalah Kebijakan Pertanian Dan Lingkungan* , *5*(1), 1–13. https://doi.org/10.29244/jkebijakan.v5i1.29781. [Bahasa]

[7] Ali, M. A., Hikmat, A., & Santosa, Y. (2016). Determining the optimal shape and size of sample plots for measuring plant species diversity in a mountain rice field forest. *Media Konservasi*, *21*(1), 42–47. [Bahasa]

[8] Firison, J., Wiryono, W., & Brata, B. (2019). Diversity of Understory Plants in Oil Palm Stands and Their Potential as Feed for Beef Cattle (Case in Kungkai Baru Village, Seluma Regency). *Naturalis: Jurnal Penelitian Pengelolaan Sumber Daya Alam Dan Lingkungan*, *8*(1), 67–76. https://doi.org/10.31186/naturalis.8.1.9168. [Bahasa]

[9] Cahyanto, M., & Lukito, E. (2019). *Flora Diversity in Area IX Conservation Forest and LNG Rhinos Natural Forest.* ITS Press. [Bahasa]

[10] Susanto, S. A., Budirianto, H. J., Maturbongs, A. C., & Putra, S. A. (2019). Potential and Diversity of Non-Timber Undergrowth in Bera Land of Womnowi Sidey District Manokwari. *Jurnal Hutan Tropis*, *3*(1), 10–18. https://doi.org/10.32522/ujht.v3i1.1878. [Bahasa]

[11] Santhyami, S., & Isnaini, A. K. (2024). The Composition, Structure, and Standing Carbon Stock of Teak Based Agroforestry Systems in Gundih, Grobogan Regency, Central Java. *Jurnal Ilmu Lingkungan*, *22*(1), 93–99. https://doi.org/10.14710/jil.22.1.93-99

[12] Resthu, M., Jamilah, M., & Zulwanis, D. (2023). Effect of Feeding Sheep Manure with Various Levels on the Growth of Kolonjono Grass (*Brachiaria mutica*). *Jurnal Ilmu Dan Teknologi Peternakan*, *11*(2), 65– 69. https://doi.org/10.20956/jitp.v11i2.30958. [Bahasa]

[13] Meriem, S., Sari, A. P., & Pasaribu, P. (2020). Prolin, Ascorbic Acid, and Relative Water Content in Drought Stressed C3 and C4 Plants. *BIOMA: Jurnal Biologi Dan Pembelajarannya*, *2*(2), 26–32. [Bahasa]

[14] Irwansah, Sugiyarto, & Mahajoeno, E. (2019). Community Structure of Mangrove Ecosystems in Serewe Bay, Lombok Island, West Nusa Tenggara. *Bioeksperimen*,*5*(1), 126–130. https://doi.org/10.23917/bioeksperimen.v5i1.2795. [Bahasa]

[15] Darwati, H., Arianti, W., Rifanjani, S., & Marwanto, M. (2023). Species Diversity of Ferns (Pteridophyta) in Lho Fat Phun Fie Nature Reserve, Monterado District, Bengkayang Regency. *Jurnal Hutan Lestari*, *11*(4), 923– 933. https://doi.org/10.26418/jhl.v11i4.63860. [Bahasa]

[16] Mokodompit, R., Kandowangko, N. Y., & Hamidun, M. S. (2022). Plant Diversity at Gorontalo State University Campus Tilong Kabila Sub-district, Bone Bolango Regency. *BIOSFER: Jurnal Biologi & Pendidikan Biologi*, *7*(1), 75–80. https://doi.org/10.23969/biosfer.v7i1.5651. [Bahasa]

[17] Sihombing, P. R., Lalupanda, M. E., & Matulessy, Y. M. (2024). Herbaceous Plant Diversity in Payeti River as Biology Learning Media in the Form of Herbarium. *BIOSCIENTIAE*, *21*(1), 20–40. https://doi.org/10.20527/b.v21i1.11882. [Bahasa]

[18] Hendrayana, Y., Sistiadi, I. F., Nurlaila, N. A., & Adhya, I. (2022). Understory Plant Diversity and Its Benefits in Cakrabuana Mountain, Majalengka. *Jurnal Penelitian Universitas Kuningan*, *13*(1), 73–84. https://doi.org/10.25134/logika.v13i01.6311. [Bahasa]

[19] Nurika, F. B. P., Wiryani, E., & Jumari. (2019). Riparian Vegetation Diversity of the Lower Long River in Ambarawa District, Semarang Regency. *Jurnal Akademika Biologi*, *8*(1), 30–34. [Bahasa]

[20] Ridianingsih, D. S., Pujiastuti, & Hariani, S. A. (2017). Inventory of Pteridophyta in Rowobendo- Ngagelan Post of Alas Purwo National Park, Banyuwangi Regency. *Bioeksperimen*, *3*(2), 20–31. https://doi.org/10.23917/bioeksperimen.v3i2.5179. [Bahasa]

[21] Susanto, S. A., Budirianto, H. J., Maturbongs, A. C., & Putra, S. A. (2019). Potential and Diversity of Non-Timber Undergrowth in Bera Land of Womnowi Sidey District Manokwari. *Jurnal Hutan Tropis*, *3*(1), 10–18. https://doi.org/10.32522/ujht.v3i1.1878. [Bahasa]

[22] Mualif, M. S., & Kusumawati, A. (2021). Impact of Soil Chemical Properties on Sugarcane (*Saccharum officinarum* L.) Productivity in Kulon Progo, Yogyakarta. *Jurnal Pengelolaan Perkebunan (JPP)*, *2*(2), 66–72. https://doi.org/10.54387/jpp.v1i1.5. [Bahasa]

[23] Mayor, J. M. K., & Wattimena, L. (2022). Utilization of Pulai Tree (*Alstonia Scholaris*) by the People of Puper Village, East Waigeo District, Raja Ampat Regency. *Jurnal J-MACE*, *2*(1), 68–81. https://doi.org/10.34124/jmace.v2i1.18. [Bahasa]

[24] Pradana, D. L. C., & Batubara, U. M. (2015). Identification of Bioactive Compounds and Antibacterial Activity Test of *Eleusine indica Infusa* and *Morinda citrifolia* Leaves on Staphylococcus aureus Bacteria In Vitro. *Prosiding SEMIRATA*, 662–669. [Bahasa]

[25] Syarifah, R. N. K. (2020). Utilization of *Mimosa invisa* Weed as a Plant Disturbing Organism Controller. *BIOFARM Jurnal Ilmiah Pertanian*, *16*(2), 59–67. https://doi.org/10.31941/biofarm.v16i2.1207. [Bahasa]

[26] Bili, D. T. (2022). Review : Pharmacological Effects of Shy Princess Plant (*Mimosa pudica* Linn). *Jurnal Beta Kimia*, *2*(2), 74–79. https://doi.org/10.35508/jbk.v2i2.12118. [Bahasa]

[27] Karim, K., Jura, R. M., & Sabang, S. M. (2015). Antioxidant Activity Test of Patikan Kebo Leaf Extract (*Euphorbia hirta* L.). *Jurnal Akademi Kimia*, *4*(2), 56–63. [Bahasa]

[28] Wati, S. S., & Wakhidah, A. Z. (2023). Kencana Ungu (*Ruellia tuberosa* L.): Botany, Phytochemistry, and Their Use in Indonesia. *Jurnal Indobiosains*, *5*(1). https://doi.org/10.31851/indobiosains.v5i1.9742. [Bahasa]

[29] Silalahi, M., & Mustaqim, W. A. (2021). *Seed plants in Jakarta. Volume 2: selected non-tree species.* UKI Press. [Bahasa]

[30] Kantor Kementerian Agama Kabupaten Barito Timur. (2024, May 15). *KEMENAG Bartim miliki rumput berkhasiat untuk obat.* https://kalteng.kemenag.go.id/bartim/berita/524047/Kemenag-Bartim-Miliki-Rumput-Berkhasiat-untuk-Obat

[31] Novita, R., Karyono, T., & Herlina, B. (2021). Nutrient Content and Physical Appearance of King Grass (Pennisetum purpureophoides) Fermented with Various Doses of Strater as Ruminant Feed Ingredients. *Jurnal Ilmu Pertanian Kelingi*, *1*(1), 18–26. [Bahasa]

[32] Novotta, Gilang, G., & Sulistyo, H. E. (2021). Genotypic diversity of local Situbondo kalopo (Calopogonium mucunoides) as a source of forage for livestock feeds. In *Thesis*. [Bahasa]

[33] Wiyancoko, D. (2014). Utilization of malela grass material for space insulation product development. *Jurnal Tingkat Sarjana Bidang Seni Rupa Dan Desain*, 1–7. [Bahasa]

[34] Hwang, D. (n.d.). *Coccinia Grandis (Buah).* Retrieved May 29, 2024, from https://biodiversitywarriors.kehati.or.id/foto/coccinia-grandis-buah-1/?lang=en

[35] Mulyani, E. (2019). In-Vitro Study: Anticholesterol Effect of Rambusa Leaf Extract (Passiflora Foetida, L). *Jurnal SuryaMedika*, *4*(2), 60–65. https://doi.org/10.33084/jsm.v4i2.606. [Bahasa]

[36] Idris, A. I., Rusmidin, & Mahfudz, M. (2023). Diversity and Utilization of Undergrowth on Agroforestry Land in Alu Village, Alu Subdistrict, Polewali Mandar Regency. *PANGALE: Journal of Forestry and Environment*, *3*(1), 38–48. https://doi.org/https://doi.org/10.31605/pangale.v3i1.2789. [Bahasa]

[37] Syahruni, R., Umar, A. H., & Asnar, N. H. (2022). Characterization of Three Antihypercholesterolemia Medicinal Plants with Anatomical, Histochemical, and Phytochemical Profile-Based Approaches. *Pharmacon: Jurnal Farmasi Indonesia*, *19*(2), 176–186. https://doi.org/10.23917/pharmacon.v19i2.19467. [Bahasa]

[38] Rahma, K. A., Nugroho, A. S., Mulyaningrum, E. R., & Hayat, M. S. (2022). Types of Medicinal Plants Utilized by the Community of RT 09 RW 04 Kayen Pati Village Central Java. *Bioeksperimen*, *8*(2), 106–113. https://doi.org/10.23917/bioeksperimen.v8i2.17141. [Bahasa]

[39] Siamtuti, W. S., Aftiarani, R., Wardhani, Z. K., Alfianto, N., & Hartoko, I. V. (2017). Potential of Tannin in Nginang Herb as an Environmentally Friendly Insecticide. *Bioeksperimen*, *3*(2), 83–93. https://doi.org/10.23917/bioeksperimen.v3i2.5186. [Bahasa]

[40] Badan Pengembangan Infrastruktur Wilayah. (2025). *International Guidelines on Urban and Territorial Planning.* Kementerian Pekerjaan Umum dan Perumahan Rakyat, Republik Indonesia. [Bahasa]

IOP Conf. Series: Earth and Environmental Science **1503** (2025) 012014 doi:10.1088/1755-1315/1503/1/012014

Composition and Structure of Poles and Sapling Community in Natural Green Open Space

Ike Ratna Widianingsih[1*], Santhyami[1,2],

[1]Department of Biology Education, Faculty of Teacher Training and Education, Universitas Muhammadiyah Surakarta, Jl. A. Yani Tromol Pos I, Pabelan, Kartasura, Surakarta 57162, Jawa Tengah, Indonesia
[2]Environmental Study Centre, Universitas Muhammadiyah Surakarta, Jl. A. Yani Tromol Pos I, Pabelan, Kartasura, Surakarta 57162, Jawa Tengah, Indonesia

*correspondence author: ikewidianingsih82@gmail.com

Abstract. Green Open Space (GOS) is a component of an initiative aimed at addressing challenges in metropolitan setting. Surakarta is located on a strategically important route, which has led to its development as a significant economic and commercial hub for the surrounding region. Urban areas will inevitably undergo growth, encompassing both tangible and intangible aspects. This phenomenon leads to the town undergoing rapid growth as a result of population dynamics, socio-economic shifts, and interactions with neighbouring regions that serve as centres of activity. The presence of GOS is crucial of persevering the biological ecology in the vicinity of Surakarta city. Plants situated at the level of poles and saplings provide numerous purposes within the ecosystem. Multiple research studies have demonstrated that pole and sapling plans possess the capacity to regulate the environment, as seen by their diverse composition. The objective of this study is to analyse the composition and structure of the poles and sapling community in the natural GOS surrounding the city of Surakarta. The data collection employed an exploratory approach utilizing the quadratic (plot) method. The plot placement was conducted by random sampling, encompassing a 0.02 hectare in pole vegetation and 0.5 hectare in sapling vegetation area along the river banks of the Bengawan Solo. It was found a total of 17 different species of poles from 14 families and 37 species of sapling from 22 families. The diversity index is mediumly high at both the pole and sapling levels. Thus, this research can be recommended by the Surakarta City government as an effort to increase the diversity in pole and sapling vegetation.

1. Introduction

A green open space (GOS) is a designated area inside urban environments that serves several purposes, including recreation, socio-cultural activities, aesthetics, urban infrastructure, and ecological functions. These spaces hold significant economic value for both humans and the development of cities [1]. GOS can be categorized into three types: natural GOS, non-natural GOS, and built GOS [2]. GOS has primary and ancillary functions. The major function of GOS encompasses ecological functions that involve contributing to clean air, controlling the microclimate, offering shade, generating oxygen, and absorbing precipitation. Furthermore, there are additional roles that encompass socio-cultural, economic, and artistic aspects. Define physical

GOS as naturally GOS, such as wild habitats, protected areas, and national parks [3]. Non-natural or built GOS refers to green areas that are created by humans, such as parks, sports fields, and flower gardens. The GOS Surakarta encompasses various green areas, including parks and urban forests spanning 187.56 hectares, road green belts spanning 21.12 hectares, and particular green open spaces such as riverbanks spanning 164.22 hectares. Surakarta boasts a variety of man-made green spaces, including city parks, greenbelts, city forests, botanical gardens, sports facilities, green lanes along important roads, lakes, and flood control retention zones. Meanwhile, the City of Surakarta possesses natural GOS in the form of river banks, rail borders, and spring protection.

The purpose of GOS in urban settings is to strategically plan the future spatial arrangement. This involves identifying regions that require preservation to protect the environment and areas that are susceptible to disasters [4]. GOS has a direct impact on climate conditions. Micro, where the microscale interacts directly with and affects living organisms [5]. Microclimate refers to a specific and restricted climate that exists in a certain area and has a crucial role in the survival and well-being of all organisms [6]. A microclimate refers to the specific climate conditions found inside a small area, which are shaped by several variables such as the presence of woods, marshes, lakes, and human activity.

The structure and composition of vegetation in a region are significantly influenced by various interrelated components. Therefore, the vegetation in that area is a manifestation of the interplay between many environmental conditions and is susceptible to fluctuations in these elements. The dominance of a certain type of plant within an ecosystem can significantly impact its stability [7]. The dimensions of the vegetation category have an impact on the resilience of the ecosystem [8]. The characteristics of vegetation diversity in an area vary based on the area's state and human activities, which are necessary for daily life. Vegetation in forest ecosystems exhibits variation in plant height, resulting in the presence of many layers within a community. Vegetation analysis can be conducted to ascertain the distribution of different species within a given area. Vegetation analysis involves the identification of species and the categorization of their growth forms.

Conducting vegetation study is crucial for gaining further insights into the factors influencing the progression of succession [9]. Vegetation analysis allows for a more comprehensive understanding of the composition, quantity, and spatial arrangement of plant species within ecosystems, as well as the influence of environmental conditions [10]. This vegetation analysis encompasses forest stands consisting of tree strata, saplings, and non-woody understory species, specifically herbaceous plants. Plants are categorized based on their size into four groups: seedling-level vegetation (saplings or young plants with a height below 1.5 m), sapling vegetation (young trees with a minimum height of 1.5 m and a diameter less than 10 cm), pole vegetation (trees with a diameter between 10 cm and less than 20 cm), and tree vegetation (diameter equal to or greater than 20 cm). Urban research focuses on poles and saplings in GOS to promote tree regeneration. Detailed information about vegetation at the pole and saplings level can help identify suitable vegetation types for non-natural GOS with limited areas. These vegetation types should have smaller canopies and roots. Additionally, pole and sapling vegetation contributes to the overall above-ground carbon calculations, making it an important carbon pool. Pole and sapling vegetation play several important roles in the development of river ecosystems, including controlling erosion and flooding and maintaining and improving river water quality from pollutants carried by water [11]. In addition, pole and sapling vegetation have a primary function in reducing sediment input into the lake so that it can maintain the biogeochemical cycle,

IOP Conf. Series: Earth and Environmental Science **1503** (2025) 012014 doi:10.1088/1755-1315/1503/1/012014

provide habitat for biodiversity, and manage river water temperature [12]. Plants thrive in protected (natural) environments due to the enhanced growth prospects they offer for vegetation [13]. Nevertheless, no previous studies have been conducted on the plants found on poles and saplings, particularly GOS of Surakarta.

2. Methods

2.1 Sapling and the Selection of Sapling Locations

This study aimed to examine the content and organization of the pole and sapling community in the GOS along the river banks of Bengawan Solo in Surakarta, Central Java. The investigation was conducted in January 2024 utilizing an exploratory technique employing a quadratic (plot) strategy. The plot placement was conducted using a random method, and random sampling was performed an area of 0.02 ha in pole vegetation and 0.05 ha in sapling vegetation. Random Sampling is used to ensure the representation of vegetation cover types by ensuring the adequacy of the area using the minimum area curve. The number of plot points taken was 20 for each pole vegetation and sapling vegetation (Figure 1). The plot area for each pole vegetation plot is 10 m x 10 m and 5 m x 5 m for the pole vegetation (Figure 2). Plots are constructed using rolling meters, raffia rope, and pegs. Thermo hygrometers, altimeters, and soil testers are instruments employed to ascertain environmental conditions. Cameras, observation tables, boardwalks, and stationery are employed to chronicle and record the findings of the measured species.

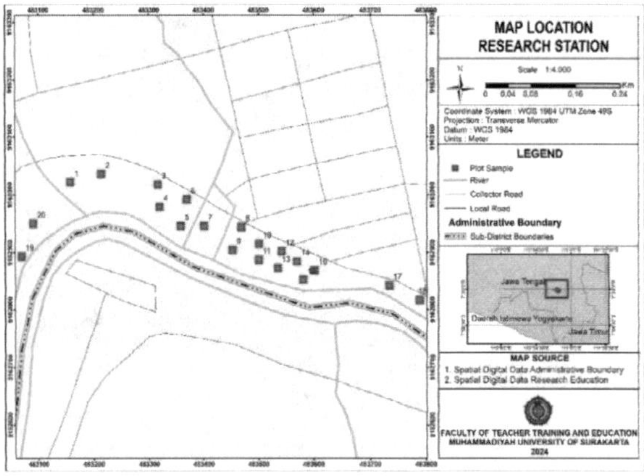

Figure 1. Map and Research Locations

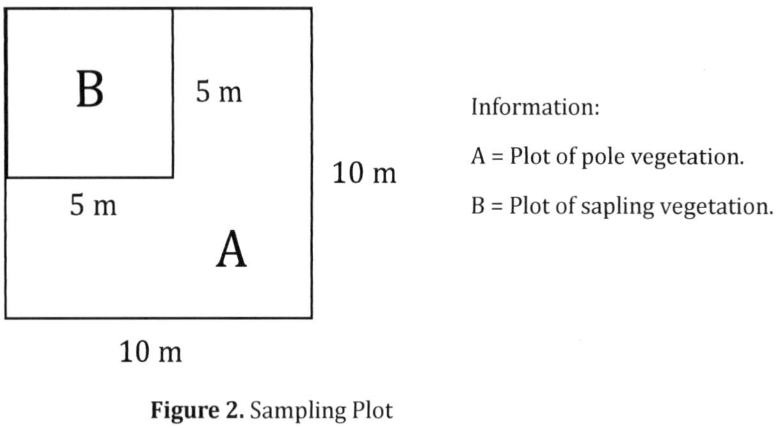

Figure 2. Sampling Plot

2.2 Identification

Plant identification and nomenclature were carried out through a literature study using the Gembong book , the Flora book, the online website POWO (Plants of the World Online), and journals related to the plants to be identified ([14], [15], [16], [17]).

2.3 Data Analysis

This study investigated various aspects, such as the species richness, population size, and degrees of plant biodiversity. Quantitative data analysis was conducted to elucidate the structure and composition of plant growth at both pole and sapling levels. Vegetation analysis involves assessing the density, frequency, dominance, important value index, diversity index, and dominance index of plant life.

3. Result and Discussion

Research conducted in the GOS near Surakarta city, specifically along the river banks of Bengawan Solo, has identified 17 types of poles belonging to 14 families. These include mango (*Mangifera indica* L.), pine (*Pinus* L.), mahogany (*Swietenia mahagoni* (L.) Jacq.), gaharu (*Aquilaria malaccensis* Lam.), water guava (*Syzygium aqueum* (Burm.f.) Alston), trembesi (*Samanea saman* (Jacq) Merr.), melinjo (*Gnetum gnemon* L.), guava (*Psidium guajava* L.), jackfruit (*Artocarpus heterophyllus* Lam.), longan (*Dimocarpus longan* Lour.), rubber (*Hevea brasiliensis* (Wills . Ex. A. Juss.) Mull.Arg.), eucalyptus (*Melaleuca leucadendra* (L.) L.), sapodilla (*Manilkara zapota* (L.) P. Royen), pule (*Alstonia scholaris* (L.) R. Br.), soursop (*Annona muricate* L.), teak (*Tectona grandis* L. f.), and frangipani (*Plumeria rubra* L.).

The findings of the calculation of the Importance Value Index (IV), Diversity Index (H'), and Dominance Index (C) of pole vegetation on GOS areas along the River Bank of Bengawan Solo are shown in Table 1. and Table 2.

Table 1. Diversity of Pole Vegetation in the Bengawan Solo River Bank Area.

No	Local name	Scientific name	Famili	RD %	RF %	RDo %	IV	IV Famili	H'	C
1.	Mangga	*Mangifera indica* L.	Anacardiaceae	73.13	3.22	74.92	187.26*	187.26	0.23	0.53
2.	Sirsak	*Annona muricate* L.	Annonaceae	2.50	3.96	2.08	8.50	8.50	0.09	0
3.	Pule	*Alstonia scholaris* (L.) R.Br.	Apocynaceae	0.63	1.96	0.83	3.41	6.42	0.03	0
4.	Kamboja	*Plumeria rubra* L.	Apocynaceae	0.63	1.96	0.44	3.02		0.03	0
5.	Karet	*Hevea brasiliensis* (Wills. Ex. A. Juss.) Mull.Arg.	Euphobiaceae	1.25	1.96	1.65	4.86	4.86	0.05	0

No	Local name	Scientific name	Famili	RD %	RF %	RDo %	IV	IV Famili	H'	C
6.	Trembesi	*Samanea saman* (Jacq) Merr.	Fabaceae	4.38	11.76	4.85	20.99*	20.99	0.14	0
7.	Melinjo	*Gnetum gnemon* L.	Gnetaceae	0.63	1.96	0.25	2.84	2.84	0.03	0
8.	Mahoni	*Swietenia mahagoni* (L.) Jacq.	Meliaceae	6.88	9.80	6.42	23.10	23.10	0.18	0
9.	Nangka	*Artocarpus heterophyllus* Lam.	Moraceae	1.88	3.92	1.00	6.80	6.80	0.07	0
10.	Jambu air	*Syzygium aqueum* (Burm.f.) Alston	Myrtaceae	2.50	5.88	1.78	10.17	10.17	0.09	0
11.	Jambu bii	*Psidium guajava* L.	Myrtaceae	0.63	3.92	0.39	2.98	6.49	0.03	0
12.	Kayu putih	*Melaleuca leucadendra* (L.) L.	Myrtaceae	0.63	1.96	0.93	3.51		0.03	0
13.	Pinus	*Pinus L.*	Pinaceae	0.63	1.96	0.57	3.16	3.16	0.03	0
14.	Kelengkeng	*Dimocarpus longan* Lour.	Sapindaceae	1.25	1.96	1.46	6.63	6.63	0.05	0
15.	Sawo	*Manilkara zapota* (L.) P. Royen	Sapotaceae	1.25	3.92	0.90	6.07	6.07	0.05	0
16.	Gaharu	*Aquilaria malaccensis* Lam.	Thymeleaceae	0.63	1.96	0.90	3.49	3.49	0.03	0
17.	Jati	*Tectona grandis* L. f.	Verbenaceae	0.63	1.96	0.62	3.20	3.20	0.03	0
		Total		100	100	100	300	300	1.23	0.54

Information: RD = Relative Density, RF = Relative Frequency, RDo= Relative Dominance, IV = Importance Value Index, * = Highest IV, H' = Diversity Index, C = Dominance Index.

The same research carried out on sapling level vegetation found 37 types of saplings from 22 families, namely red shoots (*Syzygium myrtifolium Walp*), mango (*Mangifera indica L.)*, jackfruit (*Artocarpus heterophyllus* Lam.), mahogany (*Swietenia mahagoni* (L.) Jacq.), banyan (*Ficus benjamina* L.), teak (*Tectona grandis* L.f.), frangipani (*Plumeria rubra* L.), dadab (*Erythrina variegate* L.), waru (*Hibiscus sterculiifolius* (Guill. & Perr.) Steud.), guava (*Psidium guajava* L.), bamboo (*Melocanna bamboosoideae* (Roxb.) Kurz), papaya (*Carica papaya* L.), hanjuang (*Maxillaria cordyline* (Rchb.f.) Dodson), cakra cikri (*Polyscias filicifolia* (C. Moore ex E. Fourn.) L. H. Bailey), kaffir lime (*Citrus hystrix* DC), pule (*Alstonia scholaris* (L.) R.Br.), trembesi (*Samanea saman* (Jacq.) Merr.), salam (*Syzygium polyanthum* Thwaites), sengon (*Albizia chinensis* (Osbeck) Merr.), ketapang (*Terminalia catappa* L.), moringa (*Moringa oleifera* Lam.), daluang (*Broussonetia papyrifera* (L.) LHer. Ex Vent.), soursop (*Annona muricate* L.), rambutan (*Nephelium lappaceum* L.), water guava (*Syzygium aqueum* (Burm. f.) Alston), longan (*Dimocarpus longan* Lour), chinese petai (*Leucaena leucocephala* (Lam.) de Wit), cherry (*Muntingia calabura* L.), African leaf (*Vernonia amygdalina* Delile), mentaos (*Wrightia Pubescens* R. Br.), beluntas (*Pluchea indica* (L.) Less.), bidara (*Ziziphus mauritiana* Lam.), kapok (*Ceiba pentandra* (L.) Gaertn.), eucalyptus (*Melaleuca leucadendra* (L.) L.), cassava (*Manihot esculenta* Crantz), *Chinese castor* (Jatropha multifida L.), and emery (*Ficus amelas* Burm. f.).

Table 2. Diversity of Sapling Vegetation in the Bengawan Solo River Bank Area.

No	Local name	Scientific name	Famili	RD %	RF %	RDo %	IV	IV Famili	H'	C
1.	Hanjuang	*Maxillaria cordyline* (Rchb.f.) Dodson	Agavaceae	3.07	3.49	1.37	7.93	7.93	0.11	0
2.	Mangga	*Mangifera indica L.*	Anacardiaceae	20.86	12.79	22.81	56.46*	56.46	0.33	0.04
3.	Sirsak	*Annona muricate* L.	Annonaceae	0.61	1.16	0.51	2.29	2.29	0.03	0
4.	Pule	*Alstonia scholaris* (L.) R.Br.	Apocynaceae	0.61	1.16	0.67	2.45	9.6	0.03	0
5.	Mentaos	*Wrightia Pubescens* R. Br.	Apocynaceae	0.61	1.16	0.29	2.07		0.03	0
6.	Kamboja	*Plumeria rubra* L.	Apocynaceae	0.61	1.16	3.30	5.08		0.03	0
7.	Cakra-cikri	*Polyscias filicifolia* (C.Moore ex E. Fourn.)L. H. Bailey	Araliaceae	0.61	1.16	0.19	1.96	1.96	0.03	0
8.	Daun afrika	*Vernonia amygdalina* Delile	Asteraceae	0.61	1.16	0.26	2.03	4.22	0.03	0
9.	Beluntas	*Pluchea indica* (L.) Less.	Asteraceae	0.61	1.16	0.42	2.19	2.19	0.03	0
10.	Pepaya	*Carica papaya L.*	Caricaceae	2.45	4.65	2.51	9.61	9.61	0.09	0

No	Local name	Scientific name	Famili	RD %	RF %	RDo %	IV	IV Famili	H'	C
11.	Ketapang	*Terminalia catappa* L.	Combretaceae	2.45	4.65	2.41	9.51	9.51	0.09	0
12.	Ketela	*Manihot esculenta* Crantz	Eurphobiaceae	6.75	8.14	6.31	21.19	25.45	0.18	0
13.	Jarak cina	*Jatropha multifida* L.	Euphorbiaceae	1.23	1.16	1.87	4.26		0.05	0
14.	Dadab	*Erythrina variegate* L.	Fabaceae	1.84	1.16	3.34	6.35	11.07	0.07	0
15.	Trembesi	*Samanea saman* (Jacq.) Merr.	Fabaceae	0.61	1.16	0.21	1.98		0.03	0
16.	Sengon	*Albizia chinensis* (Osbeck) Merr.	Fabaceae	0.61	1.16	0.97	2.74		0.03	0
17.	Waru	*Hibiscus sterculiifolius* (Guill. & Perr.) Steud.	Malvaceae	1.84	4.65	1.54	8.04	10.07	0.07	0
18.	Randu	*Ceiba pentandra* (L.) Gaertn.	Malvaceae	0.61	1.16	0.26	2.03		0.03	0
19.	Mahoni	*Swietenia mahagoni* (L.) Jacq.	Meliaceae	1.84	2.33	3.35	7.51	7.51	0.07	0
20.	Petai cina	*Leucaena leucocephala* (Lam.) de Wit	Mimosaceae	10.43	4.65	8.10	23.18	23.18	0.24	0.01
21.	Nangka	*Artocarpus heterophyllus* Lam.	Moraceae	4.91	3.49	3.36	11.76	28.97	0.15	0
22.	Beringin	*Ficus benjamina* L.	Moraceae	2.45	3.49	4.62	10.56		0.09	0
23.	Ampelas	*Ficus ampelas* Burm. f.	Moraceae	0.61	1.16	1.75	3.52		0.03	0
24.	Daluang	*Pluchea indica* (L.) Less.	Moraceae	0.61	1.16	1.36	3.13		0.03	0
25.	Kelor	*Moringa oleifera* Lam.	Moringaceae	3.07	5.81	4.15	13.03	13.03	0.11	0
26.	Kersen	*Muntingia calabura* L.	Muntingiaceae	1.84	1.16	0.78	3.79	3.79	0.07	0
27.	Pucuk merah	*Syzygium myrtifolium* Walp	Myrtaceaea	2.45	4.65	1.79	8.90	45.29	0.09	0
28.	Jambu biji	*Psidium guajava* L.	Myrtaceae	0.61	1.16	0.18	1.95		0.03	0
29.	Salam	*Syzygium polyanthum* Thwaites	Myrtaceae	1.23	4.65	0.53	6.41		0.05	0

No	Local name	Scientific name	Famili	RD %	RF %	RDo %	IV	IV Famili	H'	C
30.	Jambu air	*Syzygium aqueum* (Burm.f.) Alston	Myrtaceae	1.84	2.33	0.68	4.84		0.07	0
31.	Kayu putih	(*Melaleuca leucadendra* (L.) L.)	Myrtaceae	13.5	2.33	12.21	28.03*		0.27	0.02
32.	Bambu	*Melocanna bambusoideae* (Roxb.) Kurz	Poaceae	1.23	1.16	0.34	2.73	2.73	0.05	0
33.	Bidara	*Ziziphus mauritiana* Lam.	Rhamnaceae	0.61	1.16	0.86	2.63	2.63	0.03	0
34.	Jeruk purut	*Citrus hystrix* DC	Rutaceae	0.61	1.16	1.32	3.09	3.09	0.03	0
35.	Kelengkeng	*Dimocarpus longan* Lour	Sapindaceae	4.29	3.49	3.59	11.37	13.53	0.14	0
36.	Rambutan	*Nephelium lappaceum* L.	Sapindaceae	0.61	1.16	0.38	2.16		0.03	0
37.	Jati wesi	*Tectona grandis* L.f.	Verbenaceae	0.61	1.16	1.42	3.20	3.20	0.03	0
		Total		100	100	100	300	300	2.94	0.1

Information: RD = Relative Density, RF = Relative Frequency, RDo= Relative Dominance, IV = Importance Value Index, * = Highest IV, H' = Diversity Index, C = Dominance Index.

The Importance Value Index (IV) is a measure utilized to quantify the degree of species dominance within a plant community. IV can elucidate the function of a plant within a community. A plant's significance in the community increases as its IV value increases [18]. Research findings indicate that *Mangifera indica* and *Samanea saman* had the greatest IV values at the pole level, measuring 187.26% and 20.99% respectively. At the sapling level, *Mangifera indica* and *Melaleuca leucadendra* exhibited the greatest IV values, measuring 56.46% and 28.03% respectively. Therefore, this plant is classified as a tall and young plant, which holds the highest level of dominance and plays a crucial function in the surrounding ecosystem. Surakarta, situated in a low-lying area with an elevation ranging from 80 to 130 meters above sea level, experiences an average yearly temperature of 28°C and a humidity level of 64% [19]. The river banks of Bengawan Solo exhibit an air temperature of 26.7 °C and an air humidity of 90%. This indicates that the air temperature at the Bengawan Solo River Bank falls within the average temperature range of Surakarta City, however the air humidity at the Bengawan Solo River Bank is greater than the average humidity of Surakarta City. The average pH value is 7 and a substance is considered alkaline if its pH value is greater than 7, and vice versa [20]. The soil pH on the banks of the Bengawan Solo River is 6.98, which is within the range of normal soil pH. *Mangifera indica* is a plant that can live up to 100 years, its morphology consists of roots, stems, leaves, and flowers. *Mangifera indica* roots are very long and can reach a depth of 6 m, and ideally grow in rainfall of 750-2250 mm with a tropical climate [21]. *Samanea saman* has a broad crown with strong rootlets that grow 66.6% into the soil and can grow up to 100 years with rainfall of 600-3000 mm per year [22]. Meanwhile, *Melaleuca leucadendra* is a plant that is widely used in the production of essential oils, it has single roots and grows in rainfall of 1300-1750 mm per year [23]. Thus, the characteristics possessed by the three plants provide benefits to the ecosystem around the Bengawan Solo Riverbanks, one of which is in overcoming the problems of erosion and flooding that occur during the rainy season.

Abiotic variables exert an influence on the growth and development of plants. Environmental factors play an important role in determining the diversity of growth in both tree height and stem diameter [24]. The air temperature, as a constituent of climate elements, has the potential to impact the growth and development of plants. The ideal temperature is crucial for achieving optimal plant growth, although fluctuations in external temperature can also impact plant growth and development [25]. Elevated temperatures have a detrimental effect on enzymes, leading to impaired plant metabolism. Conversely, low temperatures can induce enzyme inactivity, resulting in a halt in plant metabolism [26]. The air temperature has a significant impact on the physiological processes of blooming plants, leading to alterations in the quantity of flowers, nectar, and pollen, as well as the timing of flowering [27].

In addition to air temperature, air humidity plays a crucial role in the growth and development of plants. Air humidity refers to the quantity of water vapor present in the air or atmosphere [28]. The amount of water vapor in cold air is lower compared to that in warm temperatures. There is a direct relationship between air temperature and the amount of water vapor it holds [29]. As the air temperature increases, the amount of water vapor also increases. The level of air humidity has a direct impact on the processes of photosynthesis and transpiration in plants. Photosynthesis, a vital activity in plants, takes place in specialized cells called stomata, which contain photosynthetic pigments [30]. Photosynthesis is a biological process in which plants utilize energy and sunlight to manufacture their own food and generate oxygen, which is essential for human survival [31]. Soil pH is an abiotic component that influences plant growth and development. It affects the ability of plants to absorb nutritional ions, shows the presence of substances that are harmful to plants, and influences the proliferation of microorganisms in the soil [32].

Figure 3. *Mangifera indica (A), Samanea saman (B), and Melaleuca leucadendra (C)*

The Diversity Index is an essential vegetation metric that allows for the comparison of various plant communities, particularly in terms of the impact of non-living disturbances and the degree of progression or stability within a community [33]. The Diversity Index value is obtained based on the research results. The elevation at the pole level in the GOS on the river banks of Bengawan Solo is 1.23, while at the sapling level it is 2.94. The classification of the natural GOS on the banks of the Bengawan Solo River is determined to be medium. Conversely, a lack of diversity might have a negative impact on the ecosystem's capacity to generate environmental services [34]. The concept of plant diversity always refers to two indicators, namely the number of species and distribution of plants [35]. In addition, the Dominance Index (C) value at the pole level was 0.54, while at the sapling level it was 0.1. These values suggest that the Dominance Index (C) was medium and low, respectively. Based on the vegetation seen at both the pole and sapling level in the GOS along the Bengawan Solo River, it can be inferred that there is no species that stands out as dominant, indicating that the community is in a stable state.

This research represents the diversity index of pole and sapling vegetation on the banks of the Bengawan Solo River. Therefore, this study presents a collection of data regarding research that is similar to the results of diversity indices in GOS, agroforestry, and other secondary forests in Indonesia. The following is comparative data on the Diversity Index (H') in GOS, agroforestry, and secondary forests which is presented in Table 3.

Table 3. Comparison of H' of Pole and Sapling Vegetation in Green Open Space, Agroforestry and Secondary Forest

No	Locations	Type of Vegetations	H'	Criteria	Source
	GOS				
1.	Surakarta, Central Java	Pole	1.23	Medium	This Research
		Sapling	2.94	Medium	
2.	Gunung Mas, Central Kalimantan	Pole	3.50	Medium	[13]
		Sapling	3.01	Medium	
3.	PT Kaltim Methanol Industri, Bontang, East Kalimantan	Pole	2.50	Medium	[36]
		Sapling	2.98	Medium	
	Agroforestry				
1.	Nekamese, Kupang, East Nusa Tengga	Pole	3.121	Medium	[37]
		Sapling	2.781	Medium	
2.	Leuwiliang, Bogor, West Java	Pole	1.167	Medium	[38]
		Sapling	1.022	Medium	
3.	Biih Village, Banjar, South Kalimantan	Pole	2.20	Medium	[39]
		Sapling	1.94	Medium	
	Secondary Forest				
1.	Punggaloba, West Kendari, Southeast Sulawesi	Pole	1.384	Medium	[40]
		Sapling	0.962	Low	
2.	Gayo Lues, Aceh	Pole	0.75	Low	[41]
		Sapling	0.60	Low	
3.	Alas Burno SUBKPH, Lumajang, East Java	Pole	1.77	Medium	[42]
		Sapling	1.55	Medium	

Based on the table above, shows that research regarding vegetation analysis on poles and saplings has been carried out several times. The Diversity Index (H') in agroforestry shows consistent results, namely medium diversity. This indicates that the Diversity Index (H') in agroforestry is quite stable and interactions between species in the community are quite good. The Bengawan Solo Riverbank GOS has a medium range of diversity. Diversity in the Bengawan Solo Riverbank GOS has a higher diversity value compared to secondary forests. This can be seen from the H' value of secondary forests which have a medium-low diversity range. From these data, it can be concluded that the H' value in the GOS along the Bengawan Solo River is quite stable and the interactions between species in the community are quite good. Meanwhile, in secondary forests, the H' value tends to be low, indicating that there is a dominant plant in a place. A low diversity index indicates that one species are dominant with uneven distribution [43].

4. Conclusion

The research focused on examining the structure and composition of pole and sapling vegetation in the GOS along the banks of the Bengawan Solo River. The findings revealed the presence of 17 pole types belonging to 16 families, as well as 37 sapling types from 24 families. The species *Mangifera indica*, *Samanea saman*, and *Melaleuca leucadendra* are frequently encountered plants. The highest Important Value Index (IV) is observed in *Mangifera indica* and *Samanea saman* at the pole level, with values of 187.26% and 20.99% respectively. At the sapling level, *Mangifera indica* and *Melaleuca leucadendra* exhibited the greatest IV values, measuring 56.46% and 28.03% respectively. At the pole level, *Gnetum gnemon* had the lowest Important Value Index (IV) with a value of 2.84%. Similarly, at the sapling level, *Psidium guajava* had the lowest IV with a value of 1.95%. The Diversity Index (H') is categorized as medium, with a magnitude of 1.23 at the pole level and a magnitude of 2.94 at the sapling level. The Dominance Index (C) is categorized as medium at the pole level and as low at the sapling level, with a numerical value of 0.54 at the pole level and 0.1 at the sapling level. The abiotic variables in the ecosystem along the banks of the Bengawan Solo River continue to provide favorable circumstances for plant development. Additional research is needed to investigate the community assembly between tree, pole, and sampling alongside Bengawan Solo River Bank.

5. Acknowledgments

The author expresses gratitude to the pertinent entities that have contributed to the finalization of this article, particularly the Surakarta City Environmental Service for granting permission and endorsing research sites, and the Biology Laboratory of the Muhammadiyah University of Surakarta for providing instrumental support for the author's research.

References

[1] D. Setyanto and N. S. Salahuddin, "Greenhouse Microclimate Automatic Monitoring and Control Prototype with Blynk IoT Platform," *Techno.Com*, vol. 21, no. 1, pp. 88–102, 2022, doi: 10.33633/tc.v21i1.5462. [Bahasa]

[2] N. Alifia and Y. Purnomo, "Identification of the Location and Type of Green Open Space in Urban Residential Areas," *Langkau Betang J. Arsit.*, vol. 3, no. 2, pp. 25–38, 2016, doi: 10.26418/lantang.v3i2.18329. [Bahasa]

[3] D. Mashur and Z. Rusli, "Efforts and Implications of Providing Green Open Space (RTH)," *J. Kebijak. Publik*, vol. 9, no. 1, p. 45, 2018, doi: 10.31258/jkp.9.1.p.45-52. [Bahasa]

[4] N. Budhyowati, "Study of the Availability of Non-Natural Public Green Open Spaces in Manado City," *J. Tek. Sipil Terap.*, vol. 1, no. 1, p. 8, 2019, doi: 10.47600/jtst.v1i1.220. [Bahasa]

[5] C. Aluyah and R. Rusdianto, "The Influence of Type and Number of Trees on the Microclimate in the Bukit Siguntang Archaeological Park, Palembang City, South Sumatra Province," *Sylva J. Ilmu-ilmu Kehutan.*, vol. 8, no. 2, p. 53, 2020, doi: 10.32502/sylva.v8i2.2696. [Bahasa]

[6] I. Saroh and Krisdianto, "Ecological Benefits of Tree Canopies on Microclimate in Green Open Spaces in Urban Areas," *J. Hutan dan Masy.*, vol. 12, no. 2, pp. 136–145, 2020, doi: 10.24259/jhm.v12i2.10040. [Bahasa]

[7] N. Destaranti, S. Sulistyani, and E. Yani, "Structure and Vegetation of Understory Plants in Pine Stands in Rph Kalirajut and Rph Baturraden Banyumas," *Scr. Biol.*, vol. 4, no. 3, p. 155, 2017, doi: 10.20884/1.sb.2017.4.3.407. [Bahasa]

[8] J. S. Ramadona, A. Fatiqin, and F. Azwar, "Analysis of Understory Vegetation of Burned Land Under Bambang Lanang (Michelia Champaca) Stands in KHDTK Kemampo," *Pros. Semin. Nas. Sains dan Teknol. Terap.*, vol. 3, no. 1, pp. 440–445, 2020. [Bahasa]

[9] S. Santhyami and I. Aryani, "Vegetation Analysis Of Tree And Pole Community As Part of Succession in Turgo Hill , Yogyakarta," *J. Pembelajaran Dan Biol. Nukl.*, vol. 10, no. 1, pp. 63–73, 2024, doi: https://doi.org/10.36987/jpbn.v10i1.5330.

[10] I. Aryani and S. S. Riskawati, "Analysis of Plant Vegetation in Sekipan Forest Tawangmangu District , Karanganyar District," *Int. Conf. Biol. Educ. Nat. Sci. Technol.*, vol. 1, no. 1, pp. 509–517, 2023.

[11] F. B. P. Nurika, E. Wiryani, and Jumari, "Diversity of Riparian Vegetation in the Lower Panjang River in Ambarawa District, Semarang Regency," *J. Akad. Biol.*, vol. 8, no. 1, pp. 30–34, 2019. [Bahasa]

[12] S. P. Kharisma, A. R. Fauzan, I. Permitasari, and T. Nurseha, "Riparian Vegetation Diversity: Efforts to Preserve the Aquatic Ecosystem of the Artificial Lake Air Batu Banyuasin, South Sumatra," *J. Biotropical Res. Nat. Technol.*, vol. 2, no. 2, pp. 85–92, 2024, doi: https://doi.org/10.36873/borneo.v2i2.12706. [Bahasa]

[13] N. Haryadi, "Vegetation Structure and Composition in the Telaga Kameloh Waterfall Protected Area, Gunung Mas Regency," *J. Ziraa'ah*, vol. 42, no. 02, pp. 137–149, 2017, doi: https://dx.doi.org/10.3160/zmip.v42i2.778. [Bahasa]

[14] K. A. Shah, M. B. Patel, R. J. Patel, and P. K. Parmar, "Mangifera Indica (Mango)," vol. 4, no. 7, pp. 42–48, 2010, doi: 10.4103/0973-7847.65325. [Bahasa]

[15] S. W. Ulfa, A. R. Samosir, A. Marwani, D. K. Ayunda, F. Amalia, and K. A. Fazrah, "Identification of the Fabales Order in the Area of the State Islamic University of North Sumatra and the State University of Medan," *Rev. Pendidik. dan Pengajaran*, vol. 6, no. 4, pp. 4310–4319, 2023, doi: https://doi.org/10.31004/jrpp.v6i4.23638. [Bahasa]

[16] L. Wattimena, Y. S. Serkadifat, and K. Tapaen, "Utilization of Eucalyptus Trees (Melaleuca cajuputi) by the Solal Village Community, North Misool District, Raja Ampat Regency, West Papua Province," *J. Jendela Ilmu*, vol. 2, no. 1, pp. 26–30, 2021. [Bahasa]

[17] Yarman and E. K. Damayanti, "Utilization and Conservation Efforts of Eucalyptus (Asteromyrtus symphyocarpa) in Wasur National Park," *Media Konserv.*, vol. 17, no. 2, pp. 85–93, 2012, doi: 10.29243/medkon.17.2.%p. [Bahasa]

[18] M. S. Hamidun and D. W. K. Baderan, "Vegetation Analysis of Boliyohuto Limited Production Forest, Gorontalo Province," *Artikel*, vol. 4, no. 1, pp. 1–23, 2016. [Bahasa]

[19] E. Roziaty and S. I. W. Adiningsih, "Spatial Distribution of Shade Trees in Surakarta City, Central Java," in *Seminar Nasional Pendidikan Biologi dan Saintek (SNPBS) ke-IV*, 2021, pp. 92–100. [Bahasa]

[20] F. Yusran, D. N. Ramadan, and T. N. Damayanti, "Soil pH Monitoring System and Automatic Watering of Smart Plant Based on IoT Microcontroller on Bonsai Santigi," in *e-Proceeding of Applied Science*, 2023, pp. 1–8. [Bahasa]

[21] Y. Oktavianto and A. Suryanto, "Characteristics of Mango Plants (Mangifera indica L.) Cantek, Ireng, Empok, Jempol in Tiron Village, Banyakan District, Kediri Regency," *J. Produksi Tanam.*, vol. 3, no. 2, pp. 91–97, 2015, doi: 10.21176/protan.v3i2.174. [Bahasa]

[22] D. E. Andraini, "Evaluation of the Suitability of the Physical Characteristics of Trembesi on the Green Lane of Makassar City Road to Maximize CO2 Absorption," *J. Lanskap dan Lingkung.*, vol. 1, no. 1, pp. 41–47, 2023. [Bahasa]

[23] Aswin Hendry Atmoko, Bagus Aradea, and Annisa Nanda Oktavia, "Analysis of Genetic Diversity of Eucalyptus Plant Germplasm (Melaleuca cajuputi subsp. cajuputi) Based on Morphological and Anatomical Characters of

Leaves and Oil Glands," *J. Trit.*, vol. 14, no. 2, pp. 492–507, 2023, doi: 10.47687/jt.v14i2.699. [Bahasa]
[24] M. Susanto and L. Baskorowati, "Genetic and Environmental Influences on the Growth of Sengon (Falcataria molucanna) Java Land Race," *Bioeksperimen J. Penelit. Biol.*, vol. 4, no. 2, pp. 35–41, 2018, doi: 10.23917/bioeksperimen.v4i2.6883. [Bahasa]
[25] R. Karmila and V. Andriani, "Effect of Temperature on Growth Rate of Tolo Beans (Vigna sp.)," *STIGMA J. Mat. dan Ilmu Pengetah. Alam Unipa*, vol. 12, no. 01, pp. 49–53, 2019, doi: 10.36456/stigma.vol12.no01.a1861. [Bahasa]
[26] A. Fakhrezi, R. E. Saputra, and F. C. Hasibuan, "Design and Construction of an IoT-Based Nutrient, Humidity, Soil PH and Air Temperature Monitoring System Using an ESP32 Microcontroller," *e-Proceeding Eng.*, vol. 10, no. 1, pp. 778–786, 2023. [Bahasa]
[27] K. Khotimah, E. Sudiana, and H. Pratiknya, "Impact of Climate Change on Phaseolus vulgaris L Phenology Faculty of Biology, Jenderal Soedirman University," *Bioma Berk. Ilm. Biol.*, vol. 24, no. 1, pp. 1–7, 2022, doi: 10.14710/bioma.24.1.1-7. [Bahasa]
[28] Y. S. Swarinoto and Sugiyono, "Utilization of Air Temperature and Air Humidity in Regression Equations for Monthly Rainfall Prediction Simulation in Bandar Lampung," *Pertanian*, vol. 12, no. 3, pp. 271–281, 2009, doi: https://doi.org/10.18860/ca.v3i1.2565. [Bahasa]
[29] G. M. Putra and D. Faiza, "Light in a Greenhouse for Shallot Plants Using the Internet of Things (IOT)," *J. Pendidik. Tambusai*, vol. 5, no. 3, pp. 11404–11419, 2022, doi: https://doi.org/10.31004/jptam.v5i3.2162. [Bahasa]
[30] N. Farid, A. Sarjito, and Z. Ulinnuha, "The Effect of Media Humidity on the Growth and Transpiration of Five Dendrobium Orchid Varieties," *Agromix*, vol. 14, no. 1, pp. 96–103, 2023, doi: https://doi.org/10.35891/agx.v14i1.3014. [Bahasa]
[31] Maftukhah, Ulfaturrohmah, N. I. Sholikhah, and U. Fawaida, "The Effect of Light on Photosynthesis Process in Shade Plants and Plants Exposed to Direct Light," *J. Pengabdi. Masy. MIPA dan Pendidik.*, vol. 7, no. 1, pp. 51–55, 2023, doi: https://doi.org/10.21831/jpmmp.v7i1.51510. [Bahasa]
[32] R. Dewi Hartati, M. Suryaman, and A. Saepudin, "The Effect Of Phosphate Solublizing Bacteria At Various Soil pH On Plant Growth And Yield of Soybean (Glycine max (L.) Merr)," *ournal Agrotechnology Crop Sci.*, vol. 1, no. 1, pp. 26–34, 2023.
[33] A. R. Hidayah and E. Roziaty, "Diversity of Shrubs Growing Along the Cemoro Sewu Hiking Trail, Magetan," in *Prosiding SNPBS (Seminar Nasional Pendidikan Biologi dan Saintek)*, 2022, pp. 413–419. [Online]. Available: https://proceedings.ums.ac.id/index.php/snpbs/article/view/1787 [Bahasa]
[34] Santhyami, A. Basukriadi, M. P. Patria, and R. Abdulhadi, "The Comparison Of Aboveground C-stock Between Cocoa-Based Agroforestry System And Cocoa Monoculture Practice In West Sumatra, Indonesia," *Biodiversitas*, vol. 19, no. 2, pp. 472–479, 2018, doi: 10.13057/biodiv/d190214.
[35] S. Suhendar, B. Ramdhan, and A. E. Triana, "Study of Tree Types in Kibitay Sukabumi City Forest Development," *Bioeksperimen J. Penelit. Biol.*, vol. 6, no. 2, pp. 141–153, 2020, doi: 10.23917/bioeksperimen.v6i2.8609. [Bahasa]
[36] A. A. Ridlo *et al.*, "Diversity of Vegetation Types Composing the Green Open Space of PT Kaltim Methanol Industry," *ULIN J. Hutan Trop.*, vol. 7, no. 1, p. 10, 2023, doi: 10.32522/ujht.v7i1.8107. [Bahasa]
[37] R. C. Tabun, H. P. Nastiti, and T. S. Temu, "Identification of Forage Communities in the Agroforestry Area of Oben Village, Nekamese Sub-District, Kupang Regency (Identification on Forage Communities in the Agroforestry Area of Oben Village, Nekamese Sub-District, The Regency," *J. Peternak. Lahan Kering*, vol. 2, no. 3, pp. 972–983, 2020, doi: https://doi.org/10.57089/jplk.v2i3.512. [Bahasa]
[38] Rozalina, "Profile of Mixed Gardens in Karacak Village, Leuwiliang District, Bogor Regency," *J. Akar*, vol. 8, no. 1, pp. 72–82, 2020, doi: 10.36985/jar.v8i1.116. [Bahasa]
[39] A. Davita, P. S. Kehutanan, F. Kehutanan, and U. Lambung, "The Vegetation Diversity in the Agroforestry System of Dukuh Hamlet in Biih Village, Banjar Regency, South Kalimantan," vol. 07, no. 5, pp. 717–725, 2024, doi: https://doi.org/10.20527/jss.v7i5.9383. [Bahasa]
[40] F. Satria, Rosmarlinasiah, L. O. A. S. Mando, S. Bana, and Basrudin, "Vegetation Analysis in the Protected Block of the Nipa-Nipa Grand Forest Park (Tahura) in Punggaloba Village, West Kendari District," *J. Kehutan. Indones.*, vol. 2, no. 2, pp. 83–92, 2021. [Bahasa]
[41] S. Toni, Misdi, and S. Rasnovi, "Inventory of Plant Types with Potential as Producers of Non-Timber Forest Products (NTFPs) in the Production Forest of KPH Region V Aceh, Gayo Luwes Regency," *J. Ilm. Mhs. Pertan.*, vol. 7, no. 41, pp. 785–795, 2022. [Bahasa]
[42] Rawana, S. Wijayani, and M. A. Masrur, "Importance Value Index and Diversity of Vegetation Communities Composing Forests in Alas Burno SUBKPH Lumajang," *J. Wana Trop.*, vol. 12, no. 02, pp. 80–89, 2023, doi: 10.55180/jwt.v12i02.215. [Bahasa]
[43] K. Hasanah, A. Hayati, and H. Zayadi, "Diversity of Wild Plant on Corn Field (Zea mays L.) in Bung-Bungan Village Bluto District Sumenep Regency," *Biosaintropis (Bioscience-Tropic)*, vol. 6, no. 1, pp. 54–60, 2020, doi: https://doi.org/10.33474/e-kbst.v6i1.298.

IOP Conf. Series: Earth and Environmental Science **1503** (2025) 012015 doi:10.1088/1755-1315/1503/1/012015

Spatio-Temporal Analysis Of Erosion Change Rate In 2018 And 2023 Based On Geographic Information Systems (Case Study Of The Bodri Watershed)

Muhammad Catur Prasetyo[1], Yuli Priyana[2*], Aditya Saputra[3]
[1] Geography Faculty, Muhammadiyah University Of Surakarta, Surakarta 57162, Central Java, Indonesia

*E-mail: yp121@ums.ac.id

Abstract. Indonesia has many watersheds (DAS), based on data from the Ministry of Environment and Forestry (KLHK), at least Indonesia has 17,076 watersheds where of all the existing watersheds as many as 2145 watersheds need to be restored, 108 watersheds are on the critical threshold and 15 of them have a high priority scale to be addressed immediately, damage to a watershed can be caused by several factors, one of which is erosion, Bodri watershed is one of the watersheds in Central Java Province with a relatively high erosion rate, this is evidenced by the amount of sediment accumulation, where in 2023 as much as 390. 811.95 tonnes of sediment material carried by water flow. watershed damage can have a domino effect on other aspects of life. this study aims to determine the rate of change in erosion that occurs in the Bodri watershed between 2018 and 2023, as well as the main factors that trigger changes in erosion in this watershed, which erosion will be calculated by the RUSLE equation, with rainfall data and vegetation density (NDVI) as a differentiator between the two types of data. The results obtained show that the average erosion that occurred in the Bodri watershed in 2018 was 250.21 tonnes/ha/year with the highest erosion of 18,982.01 tonnes/ha/year in a small part of the upper Bodri Hilir sub-watershed, while the average erosion in 2023 was 248.21 tonnes/ha/year with the highest erosion distribution of 17. 446.39 tonnes/ha/year spread over a small part of the Bodri Hilir sub-watershed in the upstream and central parts, these results indicate that there is a decrease in erosion rates between 2018 and 2023, the main factor in reducing the amount of erosion rate is the density of vegetation where in 2023 there is an increase in vegetation density in the dense vegetation class which affects the amount of erosion in the Bodri watershed, these results indicate that conservation measures taken by related agencies are quite effective in reducing erosion rates in the Bodri watershed.

1. Introduction

Indonesia is an archipelago that is in a tropical climate, this results in Indonesia having two seasons, namely the rainy season and summer, the tropical climate in Indonesia allows Indonesia to have a complex river flow system from this flow system then formed a watershed, according to the Minister of Forestry Regulation (Permen) No. 42 of 2009 concerning general patterns, criteria and standards for integrated watershed management [1], A watershed is a land area that connects the main river with its tributaries, and functions to collect, store, and channel rainwater naturally towards reservoirs such as lakes or seas. Generally, a watershed is bounded by a topographic arrangement influenced by terrestrial activities [2].

Indonesia has a relatively large number of watersheds, from the many watersheds that exist of course have a high potential for damage, this is due to several factors including due to human activities or natural factors such as rainfall and slope which results in a watershed has the potential for damage [3]. Based on data from the Ministry of Environment and Forestry, Indonesia has at least 17,076 watersheds, of which there are 2145 watersheds that need to be restored [4]. In the arrangement of the 4th National Medium-Term Development Plan (RPJMN IV) for 2020-2024, at least 108 watersheds are in critical condition and 15 of them have a high priority scale to be addressed immediately [5]. the restoration of conditions and handling of critical watersheds must be realised immediately because it can cause more serious impacts on a large area coverage.

The criticality of a watershed can be influenced by several factors including the high rate of surface run off, soil erosion and transport of sediment material from upstream, but in fact erosion is one of the main factors that cause a watershed to be on the verge of critical. this is because erosion can result in the degradation of agricultural land in the upstream while for the middle to downstream areas of the River, will experience the process of increasing sediment material which in the long term will result in damage to aquatic ecosystems, decreased water quality, sustainability of power plants and reduced reservoir performance [4]. To overcome the erosion problem on the island of Java alone, at least the government must spend a budget of $400 million each year [6], this figure is likely to continue to increase if the erosion problem is not addressed, therefore there needs to be a serious handling step to overcome this, one step that can be done is to analyse the distribution and magnitude of erosion that occurs in an area temporally.

Bodri watershed is located at 6° 50' 28" - 7° 15' 47" South Lattitude and 109° 56' 12" - 110° 20' 37" East Longitude. Bodri watershed is one of the watersheds that has a relatively high erosion rate so that action needs to be taken. Bodri watershed is a watershed that is part of the watershed management unit SWP Bodri watershed, with an area of 65,247.48 ha, Bodri watershed has a circumference of 171 km long and flowed by Bodri River as the main river along 80.73 km. Bodri watershed is located in the northern part of Java Island and crosses 4 regencies namely Kendal Regency, Temanggung Regency, Semarang Regency and Wonosobo Regency. Bodri watershed is included in the administrative area of Bodri Kuto PSDA, where the north serayu mountain or sindoro mountain is the upstream area of the Bodri River and the downstream Bodri sub-watershed becomes the output of the big Bodri River. Based on the calculation of sedimentation by the Department of Public Works Water Resources and Public Works of Central Java Province in 2023 there is at least 390,811.95 tonnes of sediment in a period of one year [7], this indicates that the rate of erosion that occurs in the Bodri watershed is quite large, land use change and fluctuations in rainfall can be one indicator of increased erosion and sedimentation in the Bodri watershed.

Land use change and fluctuations in rainfall that occurred between 2018 and 2023 in the Bodri watershed experienced significant changes, this could indicate changes in the amount of erosion that occurred in this area so that researchers felt the need to conduct research whose focus was to map and analyse the extent and amount of erosion that occurred in the Bodri watershed. This study aims to determine the rate of erosion that occurs in the Bodri watershed temporally between 2018 and 2023 and determine the factors that affect erosion in this watershed. The method that will be used in this research is the RUSLE (Revised Universal Soil Loss Equation) method.

2. Study Area and Data

2.1 Study Area

The research was conducted in Bodri watershed located at coordinates 6°50'28'-7°15'47' South Latitude and 109°56'12'- 110°20'37' East Longitude. Bodri watershed is located in 4 different districts namely Wonosobo Regency, Kendal Regency, Temanggung Regency, and Semarang Regency. Bodri watershed consists of 4 sub-watersheds namely Bodri Hilir sub-watershed, Logung sub-watershed, Lutut sub-watershed, and Putih sub-watershed. a spatial overview of the Bodri watershed is illustrated in Figure 1.

IOP Conf. Series: Earth and Environmental Science **1503** (2025) 012015 doi:10.1088/1755-1315/1503/1/012015

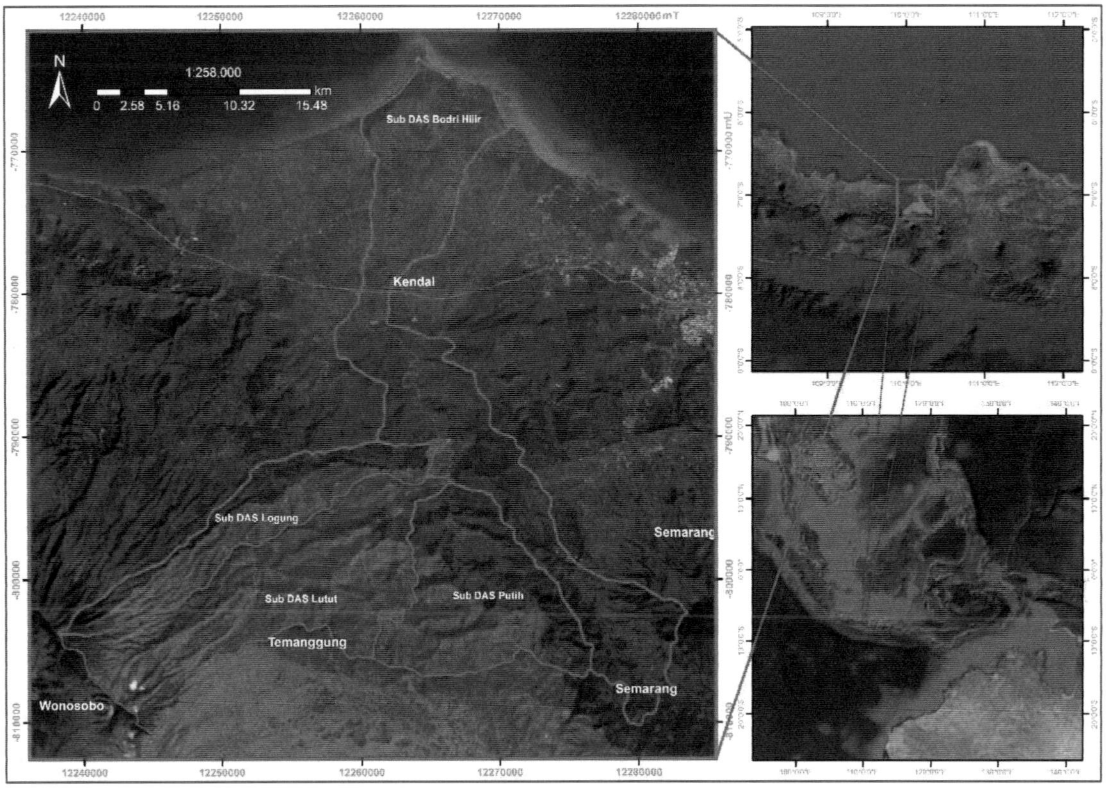

Figure 1. Map of Study Location

2.2 Data

For RUSLE data, in this study requires image processing, specific data sources and tools to process existing data, to be able to provide accurate calculation results. as for the materials and instruments needed in this study are in table 1.

Table 1. Tools and Materials

Tools & Materials	Source	Function
Data Vector of Watershed Boundaries in Indonesia	www.lapakgis.com PEPDAS 2018	As a research ROI (Region of interest)
Rainfall Data in Bodri Watershed	PUSDATARU Central Java Province	As a reference for rainfall data to obtain erosivity (R) value in 2018 and 2023
Vector Data of Java Soil Types	www.lapakgis.com Ministry of Agriculture	As data to determine the erodibility parameter (k) based on soil type at study site
Sentinel 2A/B Imagery	Copernicus Browser	As data to obtain vegetation density classification in 2018 and 2023
DEMNAS	Ina-Geoportal	To analyse level of slope factor (LS) at the sttudy site

3. Methods

3.1. Data Processing
this study aims to analyse the rate of change in erosion between 2018 and 2023 and the factors that influence the rate of change in erosion, using the RUSLE equation. the study area in this study is the Bodri watershed in Central Java province. the observation variables used in this study are variables of erosivity, erodibility, length and slope, crop management techniques, and conservation measures. these data will later be used to calculate the rate of erosion that occurs. The investigation in this study has a period of five years starting from 2018 and ending in 2023. the results of this study will later be disseminated to the general public, especially for people who live in the entire Bodri watershed area. this research is expected to be a guide for officials or the community to overcome erosion problems in the Bodri watershed and other watersheds in general.

3.2. Pre-processing
The Sentinel 2 A/B imagery used has been radiometrically and geometrically corrected so that the imagery used does'nt require any correction process. Sentinel 2 A/B must first be processed to merge separate images of the study area using the image mosaic function, after merging the image covered by clouds, the cloud removal process is performed by masking clouds to patch cloudy areas with images acquired at relatively close times to avoid distortion.

3.2.1. Image Mosaic
Image mosaic is a method or technique used to combine two or more images into one complete image, provided that each image has overlapping parts. each image has a part that overlaps so that it can be combined into a single image, so that they can be combined into a single image [8].

3.2.2. Cloud Masking
Cloud masking is a process to detect and reduce cloud cover in an image [9]. Cloud masking is performed when a study area is covered by clouds. because the presence of clouds can cause a reduction in the level of information detail of an image processing result. The working mechanism of cloud masking is to replace the cloudy part of the image with an image that is not covered by clouds. The cloud-free image must be a recognised image in terms of the requirements for the image used for patching [23]. The patches must be images acquired at close times to minimise the distortion of differences in cloud use. To minimise the distortion of differences in land use between the time the image is used and the time the image is used for patching.

3.3. Erosivity Processing
Erosivity is the sum of the average accumulation of monthly/annual rainfall at a study site which is used to determine the value I[10]. The Bols (1978)[11] equation used to analyse rainfall data and derive annual erosivity values from the rainfall data. Rainfall data and to derive annual erosivity values from each rainfall station using the following equation :

$$R = 6.119\,(CH_b)^{1.21}(HH_b)^{-0.47}(CH_{24})^{0.53} \tag{1}$$

Description :
R = Erosivity
CH_b = Monthly Rainfall (cm)
HH_b = Number of Rainy Days
CH_{24} = Maximum 24-Hour Rainfall (cm)

3.4. Erodibility Processing

Soil erodibility is the sensitivity of a soil erosion based on the type or components of which it is composed. Based on the nature or components of its composition to soil erosion, characterised by the difficulty or ease with which a soil is eroded by rainwater run-off or surface flow. Characterised by the difficulty or ease with which a soil is eroded by rainwater runoff or surface flow [12]. In this study, the erodibility value is based on the soil type at the study site, with a score value taken from the erodibility value issued by the Bogor Irrigation Research and Development Centre.

3.5. Length and Slope Data Processing

Erosion will generally increase if a slope has a certain length and steepness. This is because the longer and steeper the slope, the faster surface runoff, which causes material detachment[13]. The LS factor in this study employs a scoring system, with a scoring table sourced from the Land Rehabilitation and Soil Conservation Agency (1986)[14]. In the scoring table used, the scores are assigned based on five classes, each of which has a different slope percentage. The use of this scoring system is motivated by the output of this processing, where the results obtained are in vector data form, making scoring the preferred method for proceeding to the next data processing stage, which is the overlay phase.

Table 2. Slope Steepness Class Score Based on Land Rehabilitation and Soil Conservation Agency

Slope Classes	Slope Steepness	LS Value
A	0 - 5	0.25
B	5 - 15	1.20
C	15 - 35	4.25
D	35 - 50	9.50
E	>50	12.00

3.6. Vegetation Density Data Processing

The vegetation index or NDVI (Normalised Difference Vegetation Index) is an indicator of the density, greenness and condition of vegetation in an area [15]. NDVI is a mathematical combination of band 4 (red) and band 8 (Near Infrared)[16]. In this study, vegetation density data is used to calculate the parameters of the crop management system and conservation measures, the NDVI algorithm is used to analyse the level of vegetation density between 2018 and 2023, so that it can be known the difference in vegetation density of the study area temporally, The equation used in the calculation of vegetation density is :

$$NDVI = \frac{NIR-RED}{NIR+RED} \tag{2}$$

Description :
NDVI = Normalized Difference Vegetation Index
NIR = Near Infra Red (Band 8)
RED = Red (Band 4)

IOP Conf. Series: Earth and Environmental Science **1503** (2025) 012015 doi:10.1088/1755-1315/1503/1/012015

3.7. RUSLE (Revised Universal Soil Loss Equation)

RUSLE is a formula developed by Wischmeier (1978)[17], this equation is used to calculate the erosion rate at a research site using the formula :

$$E = R \times K \times Ls \times CP \tag{3}$$

Description :

E = Erosion
R = Erosivity
K = Erodibility
Ls = Length and Slope
CP = Soil Cover by Plants and Conservation Measures

The USDA (United States Department of Agriculture) classifies the level of erosion hazard based on the amount of erosion that occurs in tons/ha/year [10]. The erosion hazard classification by the USDA is as follows :

Table 3. Erosion Hazard Class

Erosion Hazard Class	Erosion Rate (ton/ha/Year)	Explanation
1	<15	Very Light
2	15 – 60	Light
3	60 – 180	Moderate
4	180 – 480	Heavy
5	>480	Very Heavy

3.8. Research Flowchart

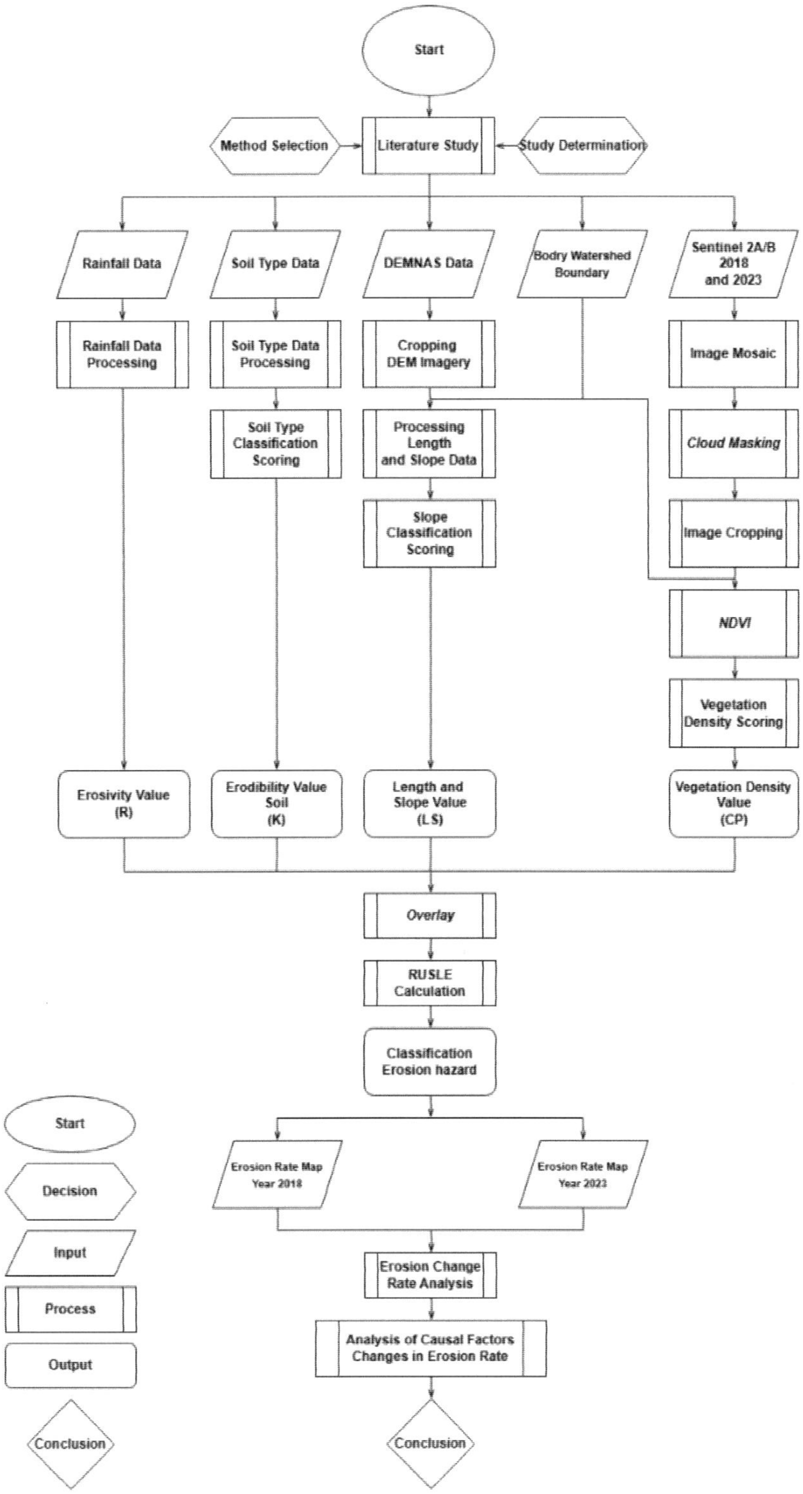

Figure 2. Research Flowchart

IOP Conf. Series: Earth and Environmental Science **1503** (2025) 012015 doi:10.1088/1755-1315/1503/1/012015

4. Result and Discussion

4.1. Erosivity I

The erosivity map was processed using the Thiessen polygon method to determine the area of rainfall distribution within the study location[18]. The calculations performed using the Bols equation for each rainfall station yielded values as shown in Figure 3.

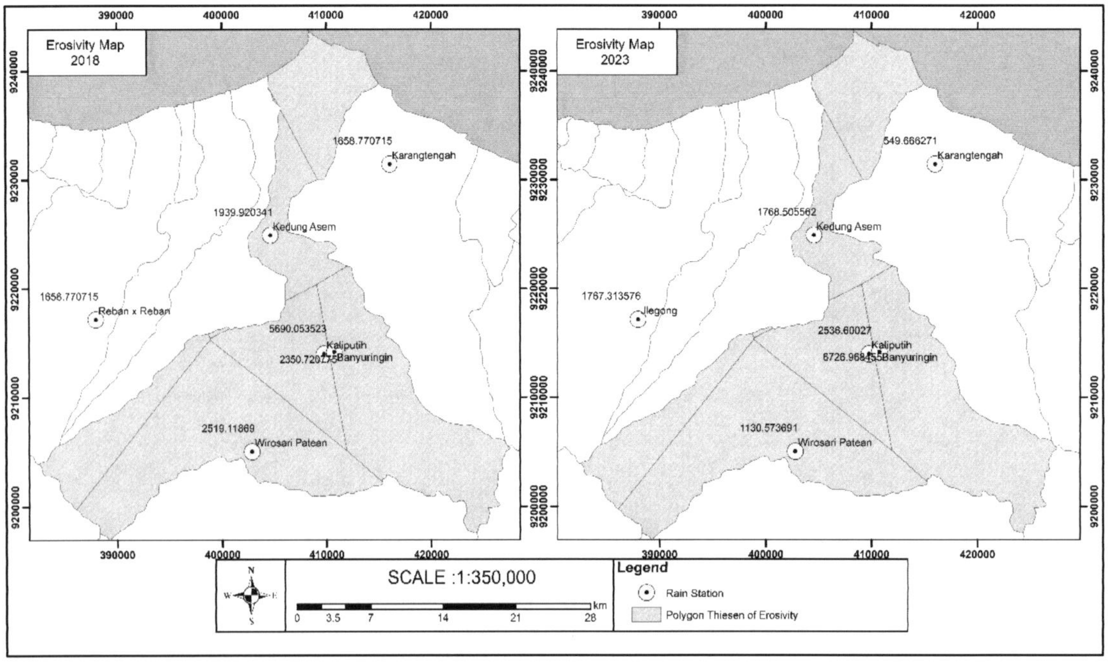

Figure 3. (a) Polygon Thiesen of Erosivity Value 2018 (b) Polygon Thiesen of Erosivity Value 2023

Based on the map, there are at least five rainfall stations in the Bodri Watershed (DAS Bodri). The map illustrates the distribution of each rainfall station's location, which was derived using the Thiessen polygon method to determine the rainfall coverage area for each station. Detailed erosivity values are presented in Table 4.

Table 4. Bodri Watershed Erosivity

Rain Station	2018 Erosivity	2023 Erosivity
Kedung Asem	1939,92	1768,506
Kaliputih	2350,721	2536,6
Banyuringin	5690,054	6726,968
Wirosari Patean	2519,119	1130,574
Karangtengah	1658,771	549,6663
Reban/Jlegong	1658,771	1767,314
Average	**2636,226**	**2413,271**

The analysis shows that the erosivity values in the Bodri Watershed (DAS Bodri) between 2018 and 2023, observed at five rainfall stations distributed across the watershed, exhibit differences. In 2018, the average erosivity value across the stations was 2636,226, with the highest value recorded at the Banyuringin station at 5690,054, and the lowest values at the

Reban/Jlegong and Karangtengah stations at 1658,771. By 2023, the average erosivity value decreased to 2413,271, with the highest value still recorded at the Banyuringin station and the lowest at the Karangtengah station at 549.666. This decline may indicate changes in erosion patterns within the Bodri Watershed between 2018 and 2023.

4.2. Erodibility (K)

The Bodri Watershed (DAS Bodri) exhibits a diverse range of soil types. According to data obtained from the Ministry of Agriculture, the watershed encompasses at least 10 different soil types. The distribution of these soil types is illustrated in Figure 4.

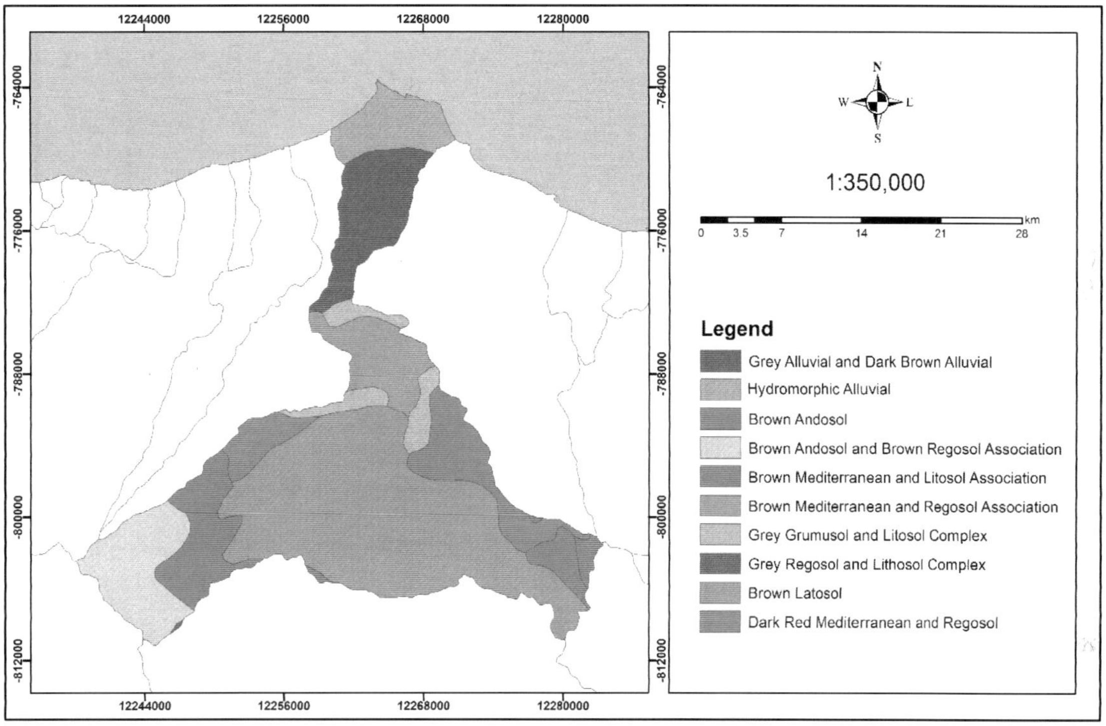

Figure 4. Soil Type in Bodri Watershed

Based on the illustration in Figure 4, it can be observed that the 10 soil types distributed across the Bodri Watershed are, in sequence: gray alluvial and grayish-brown alluvial; hydromorphic alluvial; brown andosol; an association of brown andosol and brown regosol; an association of brown Mediterranean soil and lithosol; an association of brown Mediterranean soil and regosol; a complex of gray grumusol and lithosol; a complex of gray regosol and lithosol; brown latosol; and red Mediterranean soil and regosol. The area covered by each soil type varies, with brown latosol having the largest area compared to the other soil types. Table 5 provides information about the erodibility values of each soil type, based on data released by the PUSLITBANG Irrigation Research Center in Bogor, 1985, In Table 4, the erodibility values of each soil type show different scores, with the highest erodibility found in brown andosol soil at a score of 0.273, while the lowest erodibility is observed in hydromorphic alluvial soil with a score of 0.156.

Table 5. K Factors In Each Soil Type In Study Area Based On PUSLITBANG Pengairan Bogor, 1985

Soil Type	K Factor
Hydromorphic Alluvial	0.156
Grey Alluvial and Dark Brown Alluvial	0.193
Brown Andosol	0.278
Brown Andosol and Brown Regosol Association	0.271
Brown Mediterranean and Regosol Association	0.273
Brown Mediterranean and Litosol Association	0.273
Grey Grumusol and Litosol Complex	0.187
Grey Regosol and Lithosol Complex	0.172
Brown Latosol	0.175
Dark Red Mediterranean and Regosol	0.188

4.3. Length and Slope (LS)

Figure 5 illustrates the slope gradient and slope length conditions in the Bodri Watershed after processing DEM data into a slope gradient map. The resulting data were analyzed and scored to determine the values for the LS parameter. A visual representation of the slope gradient can be observed in the following illustration.

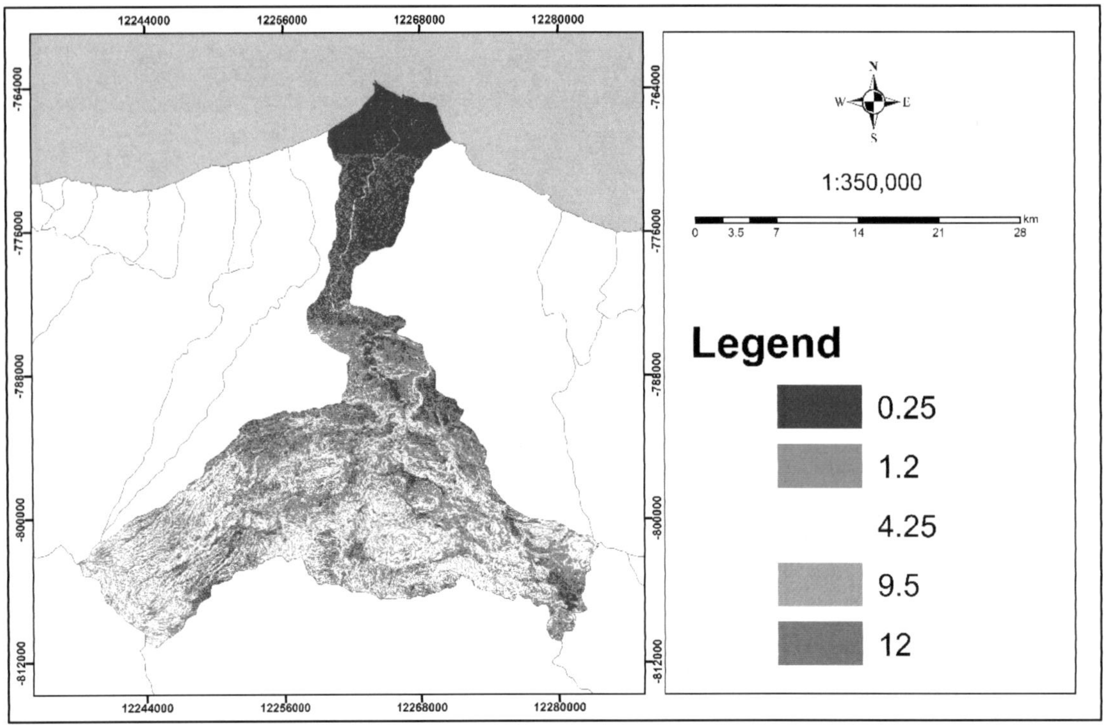

Figure 5. Map of LS Factor

Based on the processed slope and slope length data illustrated in Figure 5, it can be observed that, generally, the upper reaches of the Bodri Watershed have steep slopes with long slopes, which may indicate a high rate of surface runoff that could lead to erosion. In contrast, the middle to lower reaches feature relatively gentle to very gentle slopes, which suggests a reduction in surface runoff, resulting in a decrease in erosion rates. Visually, areas with moderate to high slope gradients, represented by yellow to red colors, dominate almost the entire Bodri Watershed. This could potentially result in a high erosion rate if calculated cumulatively.

4.4. Vegetation Density (CP)

The vegetation density in the Bodri Watershed, as illustrated in Figure 6, was obtained from the processing of Sentinel 2A/B imagery through NDVI visualization. This provided results showing the distribution of vegetation density between 2018 and 2023. Visually, the vegetation density in the Bodri Watershed can be observed in the following illustration.

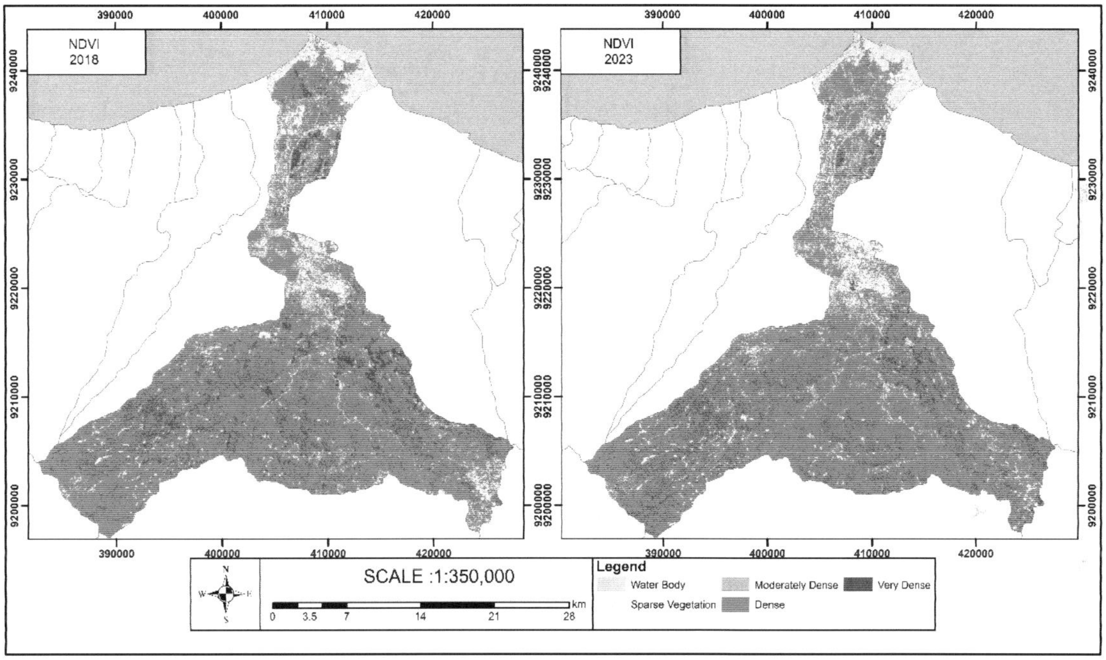

Figure 6. (a) NDVI 2018, (b) NDVI 2023

Table 6. The difference in the area of vegetation density classes.

Vegetation Density Class	Area (ha)		Area Difference
	2018	2023	
Water Body	807.6721	1000.577	192.905
Sparse Vegetation	4936.044	5176.574	240.53
Moderately Dense	5089.485	4935.334	-154.1512
Dense	48741.49	49333.06	**591.568**
Very Dense	5672.791	4801.94	-870.8518
Total Area	**65247.48**	**65247.48**	

The results of the NDVI processing in the Bodri Watershed are divided into five classes: water body, sparse vegetation, moderately dense, dense, and very dense vegetation. The area of each class can be seen in Table 5. In 2018, the water body class covered 807.6721 ha, while in 2023, it increased to 1000.577 ha. The sparse vegetation class in 2018 covered 4936.044 ha, and in 2023, it increased to 5176.574 ha, showing a difference of 240.53 ha. The moderately dense vegetation class in 2018 covered 5089.485 ha, but in 2023, it decreased to 4935.334 ha, with a reduction of 154.1512 ha. The dense vegetation class in 2018 covered 48741.49 ha, which increased to 49333.06 ha in 2023, showing an increase of 591.568 ha, making it the class with the highest area increase compared to the others. Finally, the very dense vegetation class had an area of 5672.791 ha in 2018, which decreased to 4801.94 ha in 2023, with a difference of 870.8518 ha. This decrease indicates a significant land-use change, especially in areas with very dense vegetation, which could be caused by several factors, including land-use changes, deforestation, and natural causes such as the prolonged dry season that occurred between 2018 and 2023, leading to a reduction in vegetation in the very dense class. This result provides a brief overview of the changes in erosion rates due to changes in vegetation density levels in the Bodri Watershed.

Based on previous research conducted in the Bodri Watershed to analyze land use changes in the study area from 2016 to 2020, it was found that the forest land cover class decreased by 38.039 km^2, while the open land cover class increased to 28.442 km^2[19], This study proves that the very dense vegetation class has consistently decreased from 2018 to 2023. This is in line with the results of the NDVI processing, which shows a decrease in the area of the very dense vegetation class.

4.5. RUSLE (Revised Universal Soil Loss Equation)

The results of the processing for each parameter to obtain the erosion rate using the RUSLE method, namely the RKLSCP parameters, were combined into a single map to then analyze the rate of erosion change and assign erosion classes based on the quantity of erosion that occurred. Figure 7 represents the result of merging all parameters into one complete map, providing a clear depiction of the distribution of areas with varying levels of erosion hazard in each region of the Bodri Watershed.

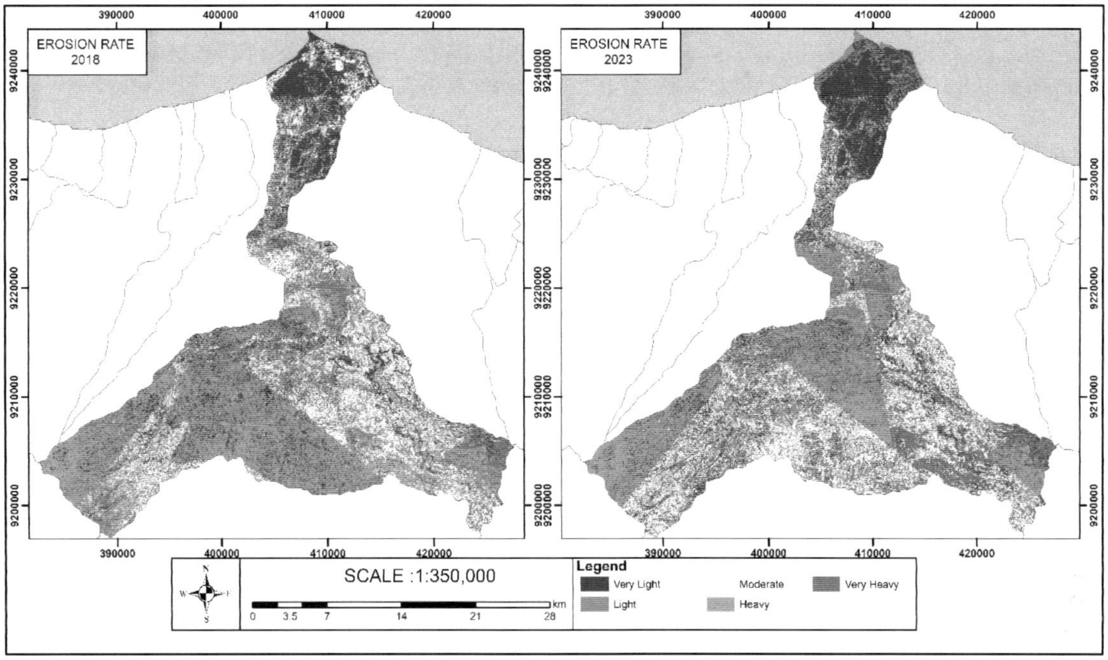

Figure 7. (a) Erosion Rate 2018, (b) Erosion Rate 2023

From the processing results, it is visually apparent that there are differences in the erosion rate between 2018 and 2023. Spatially, the distribution of areas with erosion rates falling into the moderate, heavy, and very heavy erosion hazard classes in 2018 was predominantly in the upper part of the Bodri Watershed, with the average erosion hazard in these classes being found in the Logung Sub-Watershed and the Lutut Sub-Watershed. This occurred because, in the Logung and Lutut Sub-Watersheds, the average slope and slope length were categorized as high, allowing for significant erosion accumulation in both Sub-Watersheds. This pattern also occurred in the erosion rates of 2023, but with a difference: in 2023, the erosion hazard levels in the Logung and Lutut Sub-Watersheds experienced a decrease. While in 2018, these areas were categorized under the heavy erosion hazard class, in 2023, they were classified under the moderate erosion hazard class. Overall, based on the illustration in Figure 7, the erosion occurring in the Bodri Watershed is dominated by areas in the moderate to very heavy erosion hazard classes. The area sizes for each erosion hazard class shown in Figure 7 can be seen in Table 6.

Table 7. The area of each erosion hazard class.

Erosion Hazard Class	**Area (ha)**		**Area Difference**
	2018	**2023**	
Very Light	12933.31	13990.71	1057.4
Light	15505.94	17850.59	2344.65
Moderate	12872.43	15398.69	2526.25
Heavy	18895.5	9402.59	-9492.908
Very Heavy	5040.303	8604.911	3564.61

The table shows the area of each erosion hazard class, where in the very low hazard class in 2018, the area was 12,933.31 ha, and in 2023, it increased to 13,990.71 ha, with an increase of 1,057.4 ha. In the low hazard class, the area in 2018 was 15,505.94 ha, while in 2023, it increased to 17,850.59 ha, with a difference of 2,344.65 ha. In the moderate hazard class, the coverage area

IOP Conf. Series: Earth and Environmental Science **1503** (2025) 012015 doi:10.1088/1755-1315/1503/1/012015

in 2018 was 125,398.69 ha, while in 2023, it increased to 15,398.69 ha, rising by 2,526.25 ha from the previous year. In the high hazard class, the area in 2018 was 18,895.5 ha and significantly decreased to 9,402.59 ha, showing the highest decrease, with a difference of 9,492.91 ha between 2018 and 2023. Finally, in the very high hazard class, the area in 2018 was 5,040.303 ha and increased to 8,604.911 ha in 2023, with an increase of 3,564.61 ha, making the very high hazard class the class with the highest area increase compared to the others.

The results of this study are contradictory to previous research that used the SES (Soil Erosion Status) method, where the research conducted in the Mrica Dam showed that erosion, divided into three classes slight, moderate, and severe, had the following percentages : slight erosion had a percentage of 1.8%, moderate erosion had a percentage of 70.2%, and severe erosion had a percentage of 27.9%. This differs from the study conducted by the author, where the class of very severe erosion actually showed an increase in area, indicating a rise in the erosion rate for the very severe class[20][24].

In this study, there are areas with the highest erosion rates spread across several parts of the Bodri watershed. Figure 8 illustrates the distribution of areas with the highest erosion rates that occurred in both 2018 and 2023.

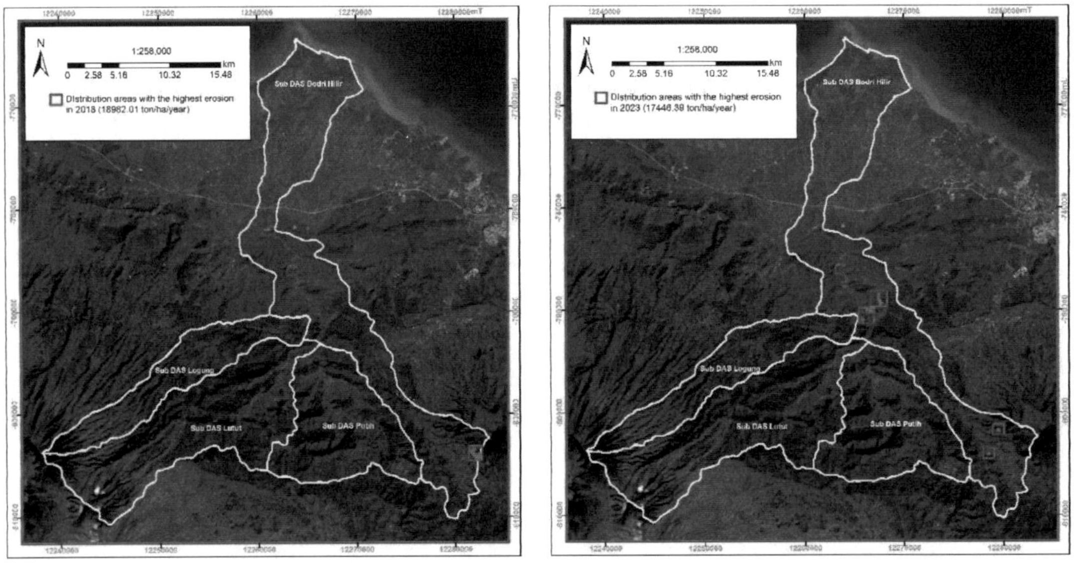

Figure 8. (a) Distribution of The Highest Erosion in 2018, (b) Distribution of The Highest Erosion in 2023

Based on the results of the overlay analysis conducted to obtain data on the distribution of accumulated erosion quantities in each erosion hazard class, the findings provide an overview of the erosion magnitude in tons/ha/year. These results allow for identifying areas with the highest erosion rates within the study location. Figure 8 illustrates the distribution of areas with the highest erosion rates in the Bodri Watershed. Both maps show that the highest erosion rate is located in the Lower Bodri Sub-Watershed. In 2018, the highest erosion area was concentrated in a single point in the upper part of the Lower Bodri Sub-Watershed, with an erosion quantity of 18,982.01 tons/ha/year. Meanwhile, in 2023, the highest erosion spread across the upper and central parts of the Lower Bodri Sub-Watershed, with an erosion quantity of 17,446.39 tons/ha/year. The distribution of high erosion, compared to the quantity of erosion distribution in the Bodri Watershed in 2018, shows a change. In 2018, the highest erosion had a centralized pattern, whereas in 2023, it spread across two areas. This occurred because, in 2023, the distribution of erosion classified as severe was relatively concentrated in the upstream and middle regions of the Bodri Hilir sub-watershed, allowing the peak erosion to spread within the Bodri Hilir sub-watershed. An illustration of this distribution can be seen in Figure 7. These results indicate that the erosion rate in the Bodri Watershed has decreased. The calculations of the total erosion quantity in the Bodri Watershed show that the average erosion in the Bodri Watershed in 2018 was 250.21 tons/ha/year, while in 2023, it was 248.21 tons/ha/year. These figures were derived from the accumulated average erosion in the Bodri Watershed as a whole, ranging from the lowest to the highest values. This figure indicates that erosion in the Bodri Watershed experienced significant fluctuations, with the average erosion in 2017 recorded at 169.47 tons/ha/year[21]. These findings contradict previous research conducted in the Garang Watershed, where erosion rates showed an increase. In 2013, the erosion rate was 200.343 tons/ha/year, while in 2018, it increased by 77.304 tons/ha/year, reaching 277.647 tons/ha/year. The results of that study indicate that the increase in erosion rate in the Garang Watershed was caused by significant land-use changes, which contributed to the higher erosion rates. [10]. The same phenomenon was observed in a study conducted in 2018 in the Bedadung and Tanggul Watersheds, where land-use changes were a significant factor influencing the changes in erosion rates in these watersheds. [22]

5. Conclusions

This study aims to analyze the changes in erosion rates in the Bodri Watershed from 2018 to 2023 using a Geographic Information System (GIS) approach and the Revised Universal Soil Loss Equation (RUSLE) method. The Bodri Watershed, which covers an area of 65,247.48 hectares in Central Java, has high erosion rates due to various factors, such as land-use changes, vegetation density, rainfall, and land slope. The results show that the average erosion rate in the Bodri Watershed in 2018 was 250.21 tons/ha/year, while in 2023, it decreased to 248.21 tons/ha/year. Although the difference appears small, there was a significant spatial shift in the distribution of erosion levels.

The main factors influencing changes in erosion rates are vegetation density, rainfall, and land slope. NDVI (Normalized Difference Vegetation Index) analysis shows that the area of dense vegetation increased by 591.57 hectares, but the area of very dense vegetation decreased by 870.85 hectares, caused by human activity and natural conditions such as prolonged dry seasons. The erosivity value also decreased on average from 2636.23 in 2018 to 2413.27 in 2023, contributing to the reduction in erosion rates. Additionally, the upstream area of the Bodri Watershed, with steep slopes, tends to experience higher erosion compared to the middle and downstream areas, which have gentler slopes.

The distribution of erosion levels shows a significant change between 2018 and 2023. In 2018, the highest erosion levels were concentrated in the upstream area of the Bodri Hilir Sub-Watershed with a maximum value of 18,982.01 tons/ha/year. In 2023, the highest erosion levels spread to the middle and upstream areas of the Bodri Hilir Sub-Watershed with a maximum value of 17,446.39 tons/ha/year. The erosion hazard classification also showed an increase in the area of light and moderate hazard classes by 2344.65 hectares and 2526.25 hectares, respectively, while the heavy hazard class decreased significantly by 9492.91 hectares. This indicates the effectiveness of conservation measures implemented in certain areas.

Conservation efforts, such as increasing vegetation density in the dense vegetation class, have proven effective in reducing erosion rates in the Bodri Watershed. However, the reduction in very dense vegetation requires further attention to prevent land degradation. Therefore, strengthening reforestation programs and protecting vegetation in critical areas are necessary. Continuous monitoring with GIS technology and public education on the importance of soil and water conservation should also be enhanced. Overall, this study successfully provides an overview of the changes in erosion rates in the Bodri Watershed, the factors behind them, and can serve as a reference for more effective watershed management and conservation in the future.

Acknowledgement

I would like to express my gratitude to the entire organizing committee of the International Conference of Geography and Disaster Management (ICGDM) for providing me with the opportunity to publish this research. This program not only serves as a platform for geography researchers but also as a medium for geography enthusiasts to expand their knowledge and deepen their understanding of geography.

References

[1] R. H. Prakoswa, 2018, "Ribuan DAS Rusak, Negara Rugi Ratusan Triliun Tiap Tahun," CNBC Indonesia [Online]. Available: https://www.cnbcindonesia.com/news/20180325172807-4-8485/ribuan-das-rusak-negara-rugi-ratusan-triliun-tiap-tahun

[2] P. D. Susanti and A. Miardini, 2017, "The impact of Land use Change on Water Pollution Index of Kali Madiun Sub-watershed," *Forum Geografi*, vol. 31, no. 1, pp. 128–137, 2017, doi: 10.23917/forgeo.v31i1.2686.

[3] Poerbandono, A. Basyar, A. B. Harto, and P. Rallyanti, 2006, "Evaluasi Perubahan Perilaku Erosi Daerah Aliran Sungai Citarum Hulu dengan Pemodelan Spasial," *J u r n a l Infrastruktur dan Lingkung Binaan*, vol. 2, no. 2, pp. 21–28.

[4] B. H. Santosa, 2024, "Ekohidrologi Untuk Restorasi DAS," Detik News, Mar. 22, 2024. [Online]. Available: https://news.detik.com/kolom/d-7254199/ekohidrologi-untuk-restorasi-das

[5] S. I. Pratama, 2021, "Hanya 108 DAS yang Diprioritaskan untuk Dipulihkan hingga 2024," Betahita.id, 2021. [Online]. Available: https://betahita.id/news/lipsus/6953/hanya-108-das-yang-diprioritaskan-untuk-dipulihkan-hingga-2024.html?v=1645264997

[6] R. P. C. Morgan, 2005, "Soil Erosion & Conservation, Third Edit" Victoria: Blackwell Publishing.

[7] F. Y. Pramono, 2023, "Laporan Ringkas Pekerjaan Penyusunan Zonasi Sungai Bodri," Semarang, 2023.

[8] S. Rahman, U. Khair, D. Siregar, and H. Harahap, 2019, "Metode 8 Parameter Prespective Transformation Menggunakan Panoramic Image Mosaic," *Semin. Nas. Teknol. Komput. Sains*, pp. 606–609.

[9] L. Sun et al., 2017, "A cloud detection algorithm-generating method for remote sensing data at visible to short-wave infrared wavelengths," *ISPRS J. Photogramm. Remote Sens.*, vol. 124, pp. 70–88, 2017, doi: 10.1016/j.isprsjprs.2016.12.005.

[10] A. S. H. Huda, A. L. Nugraha, and N. Bashit, 2020, "Analisis Perubahan Laju Erosi Periode Tahun 2013 dan Tahun 2018 Berbasis Data Penginderaan Jauh dan Sistem Informasi Geografis (Studi Kasus : DAS Garang)," *J. Geod. Undip Januari*, vol. 9, no. 1, pp. 106–114, 2020, doi: https://doi.org/10.14710/jgundip.2020.26109.

[11] P. L. Bols,1978, "The Iso-erodent Map of Java and Madura," Soil Res: Bogor.

[12] F. Isma and M. Purwandito, 2019, "Estimasi Erosi Dan Sedimentasi Lahan Pada Das Langsa Berbasis Sistem Informasi Geografis (Sig)," *Teras J. J. Tek. Sipil*, vol. 9, no. 1, pp. 29. doi: 10.29103/tj.v9i1.169.

[13] Arsyad and Sitanala, 2010, Konservasi Tanah dan Air, Bogor: IPB Press.

[14] Balai Rehabilitasi Lahan dan Konservasi Tanah, 1986, "Petunjuk Pelaksanaan Penyusunan RTL-RLKT", Jakarta: Departemen Kehutanan RI.

[15] T. Lillesand, R. W. Kiefer, and J. Chipman, 2000, "Remote Sensing and Image Interpretation", Canada: John Wiley and Son.

[16] H. Z. Hadibasyir and N. S. Firdaus, 2022, "Effect of Vegetation and Building Densities To Urban Thermal Terhadap Kenyamanan Termal Perkotaan (Studi Kasus Kota Denpasar) Effect of Vegetation and Building Densities To Urban Thermal Comfort (Case Study of Denpasar City)," *J. Purifikasi*, vol. 21, no. No 1, pp. 11–19.

[17] A. Akbar, R. Adwia, I. H. Nurnawaty, and F. Gaffar, 2023, "Metode Rusle Dan Usle Di Das Tallo," *J. Tek. dan Sains Fak. Teknol. Lingkung. dan Miner. Univ. Teknol. Sumbawa*, vol. 4, no. 1, pp. 59–73.

[18] A. N. Anna et al., 2021, "Spatial Modelling of Local Flooding for Hazard Mitigation in Surakarta, Indonesia," Int. *J. GEOMATE*, vol. 21, no. 87, pp. 145–152, 2021, doi: 10.21660/2021.87.j2306.

[19] I. D. Tisnasuci, A. Sukmono, and F. Hadi, 2021, "Analisis Pengaruh Perubahan Tutupan Lahan Daerah Aliran Sungai Bodri terhadap Debit Puncak Menggunakan Metode Soil Conservation Service (SCS)," *J. Geod. Undip Januari*, vol. 10, no. 1, pp. 105–114.

[20] B. Harjadi, 2021, "Erosion Analysis in the Mrica Reservoir Catchment Area in Indonesia using the Soil Erosion Status Method," *Forum Geografi,* vol. 35, no. 2, pp. 116–129. doi: 10.23917/forgeo.v35i2.13943.

[21] Q. B. Islami, W. Setyaningsih, and Suroso, 2017, "Estimasi Erosi Di Daerah Aliran Sungai (Das) Bodri Menggunakan Metode Universal Soil Loss Equation (Usle)," *GeoImage,* vol. 6, no. 2, pp. 115–122.

[22] M. K. Rohman, 2018, "Pengaruh Perubahan Tata Guna Lahan Terhadap Laju Erosi Menggunakan Metode Universal Soil loss Equation (USLE) Di DAS Bedadung dan DAS Tanggul," Universitas Jember [Online]. Available: http://repository.unej.ac.id//handle/123456789/96274

[23] N. Jubit, T. Masron, A. Ahmad, & R. Soda, 2024, "Investigating the Spatial Rela-tion between Landuse and PropertyCrime in Kuching, Sarawak through Location Quotient Analysis" *Forum Geografi*, vol. 38, no. 2, pp. 153-166. doi: 10.23917/forgeo.v38i2.4575.

[24] A. Saloua, A. Ahmed, A. Abdellah, F. Abdelouahed, M. Said, A. Abdelfattah, E. G. Akram, E, 2024, "Modelling of Soil Erosion Susceptibility Using the Multi-In-fluencing Factor Method in the Amizmiz Basin, Morocco" *Forum Geografi*, vol. 38, no.3, pp. 343-357. doi: 10.23917/forgeo.v381i3.6263.

Influence of Weather and Climate on the Spatial Distribution of Forest Fires in Indonesia: A Bibliometric Study

Andung Bayu Sekaranom[1*], Emilya Nurjani[1], Utia Suarma[1], Giarno[2], Sayful Amri[2], Ahmad Fadlan[2], Asshaffa Naim[1], Aurelia Zerlinda[1], and Malinda Budi Oktaviani[1]

[1] Department of Environmental Geography, Faculty of Geography, Universitas Gadjah Mada, Bulaksumur 55281 Yogyakarta, Indonesia
[2] Meteorology, Climatology, and Geophysics College (STMKG), Agency for Meteorology, Climatology, and Geophysics (BMKG), Tangerang, Indonesia

*E-mail: andung.geo@ugm.ac.id

Abstract. Forest and land fires in Indonesia represent a major environmental challenge, frequently associated with climatic and atmospheric events such as the El Niño Southern Oscillation (ENSO). This study utilizes a bibliometric methodology to examine 1,312 Scopus-indexed publications from 2000 to 2024, concentrating on the spatial and thematic distribution of research concerning fire incidents in Indonesia. The study employs bibliometric analysis based on a set of keywords, including regions and a number of weather and climate drivers, and also Latent Dirichlet Allocation (LDA) modeling. The analysis identifies ENSO as the primary atmospheric influence, especially in Kalimantan and Sumatra, where its interaction with IOD intensifies fire hazards. Through the LDA modeling, five principal study issues were discerned: data-driven fire monitoring, peatland fires, community-based fire management, air pollution and haze, and ecosystem consequences. Although Sumatra and Kalimantan receive predominant academic focus, areas such as Papua, Sulawesi, and Nusa Tenggara are less investigated, particularly due to lower occurence of forest and land fires. Based on this result, recommendations involve incorporating meteorological data into fire mitigation techniques and intensifying attention on localized atmospheric factors are importamt to enhance adaptive management of forest and land fires in Indonesia.

1. Introduction

Forest and land fires are among the most severe disasters in Indonesia, particularly in regions frequently affected by such events, such as the islands of Kalimantan and Sumatra. In addition to land characteristics—such as dry surface conditions and peatlands—forest and land fires in Indonesia are also influenced by meteorological factors. One of the key identified factors is El Niño [1]. Reduced rainfall during El Niño events is often accompanied by human activities that can trigger forest and land fires, such as land-use conversion for settlements or production activities [2].

The direct impacts of forest and land fires to the atmosphere primarily (~70%) involve the loss of forest cover. The remaining impact is dominated by peatland loss, which significantly

IOP Conf. Series: Earth and Environmental Science **1503** (2025) 012016 doi:10.1088/1755-1315/1503/1/012016

contributes to the depletion of carbon storage in the land and its release into the atmosphere as greenhouse gases [2]. Consequently, forest and land fires are among the major contributing factors to global temperature rise and climate change [3]. This impact is particularly significant because forest and land fires not only release carbon stored in biomass but also affect peatlands, which store three to five times more carbon than surface biomass [4]. As a result, their contribution to global warming and climate change is substantial.

The Indonesian government's initiatives, such as the land moratorium, have significantly contributed to the annual reduction of forest and land fire occurrences [2]. However, the potential for forest and land fires remains, further exacerbated by the growing threats of global warming and climate change. Therefore, it is crucial to emphasize that the current paradigm in forest and land fire prevention is focused on preventive measures [5]. These efforts are supported by advancements in monitoring technology, including the use of satellite imagery and the Internet of Things (IoT). In terms of fire event prediction, artificial intelligence (AI) is anticipated to play a dominant role, with meteorological factors serving as key parameters [6]. In further developments, fire prediction models may also incorporate human (anthropogenic) factors, which can act as fire triggers, along with social engineering approaches to reduce fire risks [1].

In developing a forest and land fire prediction system, the influence of each meteorological aspect must be identified in detail to ensure the accuracy of the prediction system. Previous studies have examined the effects of temperature, rainfall, humidity, and several other parameters on the occurrence of forest and land fire hotspots. These influences can span interannual, seasonal, and intraannual scales [7, 8]. In the long term, global warming and climate change are predicted to be key determinants of forest and land fire risk. Extreme conditions that rarely occur, such as droughts caused by the El Niño-Southern Oscillation (ENSO) or positive Indian Ocean Dipole (IOD) events, can significantly impact fire occurrence [9]. In the shorter term, local factors play a more dominant role and often contribute to soil moisture deficit, which serves as one of the primary triggers for forest and land fires [10]. Weather and climate are essential factors influencing fire risk and behavior across several geographical and temporal dimensions [11, 12]. This knowledge might guide methods for adaptation, landscape planning, and compositional management to alleviate increased fire activity due to anticipated escalations in fire weather severity [8].

This paper examines the major issue of forest and land fires in Indonesia, focusing on the spatial distribution of fire occurrences, seasonal and temporal patterns, and their relationships with meteorological and climatic phenomena. This study is based on a bibliometric analysis that examines published databases for research publications on forest and land fires in Indonesia. Despite the increasing quantity of literature on the subject, numerous flaws remain, particularly in integrating diverse components into an integrated framework. Addressing these inadequacies is crucial for improving our understanding of the correlation between fire incidences and environmental factors, as well as for informing more effective fire management and mitigation strategies. The article seeks to summarize existing knowledge while offering new insights to enhance future strategies for managing and mitigating forest and land fires in Indonesia.

2. Research Method

This study utilizes a bibliometric analysis to examine Scopus-indexed publication data concerning forest and land fires in Indonesia. The analysis concentrates on articles including the keywords "forest fire" and "Indonesia" in their titles, keywords, or abstracts. The chosen timeline extends from 2000 to 2024, encompassing more than twenty years of research on this significant environmental concern. The publication are limited to research journal and proceeding. A thorough filtering method revealed 1,312 pertinent papers judged appropriate for further investigation. The following study explores the thematic and spatial aspects of the research in greater detail. Additional keywords were utilized to investigate the geographic locations of fire episodes (e.g., islands with reported fires), research themes (e.g., ecological consequences, mitigation approaches, socio-economic variables), meteorological factors (e.g., precipitation, temperature, humidity), and atmospheric dynamics (e.g., El Niño, wind patterns) as outlined in the abstracts of the chosen publications. This thorough approach seeks to offer a detailed overview of the scope and trends in forest fire research in Indonesia, emphasizing both the strengths and deficiencies in the current literature.

The bibliometric analysis was conducted using text processing in Python, focusing on the occurrence of specific words in abstracts and grouping words within abstracts into specific themes using the Latent Dirichlet Allocation (LDA) model. Latent Dirichlet Allocation (LDA) is a machine learning method employed for topic modeling, aimed at uncovering concealed themes or subjects within a collection of text documents. LDA is a probabilistic generative model that characterizes each document as a combination of many themes, with each topic including a blend of words. The keywords for the study area include the major islands of Indonesia: 'Sumatra', 'Java', 'Kalimantan/Borneo', 'Sulawesi/Celebes', 'Bali & Nusa Tenggara', and 'Papua'. Keywords for meteorological parameters encompass a range of terms, including 'weather', 'climate', 'meteorology', 'climatology', 'rainfall', 'temperature', 'humidity', and 'wind'. Keywords for atmospheric phenomena include 'atmospheric phenomena', 'El Niño Southern Oscillation', 'Indian Ocean Dipole', 'Madden-Julian Oscillation', 'Borneo Vortex', 'heat waves', and 'local atmospheric phenomena'. In the final stage, a cross-tabulation of keywords for location, meteorological parameters, and atmospheric phenomena was conducted to identify their associations with forest and land fires.

3. Results and Discussion

3.1. Spatial distribution of the research topic related to forest and land fires

The bibliometric study identifies specific patterns in research emphasis and temporal trends concerning forest and land fires in Indonesia. Based on the keywords addresing the major islands of Indonesia, several information has been obtained (Figure 1). The overall number of articles has consistently risen since 2010, indicating an escalating global and national concern for this significant environmental issue. Upon regional examination, Kalimantan and Sumatra are identified as the most extensively examined regions. This is unsurprising, considering their regular correlation with extensive forest and peatland fires, which have significant ecological and air quality consequences. Both islands exhibit substantial growth in research production, especially in the years subsequent to 2015. This timeframe coincides with significant El Niño events, recognized for intensifying fire incidents in certain areas [13].

Java exhibits a steady increase in publications, however its research output is still inferior to that of Kalimantan and Sumatra. This emphasis may arise from the island's elevated population density. Conversely, Sulawesi, Papua, and Nusa Tenggara exhibit significantly less studies, however their research trends indicate a progressive increase. This indicates an increasing, albeit still restricted, focus on these areas, which face distinct environmental and climatic difficulties. The temporal patterns indicate a significant rise in research activity from 2015 to 2020, perhaps influenced by global focus and local policy changes in reaction to catastrophic fire seasons. The

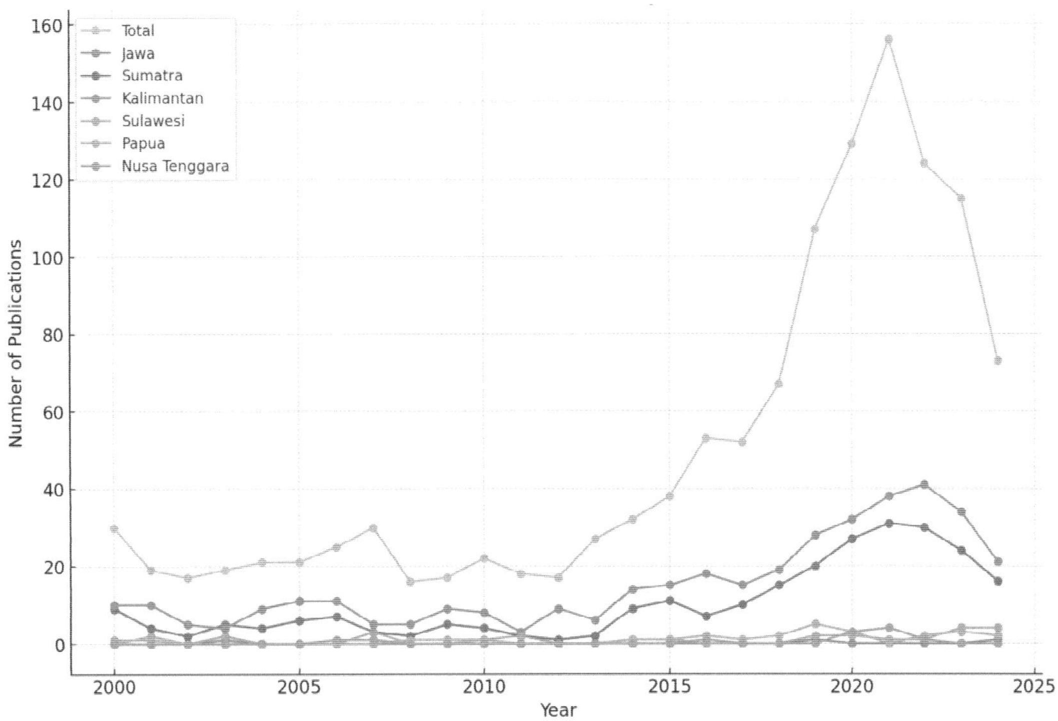

Figure 1. Publication trend and analysis based on the location of each research study (Source: Analysis Results, 2024)

comparatively diminished emphasis on lesser-explored islands like as Papua and Nusa Tenggara underscores the necessity for more equitable research initiatives to cover Indonesia's varied natural environments. It should be noted that not all publications explicitly specify the island or region of study, resulting in the number of publications for each location being lower than the total number of publications analyzed.

3.2. Spatial distribution of the research topic related to forest and land fires

The results of the Latent Dirichlet Allocation (LDA) modeling indicate that research related to forest fires can be grouped into five main topics, each reflecting a distinct focus and contribution to understanding and addressing forest fires (Table 1). The LDA analysis revealed a set of keywords classified into various broad concepts relevant to forest and land fire studies in Indonesia. The subjects covered include fire detection and monitoring, the significance of peat ecosystems and carbon sequestration, fire management and community engagement, environmental and health implications, as well as the ecological impact on biomass and biodiversity. The findings suggest that research on forest and land fires in Indonesia encompasses a wide range of disciplines, the complexity of fire dynamics, and the wider environmental and socioeconomic ramifications of these fires being brought to light.

The first topic, "Forest fires related data," highlights the use of data in studying forest fires, with an emphasis on technologies such as remote sensing and hotspot data. With 289 publications or approximately 22.03% of the total, this topic underscores the importance of data-driven monitoring to support early warning systems and impact mapping. The second topic, "Forest fires in peatland," emerges as the dominant topic with 375 publications (28.58%). The research in this

Table 1. The results of the Latent Dirichlet Allocation (LDA) modeling

No	Topic Title	LDA Keywords	Number of Publications	Percentage
1	forest fires related data	forest, fires, data, land, remote sensing, hotspot	289	22.03
2	forest fires in peatland	forest, peat, carbon, peatland	375	28.58
3	forest based community management	forest, management, community, local, conservation, development	138	10.52
4	forest fires impact on air pollution and haze	fires, forest, haze, pollution, air, health	279	21.27
5	forest fires, ecosystem, and biomass	forest, fires, areas, biomass, species, ecosystem	231	17.61

Source: Analysis Results (2024)

area primarily focuses on fires in peatlands, which have significant impacts on carbon emissions and climate change. The strong attention to this topic reflects the critical role of peatland conservation as a strategic mitigation measure. The third topic, "Forest-based community management," has fewer publications, with 138 (10.52%), but it offers a highly important approach through community-based forest management. This research highlights the role of local communities in sustainable conservation and resource management, which could provide a long-term solution for addressing forest fires. The fourth topic, "Forest fires impact on air pollution and haze," with 279 publications (21.27%), explores the impact of forest fires on air quality and public health. The haze produced by forest fires is often a national and international concern, making this research an important scientific foundation for better policy-making. The fifth topic, "Forest fires, ecosystem, and biomass," focuses on the impact of fires on ecosystems, including the loss of biomass and biodiversity. With 231 publications (17.61%), this topic emphasizes the ecological losses caused by forest fires, which affect species survival and ecosystem stability.

Overall, previous research offers a comprehensive view of various aspects of forest fires, ranging from monitoring technologies, peatland conservation, and health and environmental impacts to local community empowerment. However, there is an opportunity to strengthen research focusing on community-based management, given its relatively small contribution compared to other topics. Future research directions also encompass the establishment and implementation of collaborative governance models across different levels and situation, evaluating the long-term impact of management systems, and amending current legislation to enhance their efficacy in mitigating forest and land fires in Indonesia.

3.3. The role of meteorological variables

Prolonged climate patterns, including changes in precipitation and temperature trends, create conditions favourable to fires by withering vegetation and soil. Climate variability, particularly phenomena like El Niño, exacerbates these conditions, leading to prolonged dry spells that heighten fire risks. Short-term variations in temperature, humidity, and wind strongly impact the dynamics of fires. Arid, high-temperature environments create ideal conditions for fire ignition, yet sudden weather changes can either exacerbate or reduce fire activity.

Figure 2 depicts the number of articles addressing specific meteorological attributes related to forest and land fires, highlighting the substantial influence of these factors on fire behavior. "Temperature" is one of the most significant component, playing a large influence in drying out vegetation and increasing combustion processes. Elevated temperatures not only augment the probability of fire ignition but also exacerbate fire propagation and intensity. Similarly, "rainfall" has a twofold impact. Insufficient rainfall results in dry conditions and heightened fire vulnerability, whereas enough rainfall can mitigate flames and facilitate ecosystem restoration. Wind significantly contributes to the propagation of fires. Strong winds can transport embers over great distances, generating new outbreaks and widening fire-affected regions, making fire management measures more complex. Humidity substantially influences fire risk; low humidity levels desiccate vegetation and enhance flammability, while elevated humidity might mitigate fire activity by preserving moisture within the ecosystem.

3.4. The role of atmosphere related phenomena

The analysis of publication counts related to atmospheric dynamics reveals a clear dominance of the El Niño Southern Oscillation (ENSO) in scientific discourse, with 116 publications (Table 2). This overwhelming focus reflects the well-established influence of ENSO on global weather patterns, particularly its role in exacerbating forest and land fires. ENSO events, especially during the El Niño phase, are strongly associated with prolonged dry conditions in regions such as Indonesia, significantly heightening fire risks.

In comparison, the Indian Ocean Dipole (IOD) is mentioned in only 15 publications, indicating far less attention. While the IOD also impacts regional climate variability, particularly rainfall patterns in the Indian Ocean basin, its connection to forest and land fires appears to be less frequently studied than ENSO. The Madden-Julian Oscillation (MJO) is referenced in just one publication, reflecting minimal exploration of its influence. Despite its known role in modulating

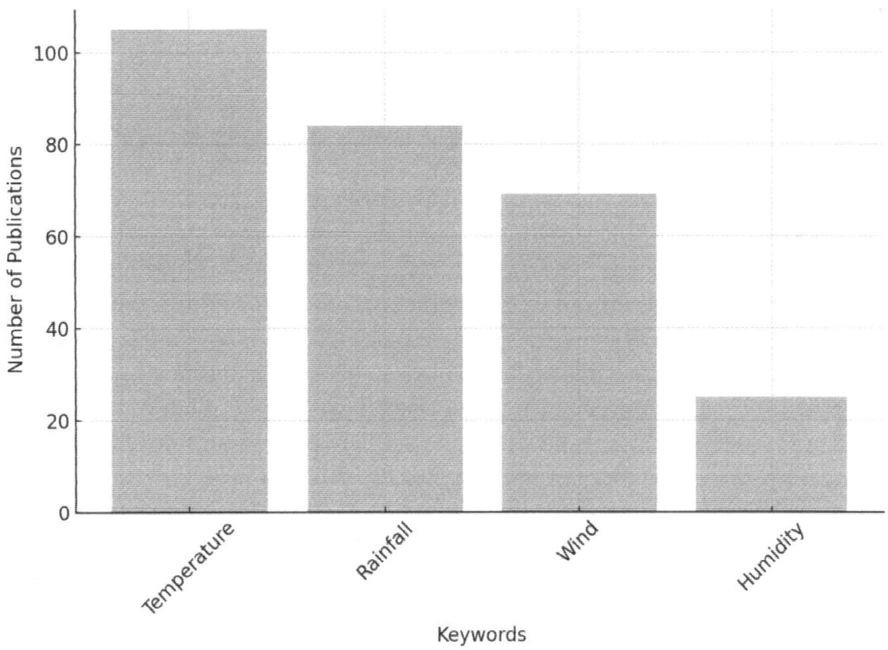

Figure 2. The role of meteorological variables in inducing forest and land fires based on bibliometrical analysis (Source: Analysis Results, 2024)

short-term weather patterns, such as tropical rainfall, its direct link to fire occurrence remains low.

Table 2. The analysis of publication counts related to atmospheric dynamics inducing forest fires

Atmospheric Dynamic	Publication Count
El-Nino Southern Oscillation	116
Indian Ocean Dipole	15
Madden Jullian Oscillation	1
Borneo Vortex	0
Heat waves	0
local atmospheric phenomena	0

Source: Analysis Results (2024)

Localized atmospheric phenomena such as the Borneo Vortex and heatwaves, along with other unspecified local dynamics, are entirely absent from the publication count. While these phenomena may have more localized or indirect impacts on fire behavior, their absence suggests they are deemed secondary in comparison to larger-scale dynamics like ENSO and the IOD. This pattern underscores a strong research emphasis on large-scale atmospheric drivers, particularly ENSO, in understanding the dynamics of forest and land fires. It also highlights that smaller-scale atmospheric drivers might have less influence to forest and land fires.

3.5. Influence of atmospheric phenomena on forest and land fires across regions in Indonesia

Table 3 reveals the significant influence of atmospheric phenomena on forest and land fires across various islands in Indonesia. The El Niño Southern Oscillation (ENSO) stands out as the most frequently recorded phenomenon, highlighting its widespread impact on fire dynamics. Kalimantan experiences the highest frequency of ENSO events, with 38 occurrences, underscoring its vulnerability to prolonged drought conditions and fire outbreaks, particularly in peatland areas. Sumatra follows with 14 ENSO-related events, reflecting a similarly high fire risk due to its extensive peatland ecosystems.

Interactions between ENSO and the Indian Ocean Dipole (IOD) also play a critical role in intensifying fire risks. These combined events are most frequently observed in Kalimantan and Sumatra, with 3 and 4 occurrences respectively, suggesting that the interaction between these phenomena exacerbates dry conditions and amplifies fire severity. Additionally, Papua and Nusa Tenggara record fewer ENSO-IOD interactions, indicating less severe impacts in these regions. While the IOD appears independently in Kalimantan and Sumatra (1 and 4 times, respectively), it contributes to regional climate variability, particularly through its influence on rainfall patterns. A rare combination of ENSO, IOD, and the Madden-Julian Oscillation (MJO) is recorded once in Sumatra, likely representing a complex and extreme climatic scenario with heightened fire risks.

Regional patterns also emerge from the data, with Kalimantan and Sumatra clearly dominating in terms of atmospheric dynamics associated with fires. This aligns with their ecological characteristics, such as extensive peatlands that are highly flammable during dry conditions. In contrast, Java, Nusa Tenggara, Papua, and Sulawesi exhibit fewer occurrences of these phenomena. Overall, ENSO remains the primary driver of fire risks across Indonesia, particularly in Kalimantan and Sumatra, where its impacts are most pronounced [13]. The interaction between ENSO and IOD further intensifies these risks, making these regions priorities for targeted fire mitigation strategies.

Table 3. Influence of atmospheric phenomena on forest and land fires across regions in Indonesia

Islands	Atmospheric Phenomenon Keywords	Frequency
Kalimantan	ENSO	38
	IOD	1
	ENSO and IOD	3
Sumatra	ENSO	14
	IOD	4
	ENSO and IOD	4
	ENSO, IOD, and MJO	1
Java	ENSO	3
Sulawesi	ENSO	3
Nusa Tenggara	ENSO	1
	ENSO and IOD	1
Papua	ENSO	3
	ENSO and IOD	2

Source: Analysis Results (2024)

4. Conclusion

The analysis illustrates the substantial influence of climatic phenomena, especially the El Niño Southern Oscillation (ENSO), on the patterns of forest and land fires in Indonesia. ENSO is recognized as the principal contributor, particularly in areas such as Kalimantan and Sumatra, which endure frequent and intense fire occurrences due to extended drought conditions intensified by El Niño phases. The interplay between ENSO and the Indian Ocean Dipole (IOD) exacerbates these effects, underscoring the cumulative influence of these events on fire severity. Nonetheless, complex atmospheric dynamics, including the Madden-Julian Oscillation (MJO) and localized atmospheric phenomena, are hardly addressed, highlighting the more important role of large-scale atmospheric phenomena.

Regional disparities demonstrate unique trends, with Kalimantan and Sumatra displaying the greatest incidence of ENSO and IOD-related events, whereas Java, Sulawesi, Papua, and Nusa Tenggara exhibit less occurrences. This underscores the necessity for focused fire management tactics, emphasizing areas with greater atmospheric susceptibility. This result also underscores the necessity of integrative research to enhance comprehension of intricate climatic interactions and their contribution to the intensification of forest and land fires. These findings highlight the imperative to create adaptable and region-specific fire prevention strategies based on a thorough comprehension of atmospheric dynamics.

Acknowledgments

This research is Supported by Grant of "Hibah Penelitian Mandiri Fakultas Geografi Universitas Gadjah Mada" in 2024. The grant entitled "*Pengaruh Variabilitas Iklim terhadap Kebakaran Hutan dan Lahan di Pulau Kalimantan*" with Dr. Sc. Andung Bayu Sekaranom, M.Sc. as the lead researcher (Contract Number: 304/UN1/GE/KPT/2024 in July 25th 2024).

References

[1] Wulandari E R, Mardiatno D, Susilastuti D H and Maryudi A 2022 *For. Soc.* **6** 609–619 https://doi.org/10.24259/fs.v6i2.21473
[2] Albar I, Jaya I N S, Saharjo B H, Kuncahyo B and Vadrevu K P 2018 *Land-Atmospheric Research Applications in South and Southeast Asia* (Cham: Springer) https://doi.org/10.1007/978-3-319-67474-2_6
[3] Aminah, Krah C Y and Perdinan 2020 *IOP Conf. Ser. Earth Environ. Sci.* **504** 012013 https://dx.doi.org/10.1088/1755-1315/504/1/012013
[4] Harenda K M, Lamentowicz M, Samson M and Chojnicki B H 2018 *Interdisciplinary Approaches for Sustainable Development Goals. GeoPlanet: Earth and Planetary Sciences* (Cham:Springer)
[5] Prayoga M B R and Koestoer R H T 2021 *J. Wil. Dan Lingkung.* 9 293–305 https://doi.org/10.14710/jwl.9.3.293-305
[6] Omar N, Al-zebari A and Şengur A 2021 *2nd International Informatics and Software Engineering Conference (IISEC)* 1-4. https://doi.org/10.1109/iisec54230.2021.9672446
[7] Abatzoglou J T and Kolden C A 2013 *Int. J. Wildland Fire* **22** 1003–1020 https://doi.org/10.1071/WF13019
[8] Drever C R, Drever M C, Messier C, Bergeron Y and Flannigan M 2008 *J. Veg. Sci.* **19** 57–66 https://doi.org/10.3170/2007-8-18313
[9] Firmansyah A J, Nurjani E, and Sekaranom A B 2022 *Arabian Journal of Geosciences* ***15*** 1746 https://doi.org/10.1007/s12517-022-11016-2
[10] Alizadeh M R, Adamowski J F and Entekhabi D 2024 *Geophysical Research Letters* **51** 1-12 https://doi.org/10.1029/2023GL105324
[11] Déandréis C et al 2020 *Climate services for forest fire risk management* (Online: EGU General Assembly 2020) https://doi.org/10.5194/egusphere-egu2020-11639
[12] Pereira M C, Parente J, Amraoui M, Oliveira A and Fernandes P M 2020 *Extreme Wildfire Events and Disasters* (Netherlands: Elsevier) https://doi.org/10.1016/B978-0-12-815721-3.00003-5
[13] Zahra R A, Nurjani E and Sekaranom A B 2023 *Quaest. Geogr.* **42** 75–86 https://doi.org/10.14746/quageo-2023-0006

Analysis of the Suitability of Salt Pond Sites Based on the Salt Suitability Index (IKG) in Bolo District, Bima Regency

P Sulistyoningsih[1], Afif Ari Wibowo[2*], Basyar Ihsan Arijuddin[3]

[1]Departement of Geography, Faculty of Geography, Universitas Muhammadiyah Surakarta, Indonesia

*E-mail: aaw346@ums.ac.id

Abstract. Salt is one of the main commodities that has a variety of uses, both as an industrial raw material and as a food additive. Bima Regency is one of the regions that has a long coastline, however, salt production in this area has not yet adequately met the demands of both consumption and industry. This is due to the lack of land extensification, lack of technology application, and erratic weather factors. Therefore, this study aims to analyze the suitability of salt pond location and analyze the limiting factors on salt ponds in Bolo District, Bima Regency, West Nusa Tenggara. This research uses the Salt Location Suitability Index (IKG) to analyze the feasibility of salt-making locations in Bolo Sub-district, which includes Sanolo, Sondosia, and Darussalam Villages. The IKG is determined using nine suitability parameters: precipitation, soil permeability, soil type, length of irradiation, air humidity, wind velocity, temperature, evaporation, and the salt production raw material saturation level used in salt production. The study's findings show that the Salt Site Suitability Index (IKG) in Sanolo Village and Darussalam Village have the same index value of 75%, indicating that both locations are classified in the Conditionally Suitable category for salt production. Meanwhile, Sondosia Village obtained an IKG value of 77%, indicating that this village is also conditionally suitable for salt production activities in Bolo Sub-district. This is due to the low value of the soil permeability parameter, which is below 6.6 x 10^{-3}, and the degree saturation level of the raw materials used for the salt making process, which is below 1.3 °Be.

Keywords: Salt, Index of Salt Suitability (IKG), Bolo District

1. Introduction

A commodity with a comparative advantage is salt, as production is abundant despite relatively low market demand. Salt plays an important role in industry for preservation and mixtures in chemical processes. In addition, salt is also often used for consumption needs [1]. Most of the salt production is done independently by salt farmers, which leads to low salt productivity with non-optimal salt quality. As a result, the salt produced does not meet the standards required by the domestic industry [2]. Indonesia is a country comprised of a vast array of islands with the fourth longest coastline globally, spanning over 95,181 km^2 [3]. Indonesia covers a sea area of about 5.9 million km2 and is a resource-rich country [4][5]. Where with these geographical conditions,

Indonesia it should be capable of fulfilling the nation's salt requirements. However, in reality until now, Indonesia is still unable to fulfill its domestic salt demands, which leads to ongoing imports from countries like Australia, India, and China [3].

Bima Regency is situated at the eastern end of West Nusa Tenggara (NTB) Province and is part of the province. It is directly adjacent to Bima City, which used to be part of Bima Regency. To the west, Bima district borders Dompu district, while to the east it borders the Sape Strait. It is bordered to the north by the Flores Sea and to the south by the Indian Ocean [6]. Bima Regency has an area of 22% of the total area of the province. The land area covers 4,016.13 km^2, while the sea waters area is 3,760.33 km^2 with a coastline length of 687.43 km^2. The sea waters of Bima Regency cover 37.71% of the total sea waters area of Sumbawa Island of 9,970.96 km^2, or 29.26% of the sea waters area of West Nusa Tenggara Province which reaches 12,852.14 km^2 [7]. Salt production typically takes place in coastal areas, as seawater is the essential component used in the process. The area of salt ponds in Indonesia currently reaches about 34 thousand ha or about 60% of the total land that is optimally managed for production [8].

Bolo District, situated in Bima Regency within West Nusa Tenggara Province, is recognized as one of the key contributors to the nation's salt production. Bolo sub-district recorded salt ponds spread across 3 villages with a total area of 760.02 ha of salt ponds in the Marine and Fisheries Service of Bima Regency (2024). The village with the largest pond area is Sanolo Village, which covers 529.48 ha and produces 42,628.82 tons of salt. In contrast, Sondosia Village has the smallest pond area, 108.78 ha, with a production of 8,260.20 tons. Meanwhile, Darussalam Village has a pond area of 121.76 ha with a total production of 11,255.14 tons. There are several salt ponds that utilize Geoisolator or Geomembrane technology, which uses HDPE plastic as a crystallization table in the salt making process. However, most still use the traditional method, where the land is directly used for the crystallization table [9][10][11]. Salt produced by farmers with geoisolator technology tends to have superior quality compared to those using traditional methods where the salt produced is often contaminated by soil. In addition, the salt produced is still unable to meet consumption standards due to its low iodine content [12]. The quality of traditionally managed salt generally has to be reprocessed to be used as consumption salt or for industrial salt [13].

The climate in Indonesia is generally influenced by the monsoon and is divided into two seasons, specifically the rainy season and the dry season. In addition, the climate in Indonesia is also influenced by global climate phenomena, such as ENSO (El-Nino Southern Oscillation) [14]. This phenomenon often causes rain in the midst of the dry season and results in a shift in the beginning and end of the dry season. In Bima District, approximately 869.15 ha of salt pond land is utilized during the dry season. However, when the rainy season arrives the land is diverted for aquaculture activities or even some are left without any activity.

The suitability of salt pond locations plays a significant role factor in maximizing salt production within a region, as the physical aspects of salt ponds act as determinants of site suitability [15]. This is the basis for measuring land resource potential and assessing land quality. In this study, the evaluation of a location's potential to serve as a salt production site was developed in the form of the Salt Suitability Index (IKG) [16]. IKG (Salt Suitability Index) is an analysis used to assess the characteristics of a location as salt production land. The assessment in the IKG analysis is based on the main parameters that affect the process of making salt by the evaporation method, also known as solar salt, such as temperature, humidity, soil type, and other environmental factors. The parameters used include precipitation, soil permeability, soil type, length of irradiation, air humidity, wind velocity, temperature, evaporation, and the salt production raw material saturation level used in salt production [16]. Building upon the conditions described above, this study is designed to assess the suitability of the salt pond location and analyze the limiting factors on salt ponds in Bolo District, Bima Regency.

2. Methods

2.1. Study location

This research was conducted in Bolo District, Bima Regency, salt ponds located in Sanolo Village, Sondosia, and Darussalam as the research location. Bolo sub-district is located at coordinates 8°30′5.83405″S 118°37′8.11265″E. Bolo sub-district it covers an area of 66.92 km^2, including a total coastal region of 19,062 ha. The Bolo sub-district area is bordered by Soromandi and Donggo sub-districts to the north, Woha sub-district to the south, Madapangga sub-district to the west, and Bima Bay to the east. The location map can be seen in **Figure 1**.

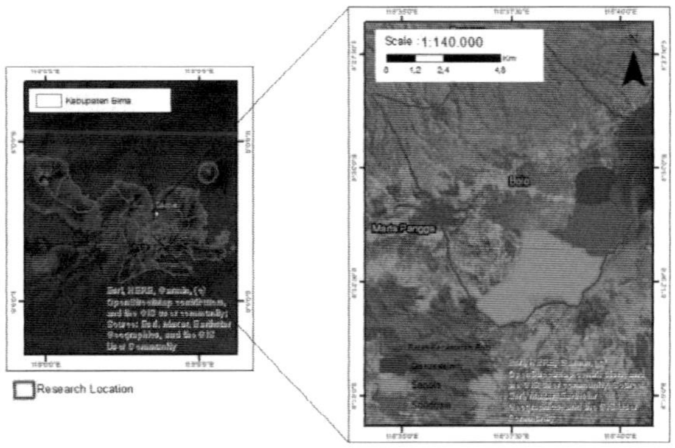

Figure 1. Study Location

2.2 Tools and Materials

This research utilizes both primary and secondary data. Primary data was obtained through field studies by conducting laboratory tests of soil samples to measure soil permeability and and on water samples to assess the water saturation levels of the raw materials used in salt production. While secondary data is obtained by collecting data from several sources, namely related agencies and thematic maps. The methods used in this research are quantitative and qualitative. The needs, tools used and data used in this study, the components are outlined as follows:

Table 1. Tools used in the study

Tool Name	Function
ArcGis 10.8	processing data in the form of maps and data analysis
GPS	determine the location of sample points
Ring	taking soil samples
Plastic Soil Samples	saving the land that has been obtained
Baume Hydrometer	measuring the water saturation level of salt production raw materials
Microsoft. Excel	data processing in the form of calculations
Microsoft. Word	report preparation

Table 2. Data used in the study

Data Name	Function	Source
Map of Bima Regency	barrier to identify the location of salt farms	DPU Kabupaten Bima
Administrative Map of Bolo Sub-district	barrier to identify the location of salt farms	DPU Kabupaten Bima
Rainfall Data, Soil Type, Duration of Illumination, Air Humidity, Wind Speed, Air Temperature, Evaporation	analyzing the suitability of salt pond locations	-Bappeda Kabupaten Bima -Stasiun Meteorologi Sultan Muhammad Salahuddin
Soil Sample	test and analyze soil permeability in the study area	*Field Study and Laboratory Test*
Seawater Sample	measuring and analyzing the saturation level of raw materials for salt production in the study area	Field Study

2.3 Stages of data processing and analysis

2.3.1 Sampling

In this study there were three points of soil sampling and measurement of saturation of raw materials for salt production. Sampling is done randomly based on certain criteria. Soil sampling was conducted at salt ponds located in Sanolo, Sonodia, and Darussalam villages. In each village, one randomly selected point was sampled. To measure soil permeability, a pipe with a diameter of 5 cm and a height of 5 cm was used. The pipe was covered using gauze, then put into a plastic sample. Measurement of the saturation level of raw materials for salt production were assessed at the three locations with two repetitions at each location.

2.3.2 Soil Permeability Measurement

Soil permeability was analyzed employing the Constant Head Permeameter method. The Constant Head permeability test is a widely used laboratory method for measuring the permeability of granular soils like sand and gravel that contain little to no silt [23].

2.3.3 Measurement of Saturation Level of Salt Production Raw Materials

Measurement of the Saturation Level of Salt Production Raw Material Water is carried out using a boumehydrometer tool by inserting the tool into a bottle containing salt production raw material water. The measurement results are determined through the number on the boumehydrometer scale indicated by the position of the water that rises in the boumehydrometer tool.

2.3.4 Secondary Data Retrieval

Secondary data were obtained from several readily available sources, such as relevant agencies and thematic maps. Data from the Sultan Muhammad Salahuddin Meteorological Station in Bima district includes parameters eg rainfall, length of irradiation, air humidity, wind speed, air temperature and evaporation. Meanwhile, data from the Bima District Fisheries and Marine Office includes information on salt production in 2023.

2.3.5 Data Analysis
The data were examined using a quantitative approach to facilitate the process of drawing conclusions about the suitability of the location, by measuring the various factors affecting salt production in an objective and measurable manner and weighted scoring done by observing the importance of a parameter to the success of a factor.

2.3.6 Calculation of Salt Suitability Index (IKG)
The calculation of salt production is based on the suitability level of nine parameters of salt pond location suitability [17]. The level of suitability of this location is calculated by applying the formula used to appraise the physical suitability class of salt ponds, which includes various factors that affect the success of production, such as soil conditions, temperature, humidity, and other environmental factors as follows:

$$\mathbf{IKG} = \sum \frac{Ni}{Nmaks} x\mathbf{100}\%$$

Description:
IKG : Production Site Suitability Index
Ni : Value of i-th Parameter (weight × score)
Nmax : Maximum value of a category
The suitability criteria based on the IKG score are as follows:
>85% : Extremely Suitable (S1)
80–84% : Suitability (S2)
75–79% : Conditionally compliant (S3)
<75% : Not suitable (N).
The Salt Suitability Index (IKG) categories are summarized in **Table 3**.

3. Results and Discussion

This study analyzes the suitability of salt pond locations in Bolo sub-district, which includes Sanolo, Sondosia, and Darussalam villages. Sanolo Village has an area of 108.78 ha of salt ponds, Sondosia Village with an area of 529.48 ha of salt ponds, while Darussalam Village has an area of 121.76 ha of salt ponds by the Department of Marine and Fisheries within Bima Regency as of 2024. Salt pond areas in the three locations are located on the east side of the village and directly facing the bay of Bima in **Figure 1**. Analysis of the feasibility of the location of salt ponds using the Salt Site Suitability Index (IKG) approach. IKG assessment is based on nine parameters that have been presented in **Table 3**. The results of the analysis of the nine parameters are presented in **Table 4** for Sanolo Village, **Table 5** for Sondosia Village, and **Table 6** for Darussalam Village.

3.1 Rainfall
Rainfall has a negative impact on changes in brine salinity in the ginning ponds. Conversely, low rainfall is good for salt production [14]. Bima Regency is a salt pond area that still relies on traditional technology, so the rainfall parameter is very important in the ongoing salt production in the area. Therefore, one of the factors that need to be analyzed to assess a location's suitability for salt production is determined by its rainfall level [22].

Rainfall data at the research site was analyzed using rainfall information obtained from the Sultan Muhammad Salahuddin Meteorological Station, Bima Regency. The average rainfall in Sanolo, Sondosia and Darussalam villages was recorded at 79.4 mm. This data was then analyzed using an index to assess the Salt Suitability Index (IKG). According to the analysis results, Sanolo,

Sondosia and Darussalam Villages obtained an index value of 15 which placed all three in the S2 or Moderately Suitable category.

3.2 Soil Permeability

Soil Permeability is one of the important parameters to analyze the suitability of salt pond sites. Soil permeability greatly affects the salt production process. This parameter is also important in determining the level of soil porosity. Soil porosity affects the speed of seawater seepage (leakage) into the soil in the peminihan or table. This can affect the success of salt production [13]. Soil permeability refers to the capacity of soil to allow water to move through its pore spaces when it is saturated [18]. Permeability depends on the pore size which is influenced by particle size, particle shape, and soil structure. The smaller the particle size, the lower the permeability.

The results of soil permeability measurements on salt ponds in Sanolo Village is 1.24×10^{-2} (k) where the value indicates an index value of 10 which puts the location into the category S3 or Conditionally Suitable, in Sondosia Village of 6.6×10^{-3} which indicates an index value of 15 which puts the location into the category S2 or Moderately Suitable, while in Darussalam Village is 1.07×10^{-2} which indicates an index value of 10 which puts the location into the category S3 or Conditionally Suitable.

3.3 Soil Type

Salt ponds used for production generally have a dense soil type or clay with little sand content [19]. This type of soil helps prevent seepage or porosity of water in the salt patch. The soil type selected for salt ponds should be nonporous with very low permeability, in order to prevent water leakage and ensure effective evaporation [20]. The type of soil also affects the color and impurities carried by the salt produced [9].

An analysis of the salt locations based on soil type conditions in salt ponds shows that the three locations are sandy loam soil. The results of the analysis of the salt suitability index (IKG) index value in the three locations amounted to 15 or Moderately Suitable for salt production.

3.4 Duration of Illumination

The length of solar irradiation is a factor in salt production that also affects the rate of water evaporation [17]. The salt-making process utilizes heat from sunlight, which allows water evaporation to occur over several days. After that, salt crystals slowly begin to form.

Length of irradiation data obtained from the Meteorological Station Sultan Muhammad Salahuddin Bima Regency which has an average of 6.6 hours / day. Based on these results, the index value obtained is 15 which indicates the three locations into the category S2 or Moderately Suitable as the location of salt pond production.

3.5 Air Humidity

Air humidity plays an important role in the salt making process [20]. Air humidity in salt ponds is no more than 50% [21]. Air humidity can also affect the evaporation process.

The average air humidity data in Sanolo Village, Sondosia, and Darussalam based on data from the Meteorological Station Sultan Muhammad Salahuddin Bima Regency is 65% where the value indicates that the three locations are included in the category S2 or Moderately Suitable for the location of salt pond production. Air humidity is a critical factor in selecting the optimal location for salt pond production.

3.6 Wind Speed

Wind speed plays an important role in determining the location of salt ponds. The higher the wind speed, the precipitation of salt will occur and the larger the size of the salt grains produced, but if the wind speed is low, the salt grains generally remain buoyant on the water's surface, resulting in a finer texture.

Wind speed data at the research site was analyzed using wind speed information obtained from the Sultan Muhammad Salahuddin Meteorological Station, Bima Regency. The data shows the average wind speed at the three locations is 8.1 m/s. Based on the wind speed conditions, the suitability analysis of the salt ponds at the three locations produced an index value of 20, placing them in the S1 category, or Extremely Suitable. Wind speed is one of the parameters that need to be considered in the level of success achieved in salt production in the region.

3.7 Air Temperature

Air temperature plays a role in boosting salt production efficiency ponds by accelerating the process of seawater in salt ponds, so as to accelerate the formation of salt [20]. To produce high quality salt, it is necessary that the location of the salt pond has an air temperature above 32°C [13]. Air temperature has a major influence on the rate of water evaporation. The higher the evaporation rate, the more salt crystals are formed and deposited [9].

Air temperature data obtained from the Sultan Muhammad Salahuddin Meteorological Station, Bima Regency. The three locations have an average of 28.4°C. The index value based on this average is 15 which puts the three locations in the S2 category or Moderately Suitable for salt production locations in Bolo sub-district.

3.8 Evaporation

Salt is produced from seawater through an evaporation process that is sped up by exposure to sunlight. An increase or decrease in water evaporation can affect the amount of salt crystals that form and precipitate.

The average evaporation at the three locations according to data obtained from the Sultan Muhammad Salahuddin Meteorological Station is 5.6 mm/day. Based on this data, the index value at the three locations is 20, which means that the three locations are in the S1 category or Extremely Suitable for the salt making process.

3.9 Salt Production Raw Material Saturation Level

Salt production raw material saturation level key factor in influence the efficiency of the evaporation process and the final result of salt produced. Water saturation plays a crucial role in the success of the salt making process.

According to the analysis results of measurements that have been taken, the saturation value of salt production raw material water in Sanolo Village has an average of 1.25 °Be where the value shows an index value of 10 which puts the location into the S3 or Conditionally Suitable category, Sondosia Village with a saturation value of 1.24 °Be the value shows an index value of 10 which puts the location into the S3 or Conditionally Suitable category, while in Darussalam Village has a saturation value of 1.26 °Be the index value in Darussalam village is 10 or Conditionally Suitable.

Table 3. Index of Salt Suitability (IKG) values

Parameters	Weight	Category 1		Category 2		Category 3		Category N	
		Value	Score	Value	Score	Value	Score	Value	Score
Precipitation (mm)	5	<10	4	10-100	3	100-200	2	>200	1
Soil permeability (k)	5	1×10^{-4}	4	1×10^{-3}	3	$<1\times10^{-2}$	2	$<1\times10^{-1}$	1
Soil Type	5	Clay	4	Sandy Clay	3	Loam	2	Silty	1

IOP Conf. Series: Earth and Environmental Science **1503** (2025) 012017 doi:10.1088/1755-1315/1503/1/012017

Parameters	Weight	Category 1		Category 2		Category 3		Category N	
		Value	Score	Value	Score	Value	Score	Value	Score
Duration of Illumination (hours/day)	5	>8,7	4	5,5-8,6	3	2,3-5,4	2	<2,3	1
Air Humidity (%)	5	<45-59	4	60-74	3	75-90	2	>90	1
Wind Velocity (m/s)	5	>5,7	4	4,1-5,7	3	2,4-4,0	2	>2,4	1
Temperature (° C)	5	>32	4	28,5-32	3	5-28,4	2	<25	1
Evaporasi (mm/day)	5	>2,0	4	1,5-2,0	3	1,0-1,4	2	<1,0	1
Salt Production Raw Material Saturation Level (°Be)	5	≥2	4	1,5-1,9	3	1,0-1,5	2	<1	1

Explanation. S1: Extremely Suitable; S2: Moderately Suitable; S3: Conditionally Suitable; N: Not Suitable

Source: [13]

Table 4. Index of Salt Suitability (IKG) Sanolo Village, Bolo Sub-district, Bima District

Parameters	Measurement Results	Category	Weight	Score	Ni= B x S
Precipitation (mm)	79,4	S2	5	3	15
Soil permeability (k)	1,24 x 10^{-2}	S3	5	2	10
Soil Type	Sandy Clay	S2	5	3	15
Duration of Illumination (hours/day)	6,6	S2	5	3	15
Air Humidity (%)	65	S2	5	3	15
Wind Velocity (m/s)	8,1	S1	5	4	20
Temperature (° C)	28,4	S2	5	3	15
Evaporasi (mm/day)	5,6	S1	5	4	20
Salt Production Raw Material Saturation Level (°Be)	1,25	S3	5	2	10

Based on table 4, the value of the salt suitability index (IKG) in Sanolo Village, Bolo District is as follows:

IKG = (∑Ni / Nmaks) ×100 %
= (135/180) ×100 %
= 75%

Table 5. Salt Suitability Index (IKG) Sondosia Village, Bolo District, Bima Regency

Parameters	Measurement Results	Category	Weight	Score	Ni= B x S
Precipitation (mm)	79,4	S2	5	3	15
Soil permeability (k)	6,6 x 10^{-3}	S2	5	3	15
Soil Type	Sandy Clay	S2	5	3	15
Duration of Illumination (hours/day)	6,6	S2	5	3	15
Air Humidity (%)	65	S2	5	3	15
Wind Velocity (m/s)	8,1	S1	5	4	20
Temperature (° C)	28,4	S2	5	3	15
Evaporasi (mm/day)	5,6	S1	5	4	20
Salt Production Raw Material Saturation Level (°Be)	1,25	S3	5	2	10

Based on table 5, the value of the salt suitability index (IKG) in Sondosia Village, Bolo District is as follows:

IKG = (∑Ni / Nmaks) ×100 %
= (140/180) ×100 %
= 77%

Table 6. Index of Salt Suitability (IKG) Darussalam Village, Bolo District, Bima Regency

Parameters	Measurement Results	Category	Weight	Score	Ni= B x S
Precipitation (mm)	79,4	S2	5	3	15
Soil permeability (k)	1,07 x 10^{-2}	S3	5	2	10
Soil Type	Sandy Clay	S2	5	3	15
Duration of Illumination (hours/day)	6,6	S2	5	3	15
Air Humidity (%)	65	S2	5	3	15
Wind Velocity (m/s)	8,1	S1	5	4	20
Temperature (° C)	28,4	S2	5	3	15
Evaporasi (mm/day)	5,6	S1	5	4	20
Salt Production Raw Material Saturation Level (°Be)	1,25	S3	5	2	10

Based on Table 6, the value of the salt suitability index (IKG) in Darussalam Village, Bolo Sub-district is as follows:

IKG = (∑Ni / Nmaks) ×100 %
= (135/180) ×100 %
= 75%

Table 7. Salt Suitability Index (IKG) Results

Location (Village)	Percentage %	Category
Sanolo	75%	Conditionally compliant
Sondosia	75%	Conditionally compliant
Darussalam	77%	Conditionally compliant

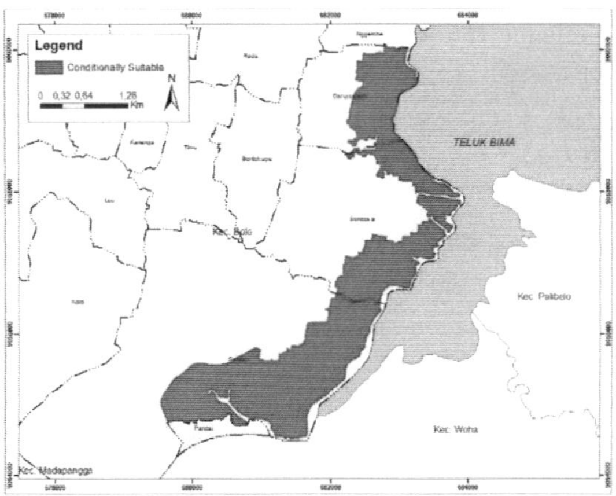

Figure 2. Salt Pond Location Suitability Map

The study's findings indicate that both Sanolo has a value of 75% and Darussalam Villages share an identical Salt Suitability Index (IKG) value of 75%, indicating that both locations are classified as conditionally suitable for salt production. Meanwhile, Sondosia Village obtained an IKG value of 77%, indicating that this village is also Conditionally Suitable for salt production activities in Bolo Sub-district. This shows that weather plays a crucial role in the sustainability of salt production, especially in these three areas. This research also shows that the main aspects that need to be considered in developing innovative technologies for salt production in Bolo Sub-district, Bima District, are soil permeability and water saturation as raw materials for salt production. These variable characteristics need to be addressed by the application of innovative technologies to boost the efficiency of salt production in this region. Alternative technologies that can be used include the installation of geomembranes or geoisolators, as well as the application of crystallization house technology such as the Greenhouse Salt Tunnel. Based on the study's findings, the salt production sites examined are deemed conditionally viable for further development, provided that innovative technologies are implemented.

4. Conclusion

The Index of Salt Site Suitability (IKG) for the three areas is as follows: Sanolo Village recorded a value of 75%, Sondosia Village also recorded a value of 75%, and Darussalam Village recorded a value of 77%. As all these values are below 80%, the three locations are classified as Conditionally Suitable (S3) for salt production.

The main aspects that need to be considered in the development of salt production technology innovations in Bolo District, Bima Regency, are soil permeability and water saturation as raw materials for salt production. The low value of soil permeability is below 6.6 x 10^{-3} and water saturation for the main resource of salt production is below 1.3 °Be it becomes one of the factors that constrain the suitability of salt pond sites in the three areas. This study presents an analysis of the Salt Suitability Index (IKG) on the location of salt ponds in Bolo sub-district, which has the potential to provide benefits for policy makers as well as a reference for further research by other researchers.

5. Acknowledgments

The author expresses his profound gratitude to Allah SWT and to everyone who has supported and contributed to this research. Special thanks go to both parents who always encourage me, Mr. Afif Ari Wibowo, S.Si., M.Si as the main research advisor who has provided guidance, input, and support, as well as the Faculty of Geography, Universitas Muhammadiyah Surakarta for facilitating, and finally to the relevant agencies, namely the Sultan Muhammad Salahuddin Meteorological Station, the Regional Research and Development Agency of Bima Regency, and the Marine and Fisheries Service of Bima Regency.

References

[1] A. S. Arifin, P. S. Geografi, F. Geografi, and U. M. Surakarta, 2019, "Kondisi Sosial Ekonomi Masyarakat (Studi Kasus Di Desa Tireman Kecamatan Rembang Kabupaten Rembang)," PhD Thesis, Universitas Muhammadiyah Surakarta.

[2] B. A. Leo, S. Suro Adhwati, and A. Asnawi, 2021, "Analisis Produktivitas dan Pendapatan Petambak Garam di Kabupaten Jeneponto Provinsi Sulawesi Selatan," *AGROVITAL J. Ilmu Pertan.*, vol. 6, no. 1, pp. 1. doi: 10.35329/agrovital.v6i1.1996.

[3] L. Baihaki, 2013, "Ekonomi-Politik Kebijakan Impor Garam Indonesia Periode 2007-2012 (Economics Politics Policy on Salt Importation in Indonesia during 2007-2012)," *J. Ilmu Sos. dan Ilmu Polit.*, vol. 17, no. 1, pp. 1–16.
[4] A. P. Indah Purwanti, Yudo Prasetyo, 2017, "Purwanti 2017," *J. Geod. Undip,* vol. 6, pp. 506–516, 2017.
[5] A. F. Novianto, H. Z. Hadibasyir, and A. A. Wibowo, 2023, "Analysis of Fish Catches of Fishermen in the Southern Waters of Java Island in 2018 and 2020 Based on Chlorophyll-A Fertility and Sea Surface Temperature" *Atlantis Press SARL,* pp. 216-232. doi: 10.2991/978-2-38476-066-4_14.
[6] BPS Kabupaten Bima, 2023, "Kabupaten Bima Dalam Angka 2023," BPS Kabupaten Bima, 2023, [Online]. Available: https://bimakab.bps.go.id/publication/2023/02/28/128b83598060f1bcceb3a6b6/kabupaten-bima-dalam-angka-2023.html
[7] R. Syarifuddin, Umar, Sukrin, Ihlas, 2022, "Strategi Muhammadiyah Disaster Manajement Center (MDMC) dalam Penanggulangan Bencana Banjir Kabupaten Bima pada Bulan April Tahun 2021," *Bima Abdi: Jurnal Pengabdian Masyarakat,* vol. 2, no. 1, pp. 29–39.
[8] Hamdani and J. Farmiati, 2022, "Analisis Indeks Kesesuaian Garam (IKG) Untuk Menentukan Kesesuaian Lokasi Produksi Garam di Provinsi Aceh," *J. Law Econ.,* vol. 1, no. 1, pp. 10–20. doi: 10.56347/jle.v1i1.36.
[9] Y. U. Hoiriyah, 2019, "Peningkatan Kualitas Produksi Garam Menggunakan Teknologi Geomembran," *J. Stud. Manaj. dan Bisnis,* vol. 6, no. 2, pp. 71–76, 2019, doi: 10.21107/jsmb.v6i2.6684.
[10] M. Salsabiela and J. Prayitno, 2022, "Pengaruh Penggunaan Teknologi Ulir Filter dan Geoisolator pada Produksi Garam Rakyat di Kabupaten Indramayu," *J. Teknol. Lingkung.,* vol. 23, no. 2, pp. 207–213, 2022, doi: 10.29122/jtl.v23i2.4976.
[11] Yunita Nurul Fatimah et at., 2022, "Analysis of Salt Production Using Geoisolator Technology," *Jurnal Mina Sains,* vol. 8, pp. 54–59.
[12] A. Wiraningtyas, A. Sandi, and R. Ruslan, 2019, "Iodized Salt Processing Technology through Solar Thermal Salt House in Sanolo Village," *MITRA J. Pemberdaya. Masy.,* vol. 3, no. 1, pp. 1–10. doi: 10.25170/mitra.v3i1.494.
[13] D. Purbani, 2000, "Proses Pembentukan Kristalisasi Garam," Pus. Ris. Wil. Laut dan Sumberd. Nonhayati, pp. 1–17 [Online]. Available: http://www.oocities.org/trisaktigeology84/Garam.pdf
[14] A. R. Kumala and Y. Sugiarto, 2012, "Analisis Pengaruh Curah Hujan terhadap Produktivitas Garam Studi Kasus: Pegaraman I Sumenep PT. Garam (Persero)," *Pros. Semin. Nas. Sains IV,* vol. 10, no. November, pp. 1–10, 2012, [Online]. Available: http://www.bom.gov.au/climate
[15] M. U. H. S. Nahar, 2019, "Analisis kesesuaian tambak garam di pesisir kecamatan trangkil kabupaten pati," PhD Thesis, Universitas Muhammadiyah Surakarta.
[16] A. Kurniawan, A. A. Jaziri, A. A. Amin, and L. N. Salamah, 2019, "Salt suitability index (Ikg) to determine the suitability of salt production sites; analysis of salt production locations in Tuban and Probolinggo Regencies," *JFMR-Journal Fish. Mar. Res.,* vol. 3, no. 2, pp. 119–127.
[17] A. Kurniawan et al., 2020, "Analisis Indeks Kesesuaian Lokasi Garam (IKG) di Kawasan Sentra Produksi Garam Jawa Timur," *Rekayasa,* vol. 13, no. 3, pp. 225–233, 2020, doi: 10.21107/rekayasa.v13i3.9130.
[18] A. A. Rudiyanto, R., Suharjo, M. S., & Sigit, 2010, "Analisis potensi bahaya tanah longsor menggunakan sistem informasi geografis (sig) di kecamatan selo, kabupaten boyolali," PhD Thesis, Universitas Muhammadiyah Surakarta..
[19] O. Jayanthi, 2021, "Karakteristik Fisik Tanah Tambak Garam Pameka," JFMR-Journal Fish. Mar. Res., vol. 5, no. 2. doi: 10.21776/ub.jfmr.2021.005.02.6.
[20] A. S. Renaldi Bahri Tambunan, Hariyadi and Jurusan, 2012, "Evaluasi Kesesuaian Tambak Garam Ditinjau Dari Aspek Fisik Di Kecamatan Juwana Kabupaten Pati," *J. Mar. Res.*, vol. 1, no. 2, pp. 181–187.
[21] D. Aldi, E. Intan, and K. Putri, 2021, "Tingkat kerentanan rumah tangga petambak garam di Desa Donggobolo akibat perubahan iklim The vulnerability of salt farmer houses in Donggobolo Village due to climate changes," vol. 11, no. 2, pp. 298–308.
[22] A. Ulfiana, M. Arsyad, & P. Palloan, 2023, " The Atmospheric Dynamics Re-lated to Extreme Rainfall and Flood Events during September-October-No-vember in South Sulawesi" *Forum Geografi*, vol. 37, no.2, pp. 107-116. doi: 10.23917/forgeo.v37i2.22339.
[23] A. Saloua, A, Ahmed, A. Abdellah, F. Abdelouahed, M. Said, A. Abdelfattah, E. G. Akram, E. G., 2024, "Modelling of Soil Erosion Susceptibility Using the Multi-In-fluencing Factor Method in the Amizmiz Basin, Morocco" *Forum Geografi*, vol. 38, no. 3, pp. 343-357. doi: 10.23917/forgeo.v381i3.6263.

Analysis of Coastline Changeson The Wedung Coast Demak District Using Multi Temporal Satellite Imagery 2019 - 2024

A N Farida[1],Danardono[2]

[1] Department of Geography, Faculty of Geography, Universitas Muhammadiyah Surakarta, Indonesia

*E-mail: dan115@ums.ac.id

Abstract. Wedung coast included in a series of coastal areas north of Java Island in Demak Regency. The coastline on the Wedung coast experiences continuous changes due to abrasion and accretion processes. Because the shoreline and relatively flat beaches are constantly impacted by ocean currents and waves, abrasion shoreline modifications take place. The accumulation of silt from the land that is deposited on beaches via river estuaries, on the other hand, is what causes accretion on beaches. One satellite that can be used to observe dynamics in coastal areas is the sentinel 2A imagery satellite. This research seeks to thoroughly assess the condition of the Wedung Coast coastline from 2019 to 2024 and to analyse alterations in the coastline resulting from accretion and abrasion during this period. The coastline extraction method employed is the Modified Normalised Difference Water Index (MNDWI). The assessment of shoreline alteration is conducted via the Digital Shoreline Analysis System (DSAS) technology, applying the Shoreline Change Envelope (SCE) and End Point Rate (EPR) methodology. The findings indicate that from 2019 to 2024, the Wedung Coast experienced 99.86 metres of abrasion and 94.89 metres of accretion, with an average abrasion rate of 13.45 metres per year and an average accretion rate of 11.29 metres per year. The area affected by abrasion on the Wedung Coast is 168.07 hectares, while the area of accretion is 167.55 hectares. Overall, the coastline changes observed on the Wedung Coast from 2019 to 2024 exhibited a predominance of abrasion over accretion.

Keywords: Coastline changes, Abration and accretion, Wedung Coast, MNDWI

1. Introduction

Indonesia is an archipelago with the second longest coastline, following Canada. Indonesia's coastline measures approximately 81,000 km [1]. The environmental state of Indonesia, characterised by an extensive coastline, is a valuable natural resource that requires optimisation. The extensive coastlines of several countries offer numerous advantages for marine tourism, aquaculture, and marine biodiversity cultivation. Another advantage of Indonesia as a marine nation is its strategic location [2].

IOP Conf. Series: Earth and Environmental Science **1503** (2025) 012018 doi:10.1088/1755-1315/1503/1/012018

Indonesian Law No. 01 of 2014, which updates Law No. 27 of 2007 for the Management of Coastal Areas and Small Islands, defines coastal areas as transitional zones between terrestrial and marine ecosystems, influenced by changes from both land and water. These maritime alterations are affected by tides, sea breezes, and seawater infiltration. The north coast of Java is located directly opposite the Java Coast, with characteristics of sloping beaches and sea level no higher than the lowest tide. The processes that occur in the coastal zone, both natural and human-made, make the coastal zone a high resource potential area. However, on the one hand, it also provides the potential for coastal disasters, such as tsunamis, tidal floods, land subsidence and high waves. The inclined features of Java's northern coastline, combined with natural phenomena (e.g. rising sea levels, land subsidence, and tidal movements) and artificial processes (land use change) make it an area vulnerable to coastal abrasion or erosion and tidal inundation [3].

In coastal areas there is a boundary between land and sea waters called the beach. The beach is a dynamic area because it is a meeting place and interaction between land, sea and air. The beach always has continuous adjustments towards a natural balance to the impacts that occur so that it affects changes in the coastline [4]. The shoreline is delineated as the interface between the water and land, affected by tidal fluctuations, and serves as the primary metric for monitoring environmental alterations in coastal regions. The coastline is a dynamic geomorphological feature that continually evolves due to natural processes, climatic circumstances, and artificial influences, including coastal construction, jetty development, and tourism areas [5].

Coastal alterations are an ongoing phenomenon resulting from both erosion and accretion processes [6]. Abrasion induces annual shoreline alteration, manifesting as the observable phenomena of retreating shorelines. Shoreline change is a multifaceted natural phenomenon driven by numerous variables, both natural and anthropogenic [7]. A beach experiences abrasion if the current year's shoreline is more inland than the previous year's shoreline. Beaches with sparse mangrove potential will experience faster abrasion due to waves and waves when compared to beaches that still have a lot of mangroves. However, beaches that have a dense mangrove ecosystem will cause an increase in land along the coast (accretion). Coastal accretion refers to the alteration of the shoreline extending seaward as a result of sediment deposition from terrestrial or fluvial sources into the ocean.

The utilization of remote sensing satellite imagery with the ability to record objects on earth in a multi-temporal and wide coverage at the same time can help in monitoring the dynamics of an area continuously. One of the satellites that can be utilized to observe dynamics in coastal areas is the sentinel 2A satellite image [21]. Satellite images can record coastal conditions periodically so that changes in the coastline will be seen if overlapping two images at different times. The advantage of sentinel 2A imagery is that it is able to carry various patches of high-resolution multispectral imagers with 13 bands. The coast of Wedung, Demak, which is prone to coastal erosion and coastal habitat change, has various environmental elements that need to be monitored simultaneously. Sentinel-2A can detect changes in vegetation, water presence, and sedimentation around the coastline. Spectral bands such as the vegetation band (Vegetation Red Edge) and infrared can provide important information to map the impact of erosion or accretion and potential environmental degradation.

Demak Regency has a coastline length of 72.14 km that stretches from Sayung Subdistrict to Wedung Subdistrict. Significant environmental degradation is currently occurring in Demak Regency's coastal area. Due to coastal erosion and seawater inundation in the pond region, which occurs for a number of years before disappearing, the coastal area of Demak Regency is experiencing a decline in land function. This has a direct impact on the decline in the quality of life of people who live around the coastal areas of Demak Regency. Communities that depend on natural resources, such as fishermen, farmers, and pond farmers, have suffered great economic losses. Alterations in environmental conditions significantly influence the socio-economic and socio-cultural dimensions of the society. Human actions, including industrial operations that lead to environmental contamination, coastal reclamation resulting in erosion and deposition, and other detrimental practices, can induce these changes [8]. Geomorphological considerations impact the rate of coastline change. The coastal geomorphology of Demak Regency, characterised by its low elevation and the annual decline in mangrove land cover, renders its coastal regions increasingly susceptible to disasters, including daily tidal floods that inundate homes and roads in villages [9].

The Wedung coastline area is situated in the northern coastal region of Java Island, within Demak Regency. Wedung Coastal is situated exactly near to the Java Sea on its western flank. Wedung Coastal is a coastal area with shallow coastal characteristics because this coastal area is located on an alluvial plain, besides that this coastal area is influenced by the activities of several rivers that flow into Wedung Coastal. Sedimentation processes influenced by fluvial (river) activities and processes due to marine activities cause erosion of land, triggering shoreline dynamics that are important to study. The study of shoreline dynamics can be used as an initiation step to determine actions and policies so as not to have further impacts that can harm the community, especially coastal communities. In light of the background data presented, the aim of this research is to ascertain the condition of the shoreline along the Wedung Coast from 2019 to 2024. Investigate shoreline modifications due to abrasion and accretion along the Wedung Coast from 2019 to 2024.

2. Method

2.1 Research Location

This study is located in Demak Regency, Central Java Province, near the Wedung Coast. The astronomical coordinates of the Wedung Coast are 6°43'26" - 7°09'43" S and 110°27'58" - 110°48'47" E. The Wedung coast constitutes a segment of the northern coastline chain of Java Island. Wedung Coast is a site in Demak Regency that directly adjoins the Java Sea. Wedung Coastal abuts the Java Sea and is contiguous with Jepara Regency in the northern area. The southern boundary borders Bonang Sub-district, and the eastern boundary borders Mijen Sub-district. The Wedung Coast is geomorphologically classified as a coastal region with shallow beaches, situated on an alluvial plain. The sedimentation process of river flow, which conveys fine material from the mainland to the coast, is the principal factor in the formation of alluvial soil in this region.

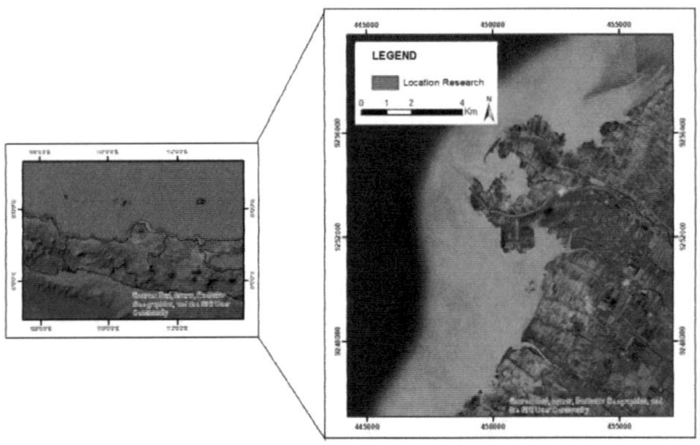

Figure 1. Research Location.

2.2 Tools and Material

The tools and materials utilised in this investigation are appended as follows:

Table 1. Tools used in research

Tool Name	Fungcition
ArcMap 10.8	Spatial data processing and map making
QGIS 3.28.7	Spatial data processing
Google Eaerth Engine	Image processing and MNDWI
DSAS	Calculating the rate of shoreline change
Microsoft Word	Compile a research journal

Table 2. Data used in research

Tool Name	Fungcition
Citra Sentinel 2A	Knowing the appearance of the coastline in 2019-2024
Peta RBI	Provide information on the boundaries of the research location

2.3 Data Processing

This research was carried out using several software or software that functions to process satellite imagery and also spatial data. The following is a flow chart in this research.

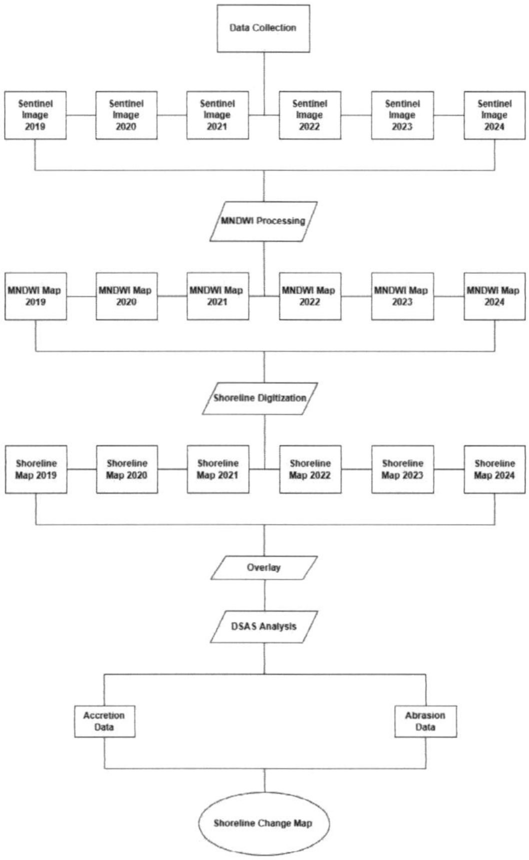

Figure 2. Flow Chart

2.3.1 Image Processing Using the MNDWI Method

The initial phase of data processing involves the utilisation of Sentinel 2A photos through Google Earth Engine (GEE). The composite image from Sentinel 2A utilises band 3 (Green) at a resolution of 10 metres and band 9 (SWIR 1) at a resolution of 20 metres. Sentinel 2A imagery, possessing a resolution of 10 metres, facilitates the mapping of shoreline alterations with considerable accuracy at the regional scale. Subsequently, the analysis of Sentinel 2A imagery is conducted via the Modified Normalised Difference Water Index (MNDWI) methodology. The Modified Normalised Difference Water Index (MNDWI) employs short-infrared channels to differentiate water from other entities, facilitating the extraction of coasts and enhancing the delineation between aquatic and terrestrial areas, hence rendering the coastline's position in the image more discernible. The MNDWI approach in image processing employs the following formula:

$$\text{MNDWI} = \frac{(GREEN - SWIR\ 1)}{(GREEN + SWIR\ 1)}$$

IOP Conf. Series: Earth and Environmental Science **1503** (2025) 012018 doi:10.1088/1755-1315/1503/1/012018

2.3.2 Shoreline Digita zation

The stage carried out after the image processing process is the coastline digitization process. Coastline digitization uses Sentinel 2A imagery which has been processed using the MNDWI method. Based on the image that has been cut according to the research area and has a new image visualization from the rationing formula process [10]. Utilise the on-screen digitisation technique to delineate the coastline in the image. Shoreline digitisation seeks to acquire vector data as lines or polylines, which are subsequently utilised for modelling through the Digital Shoreline Analysis System (DSAS) methodology.

2.3.3 Overlay

Overlay was conducted to combine the results of shoreline digitization from 2019 - 2024, prior to further processing using the Digital Shoreline Analysis System (DSAS). Overlays were performed using shoreline data from 2019, 2020, 2021, 2022, 2023, and 2024. Shoreline data from 2019 - 2024 that has been overlaid shows changes in each year. Then from the data can be seen the comparison of abrasion and accretion, after which it can be analyzed using descriptive analysis methods.

2.3.4 Digital Shoreline Analysis System (DSAS) Modeling

Digital Shoreline Analysis System (DSAS) is software developed by ESRI and USGS and is open source. The analysis system in DSAS uses a point as a reference in calculating shoreline changes, the reference point is obtained from the intersection of the shoreline and the transect line that passes through it. Shorelines from 2019, 2020, 2021, 2022, 2023, and 2024 that have been combined into one will be calculated for the rate of shoreline change. After the calculation process is complete, then the selection of EPR AND SCE data and the calculation of the area of shoreline change on the Wedung Coast are carried out. Validation was conducted using Google Earth Pro to ensure the accuracy of mapping generated from analyzing satellite imagery, such as Sentinel-2A. By utilizing Google Earth Pro's historical view feature, it is possible to compare imagery from different years to verify the extent of shoreline change. This allows us to visually see the changes in shoreline shape and utilize distance measurement tools and coordinates to ensure the accuracy of the position and rate of shoreline shift.

2.4 Data Analysis

This specific study design is part of the quantitative descriptive research methodology. The technique for extracting coastlines is the Modified Normalised Difference Water Index (MNDWI). To assess shoreline changes, the Digital Shoreline Analysis System (DSAS) tool employs the Shoreline Change Envelope (SCE) and End Point Rate (EPR) approaches. A quantitative descriptive analysis of the obtained results is conducted following the shoreline modelling procedure.

3. Result and Discussions

The results of the research obtained from this study are a map of the condition of shoreline changes on the coast of Wedung from 2019 - 2024 obtained from the results of digitizing the shoreline that has been extracted using the MNDWI method, and the second result is a map of shoreline changes on the Wedung Coast caused by abrasion and accretion.

3.1 Map of Shoreline Conditions on the Wedung Coast from 2019 - 2024.

The first result of this research is a map of shoreline conditions on the coast of Wedung from 2019 - 2024. The state of the coastline alters annually. It is affected by various causes, including both natural elements and human activity. Natural variables influencing shoreline position alterations include hydro-oceanographic processes in the sea, such as wave dynamics, variations in current patterns, tidal movements, and climate change [11]. Another cause of abrasion in coastal Wedung is land subsidence. This land subsidence occurs due to excessive pumping of groundwater for industrial and drinking water needs in coastal areas, especially if the coastal soil is dominated by clay or mud which is easily changed due to changes in water content Human activity factors in the form of beach nourishment, excavation of coastal sediments, beach filling, clearing coastal protective plants, making flood canals and regulating watershed patterns [4]. The following is a map of the condition of the coastline on the Wedung coast from 2019 - 2024.

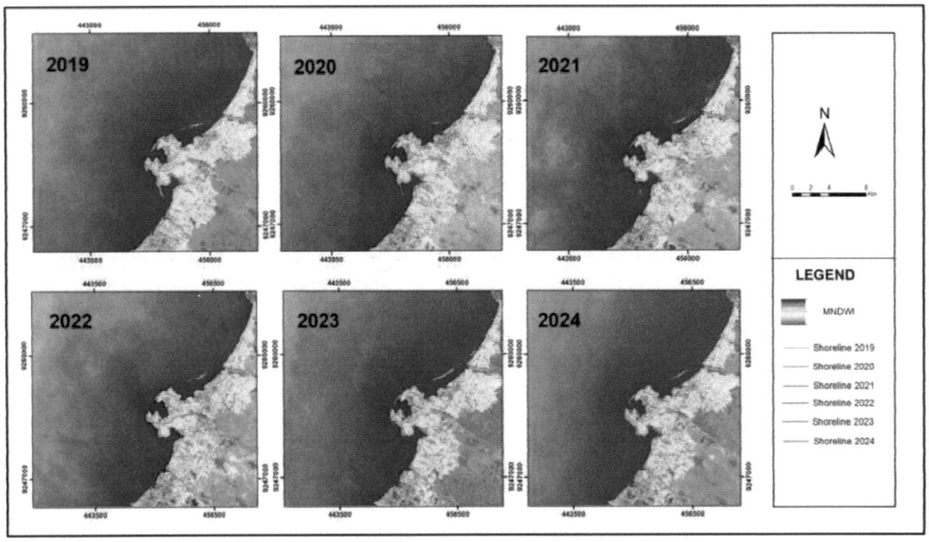

Figure 3. Shoreline condition

The figure illustrates the annual changes in the Wedung Coast's shoreline condition between 2019 and 2024. The figure illustrates the state of the Wedung coastline, which has undergone both abrasion and accretion. In 2019, the Wedung shoreline exhibited minimal abrasion, with certain land portions still visible and no evidence of land erosion in the vicinity. However, from 2020 to 2024 the area continues to experience abrasion, this occurs due to land subsidence which causes a decrease in groundwater elevation accompanied by seawater intrusion, so that the potential for land subsidence is quite large and contributes to inundation (rob) at high tide which causes land erosion [12]. From the figure, it can be seen that the abrasion that occurs on the coast of Wedung is not too significant, accretion or addition of land occurs in the delta area close to the river mouth.

3.2 Map of Shoreline Changes due to Abrasion and Accretion on the Wedung Coast from 2019 - 2024.

Shoreline alterations resulting from erosion and deposition. Abrasion refers to the displacement of sediment that leads to the recession of the shoreline [13]. coastal accretion caused by sediment movement, longshore current, and waves [14]. These maps show the shoreline changes on the Wedung coast from 2019 to 2024 as a result of accretion and abrasion.

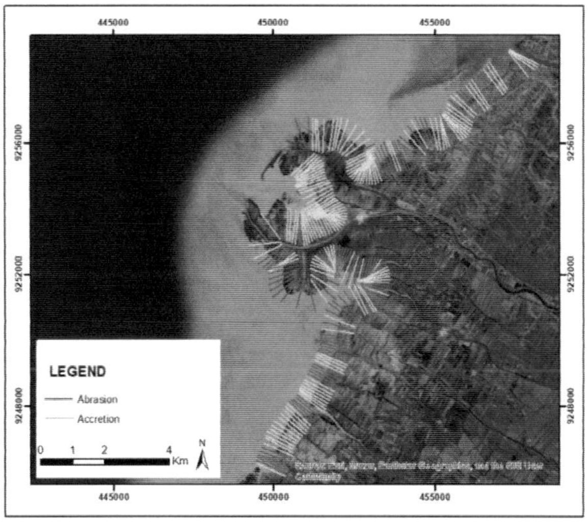

Figure 4. Shoreline change by abrasion and accretion

The Digital Shoreline Analysis System (DSAS) generates research findings that include data on Shoreline Change Envelope (SCE), End Point Rate (EPR), and the extent of shoreline modifications due to accretion and erosion. The Shoreline Change Envelope (SCE) data demonstrates the extent of shoreline modification from the initial year to the concluding year. SCE records reveal that from 2019 to 2024, the Wedung Coast underwent 99.86 meters of erosion and 94.89 meters of deposition. The End Point Rate (EPR) data is employed to calculate the average rate of shoreline change by dividing the distance between the oldest coastline and the current shoreline by the time elapsed. The End Point Rate (EPR) results reveal that the average abrasion rate is 13.45 m/year, whereas the average accretion rate is 11.29 m/year. The erosion zone at Wedung Coast spans 168.07 hectares, whereas the deposition zone covers 167.55 hectares [22]. Between 2019 and 2024, the Wedung Coast saw greater abrasion than accretion in its coastline modifications. Abrasion at Wedung Coast predominantly occurs in the northern region, whilst the southern area is characterized by accretion resulting from sedimentation of materials delivered upstream by the river. This accretion generally occurs in river estuaries, indicating the deposition of material transferred from upstream by river flow and then deposited downstream in the coastal area [15].

Shoreline change is a multifaceted phenomenon that illustrates the dynamic interplay between natural processes and anthropogenic activity in coastal regions. The research findings indicate that a notable coastline alteration is abrasion. Abrasions observed in multiple areas are affected by heightened wave intensity, ocean currents, and elevated sea levels due to climate change. Abrasion in certain areas has led to the disappearance of ponds and towns, whereas sedimentation has caused the formation of fresh land at river mouths, which communities will employ for various purposes [16].

Conversely, accretion phenomena occur in regions characterised by significant sediment activity or influenced by rivers that discharge into coastal zones, transporting sediment from upstream to be deposited on beaches or deltas. Additionally, sediment transported by ocean currents or waves can accumulate on specific beaches, particularly in areas where currents are diminished, such as bays. The coast of Wedung, Demak, comprises sediment types including silt, passive silt, sand, and silty sand [17]. The encroachment of the shoreline typically yields beneficial effects due to the expansion of land; nevertheless, coastal regions influenced by river mouths may experience disturbances from siltation, which disrupts riverine traffic [4]. The coastline has been significantly impacted by human activity over the previous decades to centuries, in addition to the dynamic effects of rivers [18].

This tidal flood is a factor that induces alterations in the shoreline, as water that inundates the land from the sea and subsequently recedes transports sediment from the mainland to the marine area [19]. Other factors that influence shoreline change are the typology of coastal constituents and geological formations that determine the type of soil in coastal areas. An increase in population in coastal areas will also affect shoreline change and land use change into built-up areas for settlements. Changes in the coastline will have an impact on changes in land use and socio-economic activities. One of them is agricultural land which is also influenced by the construction of ports that hamper irrigation lines for rice planting, besides that changes in land cover can also be utilized for mangrove planting as an effort to rehabilitate the environment and maintain ecosystems in coastal areas [20].

4. Conclusions

Research on shoreline changes in Wedung Coast shows that the phenomena of abrasion and accretion are significant dynamics in this area, it can be concluded that shoreline changes that occur are dominated by abrasion. The results showed that the shoreline in Wedung Coast has undergone major changes in recent years, characterized by the loss of most of the land in some areas due to abrasion with an average abrasion of 13.45 m/year. in addition, the accretion that occurs in Wedung Coast is generally around the mouth of the river or the bay area where the current is weak. the rate of accretion that occurs in Wedung Coast is 11.29 m/year. The area in Wedung Coast that experiences accretion and abrasion is not much different. The area of abrasion on the Wedung Coast is 168.07 ha and accretion is 167.55 ha.

5. Acknowledgments

My deepest gratitude to my supervisor, for his guidance, support and direction during the process of writing this journal article. Thank you also to my parents who have supported me all this time. thanks to friends who have provided motivation during the writing of this journal article. Hopefully the results of this research can be useful and contribute to the development of science

6. Reference

[1] J. D. Wattimena, M. R. Ayal, 2019, "Analisis Perubahan Garis Pantai Desa Rutong Kota Ambon" J Tek Sipil, vol. 14, no. 2, pp. 115–136

[2] A. Roziqin, O. Gustin, 2017, "Pemetaan Perubahan Garis Pantai Menggunakan Citra Penginderaan Jauh di Pulau Batam. Proc Ind Res Work Natl Semin" pp. 295–9, [Online]. Available from: http://jurnal.polban.ac.id/index.php/proceeding/article/view/738/591

[3] B. I. D. Astuti, A. Laksono, D. T. N. Muhammad, I. F. Nurbaiti, N. N. Hanifah, O. S. Wildiyanti, et al, 2021 "Dinamika perubahan garis pantai Kabupaten Kendal tahun 2000-2020," *Maj Geogr Indones*, vol. 35, no. 1, pp. 75.
[4] Darmiati, I. W. Nurjaya, A. S. Atmadipoera, 2020, "Analisis Perubahan Garis Pantai Di Wilayah Pantai Barat Kabupaten Tanah Laut Kalimantan Selatan" *J Ilmu dan Teknol Kelaut Trop*, vol. 12, no. 1, pp. 211–22.
[5] R. S. Dewi, W. Bijker, 2020, "Dynamics of shoreline changes in the coastal region of Sayung, Indonesia", *Egypt J Remote Sens Sp Sci.,* vol. 23, no. 2, pp. 181–93. doi: 10.1016/j.ejrs.2019.09.001
[6] D. P. Lubis, M. Pinem, M. A. N. Simanjuntak, 2017, "Analisis Perubahan Garis Pantai Dengan Menggunakan Citra Penginderaan Jauh (Studi Kasus Di Kecamatan Talawi Kabupaten Batubara)," *J Geogr.*, vol.9, no. 1, pp. 21.
[7] B. Taruna, A. A. Wibowo, H. Aziz, 2024, "Detection of Shoreline Changes with Automated Water Extraction Index (AWEI) in Galur District, Kulon Progo Regency" *IOP Conf Ser Earth Environ Sci.*, vol. 1357, no. 1.
[8] F. Istiqomah, B. Sasmito, F. J. Amarrohman, 2016, "Pemantauan Perubahan Garis Pantai Menggunakan Aplikasi Digital Shoreline Analysis System (DSAS)." *J Geod Undip.* Vol. 5, no. 1, pp. 78–89.
[9] T. P. G. Primasti, H. Hariyadi, B. Rochaddi, S. Widada, R. Widiaratih, 2021, "Pemantauan Kerentanan Fisik di Pesisir Kabupaten Demak (Studi Kasus Perubahan Garis Pantai)," *Indonesia Journal Oceanography,* vol. 3, no. 1, pp. 44–54.
[10] I. P. Akbaruddin, B. Sasmito, A. Sukmono, 2020, "Analisis Korelasi Luasan Kawasan Mangrove Terhadap Perubahan Garis Pantai Dan Area Tambak (Studi Kasus: Wilayah Pesisir Kabupaten Demak)" *J Geod Undip*, vol. 9, no. 2, pp. 217–26.
[11] Y. P. Ramadhani, I. Praktikto, C. A. Suryono, 2021, "Perubahan Garis Pantai Menggunakan Citra Satelit Landsat di Pesisir Kecamatan Sayung, Kabupaten Demak," *J Mar Res*, vol. 10, no. 2, pp. 299–305.
[12] Ima Nurmalia Permatasari, 2021 "Kajian Resiko, Dampak, Kerentanan dan Mitigasi Bencana Abrasi Dibeberapa Pesisir Indonesia" *J Ris Kelaut Trop (Journal Trop Mar Res*, vol. 3, no. 1, pp. 56.
[13] R. Aldian, E. Zuryani, A. Z. P. Ulni, 2022, "Perubahan Garis Pantai Sebagai Akibat Dari Abrasi Dan Akresi Di Kawasan Pesisir Pantai Barat Sumatera Barat" *Soc Humanit Educ Stud Conf Ser.*, vol. 5, no. 4, pp. 152.
[14] N. Kurniadin, F. Fadlin, 2021, "Analisis Perubahan Morfologi Garis Pantai Akibat Tsunami di Teluk Palu Menggunakan Data Citra Sentinel-2" *Geoid,* vol. 16, no. 2, pp. 240.
[15] B, Sasmito, 2020, "Analisis Perubahan Garis Pantai Akibat Kenaikan Muka Air Laut Pantai Kabupaten Demak," *Elipsoida J Geod dan Geomatika,* vol. 3, no.02, pp. 178–84.
[16] E. C. Hastuti, T. B. Sanjoto, P. Hardati, 2018, "Perubahan Garis Pantai dan Pengaruhnya Terhadap Status Kepemilikan dan Penguasaan Tanah Timbul di Muara Sungai Wulan Tahun 1986-2016" *Geo Image,* vol. 7, no. 2, pp. 131–140.
[17] M. F. Erfiko, S. Widada, W. Atmodjo, 2023, "Pemetaan Pola Sebaran Sedimen Dasar Di Perairan Wedung, Demak" *Indones J Oceanogr.*, vol. 5, no. 2, pp. 132–40.
[18] C. Gomez, D. E. Hart, P. Wassmer, I. Kenta, H. Matsui, M. Shimizu, 2019, "Coastal Evolution, Geomorphic Processes and Sedimentary Records in the Anthropocene" *Forum Geografi*, vol. 33, no. 1, pp. 1–24.
[19] I. A. Abdullah, Danardono, H. Z. Hadibasyir, 2023, "The Dynamics of Changes in the Coastline of the Sayung Coastal Community in Demak Regency Using Remote Sensing Imagery," *Atlantis Press SARL*, pp. 436–449, doi: 10.2991/978-2-38476-066-4_28
[20] R. Purwanti, R. H. Koestoer, 2024, "Tinjauan Perubahan dan Prediksi Garis Pantai: Studi Perbandingan Kasus di Sri Lanka dan Indonesia" *J Mar Res.*, vol.13, no. 2, pp. 203–11.
[21] B. K. K. Ahammed, I. W. G. A. Karang, I. W. Nuarsa, I. G. S. Indrawan, N. M. N. B. S. Dewi, & A. C. Pandey, 2024, Mapping of the Spatio Spectral Dynamics of Mangrove Chlorophyll Concentrations via Sentinel-2 Satellite Imagery," *Forum Geografi*, vol. 38, no. 2, pp. 244-246. doi: 10.23917/forgeo.v38i2.4518.
[22] A. Kurniawansyah, D. Susiloningtyas, & M. D. M. Manessa, 2023, "Suitability of Mangrove Tourism Areas in Cilamaya Wetan District, Karawang Regency," *Forum Geografi*, vol. 37, no. 1, pp. 10-20. doi: 10.23917/forgeo.v37i1.19852

IOP Conf. Series: Earth and Environmental Science **1503** (2025) 012019 doi:10.1088/1755-1315/1503/1/012019

Natural Disasters and Damage to Archaeological Sites in Trenggalek Regency, East Java.

Hery Priswanto[1*], Irfanuddin Wahid Marzuki[2] and Muhammad Chawari[3]

[1-3] Research Center for Prehistoric dan History Archaeology, National Research and Innovation Agency, Yogyakarta, Indonesia
*E-mail: priswanto.balaryk@gmail.com

Abstract. Trenggalek Regency has a variety of archaeological sites ranging from pre-historical, classical, and Islamic, to colonial periods. The research method used in historical archaeology combines archaeological methods with written sources (archives, old photos, and old maps) with a multidisciplinary approach. The purpose of the research is to see and test the damage to archaeological sites in the Trenggalek Regency area so that the right method can be produced for saving archaeological sites if there is a disaster in the future. The study results show that disasters that damage archaeological sites in the Trenggalek Regency area are mostly in the form of floods and landslides. This condition is caused by the geographical condition of Trenggalek and the existence of archaeological sites around the Ngasinan River. The location of these archaeological sites is in the morphology of the basin which has a sloping topography, is close to water sources, and contains fertile soil, which is ideal for choosing a site location as a settlement or erecting ritual buildings. However, the selection of the location also carries a very significant impact and a great risk of overflow (flooding) due to its sloping position and near the river basin.

Keywords: Disaster, damage, archaeological site, Trenggalek,

1. Introduction

Trenggalek Regency is an area in the southern part of East Java province. Trenggalek Regency is located at coordinates 111°24'-112°11'E and 70°63'S-80°34'S. Trenggalek Regency has an area of 1261.40 km2. The boundaries of Trenggalek Regency in the north are bordered by Ponorogo and Tulungagung Regencies, in the east by Tulungagung Regency, in the south by the Indian Ocean, and in the west by Ponorogo and Pacitan Regencies (Figure 1). Trenggalek Regency consists of 14 sub-districts, namely Dam, Dongko, Durenan, Gandusari, Kampak, Karangan, Munjungan, Panggul, Pogalan, Pule, Suruh, Trenggalek, Tugu, Watulimo. The number of residents in Trenggalek Regency is 693,104 people.[1].

Trenggalek Regency has topographical, geological, soil, and land use conditions of the Ngrowo-Ngasinan sub-watershed which is in the Gandusari and Kampak Districts. Gandusari and Kampak Districts are in a highland area along Mount Wilis which has a slope of 7°–39°. The geology of the Ngrowo-Ngasinan River sub-watershed consists of Holocene volcanic rock and tertiary sedimentary rock. The land use in the Ngrowo-Ngasinan River sub-watershed is forests, moors, irrigated rice fields, plantations, and settlements with the highest percentage being irrigated rice fields and settlements [2]. Trenggalek Regency has 28 large and small rivers with a length ranging from 2.00 km to 41.50 km. [3]. The Ngasinan River is a fairly large river estuary, namely on the north side there is the Bagong River which empties into Tamanan Village, the Prambon River which empties into Tugu District, and on the west side there is the Pinggir River which empties into Tugu District and on the south side there is the Nglongah River (Mlinjon) which empties into Trenggalek District. These rivers in the Brantas watershed are mostly used for irrigation and some enter the Niyama hydropower plant. Several tributaries in the Ngrowo Sub-watershed, namely the Tawing River and the Ngasinan River, often cause flood disasters every year in Trenggalek Regency due to high rainfall and topographic factors. Trenggalek Regency consists of 2/3 of its area in the form of mountains and 1/3 of its area in the form of lowlands with an altitude between 0 to 1,250 meters above sea level[4].

Figure. 1. Location of Trenggalek Regency, East Java Province, Indonesia.
Sumber: googleearth.com

Trenggalek Regency has seventeen mobile and immovable cultural heritage distribution points. Trenggalek has been inhabited for thousands of years from the prehistoric period to the independence period. Based on the Kamsyaka Inscription (929 AD), Trenggalek includes areas that receive autonomy rights, namely Perdikan Kampak covering Panggul, Munjungan, and Prigi which borders the Indonesian Ocean. The Kamsyaka Inscription also mentions the Dawuhan area which is one of the areas in Trenggalek Regency. In addition to the Kamsyaka Inscription, in Trenggalek there is also an inscription that is quite important, namely the Kamulan Inscription (1194 AD) which contains information about the commemoration of the king's gift to the people in the Kamulan area. The Kamulan inscription made by King Kadhiri is Sri Sarweswara Triwikramataranindita Srengga Lancana Dikwijayatunggadewa better known as Kertajaya. Information about the day, date, month, and year of making the Kamulan Inscription was later designated as the Anniversary of Trenggalek Regency. [5]. The writing of this article aims to convey information about the impact of natural disasters and their damage to archaeological remains in Trenggalek Regency.

2. Methods

This research used a historical archaeological approach combining archaeological and non-archaeological data (history, geography, and ethnography). Historical archaeology deals with the study of the archaeological remains accompanied by written notes related to the history of the remains[6]. It was expected that the use of written data, in addition to material data, could eventually reveal the background history of the archaeological remains. Historical archaeology studies cultural history as well as cultural processes documented in written sources while using historical documents to find out past life [7]. Historical archaeology studies focus on the period of history (after literation) and are defined as the study of the material remains from some periods of history. [8] [9]. Historical archaeology's sources range from written documents (archives, maps, historical notes, manuscripts) and archaeological findings to collectible materials belonging to informants used to interpret society's history and culture. The archaeological data used are artifacts and sites obtained through surveys and excavations.

3. Result and Discussion

3.1 Disaster History in Trenggalek

The types of natural disaster threats in Trenggalek Regency are floods, earthquakes, tsunamis, landslides, flash floods, droughts, tornadoes, and floods. Over the past 5 years, the health crisis that has been experienced has been flooded. [1]. Some areas of Trenggalek Regency are disaster- prone areas, especially landslides and floods, besides that in the south of Trenggalek Regency, namely in the coastal area, is an area prone to tsunami disasters and tectonic earthquakes due to plate collisions. Landslide disasters often occur in Trenggalek Regency, especially if the rainy season arrives. Two factors cause most of the Trenggalek area to be categorized as prone to medium and high-level disasters. The first is natural factors consisting of geological and soil aspects, hydrological and climatological aspects, topographic aspects, and land cover aspects (vegetation). The second is the human factor that uses nature irresponsibly. Districts that are included in the category of high vulnerability for landslide disasters are Panggul District, Munjungan District, Watulimo District, Kampak District, Gandusari District, and Bendungan District. Trenggalek Regency is one of the areas in East Java Province that often experiences disasters. On the website of the Regional Disaster Management Agency (BPBD) of Trenggalek Regency, there were 153 disaster events throughout 2018 [10].

Based on historical data, floods in Trenggalek have occurred for a long time. The Dutch colonial government even built the Neyama tunnel in southern Tulungagung to reduce the floods that hit Trenggalek and Tulungagung [11], [12]. Reports of Floods during the Dutch East Indies period can be known, one of which is from the report of the Resident of Kediri to the Governor General of the Dutch East Indies in Buitenzorg (Bogor) saying that the flood that occurred was caused by the rising water of the Brantas River in Poerwosari, so that the Ngoedjang bridge in Afdeeling Ngrowo was washed away on March 25, 1894 [11]. The Trenggalek area is included in the Ngasinan- Ngrowo watershed which has small rivers that empties into the Brantas River. To reduce flooding, the government has built several dams in the Ngasinan watershed area, namely Tugu Dam, Bagong Dam, and Kampak Dam (Table 1).

Table 1. Potential of Ngasinan Watershed Reservoir

Description	Dam Tugu	Dam Bagong	Dam Kampak
Location	1.0	1.5	2.0
River	Kali Keser	Kali Bagong	Kali Tawing
Catchment Area	43,60 km2	45,10 km2	53,50 km2
Inundation Area	1,152 km2	1,44 km2	0,68 km2
High	53, 0 m	90,0 m	40,0 m
Storage Volume	9,30 juta m3	55,20 juta m3	7,46 juta m3

A major flood disaster in Trenggalek occurred in 2016. Floods inundated Gandusari District due to the overflow of the Tawing River due to the amount of rainfall that fell due to weather anomalies and the inability of the Tawing River drainage capacity due to changes in the conditions of the river drainage area and river siltation (Figure 2). Flooding incidents only occurred in the Tawing River and did not occur in the main river Parit Raya. To control flooding in the Tawing River in Trenggalek Regency, it is necessary to carry out structural technical efforts based on several recommendations in previous studies, including river normalization, embankment construction, and realization of the Kampak Dam construction plan [4].

Figure. 2. Major Flood in Trenggalek in 2016
Source: https://www.liputan6.com/regional/read/2603266/banjir-bandang-rendam-tulungagung-dan-trenggalek

2.2 Archaeological Sites in Trenggalek
Archaeological sites in the Trenggalek Regency area are diverse, from the prehistoric period to the independence period. The existence of these archaeological sites proves the role of Trenggalek from time to time. Archaeological sites in the Trenggalek Regency area based on the results of the 2020 archaeological survey [13] in Trenggalek Regency are found in 8 sub-districts, namely Gandusari, Watulimo, Dongko, Panggul, Pogalan, Trenggalek, Durenan, and Tugu.

Archaeological remains found in Gandusari District in the form of Nglayur mounds, Pillar Fragments, Dakon Nglayur 1 and Dakon Nglayur 2 and Nglayur Stone Lumpang in Nglayur Hamlet,

Sukoarjo Village; and Lumpang Batu Sanan in Sanan Hamlet, Sukorejo Village, and Lumpang Batu Sampang in Sampang Hamlet, Wonorejo Village. Archaeological remains in Watulimo District are in the form of stone lumps, stone pestles, and Waru pillboxes. The findings of Lumpang Batu and Alu Batu are in Slawe Village. Pillbox Waru is in Slawe Village in the form of a building with a round plan made of cast cement with iron construction and equipped with an entrance and three reconnaissance holes. One of the Pillbox Waru reconnaissance holes leads to Prigi Beach which is about 5 km away. At the top of the Waru Pillbox is found a protrusion between 2-5 cm high which is suspected to be an additional construction, but the condition has been damaged.

Archaeological remains in Dongko District in the form of a stone lumpang located in Pandean Village. Lumpang Batu Krajan is believed to be the forerunner of the establishment of Krajan Hamlet. The stone lump made of andesite monolith measures 82 cm in diameter, 57 cm high with a hole diameter of 32 cm. The condition of the Krajan stone lump in the yard of the residents' houses is still intact and well maintained. Archaeological remains in Panggul District are in the form of two pillbox buildings located in the Konang Beach tourist area in Nglebeng Village and several inscriptions in Panggul Village and Wonocoyo Village. Pillbox Nglebeng 1 is a Japanese defense building made of cast cement with an iron construction equipped with an entrance to the north and three reconnaissance holes to the south facing Konang Beach. The dimensions of the Nglebeng 1 Pillbox are 283 cm long, 202 cm wide, 132 cm high, and 14 cm wall thickness. Pillbox Nglebeng 2 is hexagonal with one entrance and three reconnaissance holes with a building height of 116 cm, a bottom width of 315 cm, and a wall thickness of 14 cm. The condition of the Nglebeng 2 Pillbox is still relatively intact, but the reconnaissance hole has been covered with cement.

Archaeological remains in Pogalan District are in the form of Dakon Rinjanjang Stones, Yoni Rinjanjang, and Punden Rinjanjang, which are in Rijangan Hamlet, Ngulan Kulon Village, and Arabic inscriptions in the Jatisari Tomb located in Pogalan Village. Yoni Rinjanjang is made of andesite whose upper cross-section is worn out with a lower cross-sectional dimension of 66 cm x 66 cm, an upper cross-section of 56 cm x 56 cm with a height of 64 cm, and a spout length of 23 cm. In Pogalan District, a brick temple structure is also found, namely the Brongkah temple (Priswanto, 2017). Archaeological remains in Trenggalek District are in the form of a stone lump located in Sosutan Hamlet, Ngantru Village. The Sosutan Stone Lumpang is made of andesite stone with a diameter of 58 cm and a height of 40 cm. The condition of the Sosutan Stone Lump on the upper section has been broken.

Archaeological remains in Durenan District found four yoni and one sundial located at the Darussalam Sumbergayam Islamic Boarding School in Sumbergayam Village. This Yoni was once used as a pillar for the Islamic boarding school mosque. Morphologically, the four yoni can still be identified in shape and spout. The sundial located on the north side of the Darussalam Mosque is still functioning. In Durenan District, brick structures were also found, namely the Semarum Site which is suspected to be a fresh or ancient water reservoir, ancient brick ruins in Kendlarejo Village, and the Kamulan Site [14]. Archaeological remains in Tugu District are in the form of a brick structure which is a temple building structure called Gondang Temple.

3.3 Disaster Impact on archaeological sites

The disaster that hit the Trenggalek Regency area not only caused damage to infrastructure buildings and community agriculture but also to existing archaeological sites. Archaeological sites are vulnerable to disasters, not only during events but can be affected and damaged during emergency response and recovery. Several archaeological remains such as the Semarum Site, the Kamulan Site, and the Brongkah Temple located in the Das Ngrowo Ngasinan channel are clear

IOP Conf. Series: Earth and Environmental Science **1503** (2025) 012019
doi:10.1088/1755-1315/1503/1/012019

evidence that natural disasters have a quite destructive impact on archaeological remains. The Ngasinan River flows from Trenggalek to Tulungagung through the central Durenan District.

The Ngasinan watershed is one of the upstream parts of the Brantas River located in the Trenggalek and Tulungagung Regencies, East Java Province. The flow of the Ngasinan and Ngrowo Rivers ends in the Brantas River. The Ngasinan-Ngrowo watershed is one of the headwaters of the Brantas River, included in the watershed that has the widest rainfall catchment area when compared to other watersheds. In the northern part of Trenggalek Regency, the Rajegwesi Mountains are located, including Pringulung, Condrogeni, Gunung Cilik Ngesong, Tumpak Cenil, Watu Blandong, Mount Kekep, and one of them is Mount Bale [15], which imagines forming a basin morphology around the Ngasinan watershed.

The morphology of the basin results in alluvial deposits that accumulate very thickly, dominated by fine material with a clay texture to sand as the main matrix, with a distribution of floating structured gravel fragments (open containers). In the distribution area of this alluvial sediment, the main river flows, namely the Ngasinan River as a permanent (perennial) river, which obtains water supply from a fairly dense flow pattern, which flows upstream in the surrounding high areas with an almost centripetal radial flow pattern [16]. The river channel that tends to meander carries a considerable sediment load, some of which is deposited in the floodplain area, especially in the inner curves of the bend that form a point bar with a thick layer. The extreme difference in thickness is shown through the difference in the sediment layer of the site's stockpiling, between the Kamulan Site which is relatively far from the river flow, and the Brongkah Temple which is very close to the Ngasinan River. Both are examples of buried landscapes in the alluvial sediment distribution area in the Trenggalek region, the disclosure of which needs to take into account local geomorphological aspects (Figure 3). The landscape conditions of this alluvial sediment distribution area (basin area) also seem to be an important consideration in the selection of settlement locations in the classical period. In addition to its good environmental resources, including soil layers and water potential that ensure land fertility, its physiology also guarantees regional security factors [16].

Figure 3. Ancient Brick Ruins in Kendalrejo due to the overflow of the Ngasinan watershed (Source: Author).

Bongkah Temple is a brick temple building. The position of the temple building is now located at a depth of 3 meters below the ground level. Repeated floods in the Ngasinan River are suspected to be one of the causes of the burgeoning of the Brongkah temple at this time [5]. The building plan of Brongkah Temple is in the shape of a square with a size of 4 meters x 4 meters and an estimated building area of about 16 meters (Figure 3). Based on the decorative pattern of the relief stones, it is suspected that the Brongkah Temple building is a temple building of the Majapahit period with Hindu religious patterns [17]. The condition of this temple building still leaves the temple legs and a little part of the lower body. The building materials of the temple are in the form of bricks with a length of 32-34 cm, a width of 22-26 cm, and a thickness of 7.5-8 cm [15].

In addition to Brongkah Temple, another site that has been damaged by the flood disaster is the Semarum Site which is adjacent to the Ngasinan River. Archaeological findings at the Semarum Site are in the form of brick structures that are estimated to be temple buildings. The brick structure of the Semarum Site consists of 12 layers of bricks that are neatly arranged and without specs (no adhesive/kosot) (Figure 4). The kosot technique is a technique by rubbing the surface of the brick repeatedly with a little water. The brick is rubbed repeatedly with a little water, which then produces species, which are the glue between the bricks. The construction of the brick structure of the Semarum Site consists of leather stone (outer stone) and infill stone (inner stone) using whole bricks. The condition of the brick structure is still relatively intact on the outside from the top to the bottom layer, while the condition of the brick on the inner side has no longer been visible (destroyed). The initial suspicion of the destruction of the lower layer of brick (5 layers) is that it has been eroded or submerged in water for a long time. This is also supported by stratigraphic data that finds a layer or layer of coarse and fine sand with a thickness of about 20 cm [14].

Figure 4. Archaeological remains at the current Semarum Site (source: Author)

Based on data and monitoring results in the field, flood disasters are the main contributor to the damage to archaeological sites in the Trenggalek area. Its location is in a basin area and is traversed by several rivers and high rainfall causes frequent floods in Trenggalek Regency. The Dutch colonial government has carried out flood control efforts in Trenggalek since 1939 [4]. The flood control activities were then continued during the Japanese occupation by building the South Tulungagung Tunnel (Neyama Tunnel), and Parit Raya [12].

The vulnerability of archaeological sites is caused by various factors, such as population growth and urbanization in urban areas [18]. Resilient Cultural Heritage disasters are cultural heritage and its environment that can survive and be restored from damage caused by the impact of natural disasters or caused by human factors [18]. Realizing a disaster-resilient Cultural Heritage has become an international disaster risk reduction agenda. In 2007, UNESCO adopted the document "The Strategy for Reducing Disaster Risks at World Heritage Properties". The goal is for the 14 Disaster Resilient Cultural Heritage to adopt the principles outlined in "the Hyogo Protocol" for the preservation of world heritage. This ambition is reinforced by the publication of "the Sendai Framework for Disaster Risk Reduction 2015-2030" by the United Nations which mentions the importance of disaster risk reduction for Cultural Heritage. Archaeological sites have important values for society and the nation that are not easy to represent. Based on Article

1 of Law No. 11 of 2010 concerning Cultural Heritage, the important values contained in archaeological sites include; important historical, scientific, educational, religious, and/or cultural values. The important historical value ranks first in the Cultural Heritage Law. Archaeological sites have important value if they can provide an overview of certain historical events or figures in the past [19]. Archaeological resources have important scientific and educational value if they have the potential to be further researched to answer knowledge problems [20] and represent important stages in science and the development of science [21]. Archaeological sites have important religious value, namely as cultural products that represent a certain religion. The important value of culture is that the archaeological site can represent certain achievements, encourage the process of cultural creation, and become the identity of a certain nation or community [20].

4. Conclusions

The flood natural disaster that occurred in the Trenggalek Regency area destroyed existing archaeological sites. Evidence of this damage can be seen from the condition of archaeological findings in the form of temples and brick building structures buried in the soil resulting from the flood deposits of the Ngasinan and Ngrowo watersheds. To realize a cultural heritage that is resilient to disasters, it is necessary to take integrated and integrated steps from all elements and stakeholders. Resilient cultural heritage is a readiness to face and prevent damage to cultural heritage due to disasters; the ability to rescue Cultural Heritage in times of emergency and immediately carry out recovery. The impact of disasters on archaeological sites often results in total damage and often cannot be recovered. Therefore, management needs to pay attention to securing and saving archaeological sites that are related to disaster risk. Cultural Heritage that is resilient to disasters cannot be separated from the preservation and management of Cultural Heritage.

References

[1] Pusat Krisis kesehatan, Krisis Kesehatan. 2020.
[2] Erlina, "Analisis Banjir dan Sedimentasi Wilayah Sungai Brantas (Tinjauan terhadap Metode Pengendalian)," Jurnal Teknik Sipil, vol. XIII, no. 1, hal. 1–14, 2018.
[3] F. E. A. Ma'ruf, Kabupaten Trenggalek dalam Angka Tahun 2020. 2020.
[4] K. Rahman, "Evaluasi Dan Pengendalian Banjir Di Kabupaten Trenggalek : Studi Kasus Banjir Agustus 2016," in Conference: PIT HATHI XXXIIIAt: Semarang, 2016, no. November 2016, hal. 509–518.
[5] D. W. Utomo, "Laporan Peninjauan Struktur Bata Kuna Di Desa Semarum, Kecamatan Durenan, Kabupaten Trenggalek," Mojokerto, 2012.
[6] P. P. Funari, "Archaeology in History," in Historical Archaeology Back From the Edge, P. P. Funari, Ed. London: Routledge, 1999.
[7] N. Iriyanto, "Benteng-Benteng Kolonial Eropa di Pulau Ternate dalam Peta Pelayaran dan Perdagangan Maluku Utara," Yogyakarta, 2010.
[8] C. E. J. O. and B. . Fagan, Historical Archaeology. New York: Harper Collins, 1995.
[9] M. Hall dan S. Silliman, "Introduction: Archaeology of the Modern World," in Historical Archaeology, vol. 9, 2006, hal. 1–25.
[10] B. F. Maulana dan M. Taufik, "Pemetaan Daerah Potensi Longsor Di Kabupaten Trenggalek Menggunakan Data Citra Satelit Multi-Temporal," Geoid, vol. 15, no. 2, hal. 256, 2020, doi: 10.12962/j24423998.v15i2.7723.
[11] N. Istieni, "Banjir di Tulungagung Tahun 1955-1986," AVATARA, e-Journal Pendidikan Sejarah, vol. Vol. 6, no. No. 2, 2018.
[12] L. Kusairi, "Perang Memori dan Historiografi Studi Penyebutan Terowongan Neyama di Tulungagung Jawa Timur," Jurnal Widya Citra, vol. 1, no. 2, hal. 37–52, 2020.
[13] A. Priswanto, Hery. Muhammad Chawari, "Laporan Peninjauan Temuan Baru Benda Yang Diduga Cagar Budaya Di Kabupaten Trenggalek," 2020.
[14] H. Priswanto, "Berita Penelitian Arkeologi," 2017.
[15] H. Priswanto, "Penelitian Arkeologi Di Situs Semarum Desa Semarum Kecamatan Durenan Trenggalek," Yogyakarta, 2012.
[16] J. S. E. Yuwono, "Wilayah Trenggalek Dalam Konteks Lanskap," in Berita Penelitian Arkeologi " Bentuk & Karakter Situs Semarum dan Kamulan di Kabupaten Trenggalek," Yogyakarta: Balai Arkeologi Yogyakarta, 2017.
[17] R. Siagian, Candi Sebagai Warisan Seni & Budaya Indonesia. Yogyakarta: Yayasan Cempaka Kencana, 2002.
[18] F. Arda, G. Ghautama, P. Wijayanto, L. S. Muthmainnah, dan Y. Widiastomo, Pedoman Cagar Budaya Tangguh Bencana. 2024.

IOP Conf. Series: Earth and Environmental Science **1503** (2025) 012019 doi:10.1088/1755-1315/1503/1/012019

[19] & S. Pearson, Michael, Sullivan, Looking After Heritage Places - The Basic of Heritage Planning for Managers, Landowners and Administrators,. Melbourne University Press., 1995.
[20] D. A. Tanudirjo, "Kriteria Penetapan Benda Cagar Budaya - makalah dalam Rapat Koordinasi Pedoman Penetapan Benda Cagar Budaya," Cirebon, 2004.
[21] H. T. Wahyu, "Pelestarian dan Pemanfaatan bangunan Indis di Kawasan Kota Baru," Universitas Gadjah Mada, 2011.

IOP Conf. Series: Earth and Environmental Science **1503** (2025) 012020 doi:10.1088/1755-1315/1503/1/012020

Disaster Risk Reduction Based on Community Empowerment in Tambakrejo Village, Semarang City

Wahid Akhsin Budi Nur Sidiq[1*], Nana Kariada Tri Martuti[2], Isti Hidayah[3], Danang Dwi Saputro[4], Sri Ratna Rahayu[5] and Muhammad Adam Khatamy[1]

[1] Geography Department, Faculty of Social Sciences and Political Science, Universitas Negeri Semarang, Semarang, Indonesia
[2] Biology Department, Faculty of Mathematics and Natural Sciences, Universitas Negeri Semarang, Semarang, Indonesia
[3] Mathematics Department, Faculty of Mathematics and Natural Sciences, Universitas Negeri Semarang, Semarang, Indonesia
[4] Mechanical Engineering Department, Faculty of Engineering, Universitas Negeri Semarang, Semarang, Indonesia
[5] Medicine Department, Faculty of Medicine, Universitas Negeri Semarang, Semarang, Indonesia

[*]E-mail: akhsin1987@mail.unnes.ac.id

Abstract. Tambakrejo Village is geographically located on the coast of Semarang City with an area elevation of 1.5-2.8 above sea level which is in the downstream part of the Tenggang River and the East Flood Canal, which has an impact on the high risk of flooding and tidal disasters. Flooding has occurred in January (2023) and March (2024) which resulted in 2,000 houses flooded 50 - 100 cm. In addition, Tambakrejo Village has a high population and building density with a total of 10,429 people (2023), which has an impact on the high threat of fire. However, the high population is a potential resource in reduction on disaster risk. The purpose of this research and service activity is to the risk reduce of flood and fire disasters through community empowerment. The method of activities with participatory mapping, field surveys (GPS surveys, questionnaires and interviews), and community and regional capacity building. The results of participatory mapping show that 57.03 hectares (54%) had a high threat of flooding. As for fire disasters, all regions have a high threat. Furthermore, the level of community and regional capacity to flood and fire disasters is in the medium category. As an effort to reduce disaster risk in the region, community and regional capacity building is carried out through the abode of a Disaster Safe Education Unit at Elementary School 3 Tambakrejo with disaster education activities, facilitation of emergency response equipment and disaster emergency response simulations. Furthermore, regional capacity building is also carried out by making an Early Warning System (EWS) to monitor the water level of the East Flood Canal as a flood emergency response effort. All activities are carried out by involving the community directly as an effort to reduce disaster risk in Tambakrejo Village.

Keyword: Disaster Risk, Community Empowerment, Tambakrejo Village

1. Introduction

The phenomenon of urbanization and climate change is another one of the factors that affect the high risk of disasters in an area, including floods and fires, especially in coastal areas directly adjacent to the sea[1][2]. Climate change can cause sea levels to rise, leading to flooding and tidal disasters [3][4], In addition coastal areas with high building density can also have the potential for fires that can be caused by increased temperatures in the atmosphere as a climate change result [5]. The problem of floods and fires in coastal areas occurs in several countries in Southeast Asia that are mostly bordered by the sea, such as in the Philippines, which was affected by floods and fires in the Manila and Cebu regions (2012 and 2019)[6], subsequent flooding in the Irrawaddy Delta, Myanmar (2008) that left 77,000 dead and 55,900 missing [7], and flooding in the Mekong Delta, Vietnam (2000) due to heavy rains and sea level rise that killed 587 people. [8][9]. Based on some of these cases, it shows that coastal areas have a high threat for floods and fires, one of which is Indonesia, which most of its territory is directly adjacent to the sea with complex land use.

One of the coastal areas in Indonesia affected by disasters as a result of climate change and urbanization is Tambakrejo Village, Gayamsari District, Semarang City. Based on its geographical condition, Tambakrejo Village located in the coastal area of Semarang City which is located between 2 rivers, namely the Garang River (East Flood Canal) and the Tenggang River which empties into the Java Sea. Topographic conditions in the area are relatively flat with elevations between 2 - 4 above sea level, besides that the area also experiences the impact of land sinking/subsidence that occurs in the coastal areas of Semarang City with land sinked every year 24-36 cm [10][11]. These geographical conditions make the Tambakrejo Village area has a high threat of flooding and tidal disasters. Floods have occurred in almost all areas of Tambakrejo Village, including on December 31, 2022, floods as high as 30 - 75 cm have submerged the Tambakrejo Village area which resulted in around 2,000 residents' houses being submerged, Furthermore, in January 2023, tidal floods have submerged the Tambakrejo area with a water level of around 30 - 100 cm. Rob that occurs causes settlements to become slum and unhealthy and damages various infrastructure, such as houses, roads and various public facilities[12][13]. The disaster also disrupted economic activity and education in the area.

Furthermore, Kelurahan Tambakrejo also has a high level of building density, this is due to the large number of residents with a total of 10,429 people in 2022 who inhabit an area of 74.05 hectares. The high density of the building can be seen from the distance between houses that are less than 15 cm or even sticking together for each building. This certainly has the potential for a high threat of fire disaster because if a fire occurs it can spread quickly to other buildings [14]. In addition, in the area there are also several Micro, Small and Medium Enterprises, including milkfish processing which in its production activities uses heating and combustion equipment so that it can potentially lead to fires. The high level of fire threat is exacerbated by relatively narrow road access so that if a house fire occurs it is difficult to be quickly handled by a fire truck. Various disaster risk reduction efforts have been carried out in the region by involving various stakeholders including Tambakrejo Village, Universitas Negeri Semarang (UNNES), Regional Disaster Management Agency, PT Saka Energi Muria Limited (SEML) and several other stakeholders. Various disaster risk reduction programs in Tambakrejo Village include institutional aspects with the formation of Disaster Preparedness Groups, disaster emergency response training for kindergarten students, construction of pump houses to reduce the duration of inundation during floods and facilitation of fiberglass boats for flood evacuation.

Disaster risk reduction continues to be carried out by involving the community more as objects and subjects of various activities carried out, this is because the people in the region have a greater understanding of disasters that occur in their area, so that by involving the community to a greater extent it is hoped the more efficience on disaster risk reduction. In addition, the large population in the region is a great potential related to human resources in disaster risk reduction[15][16]. Some of the activities carried out in this research and service activity are the results of the Forum Group Discussion (FGD) with the community, the village government and the Tambakrejo Disaster Preparedness Group which are more in need of a program oriented towards increasing community capacity and facilitating equipment in supporting disaster emergency response. So that based on the needs of the community in disaster risk reduction efforts, there are several activities that are divided into two, namely research in the preparation of disaster threat maps and identification of community capacity levels, while for community service activities are implemented in the form of Establishing a Disaster Safe Education Unit and Making an Early Warning System for flooding the East Flood Canal.

2. Methods

2.1 Study Area

The location of this research and service activity is in Tambakrejo Village, Gayamsari District, Semarang City which is geographically located in the coastal area of Semarang City and close to the mouth of the Tenggang River and the East Flood Canal. Tambakrejo Village consists of 9 RW and 54 RT with an area of 74.75 hectares, where this area has a high potential threat for floods and fires, the following figure shown the area of study.

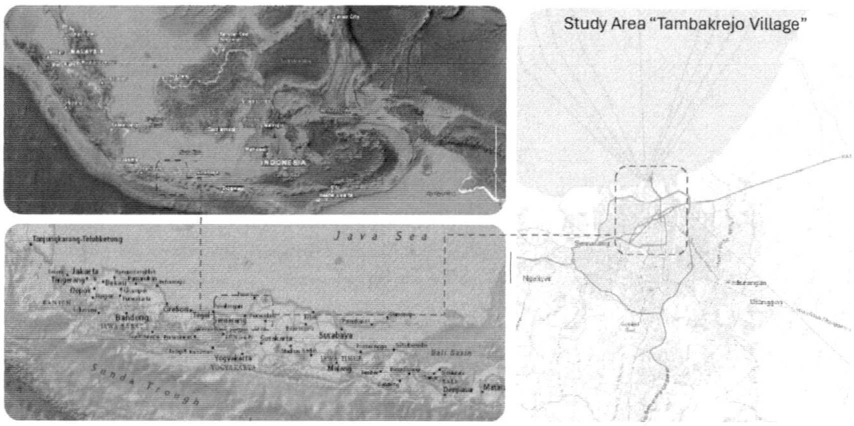

Figure 1. Study Area (Source: Arcgis base maps, 2024)

2.2 Data Source

The data source in this activity is high-resolution satellite imagery Digital Globe used in research to compile flood disaster threat maps. The satellite images were obtained from the Google Earth Enggine_https://earthengine.google.com/ provider of the 2023 recording results which have been geometrically and atmospherically corrected so that they can be directly used in participatory mapping of disaster threats[17][18]. Furthermore, other spatial data used are regional administration maps in shapefile (shp) format study area obtained from the provider Inageoportal_https://tanahair.indonesia.go.id/portal-web/ from the Geospatial Information Agency. Other data use in this research are monographic data obtained from Tambakrejo Village and disaster risk assessment data from the Regional Disaster Management Agency of Semarang City in 2023.

2.3 Data Collecting

Data collection techniques are used to obtain research data in the preparation of flood disaster threat maps and the level of community capacity in facing disasters. The following data collection techniques are used in research activities in the study area.

a) Participatory mapping

Participatory mapping was conducted to obtain data related to disaster threats in the study area which was then used as a database for the preparation of flood hazard maps. The approach used by involving the community in the identification of disaster threats, where in this study using informants Tambakrejo Disaster Preparedness Group cadres. Implementation is done in the field by presenting an image map of Tambakrejo Village with a detailed scale, then the informants delineate the flood and fire threat areas and determine the level of threat at each location. Mapping disaster threats with participatory mapping methods has several advantages, including: identification of disaster threats is done more accurately, increasing public awareness in disaster risk reduction, encouraging collaboration of various stakeholders in disaster risk reduction and providing more precise mitigation information[19][20].

b) Glogal Postioning System (GPS) Survey

The GPS survey was conducted using a 76CSX handheld GPS to obtain the coordinates of the location of the flood disaster threat. The survey was conducted accompanied by Tambakrejo disaster preparedness group cadres who have a better location orientation in the study area.

c) Questionnaire and Interview

Data collection methods with questionnaires and interviews are used to identify the level of community and regional capacity in disaster mitigation efforts. Determination of the sample in this data collection using a purposive sampling approach with respondents who have knowledge related to disasters in the study area, including Tambakrejo Village staff, Regional Disaster Management Agency, disaster preparedness group cadres and disaster-affected communities. The identification of the level of capacity is guided by the regulation of Perka BNPB No. 2 of 2012 with several indicators, including: policies, institutions, training, education, prevention, infrastructure, logistics and mitigation efforts.

2.4 Implementation Stage

Disaster risk reduction activities are carried out with research and community service activities carried out in a structured manner with the following stages of activity.

a) Secondary data collection of high-resolution satellite imagery obtained from data acquisition on google earth enggine provider, in addition to spatial data collection of shapefiles (shp) format administrative maps and disaster data from the Semarang City Regional Disaster Management Agency.
b) Flood hazard mapping using participatory mapping method.
c) GPS survey of flood and fire prone locations in the study area.
d) Identify the level of community and regional capacity for disasters.
e) Community capacity building focused on students Elementary School 3 Tambakrejo through the establishment of a Disaster Safe Education Unit.
f) Inventory of Early Warning System (EWS) for monitoring the water level of the East Flood Canal as an effort to increase regional capacity for disasters.

The technique on data analysis in this activity uses spatial analysis for disaster threat mapping to obtain the spatial distribution of flood threat levels and descriptive quantitative to identify the level of community and regional capacity against disasters in the study area.

3. Result and Discussion

The results of the activity consist of research and community service, where there are 2 research results, namely mapping the level of disaster threat and the level of capacity against floods and fires. As for community service activities, there are 2 activities carried out as an action to increase the capability of the community and the region in dealing with floods and fires, including the establishment of the Disaster Safe Education Unit Elementary School 3 Tambakrejo and the creation of EWS in the East Flood Canal.

3.1 Flood and Fire Hazard Mapping

The mapping of flood and fire disaster threats was obtained from participatory mapping by involving Tambakrejo Disaster Preparedness Group cadres. Mapping of disaster threats is carried out in each RW in Tambakrejo Village area. Based on the identification results, about 57.03 hectares (54%) have a high threat, mostly located in RW 03 and RW 09, while about 6.45 hectares have a medium threat for flood disasters. The following table shows the distribution of flood disaster threat levels in each RW in Tambakrejo Village.

Table 1. Level of Flood Threat in Tambakrejo Village[1]

No	Location	Flood Hazard Level		
		High	Moderate	Low
1	RW 01	1,23	1,56	-
2	RW 02	3,07	0,02	-
3	RW 03	6,41	1,59	-
4	RW 04	0,61	1,10	-
5	RW 05	2,34	-	-
6	RW 06	2,95	-	-
7	RW 07	2,26	-	-
8	RW 08	0,67	-	-
9	RW 09	5,93	2,18	0,60

[1]Source: Participatory mapping, 2023

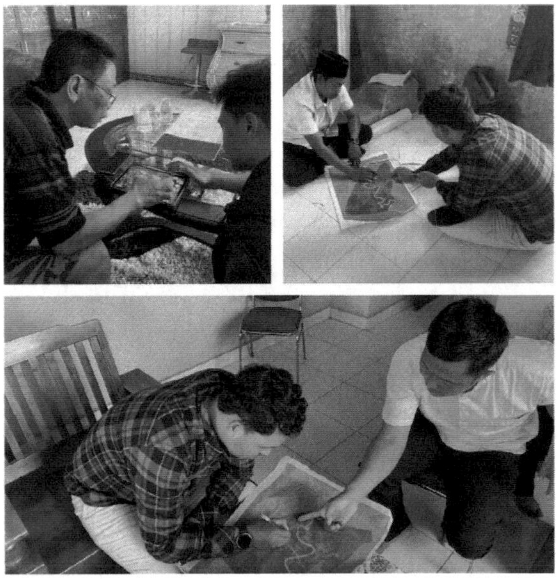

Figure 2. Participatory Mapping of Flood Hazards (Source: documentation, 2024)

The flood disaster that occurs in Tambakrejo Village is caused by the runoff of the Tenggang River and the East Canal Flood when it rains with high intensity, this condition is exacerbated by the existence of land sinking on the shoreline of Semarang City and land conversion in the upstream area of the East Canal watershed where the East Canal Flood and the Tenggang River are part of the watershed. This condition is exacerbated by the existence of land subsidence on the shoreline of Semarang City and land use change in the upstream area of the East Canal watershed where the East Canal Flood and the Tenggang River are part of the watershed. The pattern of flood disaster that occurs in the form of flooding that occurs at the peak of the rainy season, this condition is then exacerbated by the phenomenon of land subsidence which has an impact on the duration of water inundation, especially in residential areas. The greatest flood threat is found in areas around rivers and residential areas, including RW 07-09 with an area of 5.93 Ha (10%) and RW 03-06 with an area of 6.41 Ha (11%) which are adjacent to the surface flow of the East Flood Canal and Tenggang River. As for areas outside RW, there are 30.93 Ha (54%), most of which are in the area of land conversion from pond land to industrial land through the reclamation process, such as PT Galang Anugerah Sejahtera, PT Sricon Logistik Indonesia, PT Samudera Sarana Logistik, PT Unico Utama JAYA, PT Wijaya Karya, and PT Greating Fortune Logistik. The following figure shows the flood hazard map of Tambakrejo Village.

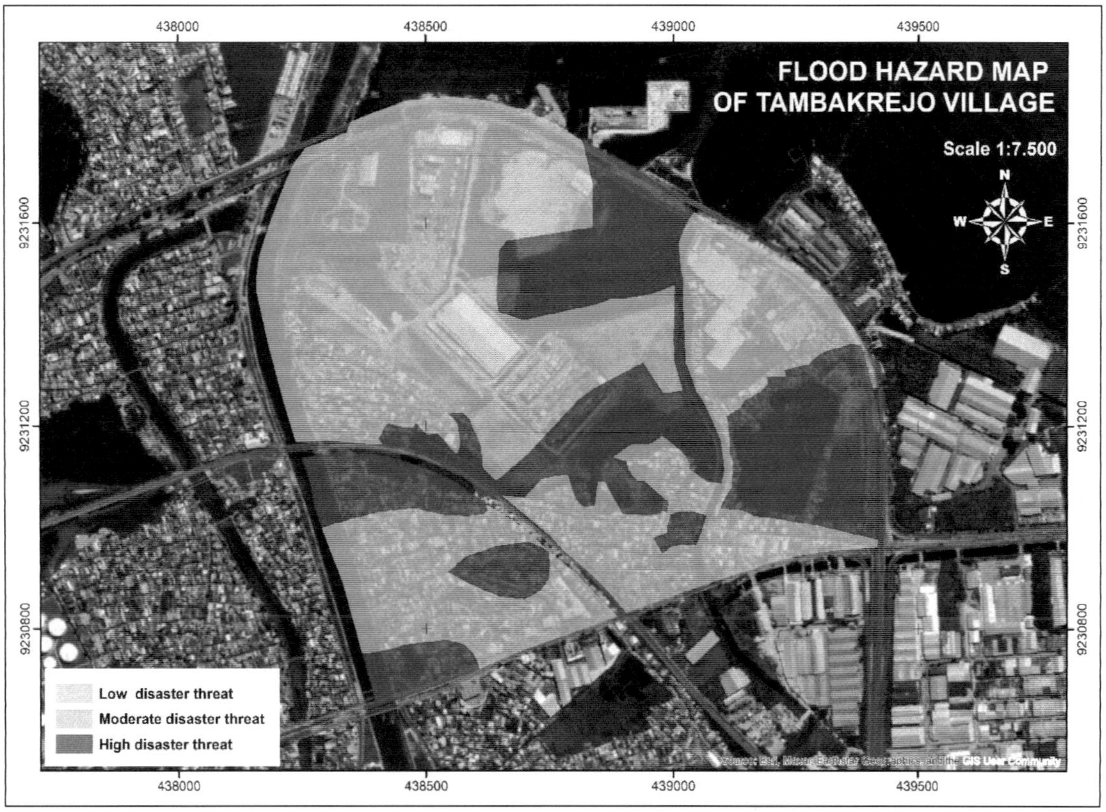

Figure 3. Flood Hazard Map of Tambakrejo Village (Source: Digital globe image and spatial analysis, 2024)

Furthermore, the results of the identification of fire threats in Tambakrejo Village have a high threat level with a total threat value of 9. There are 4 indicators used to determine the level of fire threat, including population density, settlement density, the quality of flammable buildings and the availability of emergency response road access[21]. Based on the outcome of the field survey, it seen that the parameters of population density and road class have a high influence on the threat of fire, where the high density of settlements that do not have open space between buildings, and limited access to collector roads for the fire emergency response process cause a high threat of fire. The following is a table of the results of the calculation of fire threat parameters in Tambakrejo Village.

Table 2. Calculation Result of Fire Disaster Threat Indicator[2]

No	Parameters	Weight	Disaster Threat Class Score			Total
			Low	Moderate	High	
1	Population density	35,67			3	3
2	Settlement density	47,08		2		2
3	Building quality	10,68	1			1
4	Road class	6,57			3	3
		Total				9

[2]Source: Data analysis, 2024

Figure 4. Aerial Photos of Settlements on the Edge of the Tenggang River and East Flood Canal (Source: Aerial photographs, 2024)

3.2 Community and Regional Capacity Level

Determination of disaster capacity can be started from an inventory of components consisting of policies, institutions, training, education, prevention, logistics, and mitigation efforts, where Tambakrejo Village has a moderate level of disaster management capacity for floods[21]. The assessment in this category is inseparable from the availability of capacity support in Tambakrejo Village for flood disasters that are evenly distributed in almost all RW areas, with the criteria dominated by moderate. This even distribution is one indicator that the awareness of disaster management in Tambakrejo Village is good, with various efforts to learn and improve themselves through adaptation to flood disasters in each RW. Institutions that support community capacity include the Tambakrejo Disaster Preparedness Group and Kelurahan Tangguh Bencana, which are non-governmental organizations focusing on capacity building efforts related to disaster management. In addition, there is also support from institutions such as Babinsa, Bhabinkamtibmas, and Karang Taruna that contribute to flood emergency response in Tambakrejo Village. Another supporting factor is the geographical location of Tambakrejo Village which is easily accessible, with arterial and collector roads passing through, making it easier to distribute aid to affected communities. However, the challenge in the distribution of aid is because the location of settlements is lower than the arterial road, which causes higher inundation in flood-affected residential areas, requiring boat equipment for evacuation and distribution of aid. The presestence of a disaster preparedness forum consist of Disaster Preparedness Group and Disaster Resilient Village in Tambakrejo Village is one of the social capital for emergency response and flood disaster mitigation. One of them is the existence of this group has a role to disseminate pre-disaster information or disaster threats that will occur in Tambakrejo Village. One of them is through the whatsapp group of various volunteer networks that inform disaster-related matters such as rainfall, floodgate discharge to the assistance needed during emergency response actions. In addition, structural disaster management capacity also exists in the Tambakrejo Village area with the possession of supporting infrastructure for disaster management such as 5 water pumps, river embankments, and disaster information boards in the form of gathering points, evacuation route signs, fiber boats, emergency boxes and evacuation tents that make Tambakrejo Village more resilient in facing flood disasters. The following figure shows the facilities supporting flood disaster management capacity in Tambakrejo Village.

Figure 5. Flood Mitigation Infrastructure in Tambakrejo Village (Source: documentation, 2024)

Furthermore, the calculation results of the fire capacity parameter show that Tambakrejo Village has a moderate level of capacity with a total score of 5. There are 3 indicators to measure the capacity of the region against fire disasters, including the distance of fire extinguishers, hydrant distance and socialization and disaster emergency response training[22]. Fire management capacity includes the capability of the community to face, leapfrog, and reduce the risk of fire. In the Tambakrejo Urban Village area, there are several driving factors in mitigation efforts, such as the Fire Department and other extinguishing equipment. However, improvement is still needed in terms of preparedness, rapid response, and effective recovery after a disaster occurs. Increasing capacity in this medium category is important to minimize vulnerability to loss and resilience to fire disasters. The following table shows the results of the calculation of fire capacity parameters in the study area.

Table 3. Fire Disaster Capacity Parameter Calculation Results[3]

No	Parameters	Weight	Disaster Threat Class Score			Total
			Low	Moderate	High	
1	Fire-fighting distance	61,28		2		2
2	Hydrant spacing	8,21		2		2
3	Training and socialization	30,51	1			1
		Total				5

[3]Source: Data analysis, 2024

3.3 Establishment of Disaster Safe Education Unit Elementary School 3 Tambakrejo

Disaster risk reduction in Tambakrejo Village is carried out through community and regional capacity building, one of which is in the Tambakrejo Elementary School 3 education unit with the initiation of a Disaster Safe Education Unit. There are a series of action carried out in

IOP Conf. Series: Earth and Environmental Science **1503** (2025) 012020 doi:10.1088/1755-1315/1503/1/012020

the pilot formation of Disaster Safe Education Unit Elementary School 3 Tambakrejo, including through counseling Disaster Safe Education Unit and disaster education, facilitation of emergency response equipment, formation of school disaster preparedness teams and emergency response simulations.

a. Disaster Safe Education Unit and Disaster Education Counseling

   This activity was carried out to socialize the Disaster Safe Education Unit Program which is planned to be implemented at Tambakrejo 3 Elementary School, as it is known that the education unit has a high risk level for flood disasters, besides that in this activity it was also conveyed regarding the importance of disaster education for students who are in a high risk area of disaster. The establishment of the Disaster Safe Education Unit of Tambakrejo Elementary School 3 is carried out to improve the preparedness of school community members to face disaster risks, thus creating school community members who are more prepared, responsive, and able to reduce the risks and negative impacts of disasters through understanding, coordination, and preventive action [23][24]. In general, the objectives of the Disaster Safe Education Unit include:

   - Provide information to teachers of Tambakrejo Elementary School 3 regarding awareness of potential disasters and measures to protect the school environment.
   - Build emergency response skills, so as to cope with emergency situations quickly and effectively.

b. Disaster Emergency Response Equipment Facilitation

   Disaster emergency response equipment facilitation is an important part of the formation of Disaster Safe Education Units, where complete equipment can reflect the preparedness of education units in the face of disasters. Facilitation of equipment supporting the Disaster Safe Education Unit at Tambakrejo Elementary School 3, including 2 gathering points, 9 evacuation routes, 2 first aid kits, 1 siren and 2 fire extinguishers and 5 disaster posters. It is hoped that the emergency response support equipment can support the Disaster Safe Education Unit of Tambakrejo Elementary School 3 so as to increase capacity and reduce disaster risk, and enable the evacuation and emergency response process to run more effectively in the event of a disaster in the education unit. The following figure shows the disaster emergency response equipment facilities at the Disaster Safe Education Unit of Tambakrejo Elementary School 3.

Figure 6. Disaster Emergency Response Infrastructure at Tambakrejo Elementary School 3 (Source: documentation, 2024)

IOP Conf. Series: Earth and Environmental Science **1503** (2025) 012020 doi:10.1088/1755-1315/1503/1/012020

c. Formation of School Disaster Preparedness Team and Emergency Response Simulation
The formation of the Disaster Safe Education Unit disaster alert team is important to ensure the protection and preparedness of schools in the face of disasters, where the formation of this alert team involves all school members with their respective duties. The formation of the disaster alert team is a follow-up to the declaration of the Disaster Safe Education Unit at Tambakrejo Elementary School 3. Through the disaster alert team, the action plan for further handling is detailed through the parts that will be responsible when an emergency occurs in the school environment. The composition of this team consists of the chairman, field coordinator, security team, evacuation team, first aid team and fire team, where the management structure will then be made a Principal Decree. Furthermore, after the formation of the disaster preparedness team, the next activity is a fire emergency response simulation with speakers by the Regional Disaster Management Agency of Semarang City (BPBD), where this activity is carried out because the Tambakrejo Village area has a high risk level for fire disasters with a high level of building density and narrow road access for mobilizing extinguishing equipment. Fire extinguishing training is carried out using fire extinguishers with two methods, namely with traditional fire extinguishers in the form of wet cloth and powder fire extinguishers with powder raw materials, where the practice is carried out for extinguishing small fires in stoves and medium fires. The practical activities of using fire extinguishers are carried out through the introduction of the direction of the tongue of fire, the point of fire source, and the spread of fire to find out from which direction the fire can be extinguished.

Figure 7. Fire Emergency Response Training Disaster Safe Education Unit Tambakrejo Elementary School 3 (Source: documentation, 2024)

3.4 Increased Regional Capacity with Flood Early Warning System
Flood Early Warning System (EWS) is a system designed to provide warnings before a disaster or emergency situation occurs, especially for floods in the Tambakrejo Village area[25]. The purpose

of an EWS inventory is to provide information to the public or authorities in order to take preventive or evacuation measures action quickly in the event of a flood. Flood EWS involves the use of sensors, data monitoring and communication systems to disseminate warnings to communities in the affected areas. A survey of the EWS installation location was conducted involving the village administration, community representatives and Disaster Preparedness Group cadres. The Flood EWS is specifically made to detect flood disasters from the overflowing Banjir Kanal Timur river that passes through the Tambakrejo Village area. The EWS tool design and assembly process was carried out at the UNNES Mechanical Engineering Laboratory, where the EWS system was designed with an ultrasonic sensor system that is able to read river water levels, so that when the water level from the bottom rises to 250 cm, the siren will sound. The flood EWS technology implements renewable energy as an energy source from solar panels. Furthermore, the EWS system can also be monitored via smartphone to monitor the river water level. The following is the flood Early Warning System (EWS) in the East Flood Canal of Tambakrejo Village.

Figure 8. Flood Early Warning System (EWS) in Tambakrejo Village. (Source: documentation, 2024)

4. Conclusion

Based on the research and service activities that have been carried out, the following conclusions can be conveyed.

- The results of participatory mapping show that the Tambakrejo Urban Village area has a high category of flood and fire threats, where the high flood threat has an area of 57.03 hectares, while for the high threat of fire disaster covers all built-up land areas in the study area. Furthermore, the capacity of the community and the region against disasters is at a moderate level for floods and fires.
- Disaster risk reduction in Tambakrejo Village is carried out through community and regional capacity building through the Tambakrejo Elementary School 3 Disaster Safe Education Unit Program and the creation of a flood Early Warning System (EWS) on the banks of the East Flood Canal.

5. Acknowledgments

This research and service activity is a Partnership Innovation Product Diffusion Community Empowerment Program (Partnership Innovation Product Diffusion Scheme) with Funding from the Education Fund Management Agency (LPDP) Ministry of Finance of the Republic of Indonesia. So the author would like to thank the LPDP by Ministry of Finance of the Republic of Indonesia for funding this activity, and LPPM UNNES for facilitating this service activity so that it was completed properly and smoothly.

References

[1] I. Rudiarto, "A Regional Perspective on Urbanization and Climate-Related Disasters in the Northern Coastal Region of Central Java , Indonesia," 2018, doi: 10.3390/land7010034.

[2] M. Pizzorni, A. Innocenti, dan N. Tollin, "Droughts and floods in a changing climate and implications for multi-hazard urban planning : A review," *City Environ. Interact.*, vol. 24, no. June, hal. 100169, 2024, doi: 10.1016/j.cacint.2024.100169.

[3] G. Griggs dan B. G. Reguero, "Coastal Adaptation to Climate Change and Sea-Level Rise," 2021.

[4] A. Kumar, S. Nagar, dan S. Anand, "1 - Climate change and existential threats," in *Global Climate Change*, S. Singh, P. Singh, S. Rangabhashiyam, dan K. K. Srivastava, Ed. Elsevier, 2021, hal. 1–31. doi: https://doi.org/10.1016/B978-0-12-822928-6.00005-8.

[5] J. T. Abatzoglou *et al.*, "Climate change is increasing the likelihood of extreme autumn wildfire conditions across California OPEN ACCESS Climate change is increasing the likelihood of extreme autumn wildfire conditions across California".

[6] P. J. Dickella Gamaralalage, T. Maeda, dan S. Gilby, "Enhancing Capacities for Building Climate and Disaster-Resilient Cities in Asia: Case Study of Cebu, Philippines and Nonthaburi, Thailand BT - Resilient Policies in Asian Cities: Adaptation to Climate Change and Natural Disasters," M. Tanaka dan K. Baba, Ed. Singapore: Springer Singapore, 2020, hal. 169–193. doi: 10.1007/978-981-13-8600-8_8.

[7] K. K. Lwin, I. Pal, S. Shrestha, dan P. Warnitchai, "Assessing social resilience of flood-vulnerable communities in Ayeyarwady Delta, Myanmar," *Int. J. Disaster Risk Reduct.*, vol. 51, hal. 101745, 2020, doi: https://doi.org/10.1016/j.ijdrr.2020.101745.

[8] V. Q. Thanh, D. Roelvink, M. Van Der Wegen, J. Reyns, dan H. Kernkamp, "Flooding in the Mekong Delta : the impact of dyke systems on downstream hydrodynamics," vol. 2000, hal. 189–212, 2020.

[9] N. V. K. Triet *et al.*, "Future projections of flood dynamics in the Vietnamese Mekong Delta," *Sci. Total Environ.*, vol. 742, hal. 140596, 2020, doi: https://doi.org/10.1016/j.scitotenv.2020.140596.

[10] S. Widada, M. Zainuri, G. Yulianto, A. Satriadi, dan Y. J. Wijaya, "Estimation of Land Subsidence Using Sentinel Image Analysis and Its Relation to Subsurface Lithology Based on Resistivity Data in the Coastal Area of Semarang City , Indonesia," vol. 21, no. 8, hal. 47–56, 2020.

[11] R. S. Hamdani, S. P. Hadi, dan I. Rudiarto, "Progress or Regress ? A Systematic Review on Two Decades of Monitoring and Addressing Land Subsidence Hazards in Semarang City," 2021.

[12] E. Science, "Climate Change Impact on the Coastal Settlement Quality and the Relation with the Attainment of Semarang Healthy City Climate Change Impact on the Coastal Settlement Quality and the Relation with the Attainment of Semarang Healthy City", doi: 10.1088/1755-1315/1082/1/012026.

[13] E. Science, "Coastal settlement resilience to water-related disasters in Semarang City Coastal settlement resilience to water-related disasters in Semarang City," 2021, doi:

10.1088/1755-1315/623/1/012067.
[14] F. Hermawan, N. A. Ghozali, C. E. Suci, dan F. Kistiani, "Risk Level Fire Emergency Handling among Semarang Heritage Area," vol. 02033, 2022.
[15] E. Righi, P. Lauriola, A. Ghinoi, E. Giovannetti, dan M. Soldati, "Disaster risk reduction and interdisciplinary education and training," *Prog. Disaster Sci.*, vol. 10, hal. 100165, 2021, doi: https://doi.org/10.1016/j.pdisas.2021.100165.
[16] C. Fan, C. Zhang, A. Yahja, dan A. Mostafavi, "Disaster City Digital Twin: A vision for integrating artificial and human intelligence for disaster management," *Int. J. Inf. Manage.*, vol. 56, hal. 102049, 2021, doi: https://doi.org/10.1016/j.ijinfomgt.2019.102049.
[17] E. Belcore, M. Piras, E. Wozniak, dan I.- Email, "Specific alpine environment land cover classification methodology : Google Earth Engine processing for Sentinel-2 data," vol. XLIII, hal. 663–670, 2020.
[18] A. Vollrath dan A. Mullissa, "Angular-Based Radiometric Slope Correction for Sentinel-1 on Google Earth Engine," hal. 1–14, 2020, doi: 10.3390/rs12111867.
[19] L. Cochrane dan J. Corbett, "Participatory Mapping BT - Handbook of Communication for Development and Social Change," J. Servaes, Ed. Singapore: Springer Singapore, 2020, hal. 705–713. doi: 10.1007/978-981-15-2014-3_6.
[20] B. Ryan, K. A. Johnston, M. Taylor, dan R. McAndrew, "Community engagement for disaster preparedness: A systematic literature review," *Int. J. Disaster Risk Reduct.*, vol. 49, hal. 101655, 2020, doi: https://doi.org/10.1016/j.ijdrr.2020.101655.
[21] BNBP, "Kepala Badan Nasional Penanggulangan Bencana Tentang Daftar Isi Kepala Badan Nasional Penanggulangan Bencana Tahun 2012 Tentang Pedoman Umum Pengkajian Risiko 2 . Lampiran Peraturan," 2012.
[22] V. Saputra, F. K. Marpaung, dan S. Tirawani, "The Effect Of Training And Work Involvement On Performance With Motivation As An Intervening Variable At The Fire And Rescue Service Of Medan City," vol. 2, no. 2, 2024.
[23] S. Di dan S. M. A. Negeri, "PENERAPAN SATUAN PENDIDIKAN AMAN BENCANA," vol. 2, hal. 108–117, 2021, doi: 10.5281/zenodo.4781865.
[24] Z. Palin dan F. A. Rahman, "Pemanfaatan Dashboard Inarisk untuk Evaluasi Ketercapaian Pelaksanaan Program Satuan Pendidikan Aman Bencana Provinsi Jakarta," vol. 2, no. 3, hal. 1–11, 2024, doi: 10.37817/jurnaledukasidanmultimedia.v2i3.
[25] D. Perera, J. Agnihotri, O. Seidou, dan R. Djalante, "Identifying societal challenges in flood early warning systems," *Int. J. Disaster Risk Reduct.*, vol. 51, hal. 101794, 2020, doi: https://doi.org/10.1016/j.ijdrr.2020.101794.

IOP Conf. Series: Earth and Environmental Science **1503** (2025) 012021 doi:10.1088/1755-1315/1503/1/012021

Post-Earthquake Liquefaction Potential Study on the Foundation Area of Palu IV Bridge

F Muktaf[1,2], H C Hardiyatmo[1] and S Ismanti[1]

[1] Department of Civil and Environmental Engineering, Faculty of Engineering, Universitas Gadjah Mada, Yogyakarta, Indonesia
[2] Directorate General of Highways, Ministry of Public Works and Housing, Jakarta, Indonesia

*E-mail: fikri.muktaf@mail.ugm.ac.id

Abstract. Palu IV Bridge collapsed due to the M 7.5 tectonic earthquake on September 28, 2018. Palu IV Bridge's reconstruction, located near the old bridge, was entrusted to the Ministry of Public Works and Housing (PUPR). According to the Atlas of Liquefaction Vulnerability Zones, the construction site of the Palu IV Bridge is situated within a zone of medium to high susceptibility to liquefaction. This research is conducted at the pier of the Palu IV Bridge, to investigate the likelihood of liquefaction and assess the resultant damage. The approach employed was a streamlined empirical method. Then the safety factor against liquefaction (FS) obtained was used to calculate the Liquefaction Potential Index (LPI) and Liquefaction Severity Index (LSI). Finally, the potential settlement that could occur during liquefaction was calculated. The findings indicate that the location under investigation exhibits a moderate to high level of liquefaction severity. With a potential settlement of 15.0 to 34.7 cm, further investigation is required to minimize the impact of liquefaction that could occur.

1. Introduction

On the 28th of September, 2018, a powerful 7.5 Mw tectonic earthquake struck Palu, Central Sulawesi, Indonesia, causing widespread devastation. The disaster triggered a tsunami, ground liquefaction, and the collapse of critical infrastructure, and caused over 4,000 fatalities and extensive damage to infrastructure, including buildings, bridges, and roads, including the iconic Palu IV Bridge[1]. This catastrophic event underscored the vulnerabilities of infrastructure in seismic zones and emphasized the urgent need for improved earthquake-resistant design to mitigate the impact of future disasters.

The 2018 earthquake that damaged Palu City triggered significant geological events including liquefaction, tsunami, and landslides, each of which left a profound impact on the changing landscape and geological properties of the soil[2]. These changes need to be anticipated by conducting more comprehensive soil investigations. One of the studies that needs to be conducted in post-earthquake conditions is the liquefaction potential analysis, as the history of liquefaction events can greatly influence the soil's resistance to future liquefaction[3]. In making a construction design in a potential liquefaction area, we can determine the risks that can occur and then determine the appropriate steps to mitigate them.

Palu IV Bridge's reconstruction, located 50 meters further downstream from the old bridge, was entrusted to the Ministry of Public Works and Housing (PUPR). The reconstruction project aims to improve the bridge's safety and capacity to withstand natural disasters like earthquakes, tsunamis, and liquefaction. The new Palu IV Bridge will be built using more resilient materials and modern engineering techniques to ensure its long-term durability. Once completed, the bridge will provide a vital link for the local community, facilitating smoother transportation and enhancing economic development in the region of Palu City.

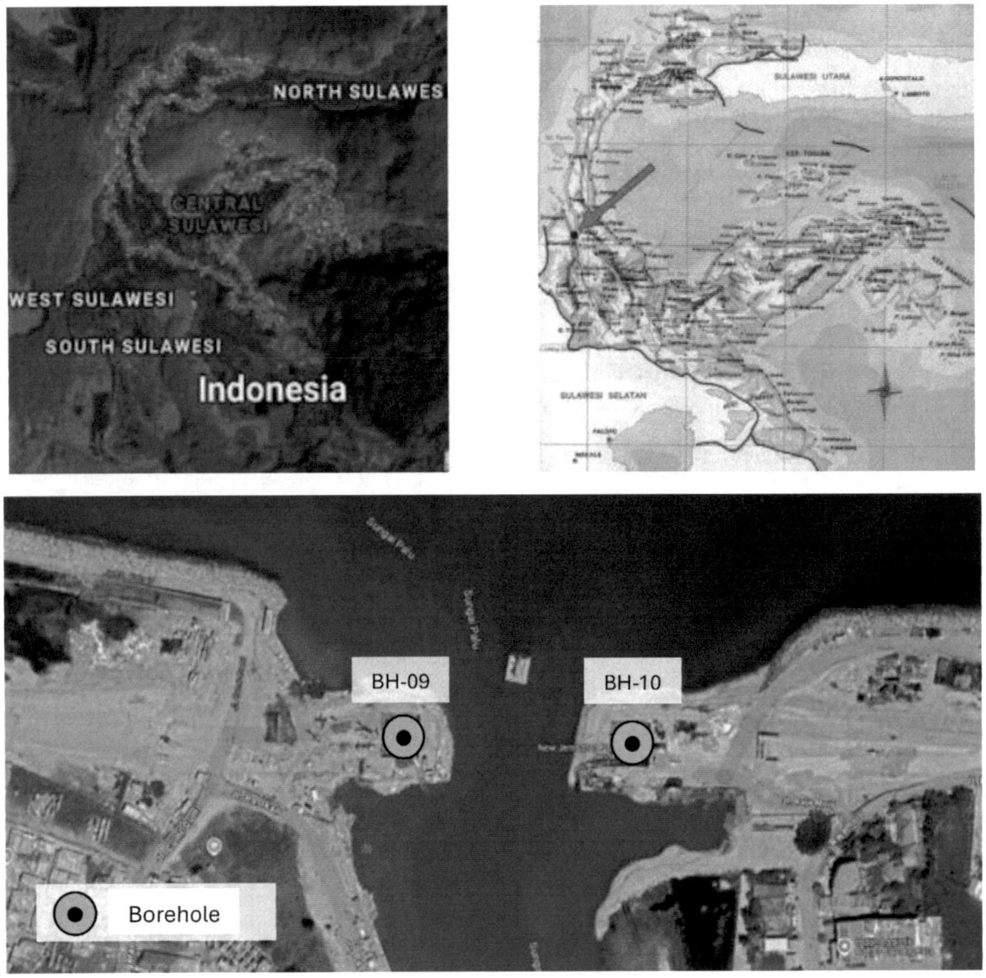

Figure 1. The study area and the borehole locations [4]

According to the Atlas of Liquefaction Vulnerability Zones, the construction site of the Palu IV Bridge is situated within a zone of medium to high susceptibility to liquefaction. Hence, examining the potential for liquefaction in the area is crucial to mitigate possible risks and maintain the bridge's stability in the face of seismic activity.

2. Site Conditions

The research area is located at the Palu IV River Estuary, Palu City, Central Sulawesi Province, Indonesia. Certain regions within Palu City are susceptible to natural calamities, particularly earthquakes, which may induce liquefaction events.[5]. Geotechnical analysis and examination of

IOP Conf. Series: Earth and Environmental Science **1503** (2025) 012021 doi:10.1088/1755-1315/1503/1/012021

geological conditions are very important in evaluating the likelihood of liquefaction occurring in a specific area[6].

Geological maps delineate rock formations and material compositions, and numerous active faults in the region have the potential to trigger earthquakes[7]. According to the geological map, Palu City is situated within the alluvium and coastal sediment formation (*Q*ap). The material of *Q*ap formation includes gravel, sand, mud, and coral limestone. This *Q*ap formation, categorized as Holocene in age, has not fully lithified, resulting in a generally unstable state that is susceptible to erosion, abrasion, landslides, and liquefaction.[8].

The study was conducted at two drilling locations, designated as Pier 1 and Pier 2 of the Palu IV Bridge, where boreholes GA-09 and GA-10 were established, spaced 116 meters apart. The stratigraphic analysis was performed based on the author's categorization of soil layers, which grouped them according to similar visual characteristics and Standard Penetration Test (SPT) values. For design purposes, it is assumed that the soil parameters of each layer are representative of those at the respective borehole location. Figure 2 illustrates the detailed soil stratigraphy observed at the research site, highlighting the diverse characteristics of the soil layers identified in the drill logs.

a. Silty Sand to Gravelly Sand (1)
   These soils range in depth from 0-12.5 meters and are loose to medium in nature. Its SPT values vary widely from 8 to 28.
b. Silty Sand to Gravelly Sand (2)
   This type of soil is in the depth range between 12.5 m to 25 m and 44 m to 59 m. The soil in this layer is loose to very loose. The soil in this layer is loose to very hard. The SPT values vary widely from 12 to more than 50.
c. Silty Clay to Clayey Silt
   This type of soil is in the depth range between 18-55 meters with very low plasticity. The SPT values vary widely from 8 to more than 50.
d. Sandy Silt to Silty Clay
   This type of soil is in the depth range between 47-63 meters with low to medium plasticity. Its SPT values vary widely from 8 to 36.
e. Dense Sand
   This type of soil is in the depth range between 59-65 meters with low plasticity. SPT values tend to be high from 35 to more than 50.

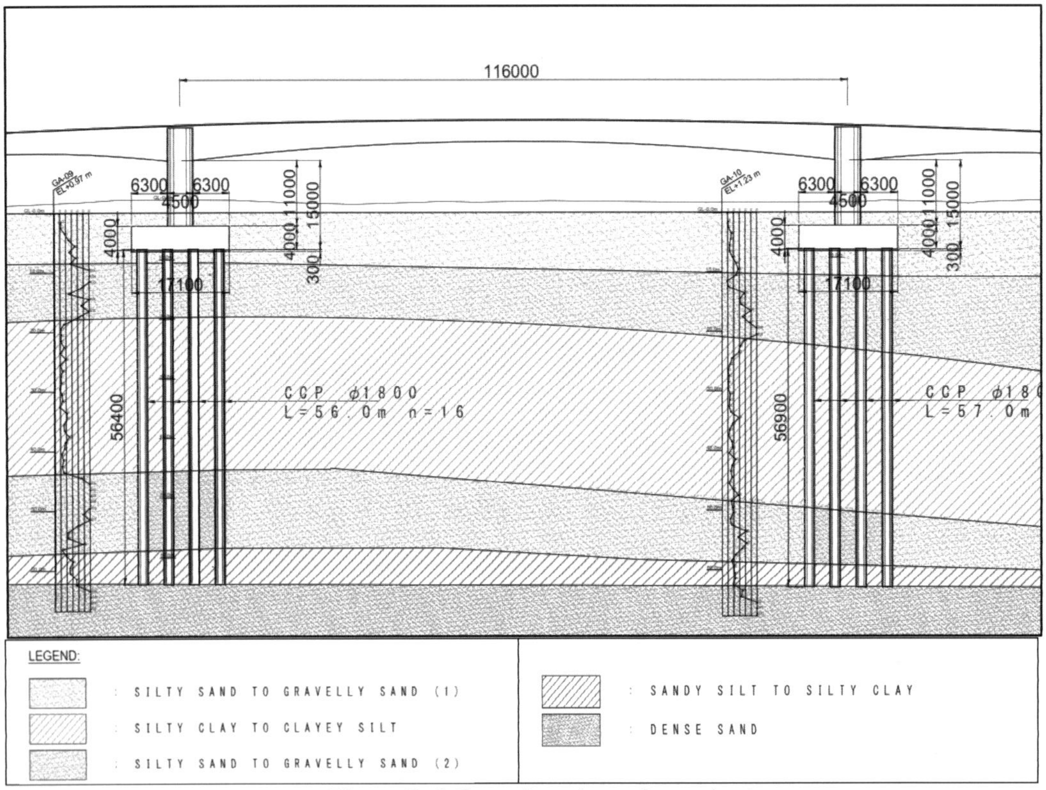

Figure 2. Soil stratigraphy at the study site.

3. Methods

The evaluation of liquefaction potential utilizes the simplified method formulated by Idriss and Boulanger, which guarantees precise and dependable outcomes[9]. The calculation involves comparing the Cyclic Resistance Ratio (CRR) with the Cyclic Stress Ratio (CSR) to ascertain the safety factor (FS) for liquefaction. If the FS value is below one, liquefaction is likely to occur. By forming a ratio between the CRR and CSR, we can determine the likelihood of liquefaction occurring during seismic events. The evaluation also employed the Liquefaction Potential Index (LPI) developed by Iwasaki [10] and and the Liquefaction Severity Index (LSI) devised by Sonmez and Gokceoglu [11]. Following that, the settlement induced by liquefaction was determined.

3.1 Liquefaction Potential Analysis using Simplified Procedure

The simplified procedure for assessing liquefaction potential is a well-established framework in geotechnical engineering, used to evaluate soil susceptibility to liquefaction under dynamic loading. This method evaluates the balance between the CSR, which denotes the seismic load applied to the soil during an earthquake, and the CRR, which signifies the soil's ability to withstand liquefaction under cyclic loading conditions. CSR is determined using equations 1 through 4.

$$CSR = 0.65\left(\frac{a_{max}}{g}\right)\left(\frac{\sigma_v}{\sigma'_v}\right)r_d \tag{1}$$

$$r_d = exp\,(\alpha(z) + \beta(z)M) \tag{2}$$

$$\alpha(z) = -1.012 - 1.126\, sin\left(\frac{z}{11.73} + 5.133\right) \tag{3}$$

$$\beta(z) = 0.106 + 0.118\, sin\left(\frac{z}{11.28} + 5.142\right) \tag{4}$$

Where a_{max} represents the peak horizontal ground acceleration, σ_v denotes the total vertical stress, σ'_v indicates the effective stress, and r_d is the depth reduction factor.

CRR is essential for evaluating the resistance of soil to liquefaction under cyclic loading conditions and Idriss and Boulanger (2014) developed equations 5-10 below to obtain the CRR value.

$$CRR_{M=7.5,\sigma'_{vc}=1} = exp\left(\frac{(N_1)_{60cs}}{14.1} + \left(\frac{(N_1)_{60cs}}{126}\right)^2 - \left(\frac{(N_1)_{60cs}}{23.6}\right)^3 + \left(\frac{(N_1)_{60cs}}{25.4}\right)^4 - 2.8\right) \tag{5}$$

$$(N)_{60} = N_m C_E C_B C_R C_S \tag{6}$$

$$(N_1)_{60} = (N)_{60} \cdot C_N \tag{7}$$

$$C_N = \left(\frac{P_a}{\sigma'_{vc}}\right)^{0.784-0.0768\sqrt{(N_1)_{60cs}}} \leq 1.7 \tag{8}$$

$$(N_1)_{60cs} = (N_1)_{60} + \Delta(N_1)_{60} \tag{9}$$

$$\Delta(N_1)_{60} = exp\left(1.63 + \frac{9.7}{FC + 0.01} - \left(\frac{15.7}{FC + 0.01}\right)^2\right) \tag{10}$$

The analysis determines the factor of safety (FS) as the ratio of CRR to CSR, where FS values below 1.0 indicate a high risk of liquefaction.

$$FS = \frac{CRR_{M=7.5,\sigma'_{vc}=1}}{CSR_{M=7.5,\sigma'_{vc}=1}} < 1 \tag{11}$$

3.2 Liquefaction Potential Index (LPI)

LPI is a method used to evaluate the probability and severity of surface effects resulting from liquefaction at a particular site. The LPI integrates the safety factor (FS) against liquefaction across various depths and applies a depth-weighting function to emphasize the influence of shallow soil layers.

$$LPI = \int_0^{20} F_L\, w(z) dz \tag{12}$$

$$F_z = 0\,; for\ FS < 1 \tag{13}$$

$$w(z) = 10 - 0.5\, z\,; if\ z < 20 \tag{14}$$

$$w(z) = 0\,; if\ z \geq 20 \tag{15}$$

LPI categorizes the severity of surface liquefaction effects into different ranges based on the calculated LPI value. These categories help in assessing the potential impact on infrastructure and land use (Table 1).

Table 1. LPI category

LPI	Category	Investigation/Countermeasures required
LPI = 0	Very Low	Detailed investigations are generally not required
0 ≤ LPI ≤ 5	Low	Further investigations are needed, especially for important structures
5 ≤ LPI ≤ 15	High	The structure requires more detailed follow-up investigations, and measures to prevent the effects of liquefaction are generally required.
LPI > 15	Very High	Detailed investigation and preventive measures of liquefaction effects are needed.

3.3 Liquefaction Severity Index (LSI)

Sonmez and Gokceoglu proposed categories for evaluating the intensity of liquefaction potential. In this way, the LSI is calculated with the equation 16-20.

$$LSI = \int_0^{20} P_L(z).w(z).dz \qquad (16)$$

$$P_L(z) = \frac{1}{1+\left(\frac{FS}{0,96}\right)^{4,5}} \text{ ; for FS} \leq 1{,}411 \qquad (17)$$

$$P_L(z) = 0 \text{ ; for FS} > 1{,}411 \qquad (18)$$

$$w(z) = 10 - 0{,}5z \text{ ; for z} < 20 \qquad (19)$$

$$w(z) = 0 \text{; for z} \geq 20 \qquad (20)$$

The LSI is divided into several risk level categories and depends on the LSI value, as shown in Table 2.

Table 2. LSI category

LSI Value	Liquefaction Severity
85 ≤ LSI ≤ 100	Very High
65 ≤ LSI ≤ 85	High
35 ≤ LSI ≤ 65	Medium
15 ≤ LSI ≤ 35	Low
0 ≤ LSI ≤ 15	Very Low
0	No Liquefaction

3.4 Post-Liquefaction Settlement Analysis

Liquefaction can also cause vertical movement, commonly referred to as settlement. Settlement is usually proportional to the volumetric strain that occurs. Ishihara and Yoshimine [12] developed a method by observing that the volumetric strain occurring during post-liquefaction reconsolidation of a sand sample is proportional to both the maximum shear strain experienced during cyclic loading without drainage and the initial relative density of the sand as shown in the following equation (21) to (27).

$$\Delta S_i = \varepsilon_v \, . \Delta H_i \tag{21}$$

$$\varepsilon_v = 1{,}5 \exp\left(-0{,}369\sqrt{(N_1)_{60CS}}\right) \min(0{,}08\,, \gamma_{max}) \tag{22}$$

$$\gamma_{max} = 0, \;\; FS_{liq} \geq 2 \tag{23}$$

$$\gamma_{max} = min\left(\gamma_{\lim}, 0.035(2 - FS_{liq})\left(\frac{1 - F_a}{FS_{liq} - F_a}\right)\right); \;\; 2 > FS_{liq} > F_a \tag{24}$$

$$\gamma_{max} = \gamma_{lim}\,, \;\; FS_{liq} \leq F_a \tag{25}$$

$$\gamma_{lim} = 1.859\left(1.1 - \sqrt{\frac{(N_1)_{60CS}}{46}}\right)^3 \geq 0 \tag{26}$$

$$F_a = 0.032 + 0.69\sqrt{(N_1)_{60CS}} - 0.13\,(N_1)_{60CS} \tag{27}$$

Where $S_{v\text{-}1D}$ represents one-dimensional surface settlement for reconsolidation, ε_v represents post-liquefaction volumetric strain, γmax represents maximum shear strain, γ_{lim} is shear strain limit, and. FS_{liq} is the liquefaction safety factor, Fa is a parameter, and $(N_1)_{60cs}$ represents the N-SPT value with fine grain content correction.

The results of research conducted by Ishihara (1996)[13] can be used to analyze the correlation between the extent of damage and the settlement that occurs, as shown in Table 3.

Table 3. Correlation between damage level and expected settlement

Level of Damage	Settlement (cm)	Manifestations on the ground
Light to no damage	0–10	Small Cracks
Medium damage	10–30	Small cracks, flowing sand
Extensive damage	30–70	Large cracks, sand bursts, large offsets, lateral movement

4. Results and Analysis

4.1 Liquefaction Potential Analysis using Simplified Procedure

Liquefaction potential analysis was conducted at BH-09 and BH-10, reaching a depth of 20 meters. The Ground Water Level (GWL) of this study is 0.0 meters because the borehole location is in the middle of the river. The PGA used derived from the Binamarga LINI application [14] and the resulting PGA at this location is 0.71 g and multiplied by an amplification factor of 1.1 so that the PGA_M used is 0.781g.

The analysis results show that all boreholes at the Palu IV Bridge reconstruction site have the potential to experience liquefaction with varying depths. A recapitulation of the analysis results at BH-09 and BH-10 is presented in Figures 3(a) and 3(b), which show the FS graph and each borehole's potentially liquefied soil layer. The GWL for all boreholes is at a depth of 0.00 meters. The potentially liquefiable soil layers are at depths of 0-4.5 meters, 5.5-6.5 meters, 7.5-8.5 meters, 12.5-13.5 meters, and 17.5-18.5 meters at Pier 1. At Pier 2, the liquefiable layers are at depths of

IOP Conf. Series: Earth and Environmental Science **1503** (2025) 012021 doi:10.1088/1755-1315/1503/1/012021

0-12.5 meters, 13.5-15.5 meters, and 16.5-17.5 meters. These results are in line with previous studies that show that liquefaction occurs in saturated sandy soils with low SPT values[15]. With a low density value, the soil tends to be porous, allowing pore water pressure to increase through the voids when shaken by an earthquake. This increase in pore water pressure reduces the effective soil pressure. Ultimately, when the effective stress reaches zero, the soil loses its strength.

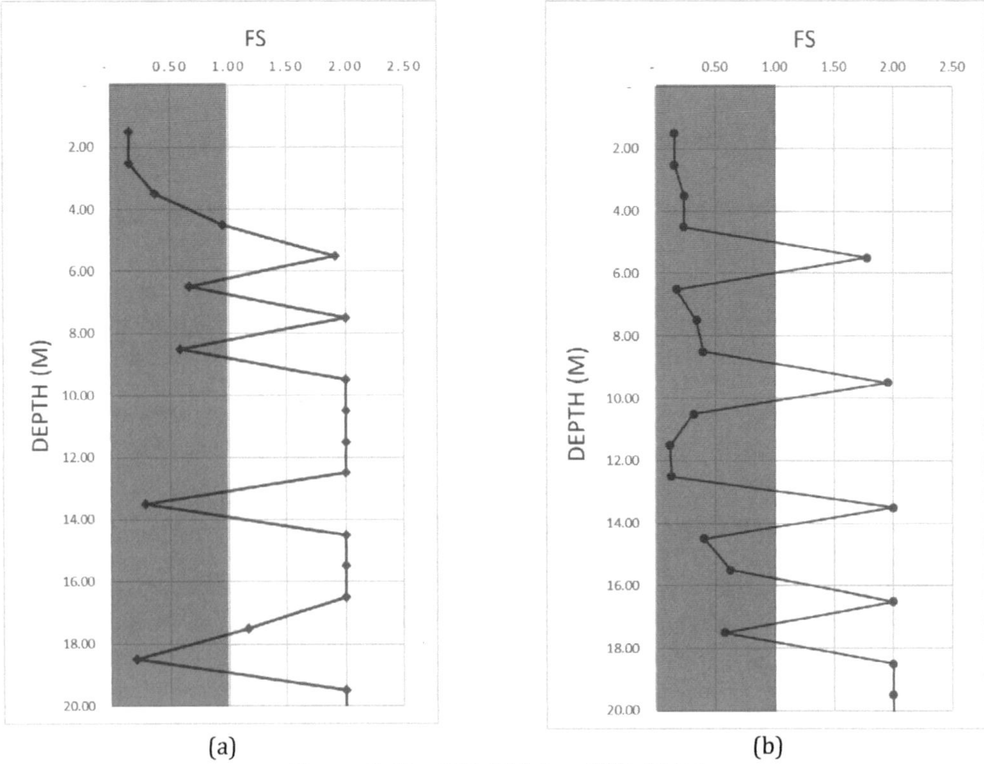

Figure 3. FS of BH-09 (a) and BH-10 (b)

4.2 Liquefaction Potential Index (LPI)

This study analyzes the degree of liquefaction potential by employing safety factor values against liquefaction to determine the LPI value and its category. The LPI analysis results at both borehole locations, which serve as indicators of structural damage risk due to liquefaction severity, reveal that the risk of liquefaction is extremely high (LPI > 15 = very high).

Table 4. Liquefaction Risk Category according to LPI value

Borehole	LPI value	Liquefaction Risk Category
BH-09/ Pier 1	32.22	Very High
BH-10/Pier 2	57.75	Very High

4.3 Liquefaction Severity Index (LSI)

The results of the LSI calculations indicated that the liquefaction severity index was classified as medium at Pier 1 and high at Pier 2. Tables 5 summarize the LSI values of BH-09 and BH-10.

Table 5. Liquefaction Severity Category according to LSI value

Borehole	LSI value	Liquefaction Severity
BH-09	45.31	Medium
BH-10	84.50	High

4.4 Post-Liquefaction Settlement Analysis

The results of the post-liquefaction settlement analysis show that the settlement at borehole BH-09 is 0.15 m for a 7.5 Mw earthquake scenario, while the settlement at borehole BH-10 is 0.347 m (Table 6). The damage observed at BH-09 is classified as medium, characterized by small cracks and the emergence of sand on the ground surface. In contrast, BH-10 experienced extensive damage, which included large cracks, lateral offsets, and noticeable lateral movement on the ground surface. This liquefaction-induced settlement typically occurs unevenly, potentially causing cracks or significant damage to bridge foundations. Such damage compromises the foundation's capacity to distribute loads effectively, thereby heightening the risk of additional damage to the bridge's superstructure. This scenario can also precipitate bridge displacement or even collapse. Therefore, the selection of foundation type is very important as an initial mitigation measure because the foundation makes direct contact with the potentially liquefied soil.

Table 6. Liquefaction Extent of Damage according to settlement value

Borehole	Total Settlement (cm)	Extent of Damage
BH-09	15.0	Moderate damage
BH-10	34.7	Extensive damage

5. Conclusions

Soil conditions in the study area show that the soil in the top 20 meters is dominated by sandy soil with low N-SPT values, and the groundwater table (GWL) is parallel to the surface because the study site is located within the river channel. The results indicate that, with a potential peak ground acceleration (PGA_M) of 0.781g, the risk of liquefaction is very high according to the Liquefaction Potential Index (LPI) and falls into the moderate to high category based on the Liquefaction Susceptibility Index (LSI).

The Post-liquefaction settlement analysis revealed that the potential settlement resulting from liquefaction ranges from moderate to extensive damage levels. Borehole BH-10 exhibited the maximum estimated settlement, measuring 34.70 cm. Additionally, it was noted that the thickness of the layer with liquefaction potential significantly influences the likelihood of settlement.

The findings of this study provide insights into the potential for liquefaction and its impact on soil integrity, thereby enabling the quantification of associated risks and the formulation of further research for mitigation planning. The empirical analysis of liquefaction potential conducted herein represents one of the foundational approaches. While not the sole method, this approach can be augmented through numerical modeling using various software tools, such as PM4Sand in Plaxis and nonlinear site response analysis with Deepsoil.

References

[1] Cilia M G, Mooney W D and Nugroho C 2021 Field Insights and Analysis of the 2018 Mw 7.5 Palu, Indonesia Earthquake, Tsunami and Landslides *Pure Appl. Geophys.* **178** 4891–920

[2] Anda M, Purwanto S, Suryani E, Husnain, and Muchtar 2021 Pristine soil property and mineralogy as the strategic rehabilitation basis in post-earthquake-induced liquefaction, tsunami and landslide in Palu, Indonesia *CATENA* **203** 105345

[3] Teparaksa J and Koseki J 2018 Effect of past history on liquefaction resistance of level ground in shaking table test *Géotechnique Letters* **8** 256–61

[4] Map of Palu City, Central Sulawesi, Indonesia." Google Maps. Google, 2024. Web. 5 Des 2024.

[5] Setyadi G P, Rifa'i A and Indrawan I G B 2024 Effects of Soil Liquefaction to Foundation Bearing Capacity Based on Empirical and Numerical Simulation *IOP Conf. Ser.: Earth Environ. Sci.* **1373** 012011

[6] Kim M, Ordu E, Lee W and Ko J 2024 Geotechnical assessment of seismicity and liquefaction potential at the Banarli landfill in Tekirdag, Turkey *Geomatics, Natural Hazards and Risk* **15** 2346148

[7] Scott C, Adam R, Arrowsmith R, Madugo C, Powell J, Ford J, Gray B, Koehler R, Thompson S, Sarmiento A, Dawson T, Kottke A, Young E, Williams A, Kozaci O, Oskin M, Burgette R, Streig A, Seitz G, Page W, Badin C, Carnes L, Giblin J, McNeil J, Graham J, Chupik D and Ingersoll S 2023 Evaluating how well active fault mapping predicts earthquake surface-rupture locations *Geosphere* **19** 1128–56

[8] Taftsoglou M, Valkaniotis S, Papathanassiou G, Klimis N and Dokas I 2022 A Detailed Liquefaction Susceptibility Map of Nestos River Delta, Thrace, Greece Based on Surficial Geology and Geomorphology *Geosciences* **12** 361

[9] Boulanger R and Idriss I 2014 CPT and SPT based liquefaction triggering procedures *Report No. UCD/CGM-14/01, Center for Geotechnical Modeling, University of California, Davis*

[10] Iwasaki T, Tokida K and Tatsuoka F Soil Liquefaction Potential Evaluation with Use of the Simplified Procedure

[11] Sonmez H and Gokceoglu C 2005 A liquefaction severity index suggested for engineering practice *Environ Geol* **48** 81–91

[12] Ishihara K and Yoshimine M 1992 Evaluation of Settlements in Sand Deposits Following Liquefaction During Earthquakes *Soils and Foundations* **32** 173–88

[13] Ishihara K 1996 *Soil behaviour in earthquake geotechnics* (Oxford : New York: Clarendon Press ; Oxford University Press)

[14] Directorate of Road and Bridge Engineering, Indonesia earthquake source and hazard map for bridges 2017, aplikasi lini (Accessed 5 Apr 2024) https://lini.binamarga.pu.go.id/

[15] Septiadi D R, Hardiyatmo H C and Faris F 2023 Study of soil liquefaction potential at Anutapura General Hospital, Palu City, Central Sulawesi Province *IOP Conf. Ser.: Earth Environ. Sci.* **1244** 012026

IOP Conf. Series: Earth and Environmental Science **1503** (2025) 012022 doi:10.1088/1755-1315/1503/1/012022

Tsunami Inundation Modeling Using High Resolution UAV-DEM in a part of the Coastal Areas of Bengkulu City

Ela Novalinda Ar[1*], Danang Sri Hadmoko[2] and Barandi Sapta Widartono[3]

[1] Master of Remote Sensing, Faculty of Geography, Universitas Gadjah Mada (UGM), Daerah Istimewa Yogyakarta, Indonesia
[2] Department of Environmental Geography, Faculty of Geography, Universitas Gadjah Mada(UGM), Daerah Istimewa Yogyakarta, Indonesia
[3] Department of Geographic Information Science, Faculty of Geography, Universitas Gadjah Mada (UGM), Daerah Istimewa Yogyakarta, Indonesia

*Email:ela.novalinda.a@mail.ugm.ac.id

Abstract. The challenge in developing a tsunami model lies in the limited availability of Digital Elevation Model (DEM) data and data related to tsunami events. According to [1], inundation modeling requires detailed topographic data, ideally with a vertical resolution of less than 0.25 m. Meanwhile, the data currently provided by the government has a contour interval of 12.5 meters. This study aims to analyze tsunami inundation modeling based on earthquake hazards in the subduction zone, supported by high-precision remote sensing techniques. This study is located in part of the coastal area of Bengkulu City, which is situated between two major megathrusts: the Mentawai-Pagai Megathrust and the Mentawai-Enggano Megathrust. This geographical setting results in a high potential for both earthquake and tsunami hazards. The objective of this study is to assess the capability of UAVs in acquiring modeling variables for tsunami inundation potential and to model the tsunami inundation potential. The results of UAV processing are capable of producing DTM and semantic information in the form of detailed land use. The tsunami propagation towards the coast of Bengkulu City had a maximum amplitude of 5.9 meters with a travel time of 25 minutes and then spread in all directions.

1. Introduction

The use of UAV (Unmanned Aerial Vehicle) technology for spatial data acquisition in tsunami hazard mapping is still relatively underutilized compared to its application in other disaster scenarios. According to a literature review of 635 scientific articles indexed in Scopus and Web of Science, UAV technology is most commonly applied to landslide-related disasters, accounting for 38% of studies. This is followed by flood-related research at 18%, earthquake studies at 14%, and cyclone assessments at 12% [2].

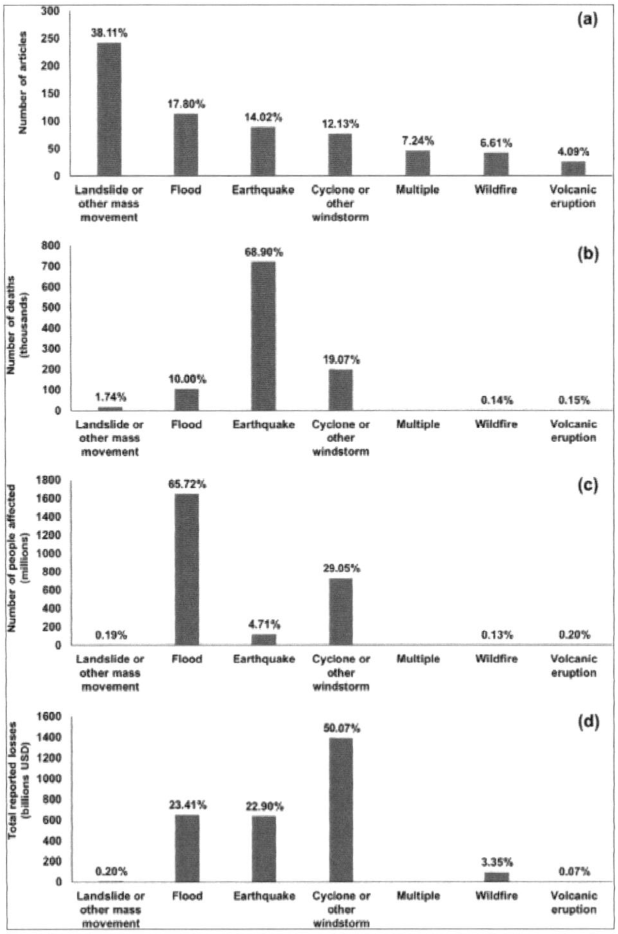

Figure 1 . (Kucharczyk & Hugenholtz, 2021).

Despite the increasing adoption of UAVs in disaster management, there is a notable lack of specific considerations regarding their application in tsunami hazard mapping, as illustrated in Figure 1. UAV-based mapping in coastal areas offers several advantages, including relatively low hardware and operational costs, high levels of photographic automation, and suitability for small-scale studies [3]. These advantages make UAVs an effective tool for detailed and efficient coastal mapping. Consequently, leveraging UAVs for tsunami-related coastal mapping presents a practical and valuable approach to enhancing disaster preparedness and hazard assessment. This study aims to develop a tsunami inundation model using high-resolution UAV-derived data combined with numerical tsunami propagation modeling through COMCOT v1.7. The integration of UAV imagery with numerical modeling is expected to improve the accuracy of tsunami disaster modeling by providing high-resolution spatial data and precise simulations of tsunami impacts. The research is conducted in a selected coastal area of Bengkulu City, which is located on the western coast of Sumatra Island and directly faces the Indian Ocean subduction zone. This region is also in proximity to the Mentawai-Pagai and Mentawai-Enggano megathrust zones, making it highly susceptible to tsunami hazards. The findings of this study are expected to contribute to more accurate and reliable tsunami hazard assessments, ultimately aiding in disaster risk reduction and coastal resilience planning.

2. Method

The overall workflow of this research is outlined in Figure 2, providing a comprehensive depiction of the methodological process.

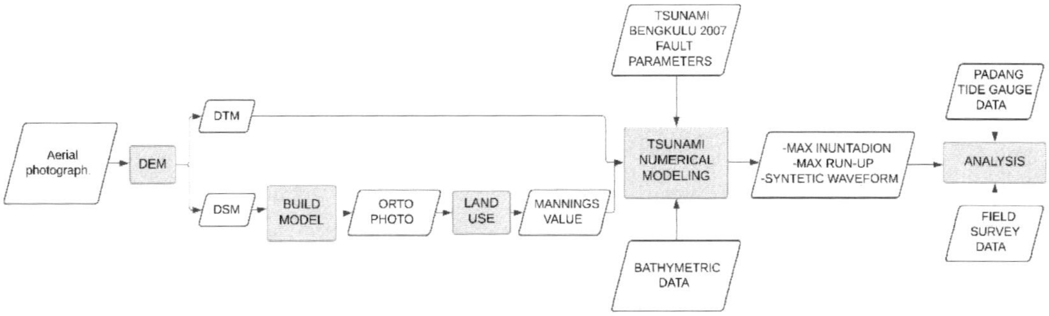

Figure 2. Comprehensive methodology that integrates high-resolution UAV-DEM alongside the BATNAS framework, coupled with parameters from the 2007 Bengkulu earthquake.

2.1 Data

The aerial imagery utilized in this research consists of photographs captured in 2023 at an altitude of 300 meters using a Farm Mapper VTOL V2 PPK (Post-Processed Kinematic) system, 24 MP resolution camera, achieving a spatial accuracy of 8–10 cm. The imagery provides a Ground Sampling Distance (GSD) of 6 cm per pixel, covering an area of 4.361,5 hectares . Based on Figure 3, where the blocks utilized in this study include blocks 4 to 8, blocks 13 to 16, and blocks 20 to 24.

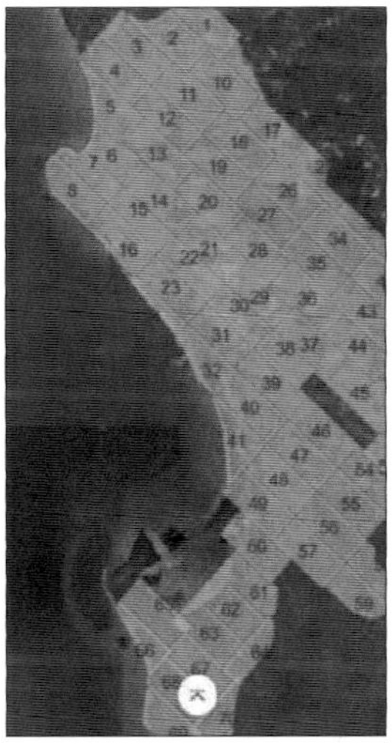

Figure 3. Block-Flight plan

To achieve seamless coverage and avoid image gaps, the endlap and sidelap were configured at 80% and 70%, respectively. The primary data utilized in this study is presented in Table 1, offering a detailed overview of the foundational inputs for the research.

Table 1. Primary Data

Acquisition Date	8-9 Oct 2023
flight altitude	300 meters
Tool	Farm Mapper VTOL V2 PPK (Post-Processed Kinematic)
Camera resolution	24 MP
Covering area	4,361.5 hectares
Independent Control Points (ICPs) measured	32 points
Land use accuracy testing	94 field survey points

For accuracy validation, a total of 32 Independent Control Points (ICPs) were measured to ensure the positional reliability of the maps generated. Distribution of ICP measurement locations based on Systematic Sampling with a 1 km distance between each point, providing a comprehensive evaluation of the positional accuracy across the study region. To assess land use accuracy, 94 field survey points were recorded [4]. As shown in Figure 4. These points were used to verify the classification and mapping of land use types, ensuring that the data accurately reflects real-world conditions. This combination of control and survey points contributes to the robustness of the accuracy validation process.

Figure 4 ICP and Land Use Field Survey

IOP Conf. Series: Earth and Environmental Science **1503** (2025) 012022 doi:10.1088/1755-1315/1503/1/012022

The secondary data utilized in this study is presented in Table 2, encompassing various external sources that supplement the primary data. These sources were carefully selected to provide additional insights and enhance the overall research framework.

Table 2. Secondary Data

Data	Source
National Bathymetry (BATNAS)	BIG
Historical Earthquake Data of Bengkulu	USGS
Tidal Wave	IOC-Sea Level Monitoring
Wave Height Data	IOC-Sea Level Monitoring
Field Survey Data	Borrero, et al., 2009

2.2 Numerical simulations

Tsunami modeling is essential for accurately simulating the complete sequence of a tsunami event, encompassing the phases of generation, propagation, and coastal inundation through the application of sophisticated mathematical equations. This area of study can be primarily divided into two categories: numerical models and empirical models [5]. In this research, we employ a numerical modeling approach that harnesses the capabilities of computer systems to accurately simulate mathematical frameworks. The numerical tsunami modeling process is methodically structured into three distinct phases: source modeling, propagation modeling, and inundation modeling. A leading model currently employed by researchers in the field is COMCOT, which was developed by [6]. This model applies shallow water equations that operate in either spherical or Cartesian coordinate systems, enabling accurate simulation of tsunami propagation and run-up. In our study, we specifically focused on key earthquake parameters that can be directly ascertained, including the epicenter location, magnitude, depth, strike, dip, and rake (slip angle) [7], as referenced in and as presented in Table 3. The model developed in this study is based on the tsunami event that struck Bengkulu on September 13, 2007, at 11:10:26 UTC, providing a critical basis for the analysis and simulation of such phenomena.

Table 3. Fault Parameters

Refer ences	Epicenter		Depth	Focal Mechanism			Dimension (Km)		Disloc (m)	Rigdity
	Lon.	Lat.		Strike	Dip	Slip	L	W		
8.4	101.367	-4.438	34	323^{0}	11^{0}	103^{0}	261.8183	109	13	30

*(Source : USGS)

IOP Conf. Series: Earth and Environmental Science **1503** (2025) 012022 doi:10.1088/1755-1315/1503/1/012022

In this research, we utilized the Blaser equation [8] to determine the fault length (L) and fault width (W) as functions of seismic event magnitudes. This empirical methodology provides a crucial framework for estimating fault dimensions, which are vital for comprehending the mechanics and potential consequences of earthquakes. The equation is expressed as follows:

$$\log10(L)=a\cdot Mw+b \quad (1)$$

$$\log10(W)=c\cdot Mw+d \quad (2)$$

Blaser defined the parameter values (a,b,c,d) based on the type of fault mechanism, which in this study is Thrust Faults. The parameter values a, b, c, and d for thrust faults, are specific coefficients used in regression scaling relationships to estimate rupture dimensions (like fault length and width) based on earthquake moment magnitude Mw. For thrust faults, the values are typically:

$a = -2.37$ $b= 0.74$
$c = -1.61$ $d= 0.41$

a and b relate to the logarithmic scaling of rupture length,
c and d correspond to the logarithmic scaling of rupture width as a function of Mw.

This modeling approach employs a Nested Grid Configuration, allowing for a flexible arrangement of computational grid points. The grid system is structured in a progressive manner, shifting from larger, coarser grids to smaller, finer ones to achieve an optimal balance between accuracy and computational efficiency, as illustrated in Figure 5.

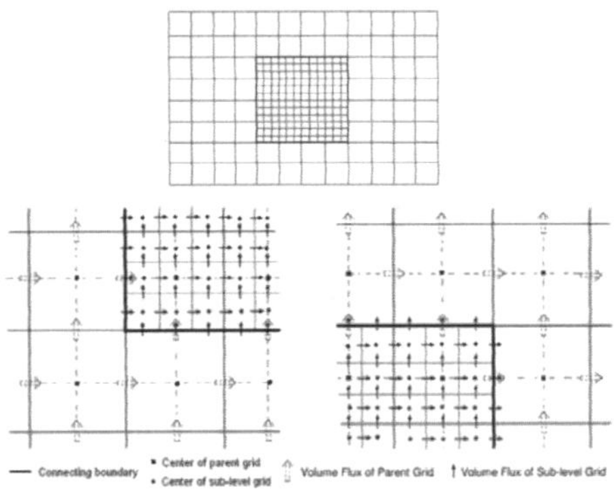

Figure 5. Sketch of Nested Grid Setup

3. Results and Discussion

3.1 UAV data processing

Aerial foto process in to photo alignement to make sparse point cloud Figure 6a and build model to dense point cloud Figure 6b using Agisoft Metashape 2.1.2 This software employs the Structure from Motion (SfM) technique to estimate 3D structures while simultaneously determining the position and orientation of the camera using overlapping photographs. Generally, SfM produces two primary datasets: a sparse point cloud, which provides a preliminary representation of the scene, and a dense point cloud, which offers a detailed and comprehensive 3D reconstruction [9]. In the subsequent stage, the density of the sparse point cloud is enhanced to produce a dense point cloud with significantly higher precision and detail. This process

IOP Conf. Series: Earth and Environmental Science **1503** (2025) 012022 doi:10.1088/1755-1315/1503/1/012022

improves the quality of the 3D reconstruction, enabling more accurate spatial analysis and visualization.

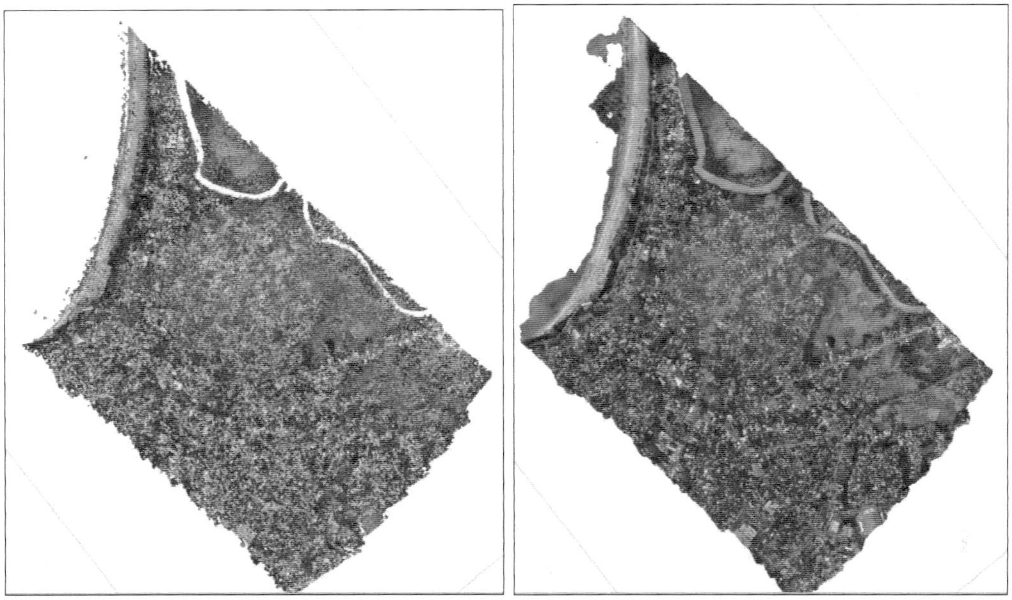

Figure 6. (a) Sparse Point Cloud (b) Dense Point Cloud

The generated dense point cloud is subsequently processed through two distinct procedures to obtain outputs in the form of orthophotos and Digital Terrain Models (DTM). To generate the DTM, the dense point cloud undergoes a ground point classification process to separate terrain points from other objects, such as vegetation or buildings. The final result of this process is a terrain model based on ground points, which is represented as a DTM, as shown in Figure 7a. To produce the orthophoto output, a Digital Elevation Model (DEM) is created from the dense point cloud, followed by the generation of the orthophoto, with the DSM source illustrated in Figure 7b.

IOP Conf. Series: Earth and Environmental Science **1503** (2025) 012022 doi:10.1088/1755-1315/1503/1/012022

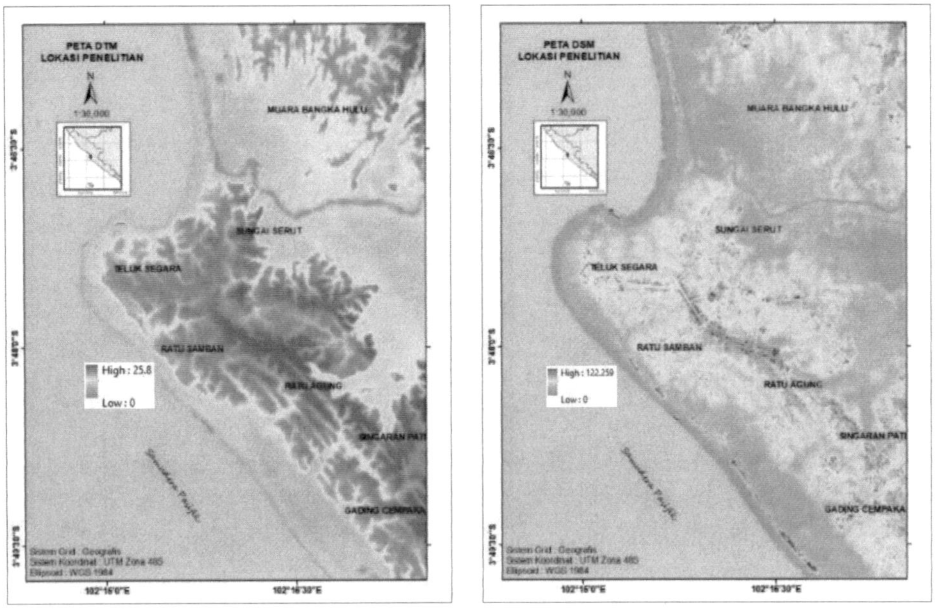

Figure 7. (a) Digital Terrain Models (b) Digital Surface Models

The orthophoto map is created from stereophotos captured at a designated altitude above the Earth's surface and is georeferenced to match the coordinates of specific points on the ground. The resulting orthophoto map, generated from aerial imagery collected in the study area, is presented in Figure 8.

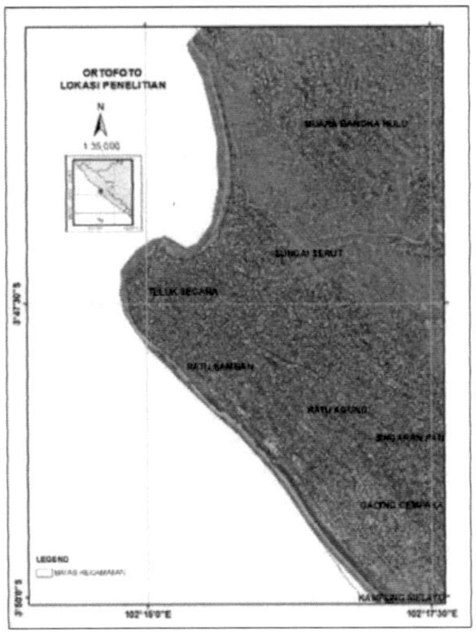

Figure 8. Orthophoto Map of the Study Area

To evaluate the quality of the generated orthophoto and DTM, accuracy or precision testing is required. This process involves comparing the produced models with known reference points or ground truth data to assess the level of error and ensure the reliability of the outputs for subsequent analysis. The distribution of ICP measurement locations is based on Systematic Sampling, with a 1 km distance between each point, using the ArcGIS application. This method ensures uniform coverage across the study area, providing a consistent and representative set of control points for accuracy testing and validation. The systematic sampling method reduces bias by ensuring that measurement locations are evenly distributed, allowing for a thorough assessment of the spatial accuracy of the generated models. The arrangement of ICP points is illustrated in Figure 9.

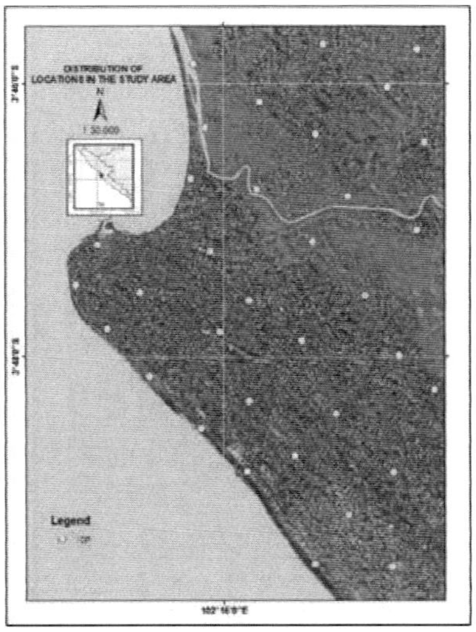

Figure 9. Distribution of ICP points

The CE90 and LE90 values can be determined using formulas based on the United States National Map Accuracy Standards (US NMAS) [10], as presented in the table. These metrics are essential for assessing the positional accuracy of the generated models, where CE90 refers to the circular error at 90% confidence, and LE90 refers to the linear error at 90% confidence, both of which are crucial for validating the precision of geospatial data in mapping applications. The formulas for determining CE90 and LE90, based on the US NMAS, are as follows:

$$\text{CE90} = 1.5175 \times \text{RMSEr}$$
$$\text{LE90} = 1.6499 \times \text{RMSEz}$$

Where :

RMSEr : Root Mean Square Error for the x- and y- coordinates (horizontal position)
RMSEz : Root Mean Square Error for the z-coordinate (vertical position)

The RMSE is calculated using the following formulas :

$$\text{RMSEr} = \sqrt{\frac{\sum(Xi-Xj)^2 + \sum(Yi-Yj)^2}{n-1}}$$

$$\text{RMSEz} = \sqrt{\frac{\sum(Zi-Zj)^2}{n-1}}$$

IOP Conf. Series: Earth and Environmental Science **1503** (2025) 012022 doi:10.1088/1755-1315/1503/1/012022

Table 3. The results of accuracy testing

No	Nama Titik ICP	Koordinat GNSS		Koordinat		dx	dy	dx2	dy2	dx2+dy2		Z Lapangan	Z Peta	dz	dz2
		X Lapangan	Y Lapangan	X Peta	Y Peta										
1	P06	62594.546	1083697.6	62594.57	1083698	-0.0191	0.0264	0.00036481	0.000697	0.001062		5.9535	5.330951	0.622549	0.387567
2	P08	64379.644	1083977.2	64379.64	1083977	0.0046	-0.0469	2.116E-05	0.0022	0.002221		15.9559	15.48904	0.466856	0.217955
3	P09	63507.956	1083195.9	63507.94	1083196	0.0138	0.0061	0.00019044	3.72E-05	0.000228		6.7022	6.107438	0.594762	0.353742
4	P10	62749.942	1082844.4	62749.9	1082844	0.0444	0.012	0.00197136	0.000144	0.002115		3.6441	3.6844	-0.0403	0.001624
5	P11	62556.804	1082157.5	62556.81	1082157	-0.0021	-0.0089	4.41E-06	7.92E-05	8.36E-05		3.4267	3.788815	-0.36212	0.131127
6	P12	61245.206	1081262	61245.18	1081262	0.0259	0.0081	0.00067081	6.56E-05	0.000736		6.8966	6.663416	0.233184	0.054375
7	P13	60942.958	1080722.3	60942.97	1080722	-0.0084	-0.023	7.056E-05	0.000529	0.0006		7.0651	6.068591	0.996509	0.99303
8	P15	65690.467	1083909.1	65690.49	1083909	-0.0205	-0.027	0.00042025	0.000729	0.001149		20.2464	19.78497	0.461427	0.212915
9	P16	65280.329	1083393.5	65280.36	1083393	-0.0276	0.002	0.00076176	4E-06	0.000766		17.5916	17.58701	0.004589	2.11E-05
10	P17	64279.174	1082765.1	64279.17	1082765	0.0022	0.048	4.84E-06	0.002304	0.002309		11.5564	11.32623	0.230173	0.05298
11	P18	63469.557	1082011.9	63469.55	1082012	0.0059	0.0143	3.481E-05	0.000204	0.000239		5.5663	5.473664	0.092636	0.008581
12	P19	62826.673	1081180	62826.72	1081180	-0.0502	-0.002	0.00252004	4E-06	0.002524		17.5946	17.4983	0.096296	0.009273
13	P20	61842.528	1080613.3	61842.56	1080613	-0.0322	-0.0206	0.00103684	0.000424	0.001461		17.7978	17.30983	0.48797	0.238115
21	P21	61384.31	1080129.1	61384.35	1080129	-0.0357	0.02	0.00127449	0.0004	0.001674		5.3674	5.259228	0.108172	0.010701
14	P24	65792.88	1082650.4	65792.88	1082650	-0.0002	0.0215	4E-08	0.000462	0.000462		16.3163	15.70663	0.609672	0.3717
15	P25	64738.163	1081913	64738.13	1081913	0.0283	-0.017	0.00080089	0.000289	0.00109		10.7429	10.57385	0.169049	0.028578
16	P26	64237.498	1081311.4	64237.52	1081311	-0.0177	0.0302	0.00031329	0.000912	0.001225		6.9292	6.636199	0.293001	0.08585
17	P27	63362.927	1080508.2	63362.92	1080508	0.0023	0.007	5.29E-06	4.9E-05	5.43E-05		15.4658	15.47341	-0.00761	-0.00579
18	P28	62960.495	1080092.5	62960.52	1080093	-0.0259	-0.006	0.00067081	3.6E-05	0.000707		19.5931	19.49818	0.09492	0.00901
19	P29	61978.19	1079480.7	61978.2	1079481	-0.0119	0.019	0.00014161	0.000361	0.000503		5.5062	5.590388	-0.08419	0.007088
20	P33	65686.443	1081468.7	65686.47	1081469	-0.0239	-0.017	0.00057121	0.000289	0.00086		7.602	6.26053	1.34147	1.799542
21	P34	64977.303	1080583.5	64977.29	1080583	0.0164	0.022	0.00026896	0.000484	0.000753		17.5309	16.35695	1.173953	1.178166
22	P35	64092.131	1079970	64092.13	1079970	0.0043	0.016	1.849E-05	0.000256	0.000274		16.7029	16.50921	0.193686	0.037514
23	P36	63363.122	1079151	63363.15	1079151	-0.0322	-0.016	0.00103684	0.000256	0.001293		12.9033	12.77195	0.131351	0.017253
24	P37	62704.099	1078786.6	62704.2	1078787	-0.1039	-0.016	0.01079521	0.000256	0.011051		6.4702	6.562438	-0.09224	0.008508
25	P42	65449.094	1079768.4	65449.07	1079768	0.0283	0.0298	0.00080089	0.000888	0.001689		6.7411	6.649614	0.091486	0.00837
26	P43	64575.407	1078980	64575.48	1078980	-0.0713	0.059	0.00508369	0.003481	0.008565		19.7261	19.49846	0.227637	0.051819
27	P44	63997.377	1078410.6	63997.4	1078411	-0.0272	-0.018	0.00073984	0.000324	0.001064		16.8583	16.5942	0.264104	0.069751
28	P45	63335.01	1078202.5	63335.11	1078202	-0.1012	0.0996	0.01024144	0.00992	0.020162		9.1729	8.733913	0.438987	0.19271
29	P52	65368.152	1078198.9	65368.19	1078199	-0.0353	0.008	0.00124609	6.4E-05	0.00131		20.5025	19.23105	1.271448	1.61658
30	P53	64753.527	1077617	64753.54	1077617	-0.0148	0	0.00021904	0	0.000219		12.0636	10.12452	1.939079	1.760027
31	P54	64276.02	1076951.4	64276.07	1076951	-0.0519	0.0226	0.00269361	0.000511	0.003204		4.7738	4.73783	0.03597	0.001294
32	P62	65352.546	1076919.9	65352.52	1076920	0.0228	0.011	0.00051984	0.000121	0.000641		14.5545	13.89882	0.655685	0.429923
									sum	0.072294				Sum	10.33989
									Average	0.002259				Average	0.31333
									RMSE	0.047531				RMSE	0.559759
									CE90	0.072128				LE90	0.923546

Based on PERKA BIG No. 6 of 2018, the calculated CE90 (Circular Error 90%) value of 0.07 meters for horizontal accuracy indicates that the aerial phot can be utilized for 1:2,500 Class 1 mapping. Meanwhile, the LE90 (Linear Error 90%) value of 0.92 meters for vertical accuracy demonstrates that the Digital Terrain Model (DTM) is suitable for 1:2,500 Class 3 mapping. It can accommodate large-scale mapping, particularly for disaster mapping, with a focus on tsunami-related disasters.

The aerial imagery, following rigorous horizontal and vertical accuracy assessments, was utilized to conduct land use classification through a visual interpretation method. This method relied on the application of an aerial photo interpretation key, as described by [11] to systematically identify and categorize land use types. The classification process was guided by the provisions of BIG Regulation No. 3 of 2016 [12], which establishes the technical specifications for the preparation and presentation of village maps. As illustrated in Figure 10, the classification revealed 24 distinct land use categories across the study area. The lowland regions and areas adjacent to riverbanks were predominantly characterized by agricultural activities, including rice fields and various forms of cultivation. Residential areas emerged as the most extensive land use type, accounting for 29% of the total study area. Conversely, the landfill (TPA) was identified as the smallest land use category, occupying a negligible 0.001% of the total area. This analysis highlights the spatial distribution of land use types, offering insights into the dominance of residential zones and the marginal allocation for waste management facilities within the study area.

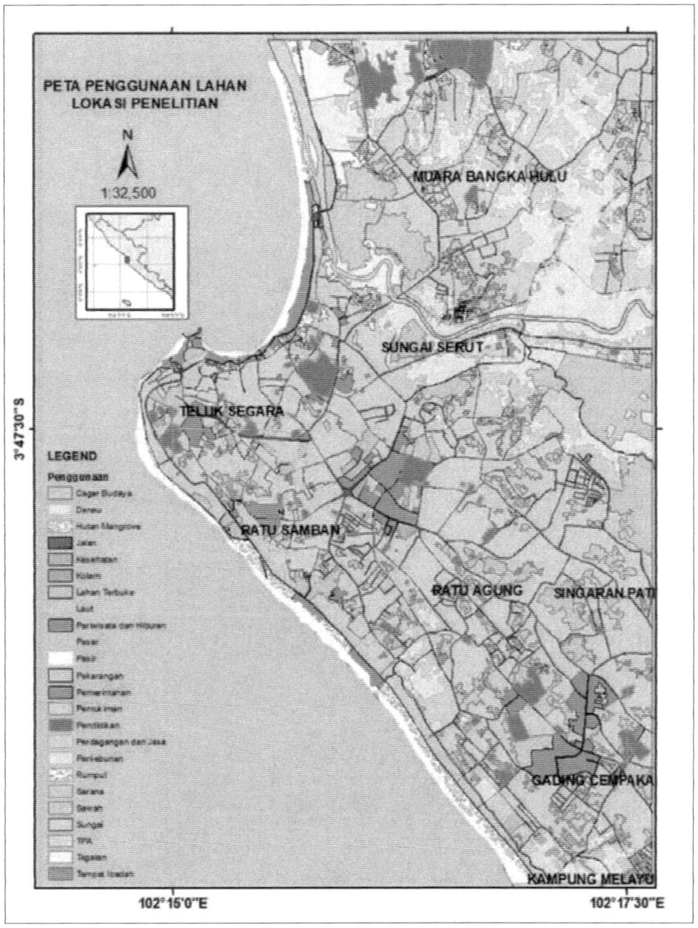

Figure 10. Land Use Map

The land cover data was then transformed into surface roughness values, which are crucial in limiting the spread and maximum inland penetration of a tsunami upon reaching the shore. Factors such as vegetation, coastal topography, and human-made structures influence surface roughness, helping to dissipate the energy of tsunami waves. Studies have shown that higher surface roughness results in less severe flooding, enhancing the resilience of coastal regions against tsunami [13]

3.2 Tsunami Hazard Mapping

3.2.1 Bathymetric Data

Figure 10 show the Bathymetric Visualization generated from BATNAS Imagery. The modeling area covers Region 1 through Region 4, with the green triangle symbolizing the location of the tide gauge station in Padang, which is used for model validation. This station is crucial for comparing observed tidal data against the model's predictions, ensuring its accuracy in simulating tsunami events. The figure shows that the lowest elevation is -6960 m, while the highest point reaches 3755 m. The Padang tide gauge station, represented by the green triangle, captured tsunami wave data during the 2007 event. This data will be used as a reference for comparing the modeled tsunami waves in this study.

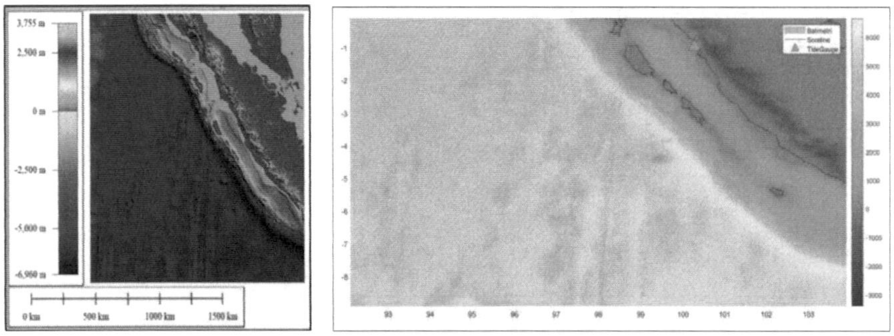

Figure 11. Bathymetric Visualization from BATNAS Imagery

3.2.2 Manning's roughness values

Figure 12 illustrates the spatial distribution of Manning's roughness values, which were obtained through land use classification in the study area [14]. The image uses grayscale to represent variations in Manning's coefficients, with darker shades indicating higher roughness values and lighter shades corresponding to lower values, as shown in the legend. These coefficients are essential for numerical tsunami modeling, as they characterize the surface resistance to water flow, which significantly influences the accuracy of simulations. This information provides critical input for assessing tsunami propagation and inundation patterns in the modeled region.

Figure 12. Manning's roughness values

3.2.3 Tsunami generation

As outlined in the research methodology, the 2007 earthquake visualization, sourced from USGS, provides critical insights into the seismic event. This visualization highlights the impact of the earthquake, facilitating an assessment of tsunami generation potential based on key event parameters, such as magnitude, location, and depth. Additionally, the layer division, ranging from Layer 1 to Layer 5, with Layer 1 representing the largest area and Layer 5 indicating the study area, is depicted in Figure 11. The yellow star on the map indicates the earthquake source that triggered the tsunami, which is subsequently modeled as a fault, as shown in the rectangular box.

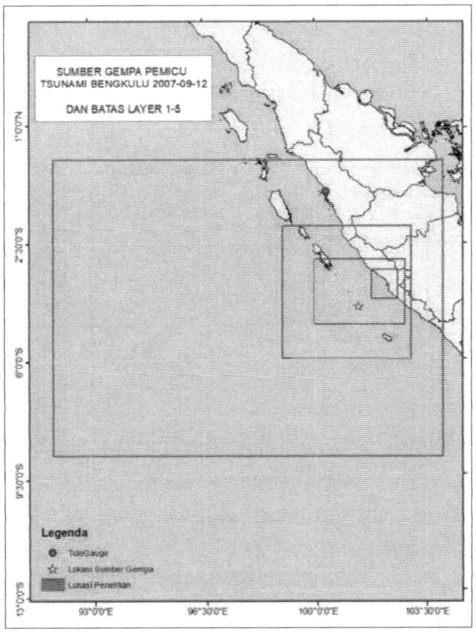

Figure 13. Earthquake Sources Triggering Tsunamis and the division of layers 1-5

One of the limitations of modeling with COMCOT is the maximum area size that can be modeled, which is restricted to 20 by 20 meters. Therefore, the study optimizes the fifth grid layer with a maximum area of 22 meters As shown in Table 4,

Table 4. Layer Boundaries

Layer	Size (Arcsec)		Size (Meter)		Layer Boundary		Area (KM2)	*Bathymetric*
	Size	Minute	Size	Meter	X	Y		
Region 1	1	minute	1850	meter	92.1	-8.9	4.9921	Batnas
					103.9	-0.1		
Region 2	0.33333	minute	616.667	meter	99.00106	-5.92511	444.3812	Batnas
					102.9011	-2.05706		
Region 3	0.11111	minute	205.5556	meter	100.62	-4.975	220.5437	Batnas
					102.8155	-3.34		
Region 4	0.03704	minute	68.519	meter	101.762	-4.215	96.7567	Batnas
					102.573	-3.34		
Region 5	0.01234	minute	22.839	meter	102.24	-3.832	8.2934	UAV
					102.294	-3.757		

*(Source : Data Processing)

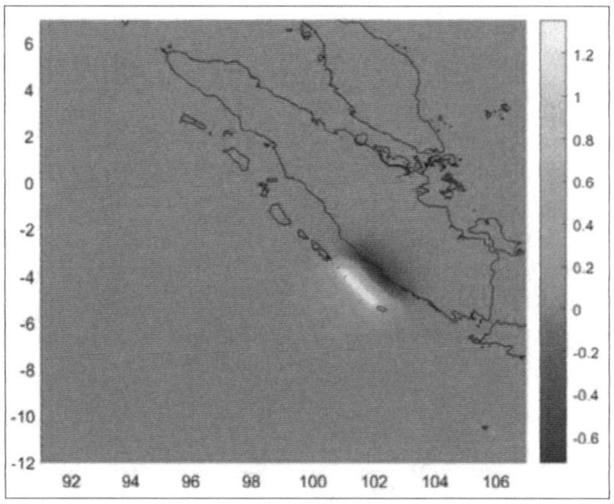

Figure 14. Tsunami-Generating Deformation

Figure 14 depicts the maximum values, indicating the vertical uplift of the oceanic crust, while the minimum values represent its downward displacement. From the figure it can be observed that the waters along the coast of Sumatra recede momentarily before the tsunami occurs a phenomenon attributed to subsidence. Meanwhile, areas near the earthquake source undergo vertical uplift. When the oceanic plate experiences a swift upward movement following an earthquake, it results in considerable variations in the water mass distribution, shifting from the ocean floor to the surface. One of the key factors that significantly influences wave height is bathymetry [15]. In shallow waters around islands, the waves generated during low tide are not as high compared to the tidal waves produced in areas near the earthquake source, where the bathymetry is deeper.

3.2.4 Tsunami Propagation and Run-up

Figure 15 presents the propagation of tsunami waves. The wave movement throughout the simulation is also available for observation in the GIF image, accessible through the provided link https://bit.ly/Tsunami2007 . The wave propagation modeling was conducted over a duration of 3 hours equivalent to 10,800 seconds, with a time interval of 1 minute. The propagation process is visualized starting from the 0th minute, representing the moment of seafloor deformation, which triggers the tsunami event. The simulation continues until the 120th minute, marking the conclusion of the modeling process.From this image, it can be observed that by the 25th second, the seawater has receded, followed by the occurrence of a tsunami. The tsunami waves begin to reach the mainland at the 35th minute.

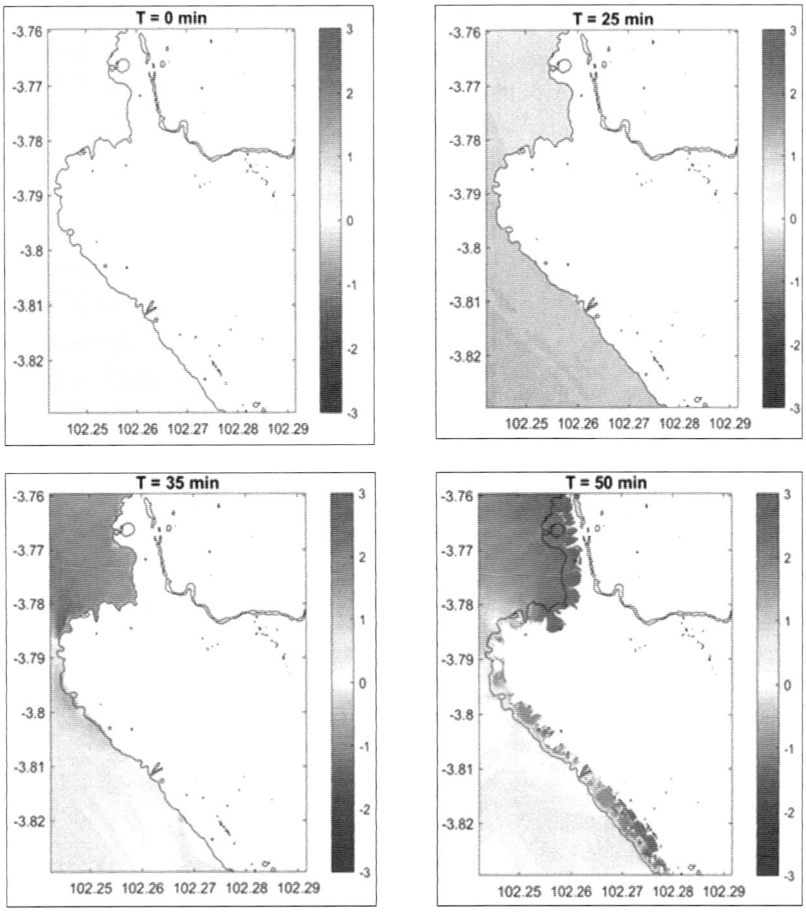

Figure 15. Tsunami Propagation

The COMCOT tsunami simulation, utilizing UAV data for the 2007 Bengkulu earthquake scenario, recorded a peak run-up of 5.9 meters, as illustrated in Figure 16. The distribution of the highest tsunami waves represents the extent of the maximum wave amplitude.

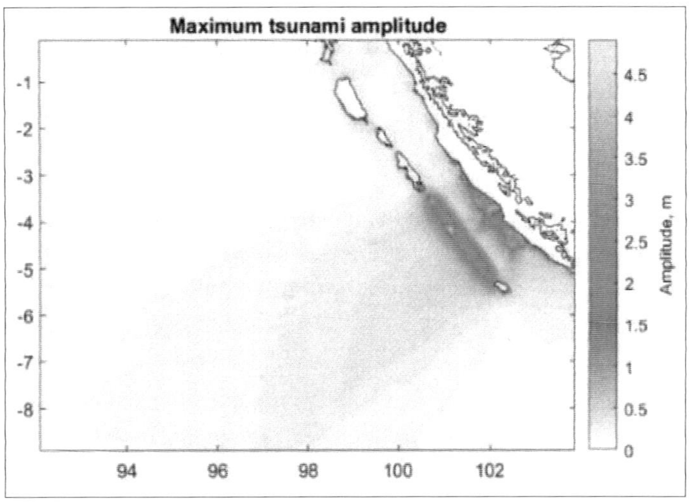

Figure 16. Tusnami Run-Up

3.2.5 Tsunami hazard map

The estimated tsunami height upon reaching the mainland, derived from the grid 5 modeling, is presented in the Tsunami Hazard Map as shown in Figure 17. This map provides a clear visualization of the potential tsunami heights across various areas, highlighting regions that are at higher risk of inundation. The classification results are represented by four potential inundation depth scales for tsunami waves: 0–1.5 meters, 1.5–3 meters, 3–4.5 meters, and 4.5–6 meters. Each depth category is visually distinguished by specific colors, as indicated in the legend of the map. This classification provides a clear and detailed representation of the varying levels of tsunami inundation risk across the affected areas, aiding in the assessment of potential impacts and the formulation of mitigation strategies. Through the tsunami hazard map, it is evident that land cover, particularly rice fields, contributes significantly to the extent of inundation, as these areas allow water to spread over a wider region. Additionally, the presence of rivers plays a crucial role in guiding the tsunami propagation, acting like natural channels that direct the flow of water. This phenomenon causes the inundation to become more concentrated and elongate along the river's path, resembling the behavior of a highway for the advancing wave.

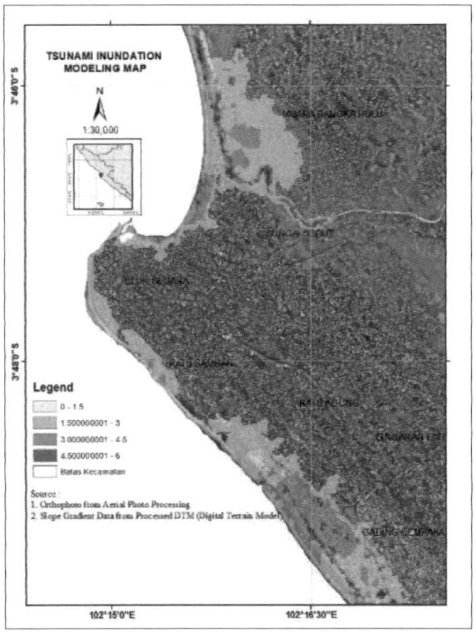

Figure 17 Tsunami Hazard Map

3.2.5 Model validation

3.2.5.1 Model validation with the Padang tide gauge observation data.

The data used to validate the waveforms generated from the modeling are tide gauge recordings from the Padang station, taken immediately after the tsunami event. The waveform is visualized in Figure 18. Initially, the waves appear regular, but as the tsunami occurs, a noticeable shift in the waveform pattern leads to irregularity.

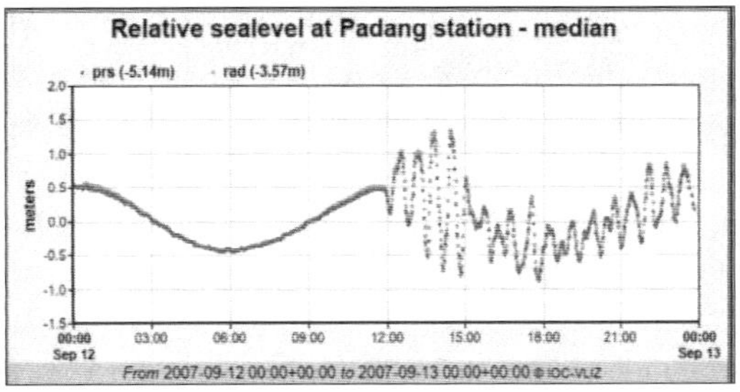

Figure 18. The waveform recorded at the Padang tide gauge station.

The image shows the waves recorded at the Padang station shortly after the tsunami event. The red line represents the tidal waves, and in order to accurately analyze the tsunami, the tidal wave component needs to be removed using a polynomial function. Validation is performed by comparing the synthetic waveform generated from the modeling with the waveforms recorded at the Padang tide gauge station. his comparison allows for an assessment of the accuracy and reliability of the model in replicating real tsunami wave behavior. As depicted in Figure 19, the black line represents the recorded data from the station, while the blue line corresponds to the synthetic data generated by the model. By examining the similarities and discrepancies between these two datasets, the effectiveness of the modeling approach can be evaluated, ensuring that the synthetic waveform closely mirrors the observed data.

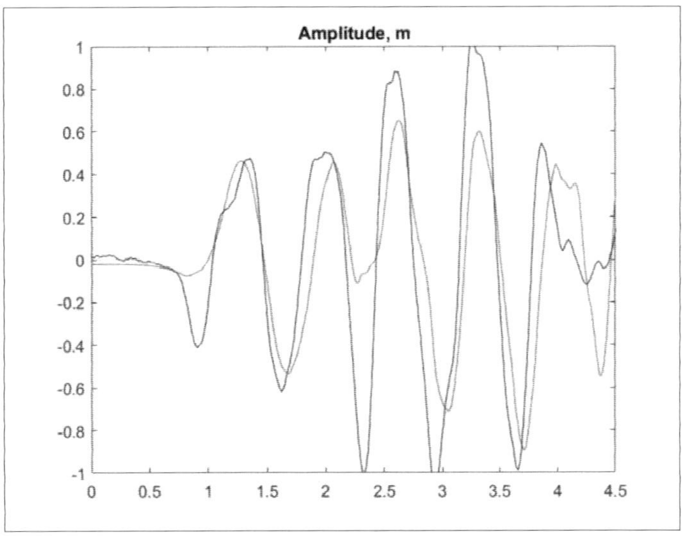

Figure 19. Synchronization of Synthetic Waveform and Observed Wavefor

From the figure, it can be observed that the modeled waveform is relatively lower than the waveforms recorded in the field. This discrepancy suggests that the model may underestimate the tsunami wave height, highlighting a potential area for further refinement in the modeling approach.

3.2.5.2 Model validation with post-tsunami field survey data.

This study utilizes field survey data derived from the research conducted by [16]. In that study, Borrero conducted a field survey soon after the tsunami event, as presented in Table 7. The survey offers essential information for understanding the tsunami's impact on affected areas, providing crucial insights for validating and analyzing the simulated tsunami waveforms. This dataset is instrumental in evaluating the model's accuracy and ensuring that the simulation accurately represents real-world conditions observed after the tsunami.

Table 5. The field survey locations by Borrero

Date/time	Longitude (°E)	Latitude (°S)	Name	Run-up (m)	Inundation distance (m)	Comment	Quality
9/17/07 17:00	101.43793	2.96731	Ipuh (Retak Ilir)	1.8	47	A maximum run-up, did not exceed	***
9/17/07 17:45	101.51154	3.07636	South Ipuh	2.2	93	Very clear inundation line. Energetic, large driftwood moved around	*****
9/17/07 15:00	101.52643	3.10175	Air Rami	1.8	172	Witness described run-up	*****
9/17/07 16:30	101.52225	3.11234	South Air Rami	2.0	150	Estimate only. Evidence of inundation across island, lagoon	***
9/17/07 12:00	101.66450	3.27505	Pantai Indah	1.9	50	Very clear run-up mark	*****
9/17/07 11:15	101.72724	3.32086	Karang Pulau	3.3	45	Very clear run-up mark	*****
9/17/07 9:00	101.90155	3.42991	Serangai	3.5 to 5.0	500	Sustained flow at 3.5 m. Splash over cliff top to 5.0 m, possibly more. Wave described as breaking bore	***
9/17/07 8:30	101.90923	3.43523	Serangai 2	3.0	140	Estimate only	**
9/15/07 17:15	102.03394	3.53072	Lais 2	3.0	35	Clear run-up mark	*****
9/15/07 16:45	102.03592	3.53163	Lais 1	3.9	30	Steep cliff coast, inundated over sea wall	*****
9/15/07 11:15	102.26022	3.78209	Bengkulu 4	1.4	151	Clear run-up mark. Houses flooded to 40 cm	*****
9/15/07 10:45	102.25414	3.78538	Bengkulu 3	1.7	130	Inundated over seawall. Did not move steel pipes	****
9/15/07 10:00	102.25116	3.78549	Bengkulu 2	1.4	0	Located in fishing port. Run-up on a vertical wall	*****
9/15/07 9:00	102.25028	3.79666	Bengkulu	1.4	40	A maximum value. No witness, run-up line not clear	***
9/18/07 11:45	102.30352	3.88556	Port North	1.2	64	Clear mark identified by witness	*****
9/18/07 8:45	102.27493	3.90038	Port South	0.9 to 1.7	90	No witness. Lower value is elevation at debris line. Larger value is needed to overtop beach berm	***
9/16/07 10:00	102.51948	4.14423	Pasar Seluma	1.3 to 2.0	100	No direct witness. Run-up mark unclear	**
9/16/07 12:30	102.79430	4.35269	Muara Maras	0.9 to 1.7	281	Clear inundation. Some tsunami damage. Flooded across road into swampy area	*****
9/16/07 14:00	102.89955	4.47753	Pasar Bawah Manna	3.5	56	Slow inundation, overtopped sea wall	*****
9/16/07 16:30	103.12446	4.58738	Teluk Biringin	1.4	20	Did not overtop beach berm	*****

Out of the 20 points in the study, 4 points are located within the research area, as shown in Figure 20. These 4 points were selected to validate the data derived from the modeling results. By utilizing these specific points, the study can effectively compare the model's output with real-world measurements, ensuring the accuracy and reliability of the tsunami simulation in the targeted region. Validation is performed using the RSR method[17], or the Run-up Standard Observation Error, which is a technique for comparing observed tsunami run-up values with those predicted by the model. A good RSR value is considered to be below 1.

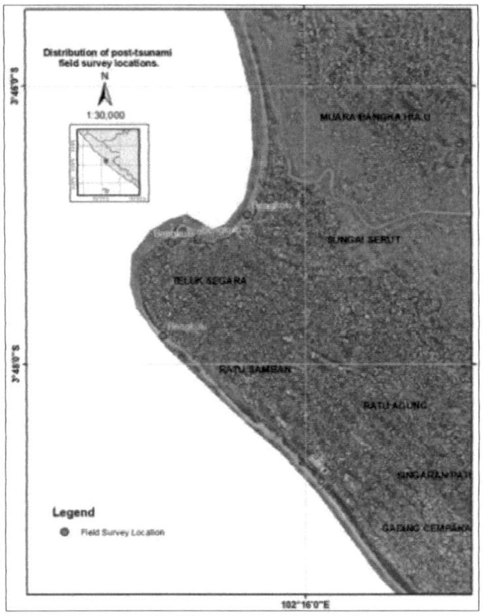

Figure 20. Distribution of Post-Tsunami Field Survey Locations

IOP Conf. Series: Earth and Environmental Science **1503** (2025) 012022 doi:10.1088/1755-1315/1503/1/012022

The formula used to calculate RSR values is as follows:

$$RSR = \frac{\sqrt{\sum_{i=1}^{n}(Oi - Mi)^2}}{\sqrt{\sum_{i=1}^{n} O_i^2}}$$

Where :
- Oi represents the observed run-up value at the i-th point.
- Mi represents the modeled run-up value at the i-th point.
- n is the total number of data points used for comparison.

The criteria14 used are:
- 0 < RSR < 0,5 very good
- 0,5 < RSR < 0,6 good
- 0,6 < RSR < 0,7 bad
- RSR > 0,7 very bad

Table 6. The calculation of RSR

Location	Coordinates		Run-Up(m)		Δhmax	$(\Delta hmax)2$	∑Oi2
	X	Y	Field Survey	Modeling			
Bengkulu 1	102.25	-3.79666	1.4	1.231	0.169	0.028561	1.96
Bengkulu 2	102.251	-3.78549	1.4	1.198	0.202	0.040804	1.96
Bengkulu 3	102.251	-3.78549	1.7	2.538	-0.838	0.702244	2.89
Bengkulu 4	102.26	-3.78209	1.4	1.822	-0.422	0.178084	1.96
					∑(Oi–Mi)2	0.949693	8.77
						0.97452193	2.961419
					RSR	=	**0.32907268**

Source : Data processing

Based on the calculated RSR (Root Square Residual) value of 0.33, the tsunami model falls into the "very good" category. This classification signifies that the model effectively represents real-world conditions observed in the field. The low RSR value demonstrates the model's high accuracy and reliability in simulating the actual phenomena, making it a valuable tool for understanding and predicting tsunami behavior in the studied area.

4. Conclusion

The COMCOT tsunami simulation using UAV data for the 2007 Bengkulu earthquake scenario resulted in a maximum run-up of 5.9 meters. Propagation observations indicated that seawater began to recede at the 25th second, followed by the occurrence of the tsunami. The tsunami waves reached the land at the 35th minute, inundating 505 hectares out of the total study area of 4,361.5 hectares. Furthermore, tsunami modeling using UAV data produced a synthetic waveform that closely resembles the observed waveform, albeit with some differences. Overall, the synthetic waveform is considered satisfactory, with its amplitude being lower than that of the observed data. Validation of the modeling results using the RSR method yielded a value of 0.3, classifying the results as very good.

5. References

[1] Berryman, K. (2005). Review of Tsunami Hazard and Risk in New Zealand.

[2] Kucharczyk, M., & Hugenholtz, C. H. (2021). Remote sensing of natural hazard-related disasters with small drones: Global trends, biases, and research opportunities. In Remote Sensing of Environment (Vol. 264). Elsevier Inc. https://doi.org/10.1016/j.rse.2021.112577

[3] González-Riancho, P., Aguirre-Ayerbe, I., Aniel-Quiroga, I., Abad, S., González, M., Larreynaga, J., Gavidia, F., Gutiérrez, O. Q., Álvarez-Gómez, J. A., & Medina, R. (2013). Tsunami evacuation modelling as a tool for risk reduction: Application to the coastal area of El Salvador. Natural Hazards and Earth System Sciences, 13(12), 3249–3270. https://doi.org/10.5194/nhess-13-3249-2013

[4] Danoedoro, P. (2015). Pengaruh Jumlah dan Metode Pengambilan Titik Sampel Penguji terhadap Tingkat Akurasi Klasifikasi Citra Digital Penginderaan Jauh. Simposium Nasional Sains Geoinformasi Ke-4.

[5] Bazykina, A. Yu., Mikhailichenko, S. Yu., & Fomin, V. V. (2018). Numerical Simulation of Tsunami in the Black Sea Caused by the Earthquake on September 12, 1927. Physical Oceanography, 25(4), 295–304. https://doi.org/10.22449/1573-160x-2018-4-295-304

[6] Wang, X., & Power, W. (2011). COMCOT: a Tsunami Generation Propagation and Run-up Model BIBLIOGRAPHIC REFERENCE.

[7] Julius, A. M., Pribadi, S., & Muzli, M. (2018). Correlation Equation of Fault Size, Moment Magnitude, and Height of Tsunami Case Study: Historical Tsunami Database in Sulawesi. IOP Conference Series: Earth and Environmental Science, 132(1). https://doi.org/10.1088/1755-1315/132/1/012012

[8] Blaser, L., Krüger, F., Ohrnberger, M., & ScherbaumFrank. (2010). Scaling Relations of Earthquake Source Parameter Estimates with Special Focus on Subduction Environment. Bulletin of the Seismological Society of America, 100(6), 2914–2926. https://doi.org/https://doi.org/10.1785/0120100111

[9] Westoby, M. J., Brasington, J., Glasser, N. F., Hambrey, M. J., & Reynolds, J. M. (2012). "Structure-from-Motion" photogrammetry: A low-cost, effective tool for geoscience applications. Geomorphology, 179, 300–314. https://doi.org/10.1016/j.geomorph.2012.08.021

[10] Standards, ed S. N. M. A. (1947). United States Bureau of the Budget.

[11] Danoedoro, P., Widayani, P., & Hidayati, I. N. (2017). Standar Pemetaan Penutup Lahan. Pusat Standarisasi dan Kelembagaan IG.

[12] BIG. (2015). PERKA BIG No 03 Tahun 2016.

[13] Grezio, A., Babeyko, A., Baptista, M. A., Behrens, J., Costa, A., Davies, G., Geist, E. L., Glimsdal, S., González, F. I., Griffin, J., Harbitz, C. B., LeVeque, R. J., Lorito, S., Løvholt, F., Omira, R., Mueller, C., Paris, R., Parsons, T., Polet, J., ... Thio, H. K. (2017). Probabilistic Tsunami Hazard Analysis: Multiple Sources and Global Applications. Reviews of Geophysics, 55(4), 1158–1198. https://doi.org/10.1002/2017RG000579

[14] Putra, R. (2008). Kajian Risiko Tsunami Tehadap Bangunan Gedung Non-Hunian dengan Skenario Variasi Ketinggian Run-up pada Garis Pantai; Studi Kasus Kota Banda Aceh Indonesia. Universitas Gadjah Mada.

[15] Pribadi, S., Afnimar, Puspito, N. T., & Ibrahim, G. (2013). Characteristics of earthquake-generated tsunamis in Indonesia based on source parameter analysis. Journal of Mathematical and Fundamental Sciences, 45(2), 189–207. https://doi.org/10.5614/j.math.fund.sci.2013.45.2.8

[16] Borrero, J. C., Weiss, R., Okal, E. A., Hidayat, R., Suranto, Arcas, D., & Titov, V. V. (2009). The tsunami of 2007 September 12, Bengkulu province, Sumatra, Indonesia: post-tsunami field survey and numerical modelling. Geophysical Journal International, 178(1), 180–194. https://doi.org/10.1111/j.1365-246X.2008.04058.x

[17] Hoover, S. V, & Perry, R. F. (1989). Simulation: a problem-solving approach. Addison-Wesley Longman Publishing Co., Inc., United States.

The Effect of Liquefaction on Carrying Capacity Loss in Palu IV Bridge Bored Pile Foundation

F Muktaf[1,2], H C Hardiyatmo[1] and S Ismanti[1]

[1] Department of Civil and Environmental Engineering, Faculty of Engineering, Universitas Gadjah Mada, Yogyakarta, Indonesia
[2] Directorate General of Highways, Ministry of Public Works and Housing, Jakarta, Indonesia

*E-mail: fikri.muktaf@mail.ugm.ac.id

Abstract. Pile design must be resilient to seismic-induced liquefaction, which can severely compromise structural stability by reducing the load bearing capacity of bridge foundations. This phenomenon typically occurs in saturated granular soils, where increased pore water pressure reduces effective stress, leading to a substantial loss of soil strength. Assessing the carrying capacity of bored piles under liquefaction conditions is essential to ensuring foundational stability and reliable geotechnical performance. This study seeks to evaluate and compare the load-bearing capacity of the bored pile group foundation of the Palu IV Bridge under both non-liquefaction and liquefaction soil conditions. The method used for liquefaction potential analysis was an empirical method with simplified procedure. Then, the Ultimate carrying capacity is calculated using the empirical approach. The result of this study showed that the soil depths vulnerable to liquefaction extend to 18.5 meters, and the decrease in the carrying capacity of the foundation under liquefaction conditions ranges from 4.45% to 15.20%. There was a reduction in skin resistance; however, the ultimate carrying capacity did not experience a significant decline, and the safety factor remains within the safe range. Based on these results, the ultimate carrying capacity of the installed Palu IV Bridge foundation is still capable of receiving loads during static and liquefaction conditions.

1. Introduction

Pile design must be resilient to seismic-induced liquefaction, which can severely compromise structural stability by reducing the load bearing capacity of bridge foundations[1]. This can be achieved through proper soil investigation and analysis to determine the potential for liquefaction in the area where the bridge will be constructed. By considering these factors in the design process, we can ensure that the bridge foundation will withstand earthquake forces and maintain stability both during and after the event.

The Palu IV Bridge is a critical infrastructure project in the post-disaster recovery effort following the 2018 Palu Earthquake that requires deep foundations to support the bridge's load over passive soil prone to liquefaction. To this end, reinforced bored pile foundations drilled into the ground to provide stability were used. By applying these techniques, Palu IV Bridge is better equipped to withstand the force of seismic events and ensure safety. Consequently, it is essential to evaluate the stability of bored pile foundations when confronted with the risk of liquefaction.

IOP Conf. Series: Earth and Environmental Science **1503** (2025) 012023 doi:10.1088/1755-1315/1503/1/012023

This study aimed to evaluate the effectiveness of Palu IV Bridge bored pile foundations in mitigating liquefaction risk and to examine the carrying capacity of bored pile foundations under normal and liquefied conditions. By thoroughly examining the engineering techniques used in the construction of the bridge, the researchers were able to provide valuable insight into best practices for building structures in seismically active regions with liquefaction-prone soils. The significance of this research lies in its contribution to informing the field of potential analysis and mitigation against liquefaction phenomena that can hopefully be applied to other projects with similar geological conditions.

2. Material and Methods

2.1 Soil and Pile Foundation Data

The study area is situated at the Palu IV River Estuary in Palu City, Central Sulawesi Province, Indonesia. The research focused on Pier 1 and Pier 2 of the Palu IV Bridge. This research location's ground water level (GWL) is 0.0 m because the borehole point is in the middle of the river. The N-SPT data of Pier 1 and Pier 2 can be seen in Figure 1.

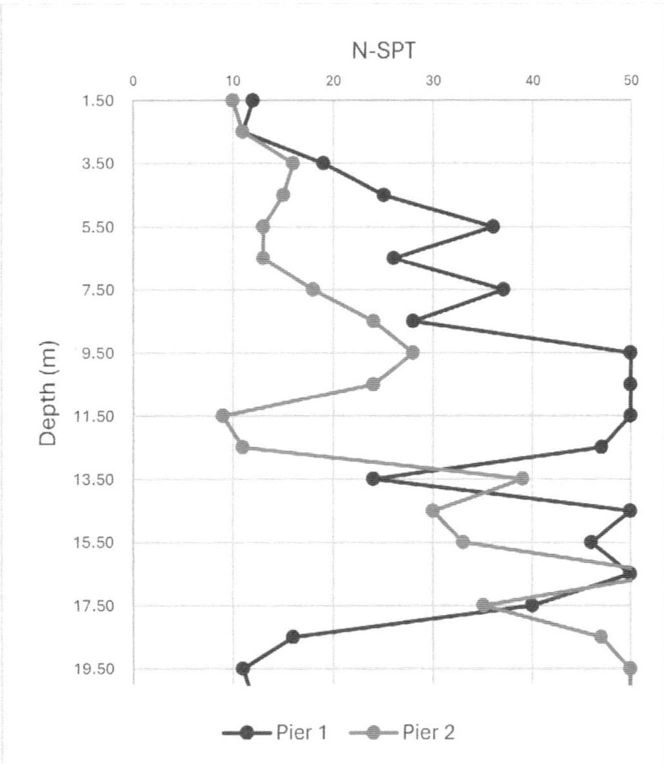

Figure 1. N-SPT value of Pier 1 and Pier 2

The foundation configuration consists of a group of 16 reinforced concrete bored piles, each with a diameter of 1.8 meters. The length of each bored pile is 56.4 meters for Pier 1 and 56.9 meters for Pier 2. The compressive strength of bored pile concrete is f_c' 35 MPa, and the reinforcement bar strength uses f_y 420 MPa. The rebar reinforcement configuration for bored piles uses S-36 and S-29 for the main longitudinal rebar and S-19 and S-22 for the transverse rebar. The specifics of the bored pile configuration are illustrated in Figure 2.

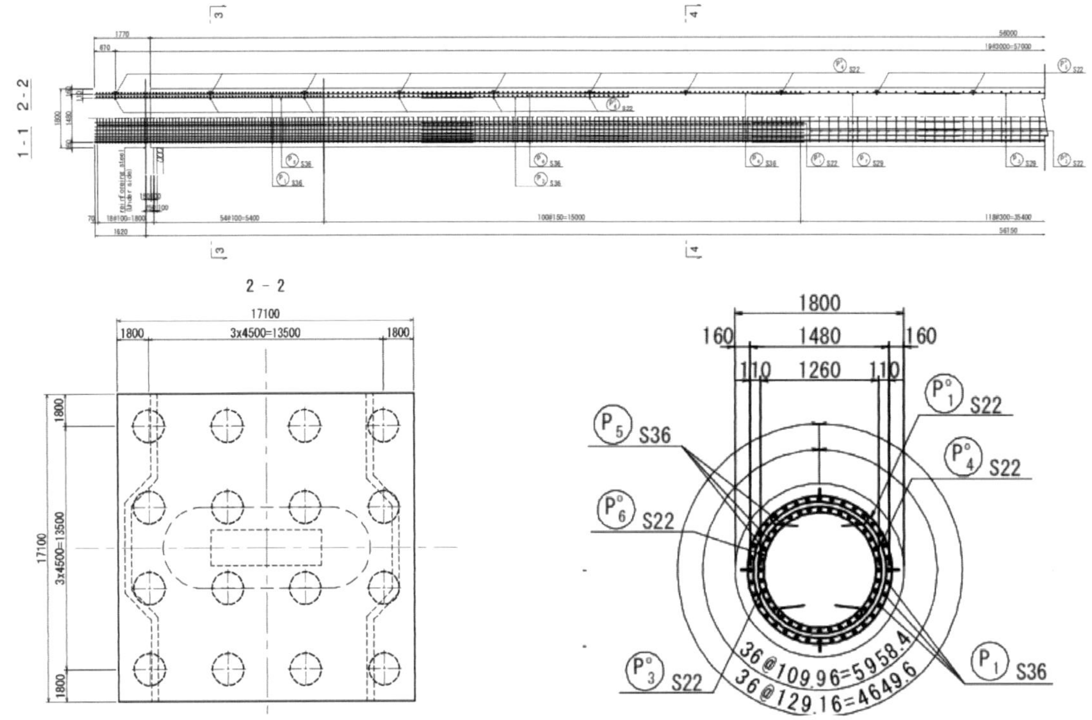

Figure 2. Configuration of Bored Pile

2.2 Liquefaction Potential Analysis using Simplified Procedure

The simplified approach for assessing liquefaction potential is a recognized framework used in geotechnical engineering to determine a soil's vulnerability to liquefaction under dynamic loads. This method examines the relationship between the cyclic stress ratio (CSR), which indicates the seismic loading applied to the soil during an earthquake, and the cyclic resistance ratio (CRR), which reflects the soil's ability to withstand liquefaction under cyclic loading conditions. CSR is computed using equations (1) to (4).

$$CSR = 0.65\left(\frac{a_{max}}{g}\right)\left(\frac{\sigma_v}{\sigma'_v}\right)r_d \tag{1}$$

$$r_d = exp\,(\alpha(z) + \beta(z)M) \tag{2}$$

$$\alpha(z) = -1.012 - 1.126\,sin\left(\frac{z}{11.73} + 5.133\right) \tag{3}$$

$$\beta(z) = 0.106 + 0.118\,sin\left(\frac{z}{11.28} + 5.142\right) \tag{4}$$

Where a_{max} is the peak horizontal ground acceleration, σ_v is the total vertical stress, σ'_v is the effective stress, and r_d is the depth reduction factor.

CRR is essential for evaluating the resistance of soil to liquefaction under cyclic loading conditions and Idriss and Boulanger[2] developed equations 5-10 below to obtain the CRR value.

$$CRR_{M=7.5,\sigma'_{vc}=1} = exp\left(\frac{(N_1)_{60cs}}{14.1} + \left(\frac{(N_1)_{60cs}}{126}\right)^2 - \left(\frac{(N_1)_{60cs}}{23.6}\right)^3 + \left(\frac{(N_1)_{60cs}}{25.4}\right)^4 - 2.8\right) \quad (5)$$

$$(N)_{60} = N_m C_E C_B C_R C_S \quad (6)$$

$$(N_1)_{60} = (N)_{60}.C_N \quad (7)$$

$$C_N = \left(\frac{P_a}{\sigma'_{vc}}\right)^{0.784-0.0768\sqrt{(N_1)_{60cs}}} \leq 1.7 \quad (8)$$

$$(N_1)_{60cs} = (N_1)_{60} + \Delta(N_1)_{60} \quad (9)$$

$$\Delta(N_1)_{60} = exp\left(1.63 + \frac{9.7}{FC + 0.01} - \left(\frac{15.7}{FC + 0.01}\right)^2\right) \quad (10)$$

The analysis determines the factor of safety (FS) as the ratio of CRR to CSR, where FS values below 1.0 indicate a high risk of liquefaction.

$$FS = \frac{CRR_{M=7.5,\sigma'_{vc}=1}}{CSR_{M=7.5,\sigma'_{vc}=1}} < 1 \quad (11)$$

2.3 Carrying Capacity of Bored Pile

The carrying capacity of a bored pile refers to the maximum load that the pile can support without failure. It is determined by both the skin friction (the resistance provided by the surrounding soil along the shaft of the pile) and the end bearing (the resistance at the tip of the pile when it reaches a stronger layer of soil or rock). The total carrying capacity is the sum of these two components, often referred to as the shaft resistance and tip resistance.

The ultimate pile tip resistance and ultimate frictional resistance for bored pile foundations are determined using the equations provided by O'Neill and Reese[3] for sandy soil and by Reese and O'Neill[4] for clayey soil. The ultimate bearing capacity of bored piles is empirically calculated using equations (12) to (14) below.

$$Q_u = Q_b + Q_s - W_p \quad (12)$$

$$Q_b = A_b f_b \quad (13)$$

$$Q_s = A_s f_s \quad (14)$$

Where Q_u = ultimate bearing capacity (kN), Q_b = ultimate pile tip resistance, and Q_s = ultimate friction resistance (kN), W_p = pile self-weight (kN), A_b = Area of bored pile (m^2), f_b = end bearing per unit area (kN/m^2), A_s=cross-sectional area of the bored pile (m^2), and f_s=skin resistance per unit area (kN/m^2).

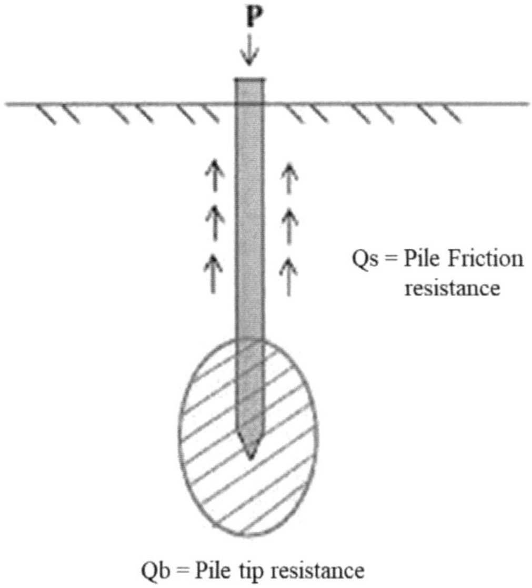

Figure 3. Pile friction and pile tip resistance

2.3.1 Ultimate frictional and pile tip resistance in sandy soil

To calculate the carrying capacity of bored piles in sandy soil, O'Neill and Reese recommend the values of frictional and pile tip resistance per unit area in Equations (15) to (18) below.

$$f_b = 0.6\, \sigma_r\, N_{60} \leq 4500\ kPa \quad (15)$$

$$\beta = 1.5 - 0.245\sqrt{z}\,;\ N_{60} > 15\,; 0.25 \leq \beta \leq 1.2 \quad (16)$$

$$f_s = \beta\, p_0' \quad (17)$$

$$\beta = \frac{N_{60}}{15}(1.5 - 0.245\sqrt{z}) \quad (18)$$

Where σ_r=reference stress (100 kN/m^2), β = frictional resistance coefficient, p_0'=the average of effective stress (kN/m^2), and z=depth (m).

2.3.2 Ultimate frictional and pile tip resistance in clay soil

To determine the carrying capacity of bored piles in clay soil, Reese and O'Neill suggest using the values for frictional resistance and pile tip resistance per unit area as outlined in Equations (19) to (22) below.

$$f_b = c_u\, N_c' \quad (19)$$

$$f_s = \alpha\, c_u \quad (20)$$

$$\alpha = 0.55\,;\ c_u < 150\ kPa \quad (21)$$

$$\alpha = 0.55 - 0.1\left(\frac{c_u}{P_r} - 1.5\right);\ 1.5 < \frac{c_u}{P_r} < 2.5 \quad (22)$$

Where c_u=undrained cohesion (kN/m^2) , α= adhesive factor, P_r=reference stress (100 kPa).

2.4 Negative skin friction (Q_{nsf})

In addition to decreasing bearing capacity in layers that can potentially experience liquefaction, the soil can also cause a negative skin friction phenomenon (Q_{nsf}) in the pile foundation. This occurs because the soil will experience subsidence when liquefaction occurs while the piles tend to remain stationary. As a result, a friction force acts downward on the pile blanket [5]. The calculation of Q_{nsf} can be seen in equation (23).

$$Q_{nsf} = \overline{P_0'}\, K_d\, tg\, \delta'\, A_s \tag{23}$$

Where $\overline{P_0'}$ average effective soil overburden pressure (kN/m^2), K_d = coefficient related to soil properties and stress conditions, and $tg\, \delta'$= shear strength factor due to the effective angle of internal friction.

3. Results and Analysis

3.1 Liquefaction Potential Analysis using Simplified Procedure

Liquefaction potential analysis was conducted at BH-09 and BH-10 to a depth of 20 meters. The Ground Water Level (GWL) of this study is 0.0 meters because the borehole location is in the middle of the river. The PGA used derived from the Binamarga LINI application [6] and the resulting PGA at this location is 0.71 g and multiplied by an amplification factor of 1.1 so that the PGA_M used is 0.781g. The analysis results show that all boreholes at the Palu IV Bridge reconstruction site have the potential to experience liquefaction with varying depths. A recapitulation of the analysis results at BH-09 and BH-10 can be seen in Figure 4, which show the FS graph and each borehole's potentially liquefied soil layer. The groundwater table for all boreholes is at a depth of 0.00 meters. The potentially liquefiable soil layers are at depths of 0-4.5 meters, 5.5-6.5 meters, 7.5-8.5 meters, 12.5-13.5 meters, and 17.5-18.5 meters at Pier 1. At Pier 2, the liquefiable layers are at depths of 0-12.5 meters, 13.5-15.5 meters, and 16.5-17.5 meters.

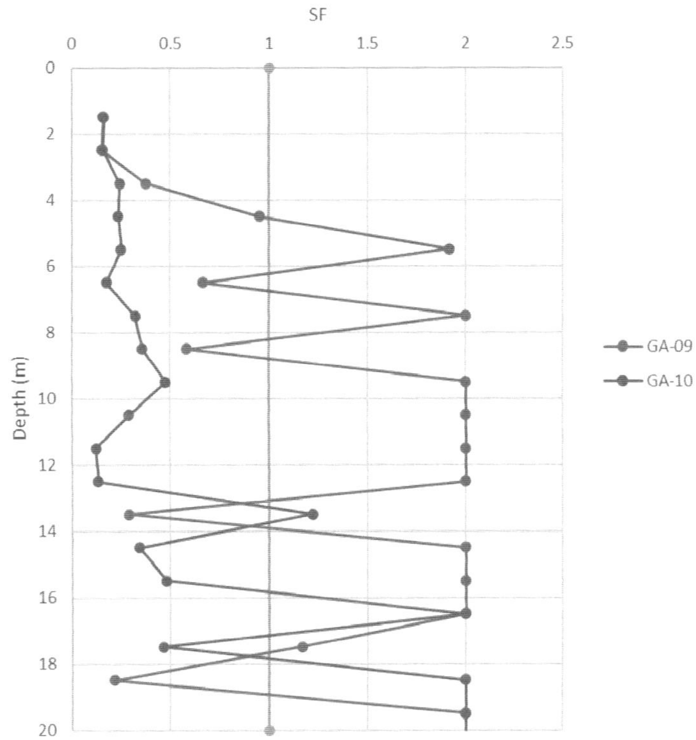

Figure 4. FS of Pier 1 and Pier 2

3.2 Carrying Capacity Losses of Single Bored Pile

The ultimate carrying capacity of the bored pile foundation, defined as the total of the ultimate tip resistance and friction resistance minus the weight of the foundation itself, was determined using the O'Neill and Reese equation for granular soils and the Reese and O'Neill equation for clay soils. This calculated ultimate carrying capacity of the bored piles served as a reference for assessing the reduction in bearing capacity during liquefaction and the factor of safety (FS) under both static and liquefaction conditions.

Table 1. Carrying Capacity of Single Bored Pile under Static Conditions

Borehole	Carrying Capacity of Bored Pile under Static Conditions			
	Ultimate skin friction Qs (kN)	Ultimate end bearing Qb (kN)	Weight of Bored Pile W (kN)	Carrying Capacity Qu (kN)
Pier 1	22,075.98	5,305.68	344.45	27.037,21
Pier 2	20,755.97	5,305.68	347.50	25.714,14

According to the liquefaction potential analysis results presented in Section 3.1, the skin resistance value is disregarded (effectively becoming zero) in the liquefied layer, as stipulated by SNI 8460:2017[7]. Additionally, negative skin friction is taken into account due to settlement following liquefaction. Consequently, the bearing capacity of the bored pile is diminished, as illustrated in Table 2. The analysis revealed a reduction in the axial bearing capacity of the bored pile foundations at Pier 1 and Pier 2. Specifically, Pier 1 experienced a decrease in bearing capacity of 1,203.31 kN (4.45%), while Pier 2 saw a reduction of 3,908.01 kN (15.20%). This occurs because when soil of low density is subjected to seismic activity, the air voids within the soil are filled by increasing pore water pressure, causing the effective stress to reach zero. Consequently, the soil assumes liquid-like properties. This transformation results in the soil losing its frictional resistance against the pile surface.

The decrease in bearing capacity at Pier 2 is greater than at Pier 1 because the soil at Pier 2 tends to be more dominant sand. This makes the FS value against liquefaction smaller, making it more susceptible to a reduction in bearing capacity. Table 2 shows the magnitude of the decrease in bearing capacity at Pier 1 and Pier 2.

Table 2. Carrying Capacity of Single Bored Pile under Liquefaction Conditions

Borehole	Carrying Capacity of Bored Pile under Liquefaction Conditions				Carrying Capacity Losses	
	Ultimate skin friction Qs (kN)	Ultimate end bearing Qb (kN)	Weight of Bored Pile W kN)	Carrying Capacity Qu (kN)	Δ Qu (kN)	Δ Qu (%)
Pier 1	20,872.67	5,305.68	344.45	25.833.90	1,203.31	4.45%
Pier 2	16,847.95	5,305.68	347.50	21,806.13	3,908.01	15.20%

According to Bina Marga Guidelines No. 02/P/BM/2022[8], in the Discussion of the Implementation of Special Bridge Safety, the safety factor during static/service loads is FS> 2.5, and during earthquake conditions is FS>1.6.

The calculation results show that the bored pile foundation can withstand the decrease in bearing capacity when liquefaction occurs both under static conditions (service load) and dynamic conditions (seismic load) and can be declared safe. Table 3 details the calculation of the safety factor (FS) of Pier 1 and Pier 2.

Table 3. Safety Factor of Bored Piled Group under normal and liquefied conditions

Point	Load Combination	Carrying Capacity Qu (kN)	Efficiency (Eg)	Number of Bored Pile (n)	Ultimate Load Pu (kN)	Safety Factor (FS)		Conclusion
						Normal Condition	Liquefaction Condition	
Pier 1	Service Load	27.037,21	0.65	16	96917.40	FS > 2,5	-	Safe
Pier 1	Seismic Load	25.833.90	0.65	16	84758.40	-	FS > 2,5	Safe
Pier 2	Service Load	25.714,14	0.65	16	96917.40	FS > 2,5	-	Safe
Pier 2	Seismic Load	21,806.13	0.65	16	84758.40	-	FS > 2,5	Safe

4. Conclusions

Analysis of liquefaction potential using empirical methods shows that there is liquefaction potential at Pier 1 and Pier 2 with varying depths. The results of empirical calculations show that when liquefaction occurs, the bored pile experiences a decrease in its ultimate capacity of 16.15% at Pier 1 and 19.24% at Pier 2. Although the ultimate capacity of the bored pile decreases during liquefaction conditions, it is still in a safe condition (SF>2.5 in static conditions and SF>1.6 in cyclic conditions). Thus, one mitigation strategy to anticipate liquefaction involves employing bored pile foundations that extend beyond 20 meters in depth and are embedded in hard soil. This approach ensures that even in the presence of a liquefied soil layer, the foundation remains capable of withstanding and effectively distributing the loads from the superstructure to the underlying hard soil layer.

References

[1] Boulanger R W, Chang D, Brandenberg S J, Armstrong R J and Kutter B L 2007 Seismic Design of Pile Foundations for Liquefaction Effects *Earthquake Geotechnical Engineering* Geotechnical, Geological and Earthquake Engineering vol 6, ed K D Pitilakis (Dordrecht: Springer Netherlands) pp 277–302

[2] Boulanger R. W., and Idriss I.M., CPT and SPT Based Liquefaction Triggering Procedures, Centre for Geotechnical Modelling, Report No. UCD/CGM-14/01, 2014, pp. 1-134.

[3] Reese L C and O'Neill M W 1989 New Design Method for Drilled Shafts from Common Soil and Rock Tests Foundation Engineering: Current Principles and Practices

[4] O'Neill, M. W., and Reese, L. C. (1999). "Drilled shafts: Construction procedures and design methods." Publication FHwA-IF-99-025, FHwA, Washington, DC.

[5] Broms, B.B, A. Fredriksson, and L. Carlsson. 1976. Land subsidence in Sweden due to waterleakage into deep-lying tunnels and its effects on pile supported structures. In Publ. 121, International Association of Hydrological Sciences. Proceedings of the Anaheim Symposium, December 1976: 375–387.
[6] Directorate of Road and Bridge Engineering, Indonesia earthquake source and hazard map for bridges 2017, aplikasi lini (Accessed 5 Apr 2024) https://lini.binamarga.pu.go.id/
[7] BSN 2017 SNI 8460: 2017 – Persyaratan Perancangan Geoteknik
[8] Directorate General of Bina Marga, 2022. Pedoman Bidang Jalan dan Jembatan : Pembahasan Penyelenggaraan Keamanan Jembatan Khusus.

IOP Conf. Series: Earth and Environmental Science **1503** (2025) 012024 doi:10.1088/1755-1315/1503/1/012024

Flood Vulnerability Analysis Using GIS Techniques: a Case Study of the Tirtomoyo Sub-Watershed, Wonogiri Regency

Rita Noviani[1*], Dicky Ardiyanto[1], Fauzan Adiatama[1]

[1] Department of Geography, Universitas Sebelas Maret, Surakarta, Indonesia

*E-mail: ritanoviani@staff.uns.ac.id

Abstract. The Bengawan Solo watershed in Wonogiri Regency consists of four sub-watersheds, one of which is the Tirtomoyo Sub-Watershed. The Tirtomoyo Sub-Watershed has a level of vulnerability to flood disasters. Areas that frequently experience flooding are Tirtomoyo District and Nguntoronadi District. Based on data from the Wonogiri Regency BPBD, in the last 5 years the flood disaster has occurred each year in the Tirtomoyo Sub-Watershed. This study aims to analyze the vulnerability of flood disaster in the Tirtomoyo Sub-Watershed. The methods used in this study are the scoring method, weighting parameters that have an influence on flooding, and spatial analysis of the Geographic Information System (GIS). The results of the classification of flood vulnerability levels in the Tirtomoyo Sub-Watershed show that the medium vulnerability level is the widest with 10,436.90 ha (34%), with a low vulnerability level occupying 6,155.84 ha (57%), and a high vulnerability level of 1,662.79 ha (9%). Tirtomoyo Sub-district is the most extensive area at each level of flood vulnerability.

1. Introduction

Flood events frequently occur in coastal areas, along riverbanks, and in densely populated areas lacking adequate drainage infrastructure [1]. The consequences of climate change are likely to exacerbate flood risks in these vulnerable areas [2]. Future projections suggest that natural disaster events are anticipated to double by 2050, increasing the number of people exposed to their adverse effects [3]. A flood can be characterized by the occurrence of water exceeding its usual level, resulting in inundation of low-lying areas adjacent to riverbanks [4]. A flood is a natural event in which excess water cannot be effectively managed by the drainage system in a particular area, resulting in detrimental inundation. The etiology of these flood disasters is largely due to meteorological and climatic factors. During these events, extreme rainfall with significant intensity often occurs, which is a consequence of climate change [5]. Floods can happen as a result of several factors, including the characteristic of catchment area, duration and intensity of rainfall, land cover, topographic features, and effectiveness of the drainage network [6]. Since the last decade of the 20th century, global estimates show that natural disasters, including floods, have resulted in the loss of more than one hundred thousand lives and have affected approximately 1.5 billion people [3].

A considerable number of low-income nations exhibit heightened susceptibility to flooding, as they frequently lack the requisite resources for each stage of the disaster management cycle: preparedness, mitigation, response, and recovery [7]. Indonesia is a tropical country with two seasons, one of which is the rainy season. During this season, rainfall in Indonesia is classified as very high and occurs in almost all regions [8]. The persistent phenomenon of climate change is leading to a heightened occurance of hydrometeorological disasters in Indonesia, one notable instance being flooding [9]. A flood is an overflow of water that exceeds normal levels, resulting in water overflowing the river channel and inundating the surrounding low-lying land [10]. Excessive rainfall intensity leads to an increase in the magnitude of the current flow created in a water body with an impact on the inability to withstand the heavy flow, causing overflow to lower areas around the river [11].

According to the findings from the mapping conducted by the Office of National Unity Politics and Community Protection (Kesbangpolinmas), around 90% of Wonogiri Regency is a disaster-prone area, including floods, landslides and hurricanes. At the national level, Wonogiri Regency has an index score of 146 and is a high category area in terms of natural disaster vulnerability in Indonesia that occurs every rainy season [12]. The rise in river levels caused by extreme rainfall due to the impacts of climate change is a major factor that increases the potential for flooding in an area [5]. River (watershed), rain plays an important role as an input to the hydrological system. Each watershed has a different response to rain with a certain amount and intensity [13]. Watersheds are catchment areas that serve as pathways for water runoff, making them prone to significant flooding events [2].

The Tirtomoyo Sub-Watershed is the largest among the watersheds within the Gajah Mungkur Reservoir area in Wonogiri. However, the contribution of water from the Tirtomoyo Sub-Watershed to the Wonogiri Reservoir has not yet been assessed using the most appropriate method [14]. The Bengawan Solo watershed in Wonogiri Regency is divided into for sub-watersheds, one of which is the Tirtomoyo Sub-Watershed. This sub-watershed passes through 18 villages spread across 4 sub-districts in the region. Tirtomoyo Sub-Watershed has a high vulnerability to flooding, especially in the Tirtomoyo and Nguntoronadi sub-watersheds that often experience flooding. Based on data from the Wonogiri Regency BPBD, in the last 5 years there has always been a flood disaster every year in the Tirtomoyo Sub-Watershed. As reported by Murianews, heavy rainfall with high intensity caused flooding in Tirtomoyo Sub-district, Wonogiri Regency, on Sunday 15 December 2024. The Dimoro River overflowed, submerging dozens of houses and a healthcare facility. The study analyzed the annual flood events in the Bengawan Hulu sub-watershed [15]. The findings indicated that rainfall lasting two days has the potential to trigger floods. [16] Flood tracing is defined as a method used to determine the timing and flow discharge (hydrograph) at a specific point in a river, based on the upstream hydrograph.

Knowing the flood vulnerability level in region is an essential preventive measure to reduce the potential losses during a flood event. Geographic Information System (GIS), as both a science and technology, enables the management and analysis of extensive spatial data. GIS can be used to develop models based on selected data, enhancing users' understanding and perspective of the issues to better reflect real-world conditions [17]. Geographic Information Systems (GIS) are one of the tools used to identify food-prone areas spatially [18]. Many methodologies are available for the preparation of flood vulnerability maps, for instance, the Spatial Multi-Criteria Evaluation (SMCE) technique, which assesses multiple variables comprehensively, has been implemented in several studies due to the observation that 80% of the information utilized by decision-makers is link to geographical factors, supported by the integration of Geographic Information Systems (GIS). GIS enhances the quality and quantity of data available for scenarios requiring decision-making [9]. This study aims to analyze the vulnerability of flood disaster in the Tirtomoyo Sub-Watershed. The methods used in this study are scoring and weighting methods to assess the parameters that affect flooding, as well as spatial analysis with the help of Geographic Information Systems (GIS). The variables considered consist of slope, rainfall, land elevation, land use, distance from the river, and soil type. The result of the research is a map of flood vulnerability in Tirtomoyo Sub-watershed, which is obtained through the process of overlaying maps of each of these parameters.

2. Methods

This research is located in the Tirtomoyo Sub-Watershed, Wonogiri Regency. The Tirtomoyo Sub-Watershed passes through 4 sub-districts, namely Batuwarno Sub-district, Baturetno Sub-district, Tirtomoyo Sub-district, and Nguntoronadi Sub-district. Astronomically, the Tirtomoyo Sub-Watershed is located between 7°52'45.36" - 8° 0'45.04"S and 110°54'31.81" - 111° 8'57.89"E with an area of 188.83 km2. 7°52'45.36"S-8°0'45.04"S and 110°54'31.81"E-111°8'57.89"E with an area of 188.83 km2. The data used in this study consisted of land use data sourced from the Indonesian Terrain Map (RBI) of Wonogiri Regency on a scale of 1:50.000 – Badan Informasi Geospasial, buffer analysis to process distance data from the river, National DEM data for the Tirtomoyo watershed area to analyze the elevation and slope, rainfall data obtained from BBWS Bengawan Solo, and soil type data for Wonogiri Regency obtained from the Central Java Province Land Type Map. The application utilized for the data processing and analytical procedures in this research is ArcGIS 10.8.2.

This study measured the level of flood disaster vulnerability using the scoring approach and weighted of parameters that affect floods [20] and the analysis of geographic information systems with overlay techniques. The parameters that affect flood vulnerability in this study include slope, rainfall, soil type, land height, land use, and distance to the river. The scoring on each of these parameters is then overlayed with the help of Geographic Information System (GIS) software, namely ArcGIS to obtain flood vulnerability maps [19].

Table 1. Score and Weighting of Flood Vulnerability Parameters

Parameter	Classification	Score	Weighted
Slope	0 – 8% (Flat)	9	
	8 – 15% (Slopes)	7	
	15 – 25% (Slightly Steep)	5	10
	25 – 40% (Steep)	3	
	>40% (Very Steep)	1	
Land Use	Water body, open land, ponds	9	
	Rice fields, settlement	7	
	Moor, plantation	5	25
	Mixed garden, shrubs	3	
	Forest	1	
Rainfall	> 2500 mm	9	
	2001–2500 mm	7	
	1501–2000 mm	5	15
	1000–1500 mm	3	
	<1000 mm	1	
Soil Type	Oxisol, vertisol	9	10

Parameter	Classification	Score	Weighted
	Ultisol, molisol, alfisol	7	
	Inceptisol	5	
	Histosol, entisol	3	
	Andisol, spodosol	1	
Elevation	0–20 mdpl	9	
	21–50 mdpl	7	
	51–100 mdpl	5	20
	101–300 mdpl	3	
	>300 mdpl	1	
Distance from river	0 - 25 m	9	
	25 - 50 m	7	
	50 - 75 m	5	20
	75 - 100 m	3	
	>100 m	1	

Source: [4][20]

Secondary data to create a map of each parameter is obtained through various sources by processing using the ArcMap 10.8.2 application. The weight of each parameter is based on the amount of influence of the parameter on the flood disaster. After each parameter is given a score or value and weight according to the reference table, the next step is to overlay the entire parameter map and continue by calculating the vulnerability index to find out the level of flood vulnerability based on the obtained value.

Flood vulnerability level intervals are classified using the following formula.

$$I = \frac{R}{C}$$

I = interval width

R = range of different value

C = constanta

Table 2. Classification of Flood Vulnerability Levels

Interval	Levels
270-443	Low Vulnerability
444-616	Medium Vulnerability
617-790	High Vulnerability

Flood vulnerability level in the Tirtomoyo Sub-watershed is classified into 3 category: low, medium, and high. The higher the value produced, the higher flood vulnerability level in an area. If the lower the value produced, the lower the level of flood vulnerability.

3. Result and Discussion

3.1 Flood Vulnerability Variables

The results obtained from this research conducted on several factors that cause floods, there are 6 variables that have the highest influence on the occurrence of flood disasters in the Tirtomoyo Sub-Watershed. Rainfall affects the occurrence of flood disasters in an area with increasing precipitation, the higher possibility of flooding [21]. The rainfall data in this study is used in one classification because the source of the coordinate point of the rain observation center in the Tirtomoyo Sub-watershed cannot be found, so it uses rainfall data from the nearest post, namely the Batuwarno Post. The average rainfall in the Tirtomoyo Sub-Watershed is 1591 mm/year which is classified as medium with a fairly vulnerable level. The rainfall map is shown in Figure 1.

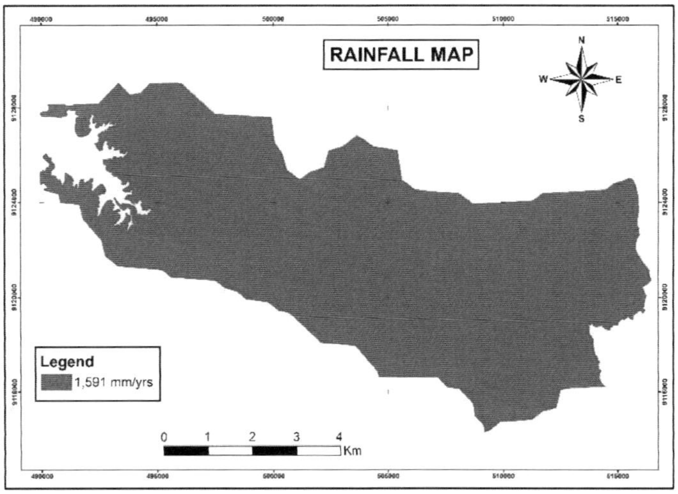

Figure 1. Rainfall Map

The distance parameter from the river has an influence in determining the chance of an area experiencing flooding or not. The closeness of a geographical area to the river significantly increases the potential for flooding events in that area. The results of distance classification in the Tirtomoyo Sub-watershed show that the area >100 m from the river has the largest area of 16836.36 ha, followed by the area 0-25 m from the river encompasses a total area of 528.01 ha, the area 25-50 m from the river comprises 520.69 ha, the area 50-75 m from the river occupies a land area of 510.84 ha, and the area 75-100 m from the river covers 495.37 ha. Rivers in the Tirtomoyo Sub-Watershed often have evaporation because the high water discharge is not able to be accommodated by the river. Areas that often occur floods are areas around river channels [20]. The distance map from the river is shown in Figure 2.

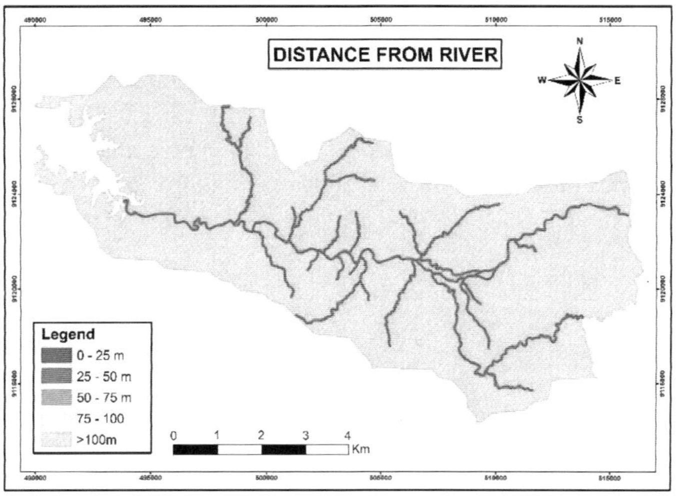

Figure 2. Distance From River Map

The condition of the slope of an area can be a triggering factor for landslides and floods. The slope slope class in the Tirtomoyo Sub-Watershed is divided into 5 classes with a flat area (0-8%) covering an area of 3302.71 ha, a sloping area (8-15%) with an area of 2977.01 ha, followed by a slightly steep area (15-25%) with an area of 4051.40 ha, a steep area (25-45%) comprises 6197.22 ha, and a very steep area (>45%) with an area of 2267.85 ha. The slope slope in the Tirtomoyo Sub-Watershed, although dominated by steep slopes (25-45%), slopes with a flat slope (0-8%) are also quite wide, making the flow of surface runoff in some areas slow and the possibility of inundation or flooding. This is because the slope of the slope affects the direction, rate, and also the concentration of rainwater. The distribution of slopes in the Tirtomoyo Sub-watershed is shown in Figure 3.

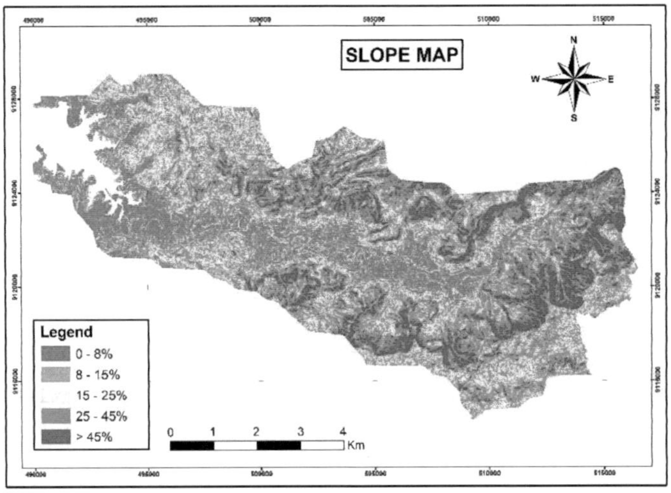

Figure 3. Slope Map

Elevation parameters affect the area of flooding that occurs. This is because low lands tends to the occurrence of water runoff during times of heavy rainfall, especially when floods hit. The Tirtomoyo Sub-Watershed has 4 elevation classes with an elevation of 50-100 meters above sea level, covering a total area of 1396.61 ha and an elevation of 100-200 meters above sea level which encompasses a total area of 1.68 ha. The dominant elevation in the Tirtomoyo Sub-Watershed is 20-50 meters above sea level encompassing a territory of 9092.39 ha and followed by an altitude of 0-20 meters above sea level which includes an expanse of 8306.19 ha. Low elevation is scattered around the Tirtomoyo river, so often river runoff has the potential to inundate the houses of local residents [20]. Related to the distribution of place height or elevation is shown in Figure 4.

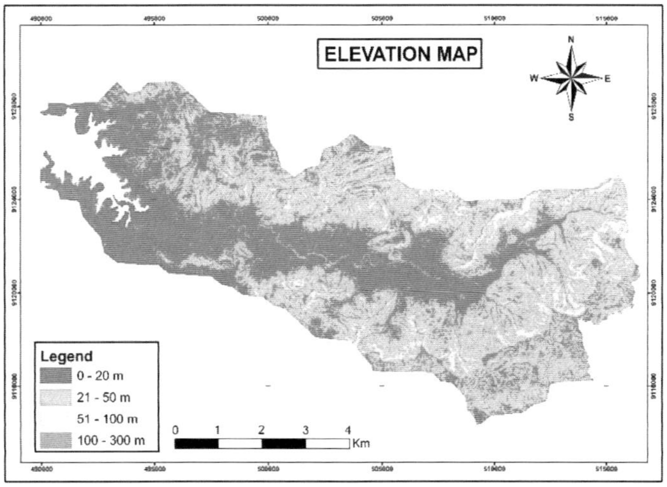

Figure 4. Elevation Map

Land use parameters have a role in the amount of surface water runoff that causes flooding. Each land use has its own ability to drain water. Land use in the Tirtomoyo Sub-watershed is generally dominated by rice fields comprising an area of 6066.33 ha, moors occupies of 4791.92 ha, and settlements encompassing an area of 3543.43 ha. When viewed through the dominance of land use, the Tirtomoyo Sub-Watershed has the potential for flooding. This is influenced by the fact that moorland and rice paddy plants have soil that has a low capacity to withstand surface water movement, while settlements have a considerable soil load and tend not to have good infiltration. The distribution of land use is shown in Figure 5.

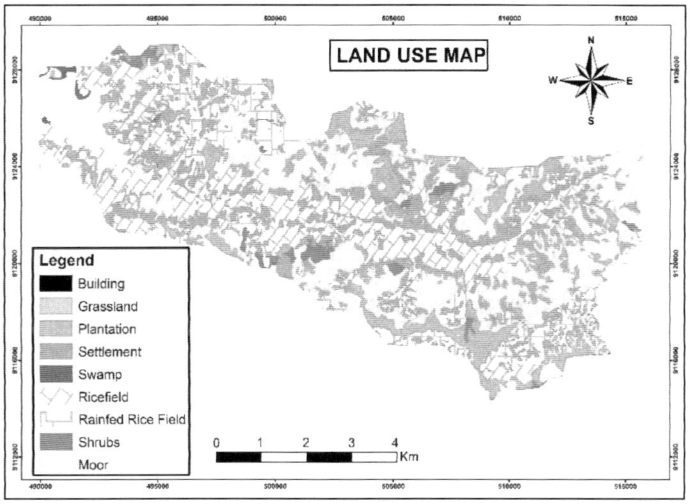

Figure 5. Land Use Map

There are 3 types of soil in the Tirtomoyo Sub-watershed, namely lithosol, grumusol, and latosol. The dominant soil type in the Tirtomoyo Sub-watershed is lithosol with an area of 15700.64 ha and is spread from the eastern to the central part of the sub-watershed. Litholomic soils are generally located in mountainous areas with large grains because they have not undergone complete weathering. Lithosol soil has low permeability because it has a tendency to dry so that runoff from rain and rivers cannot flow properly. Areas that have this land tend to be prone to landslides and if they are saturated with water, flash floods can occur. The soil type of grumusol is spread lengthwise from the middle to the west of the sub-watershed with an area of 1732.45 ha, while the latosol soil type is in the southeastern part of the sub-watershed covering an expanse of 1449.98 ha. Classification is based on the classification of orders. Litosole soils include the order Entisol, latosol soils include the order Ultisol, and grumusol soils include the order Vertisol. The distribution of soil types is shown in Figure 6.

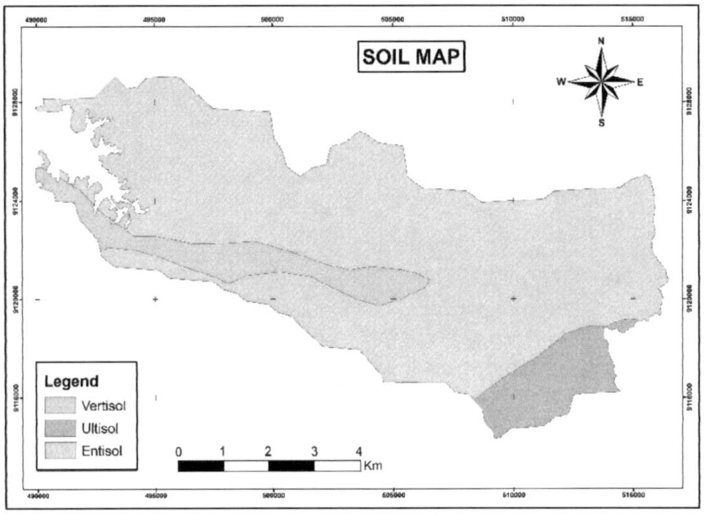

Figure 6. Soil Map

3.2 Flood Vulnerability Level

Flood vulnerability is a land area that is usually dry, can be flooded due to high rainfall and an increase in water capacity within certain area [4]. From the six flood parameters, an overlay analysis was then carried out to classify the total score with a formula to divide it into three levels of flood vulnerability: low, medium, and high.

Based on the results obtained from the classification of areas prone to flooding within Tirtomoyo Sub-Watershed, areas that are in low vulnerability are said to be the safest areas against the possibility of flooding. The low level of vulnerability covering an area of 6155.84 ha with the topography of the area which is bumpy and hilly, dominated by moorland cover and gardens, elevation above 20 meters above sea level, and a long distance from the river makes it difficult for flood runoff to inundate in this area. The classification results show that Tirtomoyo Sub-district is the largest sub-district with a low vulnerability level of 4201.90 ha, followed by Nguntoronadi Sub-district which has an area of 777.64 ha, and Batuwarno Sub-district with an area of 610.25 ha. Tirtomoyo Sub-district can be the area with the widest level of low vulnerability because this sub-district has the largest area in the Tirtomoyo Sub-Watershed and most of the topography of the area is in the form of hills with an altitude of more than 50 meters.

Medium vulnerability level, which is an area that has the potential to be critical to flooding. The area of medium vulnerability in the Tirtomoyo Sub-watershed is 10436.90 ha with an average of being in the lowlands, dominated by the use of residential land, rice fields and gardens. Settlements that do not have qualified drainage channels can experience surface runoff if there is high intensity rain. This is supported by plants from gardens and rice fields which are included in plants that are difficult to maintain surface water so that it causes water to stagnate. The level of vulnerability is medium on the slope of the slope which is classified as sloping and the elevation is between 0-20 meters above sea level, indicating that if the drainage is not managed properly, flooding can occur. Based on the size of the area, Tirtomoyo Sub-district has the widest medium vulnerability level with 4701.83 ha, then Nguntoronadi Sub-district encompasses a land area of 3236.66 ha, and Baturetno Sub-district is characterized by an area of 1146.90 ha. Most of the floods that occur in the Tirtomoyo Sub-Watershed are inundation floods due to poor drainage so that stagnant water is difficult to flow back into the river.

The high level of vulnerability is referred to as an area that is critical to potential flooding. The high level of vulnerability has an area of 1662.79 ha and is found in areas located at a distance of 0-100 meters from the river and low elevation so that if there is high intensity rain, it will be easier to experience excess water discharge. Tirtomoyo Sub-district is the largest area that has a high level of flood vulnerability compared to other sub-districts, covering 715.75 ha, followed by Baturetno Sub-district which cover 421.80 ha, and Batuwarno Sub-district included an area of 318.52 ha. Floods in the Tirtomoyo Sub-Watershed are more often caused by high rainfall intensity and supported by a small river carrying capacity. Another factor influencing higher flood vulnerability is the type of land use in the form of rice fields and settlements that do not have good water absorption capabilities. The dominant soil type aspect is also influential, namely lithosol soil which has low permeability so that inundation occurs. The representation of flood vulnerability level is shown in Figure 7.

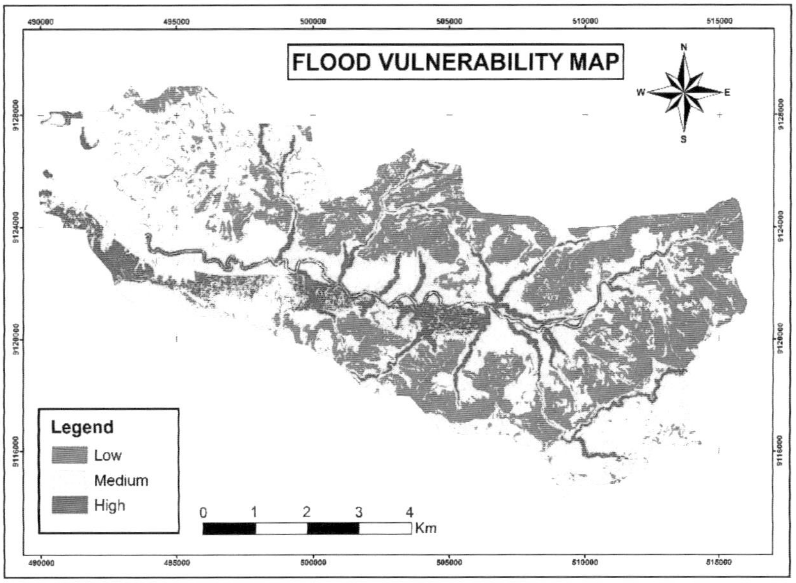

Figure 7. Flood Vulnerability Map

4. Conclusion

According to the results of research on flood-prone areas in the Tirtomoyo Sub-Watershed, Wonogiri Regency using 6 parameters showed that areas located at low elevations of <20 meters above sea level, slopes of sloping slopes, high rainfall intensity, and proximity to rivers increase the vulnerability to flooding. An area of 1662.79 ha (9%) is classified as high flood vulnerability in Tirtomoyo Sub-Watershed, followed by medium vulnerability level with 10,436.90 ha (34%), and low vulnerability level occupying 6,155.84 ha (57%). Residential areas and rice fields around the Tirtomoyo river are vulnerable to flood disasters because they have a flat and sloping topography, as well as high rainfall. Tirtomoyo Sub-district has the largest area at each level of flood vulnerability in the Tirtomoyo Sub-Watershed. Other sub-districts that reveal the potential for flooding with a high level of vulnerability are Baturetno and Batuwarno Sub-districts, while Nguntoronadi Sub-district is at a medium level of vulnerability.

The results obtain from the research are expected to support the government of Wonogiri Regency and related agencies in efforts to mitigate flood disasters and spatial planning based on disaster mitigation, especially for areas where floods often occur. Recommendations that can be given to prevent this are to prepare permanent drainage channels around settlements and rainwater catchment areas to cope with high rainfall.

References

[1] T. Q. Dang *et al.*, "Integrating Intelligent Hydro-informatics into an effective Early Warning System for risk-informed urban flood management," *Environ. Model. Softw.*, vol. 183, no. August 2024, p. 106246, 2024, doi: 10.1016/j.envsoft.2024.106246.

[2] M. F. Wiraatmaja *et al.*, "Flood vulnerability assessment trough overlay-scoring data method based on Geographical Information System (GIS) in Giriwoyo, Wonogiri, Indonesia," *IOP Conf. Ser. Earth Environ. Sci.*, vol. 1314, no. 1, 2024, doi: 10.1088/1755-1315/1314/1/012109.

[3] S. K. Abid, N. Sulaiman, C. S. Wei, and U. Nazir, "Flood vulnerability and resilience: Exploring the factors that influence flooding in Sarawak," *IOP Conf. Ser. Earth Environ. Sci.*, vol. 802, no. 1, 2021, doi: 10.1088/1755-1315/802/1/012059.

[4] H. Rakuasa, D. A. Sihasale, M. C. Mehdila, and A. P. Wlary, "Spatial Analysis of Flood Vulnerability Level in Teluk Ambon Baguala District, Ambon City," *J. Geosains dan Remote Sens.*, vol. 3, no. 2, pp. 60–69, 2022, doi: 10.23960/jgrs.2022.v3i2.80.

[5] T. N. Widodo, H. Zubair, and R. Padjung, "Land use change study and the increased risk of floods disaster in Jeneberang watershed at Gowa Regency, South Sulawesi, Indonesia," *IOP Conf. Ser. Earth Environ. Sci.*, vol. 824, no. 1, 2021, doi: 10.1088/1755-1315/824/1/012045.

[6] Nurhadi, D. R. S. Sumunar, and N. Khotimah, "Analysis of Flood Vulnerability in the Code Watershed of Yogyakarta City," *J. Penelit. Saintek*, vol. 21, no. 2, p. 75, 2017, doi: 10.21831/jps.v21i2.10553.

[7] K. Alderman, L. R. Turner, and S. Tong, "Floods and human health : A systematic review," *Environ. Int.*, vol. 47, pp. 37–47, 2012, doi: 10.1016/j.envint.2012.06.003.

[8] C. Y. Novian, S. Maarif, Kusuma, and Wilopo, "Flood Disaster Risk Management In Manado City," vol. 3, no. 6, pp. 3309–3313, 2024.

[9] I. G. Klipper, A. Zipf, and S. Lautenbach, "Flood Impact Assessment on Road Network and

Healthcare Access at the example of Jakarta, Indonesia," *Agil. GIScience Ser.*, vol. 2, pp. 1–11, 2021, doi: 10.5194/agile-giss-2-4-2021.

[10] J. S. Cabrera and H. S. Lee, "Flood risk assessment for Davao Oriental in the Philippines using geographic information system-based multi-criteria analysis and the maximum entropy model," *J. Flood Risk Manag.*, vol. 13, no. 2, pp. 1–17, 2020, doi: 10.1111/jfr3.12607.

[11] C. F. Candraningtyas, L. P. W. Syahrani, and Luthfia, "The Effect of Land Cover Changes in the Bengawan Solo Watershed on the Flood Control Function in Surakarta," no. July, 2023, doi: 10.54443/sibatik.v2i8.1272.

[12] P. M. Melati and M. T. Jatipaningrum, "Prediction of Natural Disasters in the Wonogiri Regency Area with the Concept of Markov Chains," *J. Stat. Ind. dan Komputasi*, vol. 3, no. 1, pp. 63–70, 2018.

[13] D. Auliyani and N. Wahyuningrum, "RAINFALL PATTERNS IN THE UPSTREAM PART OF THE BENGAWAN SOLO RIVER BASIN IN WATER RESOURCE UTILIZATION PLANNING," pp. 53–62, 2020.

[14] M. F. A. Safrida, Sobriyah, and A. H. Wahyudi, "Hydrographic analysis of the Tirtomoyo Watershed with several synthetic unit hydrograph methods," *Matriks Tek. Sipil*, pp. 534–544, 2014.

[15] A. P. Nova, "Analysis of Annual Floods in the Bengawan Solo Hulu Watershed," Universitas Sebelas Maret Surakarta, 2012.

[16] B. Triatmodjo, Applied Hydrology. Second Print. Yogyakarta: Beta Offset, 2009.

[17] F. Afdhalia and R. Oktariza, "Level of Physical Vulnerability to Flooding in Martapura Sub-Watershed, Banjar Regency," in *Prosiding Seminar Nasional Geotik*, 2019, no. ISSN :2580-8796, pp. 44–54.

[18] Fitrotunnisa, A. A. Wibowo, and D. Surachman, "Prediction of Flood Vulnerability Due to Land Cover Change in 2045 in Tegal Regency, Central Java, Indonesia," *Univ. Muhammadyah Surakarta*, 2023.

[19] C. Fadila, "Mapping of Flood-prone Areas in the Merangin Tembesi Watershed," Universitas Jambi, 2023.

[20] S. N. Aziza, L. Somantri, and I. Setiawan, "Analysis of Flood Vulnerability Mapping in West Bontang District, Bontang City, Based on Geographic Information System," *J. Pendidik. Geogr. Undiksha*, vol. 9, no. 2, pp. 110–120, 2021, doi: http://dx.doi.org/10.23887/jjpg.v9i2.35173.

[21] B. K. Osei, I. Ahenkorah, A. Ewusi, and E. B. Fiadonu, "Assessment of flood prone zones in the Tarkwa mining area of Ghana using a GIS-based approach," *Environ. Challenges*, vol. 3, no. January, p. 100028, 2021, doi: 10.1016/j.envc.2021.100028.

Disaster Management and Settlement Arrangement during 1800-1945 in Minahasa, Indonesia

Irfanuddin Wahid Marzuki[1*], Vivi Sandra Sari[2], Ipak Fahriani[3], Henki Riko Pratama[4], Anindya Puspita Putri[5].

[1-3] Research Centre for Prehistoric and History Archaeology, National Research and Innovation of the Republic of Indonesia (BRIN), Jakarta, Indonesia.
[4] Research centre for Archaeometry National Research and Innovation of the Republic of Indonesia (BRIN), Jakarta, Indonesia.
[5] Geography Study Program, Faculty of Social Science and Law, Manado State University, Tondano, Indonesia.

*E-mail: irfan.balarmdo@gmail.com.

Abstract. Minahasa is an area on the northern peninsula of Sulawesi flanked by two seas with a hilly contour. The location of the Minahasa community settlement in the past was in an inland area far from the coast. The Minahasa region has experienced several disasters in the past, both natural disasters in the form of earthquakes, volcanic eruptions, landslides, and floods, as well as disease outbreaks. The purpose of this study is to explore disaster management carried out in Minahasa and settlement arrangements from 1800 to 1945. The research method uses the Literature study method, with a multidisciplinary approach (geography, archaeology, and history). The results of the study show that the Dutch colonial government carried out disaster management in the Minahasa region by moving settlements to locations that were considered safer and building smaller, disaster-resistant houses. These activities were not only carried out by the Dutch colonial government but also by missionaries and *zending* in Minahasa. The results of this study are expected to provide information about disaster management in the past which can be an insight to overcome current and future disaster management problems.

Keywords: Minahasa, disaster management, Dutch colonial era.

1. Introduction

Minahasa is an ethnical area located in the far north of Sulawesi, between 0°51'00" - 1°51' 40" N and 124°18'40"-125°21'30" E [1]. Bordered by the Celebes Sea and the Philippines in the northern part, to Pacific Ocean in the East, to Tomini Coast in the South, and Kabupaten Bolaang Mongondow in the West, this region first fascinated Spain for its rice production provided for Moluccas and Cebu [2]. Minahasa was then divided into seven regencies, comprised of Kabupaten Minahasa, Kabupaten South Minahasa, Kabupaten Southeast Minahasa, Kabupaten North Minahasa, Manado City, Bitung City, and Tomohon City (Figure 1).

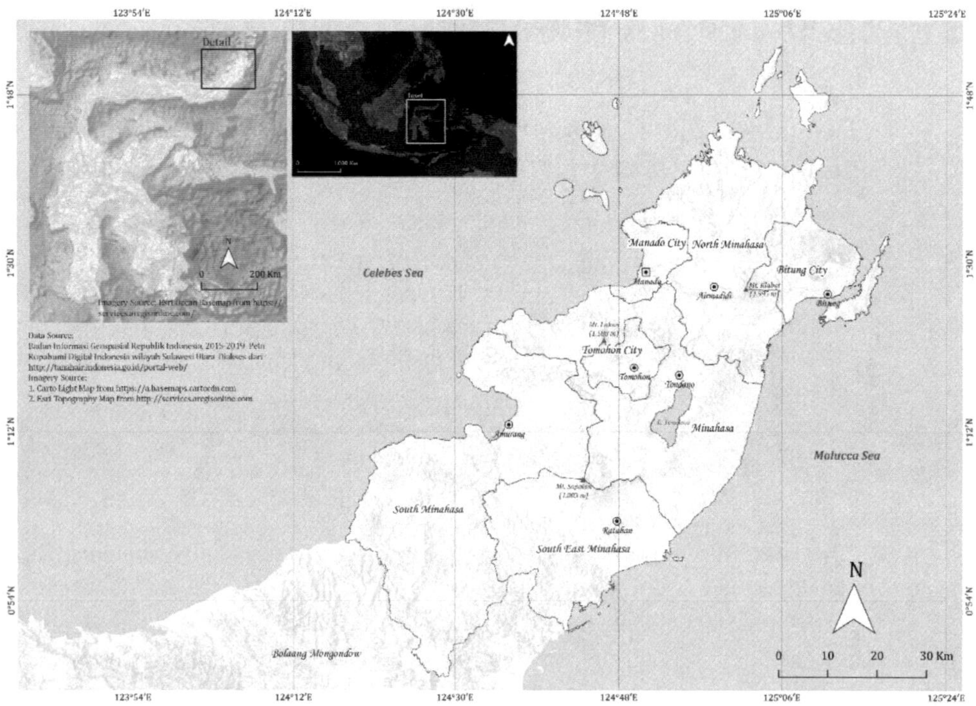

Fig. 1: Map of current Minahasa Area, Sulawesi Utara Province (Source: *Badan Informasi Geospasial Indonesia* 2015-2019, Carto Light Map, Esri Topography Map, accessed November 11th, 2023, modified by Henki Riko Pratama 2024).

The word Minahasa was first used in an official document of a report by resident J.D. Scierstein in October 1789, in which *Landstreek van Manado* was previously used [3]–[6]. The word Minahasa stems from the word "*mina*" and "*esa*" which means a united ethnic/become one [4], [5], [7]. Minahasa settlements before Europeans were in the hinterland, such as Tondano, Tomohon, and Langowan [8], [9]. Minahasa settlements in the past consisted of a big houses with some households that were irregularly located and less healthy. There were two forms of Minahasan settlement pattern in the past, i.e. *pakasaan* (a settlement of a certain ethnic) and *lukarz* (scattered small villages) [10]. *Pakasaan* was a main settlement surrounded by forts or small hamlets [11]. The pattern of Minahasan villages was static, which means that the number of people in the village tended to be stable and they refused to leave their home base village just because their farms were moved further away [12]. Traditional Minahasans chose their settlement for some reasons: the potential and character of the landscape, natural resources especially water, food ingredients, medicines or herbs, and building materials plus the freedom from any kinds of troubles (enemies and wilds) and natural disasters ([8], [10], [13], [14].

In their socio-cultural life, the Minahasa community is guided by the value of *Si Tou Timou Tumou Tou*, and *Mapalus* [15]. *Si Tou Timou Tumou Tou* implies that human life must be able to humanize each other's, which is Sam Ratulangi's philosophy. This philosophical thought inspires the life of the Minahasa people to live in harmony, humanize others and not become enemies of others (*homo homini lupus*)[15], [16]. Mapalus is an indigenous Minahasa custom, particularly when it comes to labor, where any task is completed in a collaborative or group setting (e.g., gardening). Mapalus is a cooperative volunteer activity that has been passed down through the generations and is still carried out now. In Minahasa, Mapalus is sometimes referred to as the native spirit and knowledge. Mapalus is the embodiment of the lifestyle of the Minahasa Community in solving problems that occur in the Community and is manifested in the system of rules of life order and community empowerment [15], [17], [18]. Beginning in the 1680s, Mapalus culture as a working system expanded until the 1970s and continues to this day. Since then, it has moved from the agricultural sector to a more contemporary model that encompasses the social, economic, governmental, and public health sectors [17].

Disaster management during 1800-1945 is an interesting subject to study as there has been no research related to Minahasa though it is the region in Indonesia whose Dutch influence was significantly strong. Research on disasters that occur in Indonesia is generally carried out by geographers and geologists. The research questions that arise are: "How efforts Disaster Management and settlement arrangement during 1800-1945 in Minahasa?" Research on disaster management and settlement arrangements during 1800-1845 in Minahasa aims to reveal how the disaster management and settlement arrangements were carried out by the Dutch colonial government in Minahasa and its influence on the Minahasa community further. Research targets include identifying disasters that happened in Minahasa in the past based on historical data and finding out how disaster management was carried out during 1800-1945 in Minahasa,

2. Method

This research uses a History and Literature study approach. An overview of ancient people's awareness of natural catastrophes is given by using a wide range of literature that contains earlier studies on disaster and mitigation. It is possible to suggest that new ideas for catastrophe mitigation, both locally and nationally, incorporate values that are pertinent to the current circumstances. These principles can be adjusted to meet the demands of the contemporary world. Data collection, analysis, and interpretation of findings are all part of the research phase. Data collection was carried out by literature studies, interviews, observation, and surveys. Literature studies It is done by reading and studying books, journal articles, Dutch colonial reports, and newspapers about disasters that occurred in the past. Literature studies were also conducted by searching online for old maps and old photos.

The interview was conducted to gather information from the community, especially the elderly, about disasters in the past, and the existence and dynamics of traditional Minahasa settlements. The interview was conducted using semi-structured in-depth interview techniques[19]. The topic of the question has been predetermined, namely, the history of disaster in the past and the existence, movement, arrangement, and development of Minahasa settlements. Field observation and surveys involve observing data directly in the field so that the context of the data findings can still be known. Field observations and surveys are carried out to determine the current state of the site and area.

Data Analysis and Interpretation, all data collected are arranged according to the historical context and environment of Minahasa settlements chronologically (historical diachronic) until a common thread is obtained. Furthermore, the analysis results are discussed with various theories or concepts to obtain interpretation from the research. Historical analysis is carried out by identifying and interpreting historical sources, reconstructing and measuring disasters, and analysing them. The study of past disasters is not just the study of historians, archaeologists, anthropologists, geographers, and climate scientists enthusiastically embrace the potential of 'natural archives' to reconstruct, date, and interpret the role of past catastrophic events and their changes [20]. Data analysis activities are carried out by sorting existing library data and cross-checking with other sources. Findings and preparation of research conclusions. The last step involved the preparation and formulation of a qualitative description of the phenomenon of disaster management and settlement arrangement in Minahasa during 1800-

1945. This research contributes to the current knowledge of the history of disasters, management, and settlement arrangements in Indonesia reviewed from a historical perspective.

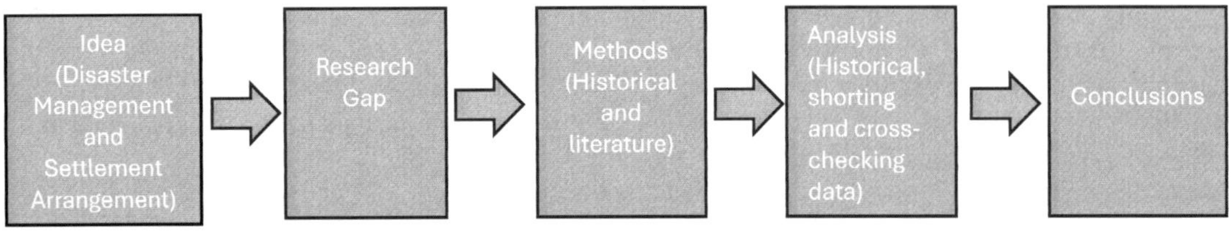

3. Result

3.1. History of Disaster in Minahasa

From 1815 to 2014, Minahasa area was hit by 169 natural disasters [21], the most severe earthquake [9] because located at the meeting of the earth's microplates the Sunda, Indo-Australian, and Philippine Sea Plates [22]–[24]. One of the biggest earthquakes in Minahasa was a big earthquake in 1845. According to Winchman (1922), the earthquake on February 8, 1845, was devastating and spread throughout the Minahassa peninsula. The storm resulted in damage to the Amsterdam Fort and the Amurang Fort. The disaster resulted in 56 deaths, and many were injured. The earthquake resulted in many cracks forming in the ground and landslides, and the seawater in the port receded, leaving fish to gather from the seabed [25]. Meanwhile, based on Henley (2005) the earthquake that occurred in 1845 killed about 100 people and was hit by the rubble of buildings.

Earthquakes and tsunamis occurred again in 1856 & 1857 which damaged the walls of the fort, and several houses on the coast of Kema (the southern coast of Minahasa). A major earthquake struck Minahasa again in 1858, an 'unusually strong and long-lasting earthquake' throughout Celebes. The impact of largest earthquake occurred in Tondano resulting in 15 houses collapsing and landslides in several locations. In addition, the earthquake caused a tsunami which caused damage to villages along the east coast of Sulawesi to Banggai [25]. In 1859, on 29 July, Minahasa, at 1:30 p.m., there was a strong tremor lasting for 5 minutes 'which threatened to knock everything down'. The tremors repeated at 4:30 a.m. and throughout the night and night. The tsunami trapped Kema and swept away a warehouse. Earthquakes and tsunamis occurred again in 1859, December 17 and 25. The shock was accompanied by a tsunami that hit the southern coast of Minahasa. The tsunami wave reached the roof of a coal warehouse and partially hit the fence (protective fence) along the road in Kema [26].

In addition to the earthquake and tsunami disasters, the Minahasa region was also hit several times by the eruption volcano disaster, because it was included in the belt of active volcanoes in Indonesia (Figure 2). Active volcanoes in Minahasa include Soputan, Lokon, Mahawu, and Tangkoko (Dua Saudara). The Minahasa volcano makes a great contribution to soil fertility [13]. Mount Soputan is different from other volcanoes in Minahasa, because it most often erupts both explosively and effusively. Mount Soputan first erupted in 1785. The eruption period ranges from one to two years. When it erupted in 1907, the volcano erupted again one year later in 1908, 1909, 1910, 1911, 1912, 1913, 1915, 1917, 1923, 1924, and 1947 [27], [28]. In November 1923 there was a powerful eruption, accompanied by ash, while the stones removed fell as dense as snowflakes in the Netherlands [29]. The longest period is around 45 years after the eruption of Mount Tangkoko (Dua Saudara) in northern Minahasa erupted in 1694, whose eruption was heard as far as Ambon [30], 'throwing out over all the lands of Manado such a quantity of ash that the roofs of many houses collapsed' [3] but the amount of damage and casualties due to the event is unknown. In addition to the two mountains, Mount Lokon in Tomohon has also erupted several times. Unlike other mountains, the location of the Lokon is close to the center of the Tomohon settlement.

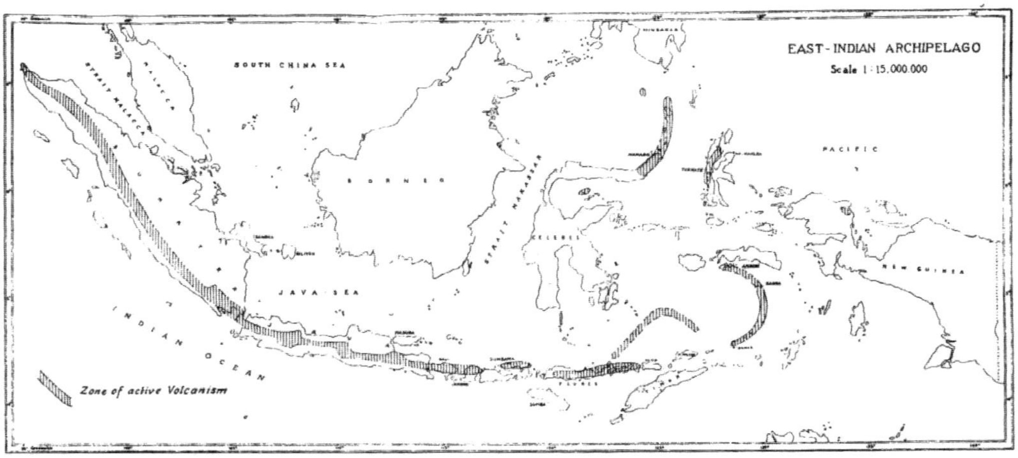

Figure 2. Belt of active volcanoes in Indonesia (Source: [31]).

Other disasters that often occur in Minahasa are floods and landslides. This is due to high rainfall and the influence of the slope of the land from the hilly Minahasa area[9]. Several flood and landslide disasters in the past can be known from the archives of Dutch newspapers. A major flood hit Minahasa on December 29, 1882. Continuous heavy rains for several days have caused floods and landslides. Flood water in several villages reached 2 meters and landslides occurred in many places in the Minahasa mountainous area resulting in 4 deaths (*Sumatra-courant* edition in February 1883). The March 25, 1927, *the Bataviaasch Nieuwsblad* newspaper also reported the flood that occurred in Tomohon. Major damage occurred and damaged the power generation engine. The floods also damaged houses and government offices in Manado. The *Het Nieuws* newspaper on November 19, 1934, reported on the flood disaster in Menado and in the Minahasa mountains which resulted in a building and a rice milling machine being washed away by water, six people drowned.

The biggest flood disaster occurred on March 15, 1936. Heavy rain has fallen since morning in the mountainous areas of Minahasa and the lowlands of Manado, resulting in landslides in several places in the Minahasa highlands and heavy damage to several villages. Flash floods hit Manado on March 17; brown water filled with material came at high speed. Quickly the water hit everything in its path. "People from the area have fled to the highland areas, so there have been no accidents and casualties," wrote the correspondent of *Soerabaijasch Handelsblad,* 28 Maret 1936. In addition to damaging settlements, the flood disaster that occurred damaged road and bridge infrastructure in Minahasa.

In addition to natural disasters, disasters in the form of diseases have hit Minahasa several times. Based on historical notes, the following were some of the plagues in Minahasa: dengue, dysentery, diarrhea, smallpox, malaria, and cholera[5], [8], [9], [32]. Among them, cholera occurred most of the time (years 1852-1854, 1866-1867, 1874, 1885-1886, 1901-1903, and 1914-1915), and the greatest number of tolls. The cholera occurred in Minahasa in 1830 and killed more than 480 people from various regions (*algemeen Verslaag of* Manado resident to the Governor of Moluccas dated 20-8-1830, a collection of ANRI Manado [9]. Also, history noted that cholera attacked Minahasa in 1852-1854, 1866-1867, 1874, 1885-1886, 1901-1903, and 1914-

IOP Conf. Series: Earth and Environmental Science **1503** (2025) 012025 doi:10.1088/1755-1315/1503/1/012025

1915 [9], [33]–[37]. The fast spread of cholera in Minahasa happened as the people ignored dwelling sanitary and cleanness, not to mention that they were also very dirty [11], [38]. The bottom of the house is used as a livestock pen (pigs, chickens, and dogs), as well as garbage disposal [5]. Further, they did not wash their clothes regularly and did not clean themselves after using the toilet either. Commonly, the traditional settlement of Minahasa was unhealthy, humid, messy, and dirty especially because pigs were also caged under the house [11] and thus potentially infectious dirt accumulated beneath the house [38]. Things were worse as Minahasans also used the piggy cage or bushes around the house for their toilet. Therefore, this habit accelerated the spread of the plague [39]. Cholera also spread through the burial system in which *waruga* was used (Figure 4). *Waruga* was a cubical stone with a hole in the middle and its lid was pyramid-shaped. *Waruga* was placed around the house and not buried to minimize bad smell during decomposition, some mud covered the *waruga*'s body and lit it. The liquid from the decomposition process was possibly absorbed by the soil surrounding the houses. Nowadays, the *warugas* in Minahasa are still on site though some have been moved to one area. The *warugas* that are still on site are the indicators of past settlement as usually *warugas* were situated around houses.

To properly develop mitigation measures, catastrophe history can provide insight into the nature of disasters in each area. Archaeology, anthropology, and history are examples of social sciences with a humanities perspective that are anticipated to be involved in the new disaster management concept. These disciplines can assist uncover the history of catastrophes in the area and how people have responded to them. Depending on its geographic circumstances, each region will display varying outcomes. Studies of this type can help local governments develop laws and regulations for disaster mitigation that consider the unique features of each [40].

3.2. Disaster Management in Minahasa 1800-1945

The Dutch colonial government began to handle the disaster by raising funds after the eruption of Mount Awu which was accompanied by an earthquake and tsunami in Sangihe, which became the residential area of Manado in 1856. This activity was the beginning of humanitarian activities carried out by the Dutch East Indies government in the nineteenth. Some of the activities carried out were in the form of assistance in the flood disaster in Central Java in 1861 and the Yogyakarta Earthquake in 1867 [41]. This is driven by the relationship and cooperation between the local community and the Dutch colonial government. Refugee camps were built after the eruption of Mount Merapi in 1930, and Europeans in Batavia held a fundraiser for European organizations affected by the eruption of Mount Krakatoa [41]. However, rules or laws regarding disaster management or refugee management have not been made by the Dutch government [42].

The people of Minahasa in responding to disasters (especially natural disasters), help each other, based on the existence of the *mapalus* culture. Mapalus comes from the word *palus (pa-elus)* and the prefix *ma*. *Elus* refers to a type of plant whose leaves are used to wrap boiled rice. The word *palus* is also known in the Tondano language with the meaning of 'pouring', so *mapalus* can also carry the meaning of 'pouring the contents of one container into another' [54]. The life of the Minahasa people is regulated according to *kanaamen* (customary law) [44] and *ta'ar* (ancestral will) [43] The *kanaamen* and *ta'ar* define the rules that govern the mapalus and the practice of the mapalus. The Mapalus Institute comes from "the oldest times", and comes from "mutual aid" while working in the field [35].

The colonial government also carried out disaster mitigation for the Minahasa people. The handling of refugees underwent significant changes during the Dutch colonial rule, by regulating post-disaster migration through employment assistance and shelters. The guiding principle of disaster management remains to control the movement of the people [41]. This was also done in Minahasa, by moving and arranging settlements for victims of natural disasters. The relocation of the settlements was carried out, among others, in Kakas, Tomohon [45]. The settlement arrangement was carried out by the Dutch colonial government after the disasters in Minahasa. The eruption period of a volcano has a time range or duration of eruption time that tends to be the same as the previous period [46]. However, the number of damage and deaths may increase along with the increase in the number of people, the expansion and change of the location of settlements and environments in mountainous areas, as well as the lack of awareness and disaster response (risk) from the community. The mountainous environment is one of the areas that is attractive for the community to develop the agricultural economy to other resources such as ecotourism. The increasing number of people and infrastructure occupying the mountainous area, causing the risk of death and damage is also greater during volcanic eruptions.

The Minahasa community has local knowledge as part of adaptation and environmental management in dealing with disasters that occur [47]. The importance of local knowledge in disaster risk management and climate change has been recognized in real terms, but until now it has not been applied in policy making and disaster science by the government [48]. Indigenous knowledge is a learning of resilience to earthquake risk, especially in Indonesia [49]. The preservation of Indigenous knowledge is useful for risk literacy because education, as stated by individuals and communities [50] can increase their capacity to reduce disaster risk. Observation and long interaction with disasters, local knowledge becomes the raw material of their prevention culture in daily life, especially in the face of natural hazards [51]. For example, local knowledge emphasizes the materials for houses should be flexible and recommended such as wood, bamboo, coconut rope, and coconut leaves to resist earthquakes [61]. Local knowledge is very important to be preserved, because it enriches people's knowledge in dealing with disasters which can increase community resilience to disasters. The preservation of local knowledge can be done, among others, by incorporating it into school and government agency education materials. Disaster resilience can connect urban planning and disaster management in modern concepts. [49].

Disaster response is characterized by cross-cultural communication that is interrelated between politics and religion. Religion informs humanitarian actions and provides causal explanations to the rulers over the disaster. The report on disasters during the colonial period is proof of how natural disasters in the archipelago are experiences that need to be well publicized to add new knowledge about different disasters. The impact of the disaster was not only on the safety of the community, but also on the political policies that occurred at that time. The policy changed over time with the addition of the colonial government's role in disaster management efforts [41].

The new settlements were built in better order with smaller, and healthier houses facing the road. The colonial government believed that any kind of illness could turn into an infectious disease if people lived in a big, dense, and unhealthy house. The Minahasan houses were rebuilt with some changes following the government and Christian's priest suggestion. The houses were transformed into smaller ones sustained by more beams [53]. An architectural timber house stands on stilts and is constructed without nails but uses a wooden plug (pasak). This house has been proven to successfully withstand earthquakes [54]. As cited, Dutch colonial and their mission to spread religion (*zending*) brought tremendous changes among Minahasan [55].

Smaller homes increasingly replaced the original multiple-hearth residences in Minahasa starting in the 1840s, in part due to Dutch orders [9]. It's possible that resident Jansen was correct when he said that replacing the ancient homes with "small hygienic dwellings" would encourage population increase [56]. There may possibly have been important steps taken to maintain the cleanliness of the new towns. From an outward perspective, the outcomes were undoubtedly remarkable [9]. The messy patterns of settlement with its big houses were transformed into a better-ordered settlement with small houses for only one family, road net, fenced, and healthy. The settlement arrangement by the Dutch colonial almost in all areas of Minahasa. The Dutch colonial then considered some compatibility factors for dwelling areas. The traditional settlement was relocated and managed better. The condition of the Minahasa settlement after being organized by the colonial government.

4. Conclusion

Disaster mitigation activities based on strengthening community capacity have been carried out by the Minahasa community since the colonial period until now. The Minahasa people have a local wisdom in dealing with disaster situations which is often referred to as the Mapalus Culture. Mapalus culture is an entity of the culture of cooperation or cooperation in dealing with joint work (disaster situations) in the Minahasa people.

The intervention of the Dutch colonial government has had a significant impact on people's resilience to disasters. Disaster risk reduction was the responsibility of the colonial government at that time. Where the Dutch colonial government participated in the post-disaster spatial planning, the provision of jobs, and evacuation shelters. The relocation and arrangement of settlements for disaster victims were carried out in Kakas and Tomohon is the purpose of arranging smaller and healthier settlements so as to reduce the impact of disease outbreak disasters.

In addition, the community also participates in disaster risk reduction through zending and missionary by establishing polyclinics and medical assistance when disasters occur. In addition, zending and missionaries also teach a healthy lifestyle and environmental cleanliness so that disease outbreaks do not recur. The people of Minahasa who were not affected by the disaster also played an active role in disaster mitigation efforts in Minahasa by providing energy and food assistance.

An overview of the history of disasters and the form of disaster mitigation activities are expected to provide a strategy for community development that is resilient and has local capacity to deal with disasters. Disaster risk reduction in the Minahasa community can at least be realized by the community that already has local capacity. Mapalus culture and the history of post-disaster settlement arrangements in Minahasa must be used as social capital for the government in development activities based on disaster risk reduction in Minahasa.

References

[1] T. Penyusun, *Adat Istiadat Daerah Sulawesi Utara*. Jakarta: Proyek Penelitian dan Pencatatan Kebudayaan Daerah Departemen Pendidikan dan Kebudayaan, 1978.
[2] B. Supit, *Minahasa dari Amanat Watu Pinawetengan sampai Gelora Minawanua*. Jakarta: Sinar Harapan, 1986.
[3] E. . Godee Molsbergen, *Geschiedeniss van de Minahassa tot 1829.* Weltevreden: Landsdrukkerij., 1928.
[4] D. Henley, "Nationalism and Regionalism in Colonial Indonesia : The Case of Minahasa," *Indones. No. 55, East Indies Dutch*, no. 55, pp. 91–112, 1993.
[5] M. J. Schouten, *Leadership and Social Mobility in A Southeast Asian Society Minahasa 1677-1983*. Leiden: KITLV Press, 1998.
[6] B. Pantouw, "Beberapa Perubahan Kebudayaan di Minahasa Tengah 1829-1859 Suatu Kajian Sejarah Berdasarkan Tulisan-tulisan J.G.Schwarz," Jakarta, 1994.

[7] H. . Palar, *Wajah Lama Minahasa*. Bogor: Yayasan Gibbon Indonesia, 2009.
[8] N. Graafland, *Minahasa: Negeri, Rakyat, dan Budayanya*. Jakarta: Pustaka Utama Grafiti, 1991.
[9] D. Henley, *Fertility, Food, and Fever Population, Economy, and Environment in North and Central Sulawesi 1600-1930*. Leiden: KITLV Press, 2005.
[10] P. R. Renwarin, *Matuari Wo Tonaas Jilid I Mawanua*. Jakarta: Penerbit Cahaya Pineleng, 2007.
[11] J. N. F. M. à Campo, *Koninklijke Paketvaart Maatschappij; Stoomvaart en staatsvorming in de Indonesische archipel 1888-1914*. Hilversum: Verloren (Publikaties van de faculteit der Historische en Kunstwetenschappen 3), 1992.
[12] N. S. Kalangie, "Kebudayaan Minahasa," in *Manusia dan Kebudayaan di Indonesia*, Keempat., Jakarta: Penerbit Djambatan, 1979.
[13] D. Henley, *Fertility, Food, and Fever*. Leiden: KITLV Press, 2005.
[14] M. J. . Bhatta, "Tinjauan Tentang Tempat Kediaman Orang Jang Beragama Islam di Minahasa," Jakarta, Indonesia, 1958.
[15] T. Pangalila and J. Mantiri, "The role of Tomohon society's local wisdom in developing tolerance attitudes," *Int. J. Eng. Adv. Technol.*, vol. 8, no. 5, pp. 366–372, 2019.
[16] N. F. Wuryaningrat, D. W. Mandagi, and D. I. Rantung, "Mapalus as a Knowledge Transfer Practice to Improve Innovation Capability: Success or not Success?," *Rev. Gestão Soc. e Ambient.*, vol. 17, no. 1 SE-, p. e03162, Feb. 2023.
[17] J. E. Nelwan, E. Widjajanto, S. Andarini, S. Djati, and O. J. Sumampouw, "The Role of Mapalus Culture by Minahasa Ethnic in North Sulawesi to the Coronary Heart Disease Incidents," *Int. J. Sci. Res. Publ.*, vol. 8, no. 3, pp. 45–52, 2018.
[18] C. A. Rumbay, H. Lumapow, P. E. A. Tuerah, E. J. Usoh, V. N. J. Rotty, and J. S. J. Lengkong, "Embracing mapalus traditional management values for Christian religious education," *HTS Teol. Stud. / Theol. Stud.*, vol. 78, no. 1, p. 7986, Dec. 2022.
[19] B. Mikkelsen, *Metode Penelitian Partisipatoris dan Upaya-Upaya Pemberdayaan Sebuah Pegangan bagi Para Praktisi Lapangan*. Jakarta: Yayasan Obor Indonesia, 2001.
[20] B. van Bavel *et al.*, *Disasters and History*. Cambridge University Press, 2020.
[21] F. P. Makalew, S. W. M. Supit, N. E. Kondoj, and H. G. Mantiri, "Planning area of resist natural disaster in Bentenan Village South Eastern Minahasa," *J. Phys. Conf. Ser.*, vol. 1450, no. 1, p. 12025, 2020.
[22] A. Cipta, R. Robiana, J. Griffin, N. Horspool, S. Hidayati, and P. R. Cummins, "A probabilistic seismic hazard assessment for Sulawesi, Indonesia," *Geol. Soc. London, Spec. Publ.*, vol. 441, no. 1, pp. 133–152, Jan. 2017.
[23] D. Lei, J. Wu, and G. Yang, "Coseismic Coulomb stress changes imparted by the 1996 Minahasa Mw7.9 earthquake on the 2018 Palu Mw7.5 earthquake and expected seismicity rate changes," *Terra Nov.*, vol. 32, no. 1, pp. 44–52, Feb. 2020.
[24] C. Liu and Y. Shi, "The role of fault interaction in earthquake migration in Central Sulawesi, Indonesia," *Tectonophysics*, vol. 839, p. 229530, 2022.
[25] R. Harris and J. Major, "Waves of destruction in the East Indies: the Wichmann catalogue of earthquakes and tsunami in the Indonesian region from 1538 to 1877," in *Geohazards in Indonesia: Earth Science for Disaster Risk Reduction*, Special Pu., P. R. Cummins and M. I, Eds. London: Geological Society, 2016.
[26] C. E. A. Wichmann, *Die Erdbeben des Indischen Archipels von 1858 bis 1877*. Amsterdam: Koninklijke Akademie van Wetenschappen, 1922.
[27] A. N. Pratama and B. Galih, "Letusan Gunung Soputan dari Masa ke Masa...," *Kompas*, Jakarta, 2018.
[28] B. A. Wicaksono, "Menguak Sejarah Letusan Soputan," *viva.co.id*, 2015.
[29] L. F. van Gent, W. A. Penard, and D. Rinkes, *Indie in Woord en Beeld*. Weltevreden:

Topografische Inrichting, 1924.
[30] M. Neumann van Padang, "History of the volcanology in the former Netherlands East Indies," *Scr. Geol.*, vol. 71, pp. 1–76, 1983.
[31] R. W. van Bemmelen, "Register of the localities of volcanic activity in the East Indian Archipelago," *Bull. East Indian Volcan Surv.*, vol. 95–98, 1943.
[32] S. Hickson, *A Naturalist in North Celebes*. London: John Murray., 1889.
[33] C. Bosscher and P. . Matthijssen, "Schetsen van de rijken van Tomboekoe en Banggai, op den oostkust van Celebes," *Tijdschr. voor Indische Taal en Volkenkd.*, vol. 2, pp. 63–107, 1854.
[34] P. van der Crab, "Memorie van overgave van de residentie Menado," 1875.
[35] N. Graafland, *Minahasa Masa Lalu dan Masa Kini. De Minahasa, Haar Verleden en Haar Tegenwoordige toestans. De Erven F. Bohn. Haarlem.* 1898.
[36] A. Pennings, "Mededeelingen van de Talauer-eilanden," *Meded. vanwege het Com. tot Voorziening Godsdienstige Behoeften van Gevestigde Inlandsche Protestantsche Christ. op Sangi- en Talauer Eilanden*, vol. 4, pp. 13–20, 1908.
[37] F. A. E. Wubben, "Naar de Talaut-eilanden," *Tijdschr. van het K. Ned. Aardrijkskd. Genoot.*, vol. 6, pp. 201–212, 1889.
[38] A. C. N. Bouvy, "Uit en over de Minahassa I. De Minahassa en de geneeskunst," *Bijdr. tot Taal en Volkenkd.*, vol. 80, pp. 365–369, 1924.
[39] H. F. Tillema, *"Kromoblanda"; Over't vraagstuk van "het wonen" in Kromo's groote land.*, 5th–1st ed. Gravenghage: Masman, 1922.
[40] A. Winaya and A. Murdihastomo, "How Indonesian People in the Past Deal with Disaster Mitigation? An archaeological perspective," *Kapata Arkeol.*, vol. 17, no. 1, pp. 13–20, Jan. 2022.
[41] A. Schrikker, "Disaster Management and Colonialism in the Indonesian Archipelago, 1840-1920," in *Natural Hazards and Peoples in the Indian Ocean World Bordering on Danger*, G. Bankoff and J. Christensen, Eds. Pallgrave Mac Millan, 2016, pp. 225–254.
[42] R. Djalante and M. Garschagen, "A review of disaster trend and disaster risk governance in Indonesia: 1900–2015," in *Disaster risk reduction in Indonesia: progress, challenges, and issues*, R. Djalante, M. Garschagen, F. Thomalla, and R. Shaw, Eds. Springer International, 2017.
[43] F. E. W. Parengkuan, "A Contribution to the History of Mapalus in the Minahasa, North Sulawesi," *J. Masy. dan Budaya*, vol. 8, no. 2, pp. 1–18, 2006.
[44] H. . Taulu, *Sejarah Minahasa*. Jakarta: Departemen Pendidikan dan Kebudayaan, 1978.
[45] A. Kojongian, *Tomohon Kotaku*. Tomohon: Dinas Pendidikan Nasional, 2006.
[46] E. Tantri, "Narasi Dampak (Alam dan Sosial) Letusan Gunung Tambora 1815," *Patrawidya*, vol. 20, no. 2 (Agustus), pp. 131–148, 2019.
[47] O. Anisimov, D. Vaughan, T. Callaghan, C. Furgal, H. Marchant, and T. Prowse, "Polar regions (Arctic and Antarctic)," in *Climate Change 2007: Impacts, Adaptation and Vulnerability, Contribution of Working Group II to the Fourth Assessment Report of the Intergovernmental Panel on Climate Change*, M. L. Parry, O. F. Canziani, J. P. Palutikof, P. J. van der Linden, and C. E. Hanson, Eds. Cambridge: Cambridge University Press, 2007.
[48] W. N. Adger, J. Barnett, F. S. Chapin, and H. Elmor, "This Must Be the Place: Underrepresentation of Identity and Meaning in Climate Change Decision-Making," *Glob. Environ. Polit.*, vol. 11, no. 2, pp. 1–25, 2011.
[49] H. Kurnio, A. Fekete, F. Naz, C. Norf, and R. Jupner, "Resilience Learning and Indgebous Knowledge of Eartquake risk in Indonesia," *Int. J. Disaster Risk Reduct.*, vol. 62, pp. 1–11, 2021.
[50] W. Wicaksono, *Pedoman Menghadapi Bencana Gempa Dan Tsunami*. Jakarta: ISDR-UNESCO, 2007.

[51] Zulfadrim, Y. Toyoda, and H. Kanegae, "The Integration of Indigenous Knowledge for Disaster Risk Reduction Practices through Scientific Knowledge: Cases from Mentawai Islands, Indonesia," *Int. J. Disaster Manag.*, vol. 2, no. 1, pp. 1–12, 2019.
[52] B. Rifa'i, "Rumah Adat Baduy Tetap Kokoh Meski Ada Gempa M 6,9, Apa Rahasianya?," 2019.
[53] O. H. A. Rogi and W. Siswanto, "Identifikasi Aspek Simbol dan Norma Kultural Pada Arsitektur Rumah Tradisional di Minahasa," *EKOTON*, vol. 9, no. 1, pp. 43–58, 2009.
[54] Marwati, "Studi rumah Panggung Tahan Gempa Woloan Di Minahasa Manado," *Teknosains Media Info Sains dan Teknol.*, vol. 8, no. 1, pp. 95–108, 2014.
[55] C. E. . Wuisang, "Defining Genius Loci and Qualifying Cultural Landscape," The University of Adelaide, Adelaide, 2014.
[56] ANRI, "Algemeen Verslag 1857," Jakarta.

Enhancing Community Knowledge for Tsunami Risk Reduction in Pecaron Beach, Kebumen

Sofi Yuni Pramesti[1], Anang Widhi Nirwansyah*[1,2] and Suwarno[1]

[1] Department of Geography Education, University of Muhammadiyah Purwokerto 53182, Banyumas, Indonesia
[2] Department of Social Science, University of Muhammadiyah Purwokerto, Banyumas 53182, Indonesia

*E-mail: anangwidi@ump.ac.id

Abstract. A total of 42.39% of the Kebumen area is categorized as an area that is very vulnerable to tsunamis, including the Pecaron Beach which is directly adjacent to the Indian Ocean, Sunda Trench, and active faults in Southern part of Java. This region has the potential for earthquakes with a magnitude above 7.0 that can trigger a tsunami and has a history of tsunami in 2006 with a height of 5-15 meters. Despite having a high risk, the results of preliminary interviews show that merchant of Pecaron Beach have a low level of knowledge and preparedness in dealing with tsunami. This study aims to develop and evaluate the effectiveness of efforts to increase community knowledge in reducing the risk of tsunami in Pecaron Beach. The research used a qualitative approach with data collection through observation and in-depth interviews. The interviews were conducted to explore the public's understanding of tsunami risk, natural signs, evacuation routes, disaster experiences, and views on capacity building efforts. Meanwhile, observations were focused on evacuation readiness, routes to gathering points, discussion, adaptation, and community involvement in mitigation activities. The research informants amounted to 10 people (n=10), consisting of government officials, chairman of local leader and DESTANA, merchant in Pecaron Beach and Pecaron beach manager. Data analysis was carried out through the triangulation method. The study shows that the dissemination and tsunami mitigation training have a positive impact in capacity of knowledge and preparedness of the community. The increase can be seen from the change in public understanding from low to high, including public understanding of tsunami risk, natural signs, evacuation routes, disaster experiences, evacuation alertness, adaptation, and mitigation activities participation. Practical dissemination strategies and evacuation exercises are identified as important strategies in reducing tsunami risk, while at the same time building sustainable resilience of coastal communities.

1. Introduction

Disaster mitigation encompasses a sequence of strategic actions designed to minimize the impact and risk of disasters through effective planning, preparation, and reduction of potential damage (1). Indonesia, located along the Ring of Fire in the Pacific, is particularly susceptible toward natural disasters, with the National Disaster Management Agency (BNPB) recording over 4,000 such events in 2023, of which 0.65% were geological in nature (2). As reported by Maiti and Kayal in 2024, on a global scale, natural disasters in 2022 resulted in economic losses amounting to 313 billion dollars. The 2004 tsunami in Aceh, regarded as one of Indonesia's most catastrophic disasters, was caused by seismic activity that produced enormous ocean waves, as noted by Rahiem, Rahim, and Ersing in 2021 in their investigation of child survivor narratives from the disaster. This event was precipitated by an earthquake with a magnitude of 9 Mw or IX MMI, indicating extreme magnitude (3). The tsunami claimed over 230,000 lives and caused material losses estimated at 4.45 billion US dollars, affecting 12 countries bordering the Indian Ocean (4). The substantial loss of life was largely attributed to inadequate preparedness and the limited ability of communities to effectively react to such calamities (5).

In light of the escalating threat of disasters, particularly in vulnerable regions such as the west coastline of Sumatra and the southern shoreline of Java, Supendi et al. in 2023, emphasize the importance of understanding the potential for megathrust earthquakes and tsunamis in these areas, enhancing Indonesia's capacity for disaster mitigation is crucial. Capacity building includes providing individuals and communities with essential competencies, knowledge, and resources to respond more effectively to disasters (6). As highlighted by Righi et al. in 2021, evidence suggests that trained individuals have a significantly higher likelihood of survival during emergencies. Nevertheless, in Indonesia, awareness of tsunami evacuation procedures remains limited, with fewer than 50% of coastal residents possessing adequate understanding (7). Therefore, it is crucial for the government to enhance community capacity through various initiatives, including dissemination of information and tsunami mitigation training programs.

Preparedness involves a systematic effort to equip communities to respond to disasters effectively and efficiently (8). This encompasses recognizing early warning signs, such as the sudden retreat of seawater preceding a tsunami (9). In Pangandaran, survival during a tsunami was attributed to the transmission of local knowledge regarding natural indicators across generations (10). Similarly, Manado has established clearly marked evacuation routes and signage in coastal areas to enhance preparedness (11). Effective preparedness enables communities to respond swiftly to disasters, thereby reducing casualties and losses (12). Besides preparedness, community resilience is essential for recovering from disaster impacts. According to 2023 data from the BNPB, Indonesia's disaster resilience index is 3.6 out of 5 by Copeland et al. in 2020. Strong resilience facilitates quicker recovery and mitigates long-term disaster impacts.

Enhancing community knowledge and capacity is increasingly vital in disaster mitigation, especially given the tsunami threats facing Indonesia's coastal regions (13). BNPB reports that over 4 million individuals residing in these areas are at significant risk from tsunamis, as examined by Sufri and Lassa in 2024. Tsunami vulnerability data indicate that approximately 42.39% of Kebumen coastal regions fall into the highly vulnerable category, with an additional 10.03% classified as moderately vulnerable (14). Despite the substantial threat, community preparedness for tsunamis remains inadequate (15). This underscores the necessity of reducing tsunami risk through enhancing community knowledge and capacity, a critical measure when disasters occur.

Increasing community knowledge is pivotal in mitigating tsunami risk (16). Disaster knowledge capacity refers to the level of understanding, awareness, and response capability that individuals, groups, or communities possess concerning potential disasters in their environment. This encompasses a thorough knowledge of disaster categories, causes, impacts, and appropriate mitigation and response strategies, as outlined by Monte et al. in 2021. Enhancing disaster knowledge capacity is crucial for preparing communities, reducing risks, and expediting recovery post-disaster. It also involves accessing relevant information, participating in training, and

IOP Conf. Series: Earth and Environmental Science **1503** (2025) 012026 doi:10.1088/1755-1315/1503/1/012026

engaging in outreach programs that promote disaster preparedness and awareness. With robust knowledge capacity, communities can become more resilient in facing disasters and minimize resultant losses (17). For a complete overview regarding the knowledge capacity parameters, please refer to Table 1.

Table 1. Knowledge Capacity Parameters

No	Knowledge Parameters	Description
1	Definition of Tsunami	Understanding that a tsunami is a large wave caused by tectonic, volcanic, or underwater landslides.
2	Causes of Tsunamis	Knowledge of triggering factors such as earthquakes in subduction zones, underwater volcanic eruptions, or landslides.
3	Signs and Symptoms of Tsunami	The ability to recognize early signs, such as a strong earthquake, a sudden receding of sea water, or a rumbling sound from the sea.
4	Tsunami Impact	Understanding of physical damage (buildings, infrastructure), casualties, and the resulting economic and environmental impacts.

Source: Ester Krisnawati 2022

Based on observations and interviews with merchants at Pecaron Beach, along with an assessment of the knowledge capacity parameters mentioned earlier, it is evident that the local population possesses limited understanding of tsunamis. This is reflected in their insufficient comprehension of tsunamis as large waves induced by tectonic shifts, volcanic activity, or underwater landslides, as well as a lack of awareness regarding causative factors such as earthquakes in subduction zones and underwater volcanic eruptions. Moreover, their ability to identify early tsunami warnings—such as strong earthquakes, sudden sea water recession, or unusual rumbling sounds from the ocean—is also constrained. The community's grasp of the tsunami's impacts, including physical destruction, casualties, and economic and environmental consequences, remains incomplete. Prior research highlights that some residents in Kebumen have never engaged in disaster mitigation training or dissemination (19). Consequently, this study seeks to develop and assess the effectiveness of initiatives aimed at enhancing community knowledge to mitigate tsunami risks at Pecaron Beach, Kebumen.

2. Methods

2.1 Geographical Settings

This study was conducted in Srati Village, Ayah District, Kebumen (coordinates: 7°45'-7°83' S and 109°38'-109°83' E). Situated at an elevation ranging from 0 to 200 meters above sea level, Srati spans approximately 352.03 hectares, encompassing hilly terrain extending to Pecaron Beach. The research primarily focuses on the Pecaron Beach tourist area (20). For a detailed depiction of the research site, refer to Figure 1. According to data from the Central Statistics Agency (BPS) as of December 2023, Srati Village has a population of 4,071, comprising 51.33% males (2,090 individuals) and 48.64% females (1,981 individuals). The population distribution is illustrated in Figure 2.

IOP Conf. Series: Earth and Environmental Science **1503** (2025) 012026 doi:10.1088/1755-1315/1503/1/012026

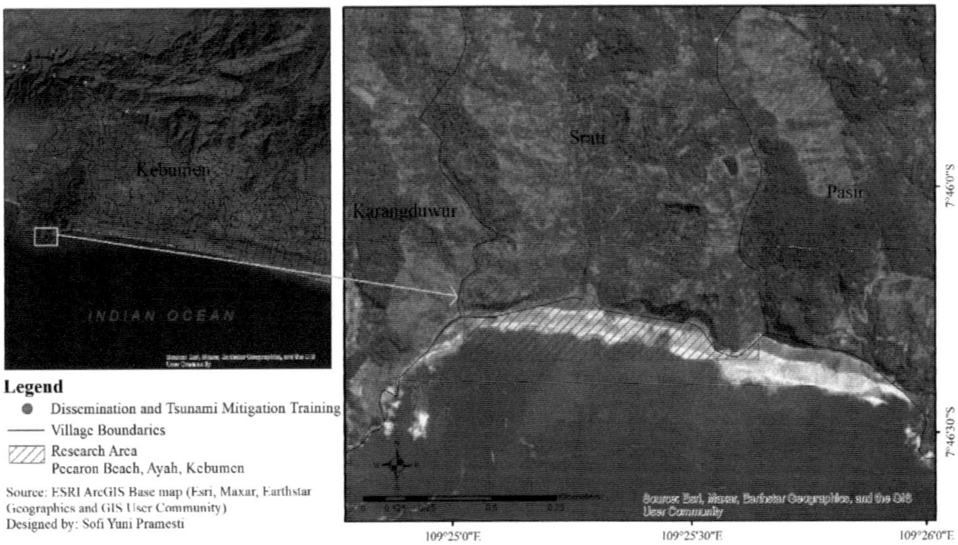

Figure 1. Case Study Area Map

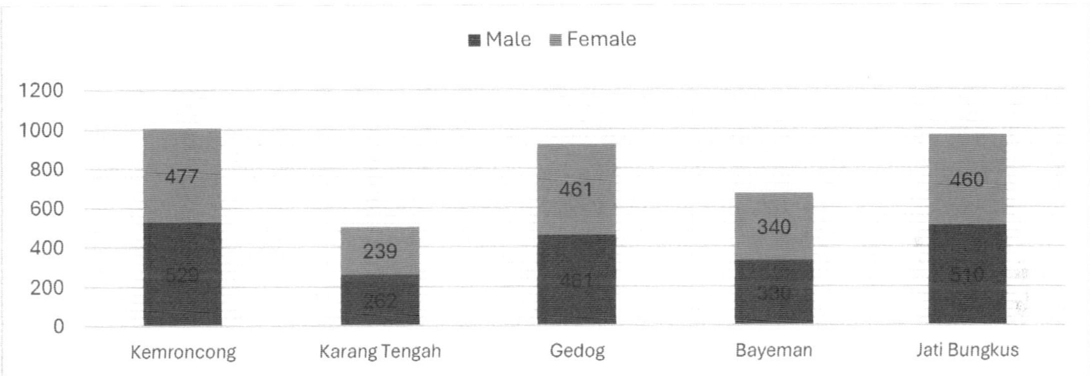

Figure 2. Population in 2023 by sex group

Geologically, Pecaron Beach is located within the South Serayu Mountains Zone, characterized as a karst region formed from Miocene limestone (21). The beach is oriented toward a considerable tsunami risk because of its closeness to the Indian Ocean, the Sunda Trench, and the active fault line south of Java (22). Approximately 42.39% of Kebumen's coastal area is highly susceptible to tsunamis. Additionally, Pecaron Beach's status as a river estuary heightens this risk. Data from the Meteorology, Climatology, and Geophysics Agency (BMKG) show the possibility for earthquakes exceeding a magnitude of 7.0, which could trigger tsunamis. Historically, the 2006 Pangandaran tsunami affected the southern coast of Java, as pointed out by Nijman in 2021, impacting areas like Ayah District in Kebumen with wave heights between 5 and 15 meters. Further historical tsunami events in Kebumen are detailed in Table 2.

Table 2. History of Tsunami in Kebumen

No	Location	Year	Cause
1	Petanahan Beach, Karangbolong Beach, and surroundings	1904	Triggered by a powerful tectonic earthquake within the subduction area of the Indo-Australian Plate off the southern coast of Java.
2	Ayah Beach, Karangbolong Beach, and Menganti Beach	1921	Triggered by large tectonic earthquake activity in the southern subduction zone of Java Island.
3	Karangbolong Beach, Petanahan Beach, and areas near the Lukulo River Estuary	1957	Caused by a significant tectonic earthquake in the Indian Ocean megathrust zone.
4	Karangbolong Beach, Suwuk Beach, and Sand Beach	1994	Triggered by a magnitude 7.8 earthquake in the subduction zone near Banyuwangi.
5	Suwuk Beach, Ayah Beach, and Menganti Beach	2006	Initiated by a 7.7 magnitude quake in Java's subduction.

Source : BPBD Kebumen

Efforts to reduce disaster risk extend beyond technical and infrastructural measures to include enhancing community capacity for understanding and addressing disaster threats, as emphasized by Zuccaro et al. in 2020 in their exploration of integrated disaster risk management strategies. The Hyogo Framework for Action (HFA) provides a comprehensive strategy through five key priorities: institutional strengthening, early warning systems, education and innovation, mitigation of root risk factors and enhancement of preparedness. Implementing this framework aims to cultivate a disaster-resilient community by integrating policy and public awareness. At Pecaron Beach, applying HFA principles is crucial for enhancing community knowledge capacity, serving as a strategic initiative to mitigate tsunami risk, aligning with the insights of Sufri and Lassa in 2024 on the integration of disaster risk reduction and climate change adaptation, as outlined in Table 3.

Table 3. HFA Achievement Variables and Indicators

No	Program Priorities	Program Priorities Disaster Risk Reduction Achievement Indicators
1	Risk reduction becomes a strong institutional foundation	There is a national/local policy and legal framework; there are resources that are specifically allocated at all levels of government; there is participation and equity through the distribution of authority and resources at the local level; and functional local-only network forums
2	Risk evaluation and early alert	There is an analysis of hazards, vulnerabilities and disaster risks; the existence of a potential disaster monitoring system; availability of an operative wide-scale early warning system; there is a disaster risk assessment that considers cross-regional risks to work together
3	Learning, innovation and disaster awareness	Disaster information accessible at all levels; curriculum materials and training on hazard risk mitigation; the availability of multi-disaster risk assessments; implementation of approaches to build awareness of disaster resilience in the community
4	Mitigating underlying risk factors	Policies and plans for disaster risk in environmental, social and economic sectors; planning and management of settlements related to permits and conditions for erecting buildings; there is a guide to the rehabilitation and recovery process; availability of procedures that are ready to initiate the impact of disaster or development risk
5	Preparedness advancement	Existence of disaster emergency response policies, capacities, and procedures; there are disaster contingency plans at all levels and emergency response exercises carried out; the readiness of funds and anticipatory logistics during disaster emergency response; there is a procedure for exchanging information during the emergency response period

Source: Sufri and Lassa 2024

2.2 Method

This study uses qualitative analysis of data gathered through in-depth interviews and observations. The semi-structured interviews were carried out to explore the public's understanding of tsunami risk, natural signs, evacuation routes, disaster experiences, and views on capacity building efforts, Rafliana et al. in 2022. Meanwhile, observation is focused on evacuation readiness, including routes and gathering points, as well as community involvement in mitigation activities (23). Data collection was conducted from June to October 2024, with a focus on data exploration through in-depth interviews with 10 informants (listed in Table 4 and DESTANA) groups involved in increasing community capacity and knowledge in reducing tsunami risk in Pecaron Beach.

Table 4. Research Informants

No	Subject	Code	Number
1	Government Agencies	GA	2
2	Village Head	VH	1
3	Chairman of Local Leader (RW)	CR	1
4	Chairman of Destana	CD	1
5	Merchant in Pecaron Beach	SR	3
6	Pecaron Beach Manager	PBM	2

The research was conducted directly to observe and interview the community (24). The data analysis adhered to the methodology outlined by Miles and Huberman, involving data collection, data reduction, and conclusion drawing (25). To increase the credibility of qualitative research, data triangulation analysis is used by verifying findings through various sources and perspectives. This approach helps to reduce bias and ensure more valid and reliable results, as mentioned in (26).

This research began with observation including topographic condition (as can be seen in Figure 4a), followed by surveys and interviews with the community, including merchant in the Pecaron Beach area (presented in Figure 4b). These measures aim to collect preliminary information on the condition of its vulnerability to tsunami risk and to assess the level of public understanding regarding tsunami mitigation efforts (27). The identification results show that the coastal area of Pecaron Beach has high vulnerability and tsunami risk, while the capacity of community knowledge related to tsunami mitigation is at a low level. Subsequently, the Student Organization Capacity Strengthening Program (*Program Peningkatan Kapasitas Organisasi Mahasiswa* or PPK Ormawa) was implemented to enhance community knowledge capacity through dissemination and tsunami mitigation training. The study's data were gathered via in-depth observations and interviews with ten informants. These informants included representatives from government agencies, village heads, local leaders (specifically *Rukun Warga* or RW), the chairman of DESTANA, traders in tourist areas, and managers of Pecaron Beach. They are considered to have better abilities and understanding than other informants, so they are able to provide rich and in-depth data related to the topic being researched. The data obtained was then examined through the triangulation method which included the process of data collection, data reduction, and drawing conclusions (28). The analysis indicates a substantial enhancement in public knowledge capacity, positively influencing efforts to mitigate tsunami risk in the coastal region of Pecaron Beach. This capacity building can be seen in Figure 3, which illustrates the changes in community knowledge before and after the dissemination and tsunami mitigation training program is implemented.

IOP Conf. Series: Earth and Environmental Science **1503** (2025) 012026 doi:10.1088/1755-1315/1503/1/012026

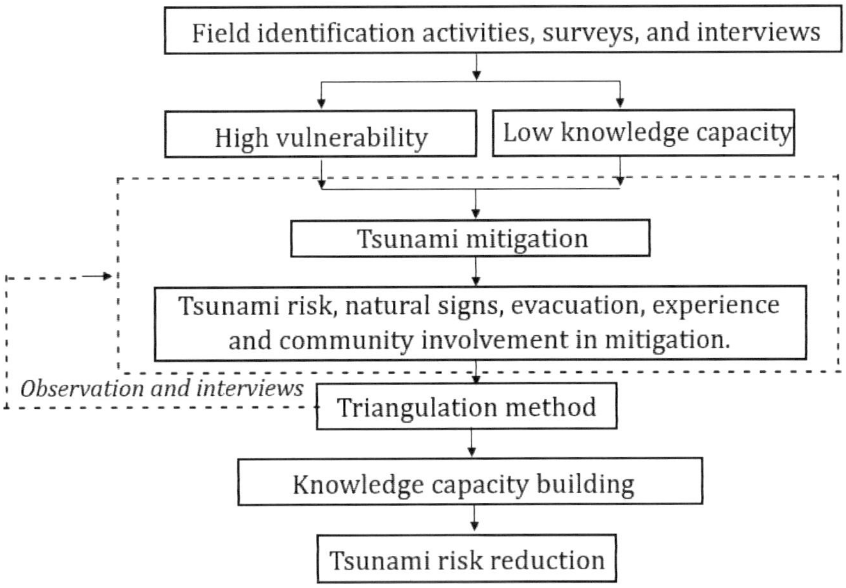

Figure 3. Research flowchart

Figure 4. (a) Observation, (b) Surveys and Interviews with the Community

3. Result and Discussion

3.1 Result Dissemination and Tsunami Mitigation Training

The initiation to increase community knowledge capacity through dissemination and tsunami mitigation training was held on July 20, 2024, which was attended by 30 participants from the Srati residents including the DESTANA group. This activity aims to increase public understanding of the threat of tsunamis, natural signs as early warnings, and evacuation actions that need to be taken. In this training, the community is taught to recognize the early signs of a tsunami, such as an earthquake or a drastic drop in sea level, so that they can immediately take appropriate evacuation actions. In addition, the program also includes a tsunami mitigation training that involves evacuation drills in realistic emergency scenarios, presented in Figure 5 and Table 5. Through this training, the community is trained to move quickly, follow a safe evacuation route, and respond to the situation systematically, aligning with the approach by Putra et al. in 2021 in enhancing disaster response capabilities.

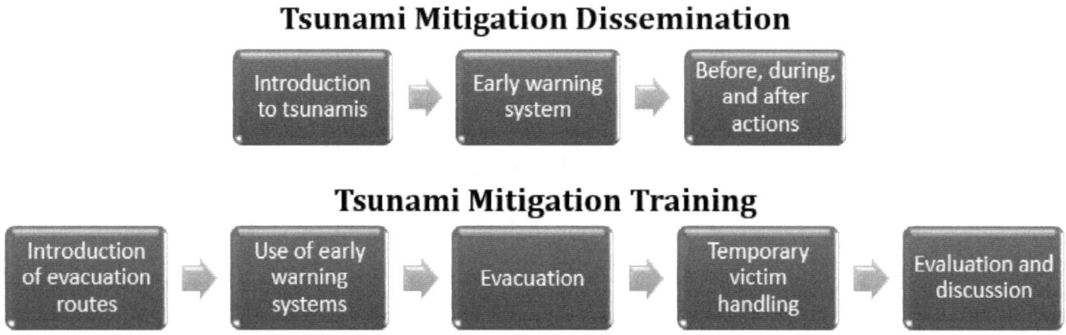

Figure 5. Dissemination and Tsunami Mitigation Training Curriculum

Table 5. Dissemination and Tsunami Mitigation Training Materials

No	Materials	Purpose
1	Tsunami concept	Provides an understanding of the characteristics of tsunamis, causes, and potential risks for affected areas
2	Early warning system	Train the public to recognize early warning signs, both from natural phenomena and official warning systems
3	Evacuation route	Identifying and preparing safe and efficient evacuation routes to be used in the event of a tsunami
4	Realistic evacuation	Carry out realistic evacuation training to ensure community preparedness for tsunami emergencies
5	Active role	Improve their understanding and active role in reducing risks and better dealing with disasters
6	Adapt	Reducing potential losses and strengthening community resilience to disaster impacts

Source: Asrizal 2023

The interviews revealed changes in public understanding of tsunami risk after participating in dissemination (as can be seen in Figure 6a) and tsunami mitigation training (presented in Figure 6b). Some residents who previously did not understand the dangers of tsunamis are now beginning to realize the potential threats and the importance of quick action when there are signs of danger. The majority of residents now better understand that natural signs, such as a strong earthquake or a sudden receding sea level, are early indications of a tsunami that requires an immediate response. The informant stated, "*... Previously, many residents did not know what to do if a tsunami occurred. After the* dissemination, *they now understand and know the correct evacuation steps*" (Interview VH-1). Increasing this knowledge capacity, the community is better prepared mentally and practically in facing potential disasters. Meanwhile another informant mentioned *"Activities like this really help the community to understand disaster risk, not only* *recognize emergency signs, but also have the readiness to act according to evacuation procedures"* (Interview GA-1). This capacity building not only leads to an understanding of natural signs, but also strengthens community readiness to respond and carry out correct evacuation procedures. This capacity building is an important foundation for the community to face the threat of tsunami with appropriate and organized measures.

Figure 6. (a) Dissemination Tsunami Mitigation, (b) Tsunami Mitigation Training

From the interviews conducted, it was found that most people now know about the existence of evacuation routes related to the tsunami disaster after participating in the dissemination, but the level of utilization is still limited due to insufficient information and the absence of previous tsunami mitigation training. One of informant stated, *"We know there is an evacuation route, but we have never been given instructions on how to use it properly when a tsunami occurs"* (Interview SR-1). Another informant added, *"We continue to encourage* tsunami mitigation training *so that the public not only knows the location of the evacuation route, but is also able to use it efficiently when a tsunami disaster occurs"* (Interview GA-2). The increase in community capacity in this case can be seen in the change in their understanding of evacuation routes and the importance of training to maximize the use of these routes during disasters. Another informant revealed, *"The threat of this tsunami is just beginning to be realized by residents, so we must work with various parties to provide further education and ensure preparedness when a disaster occurs"* (Interview CR-1). This shows that community capacity building related to disaster mitigation needs to be carried out continuously through education and training so that they not only know the theory but are also ready in practice to face the threat of tsunamis.

The people of Srati Village generally welcome efforts to increase capacity related to disaster mitigation, especially the tsunami disaster. Residents showed enthusiasm for training and the provision of better evacuation facilities. A informant stated, "*This training really helped us understand the steps to take when a tsunami occurs. We also became more mentally prepared"* (Interview SR-2). Another informant added, "*This initiative not only enhances community understanding, but also reinforces solidarity among residents. We are increasingly convinced that if we work together, disaster risk can be minimized"* (Interview CD-1). This capacity building is seen not only as important for personal safety, but also to empower community unity when dealing with disaster threats. The tsunami mitigation training provided have increased the readiness of

IOP Conf. Series: Earth and Environmental Science **1503** (2025) 012026 doi:10.1088/1755-1315/1503/1/012026

the community both intellectually and mentally, as well as strengthened the sense of togetherness to face potential risks.

In the dissemination and tsunami mitigation training carried out in Srati Village and the Pecaron Beach area, it can be seen that evacuation routes and gathering points in several locations are still poorly maintained and do not have clear signs. This has the potential to confuse the public during a tsunami emergency. A informant revealed, *"We have participated in the tsunami mitigation training, but sometimes evacuation routes are difficult to find because of the lack of clear signs, especially in coastal areas"* (Interview SR-3). However, there are community efforts to maintain readiness by cleaning evacuation routes regularly. The Pecaron Beach manager added, *" Tsunami mitigation training are very important for us, but evacuation routes and gathering points must be more emphasized and well maintained so that they are more accessible when needed"* (PBM-1 Interview). Improving the maintenance of evacuation routes and strengthening signs and information in strategic locations is key so that people can quickly and appropriately move to a safe place during a disaster.

Community involvement in disaster mitigation activities is still sporadic, depending on the initiative of certain individuals or groups. Some communities have embarked on joint efforts, such as holding simple training or installing evacuation signs. Another informant revealed, *"We are starting to realize the importance of preparation and have done some small tsunami mitigation training, but they are still very limited"* (Interview SR-2). Another informant also added, *"We are trying to raise awareness by making evacuation signs, but without further support from related parties, this effort has not been maximized"* (Interview PBM-2) The lack of coordination and support from related parties is the main challenge in expanding the scope of mitigation activities. This emphasizes the importance of synergy between the community, coastal managers, and the government to strengthen preparedness in dealing with a potential tsunami. With more intense collaboration between various parties, mitigation activities can be more widespread and well-coordinated, as well as improve the ability of the community to respond to tsunami threats effectively.

3.2 Discussion

The results show that the communities in Pecaron Beach is highly susceptible to tsunami, especially related to the lack of understanding. Prior to the dissemination and tsunami mitigation training, the majority of individuals lacked a comprehension of the initial indicators of a tsunami, such as a severe earthquake or a sudden retreat of sea level. Poorly maintained evacuation routes and gathering points and the absence of clear signs are one of the factors hindering their readiness. However, the dissemination activities and tsunami mitigation training provided have succeeded in increasing public understanding and awareness of tsunami risks, so that they are more mentally and practically prepared to face potential threats (29). Community capacity building is not only focused on theoretical knowledge, but also on practical aspects such as understanding evacuation routes and quick action in case of emergencies. This is shown through the enthusiasm of the community in participating in the tsunami mitigation training that was held, although its use still needs to be maximized through more organized support. The dissemination carried out also strengthens community solidarity which is the main force in facing the threat of disasters. The involvement of relevant stakeholders, including BPBD and the management of Pecaron Beach, further enhances the collaborative efforts among various parties to ensure the sustainability of this capacity development initiative. Enhancing community knowledge is a strategic step in reducing tsunami risk (30). With a better understanding, people are not only able to recognize the early signs of a tsunami, but also become more skilled in carrying out organized evacuations (31). Maintenance of evacuation routes and gathering points, although not optimal, has begun to be carried out regularly by the local community. This shows that there is an increase in collective awareness that supports disaster risk reduction (32). The success of tsunami mitigation training also builds public confidence in dealing with the threat of tsunamis (33), thereby reducing the potential for panic which is often the cause of increased casualties when disasters occur. The risk reduction can also be seen from the coordination of the community and beach managers in installing additional evacuation signs and improving mitigation infrastructure. Continuous assistance from the government and related institutions is needed to ensure that the mitigation efforts continue to develop and are able to reach all levels of society. With more intense collaboration, tsunami risk can be significantly reduced through sustainable community knowledge (34). This success is in line with the example in Japan, where tsunami simulations and training have succeeded in improving community preparedness, such as the post-2011 tsunami that reduced casualties thanks to evacuation drills that directly involved the community (35)

As part of mitigation efforts, the existence of a tsunami evacuation map on Pecaron Beach is an important element in increasing community preparedness. This map offers detailed information regarding evacuation routes, assembly points, and safe locations accessible during a tsunami event. The evacuation path is depicted by a blue arrow, illustrating the route from the beachfront to designated safe areas and further evacuation points. For a comprehensive view of the Tsunami Evacuation Map for Pecaron Beach, please refer to Figure 7. With evacuation maps installed at various strategic points, the public and tourists can more easily understand the rescue route, thereby minimizing confusion and panic during emergency situations (36). The existence of this evacuation map also strengthens coordination between residents, volunteers, and related parties in building a more effective and sustainable mitigation system (37).

IOP Conf. Series: Earth and Environmental Science **1503** (2025) 012026 doi:10.1088/1755-1315/1503/1/012026

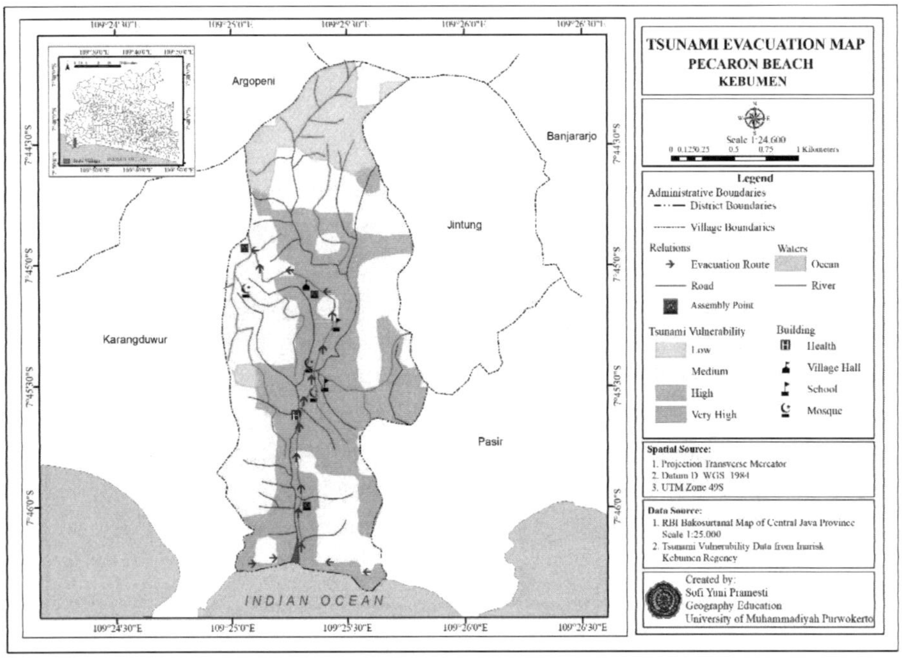

Figure 7. Tsunami Evacuation Map Pecaron Beach

This research makes a new contribution by integrating a participatory approach in enhancing community knowledge-based communities in coastal areas. Unlike previous studies, such as López-Caudana et al. in 2022, which focused on the development of technology-based evacuation routes without directly involving the community, this research highlights the significance of active community participation in mitigation planning and implementation. This approach has proven effective in increasing community solidarity and sense of ownership towards the mitigation efforts carried out, aligning with the findings of Hidayat et al. in 2022 on the role of cultural values in enhancing community resilience. This research also complements the findings from (38) which shows the importance of formal education related to the tsunami disaster. However, this study adds a new dimension by showing that non-formal education through tsunami mitigation training and direct training in the field can have a more significant impact on changing people's behaviour. The training provided not only adds insight, but also builds mental readiness and social solidarity which is the foundation in dealing with disasters. In addition, compared to studies that highlight the crucial role of information technology in the dissemination of mitigation information, this study presents aspects of physical completeness in the field, such as improving evacuation route signs and gathering points. This provides a practical solution that is often overlooked, but is crucial in emergency situations. Thus, this study offers a holistic and community-based mitigation model that can be replicated in other regions with similar disaster risks.

Before the intervention was carried out, people in coastal areas showed low knowledge capacity related to the tsunami disaster. Their knowledge of tsunamis as large waves triggered by geological activity, such as earthquakes in subduction zones, underwater volcanic eruptions, or underwater avalanches, is still very minimal. In addition, the public does not have an adequate awareness of the initial indicators of a tsunami, such as a powerful earthquake, a sudden receding of sea water, or an unusual rumbling sound from the sea. These limitations result in a lack of awareness of the major risks that tsunamis can pose, including their impact on lives, infrastructure, economies, and the environment (39). This lack of knowledge also impacts the community's inability to plan mitigation measures or respond quickly when signs of danger appear. The findings of the research show that the intervention carried out have had a significant impact on increasing the capacity of community knowledge. The community has become more aware of scientific aspects related to the causes of tsunamis, such as tectonic, volcanic, and underwater avalanche mechanisms (40). In addition, they were able to better identify the initial indicators of a tsunami, like a strong earthquake or an abrupt withdrawal of sea water, and a rumbling sound coming from the sea. This increase in understanding is accompanied by increased awareness of the impact of the tsunami disaster, including physical damage, casualties, as well as economic and environmental impacts. The intervention also equips the community with practical skills to take mitigation measures, such as recognizing and using evacuation routes, utilizing safe gathering points, and coordinating with local authorities. In addition to increasing knowledge, this training has proven to be able to build mental readiness and strengthen social solidarity among the community. Thus, the community becomes more responsive and adaptive in dealing with the threat of tsunamis, which directly supports efforts to reduce disaster risk holistically.

4. Conclusion

The tsunami dissemination and mitigation training program at Pecaron Beach has successfully in enhanced the knowledge capacity and preparedness of coastal communities. The program is able to change people's understanding from low to high, so that they can identify tsunami risk, recognize natural warning signs, understand safe evacuation routes, follow evacuation procedures correctly, and actively participate in mitigation efforts. Participants also showed increased accuracy in responding to disaster scenarios. However, this study has limitations because it only covers the coastal area of Pecaron Beach, so it has not taken into account the impact of the tsunami on other areas such as agricultural land and livestock. Therefore, further research is recommended to reach a wider area, including agricultural and livestock areas, in order to develop more comprehensive mitigation guidelines for communities that depend on these sectors.

5. Acknowledgements

Acknowledgment is given to the Directorate of Learning and Student Affairs (BELMAWA) under the Ministry of Education, Culture, Research and Technology of the Republic of Indonesia for funding the community service activities under the Student Organization Capacity Strengthening Program (PPK Ormawa) scheme, as stated in reference number 3508/E2/DT.01.01/2024 of 2024. Appreciation is also extended to the residents of Srati Village, Kebumen, and the village officials for their comprehensive support throughout the program. All essential data, visuals, and documentation were carefully collected during the implementation of the program.

References

(1) Fuady M, Munadi R, Fuady MAK. Disaster mitigation in Indonesia: between plans and reality. IOP Conf Ser Mater Sci Eng. 2021;1087(1):012011.
(2) Rosyida A, Aziz M, Firmansyah Y, Setiawan T, Pangesti KP, Kakanur F. Buku Data Bencana Indonesia 2023. Vol.

3, Buku Data Bencana Indonesia. 2023. 3–11 p.
(3) Rusydy I, Idris Y, Mulkal, Muksin U, Cummins P, Akram MN, et al. Shallow crustal earthquake models, damage, and loss predictions in Banda Aceh, Indonesia. Geoenvironmental Disasters. 2020;7(1).
(4) Irwansyah M, Nursaniah C, Qadri L, Mariana M. City of prone natural disasters: Mitigating post-tsunami on the coastal of Banda Aceh, Indonesia. Environ Challenges [Internet]. 2024;15(November 2023):100925. Available from: https://doi.org/10.1016/j.envc.2024.100925
(5) Ayuningtyas D, Windiarti S, Sapoan Hadi M, Fasrini UU, Barinda S. Disaster preparedness and mitigation in indonesia: A narrative review. Iran J Public Health. 2021;50(8):1536–46.
(6) Cvetković VM, Tanasić J, Ocal A, Kešetović Ž, Nikolić N, Dragašević A. Capacity development of local self-governments for disaster risk management. Int J Environ Res Public Health. 2021;18(19).
(7) Harnantyari AS, Takabatake T, Esteban M, Valenzuela P, Nishida Y, Shibayama T, et al. Tsunami awareness and evacuation behaviour during the 2018 Sulawesi Earthquake tsunami. Int J Disaster Risk Reduct [Internet]. 2020;43:101389. Available from: https://doi.org/10.1016/j.ijdrr.2019.101389
(8) Ryan B, Johnston KA, Taylor M, McAndrew R. Community engagement for disaster preparedness: A systematic literature review. Int J Disaster Risk Reduct. 2020;49:0–29.
(9) Mavrouli M, Mavroulis S, Lekkas E, Tsakris A. Respiratory infections following earthquake-induced tsunamis: Transmission risk factors and lessons learned for disaster risk management. Int J Environ Res Public Health. 2021;18(9).
(10) Amahoru AH, Ridho A, Lukafiardi R, Bisri MBF. Indonesian "tsunami-generation" in a nutshell: Systematic literature review. IOP Conf Ser Earth Environ Sci. 2023;1245(1).
(11) MacAfee E, de Jong E, Lohr AJ. Leveraging local knowledge for landslide disaster risk reduction in an urban informal settlement in Manado, Indonesia. Int J Disaster Risk Reduct [Internet]. 2024;111(March):104710. Available from: https://doi.org/10.1016/j.ijdrr.2024.104710
(12) Perera D, Agnihotri J, Seidou O, Djalante R. Identifying societal challenges in flood early warning systems. Int J Disaster Risk Reduct [Internet]. 2020;51:101794. Available from: https://doi.org/10.1016/j.ijdrr.2020.101794
(13) Alhadi Z, Sasmita S, Yulfa A, Fatimah S, Syafrini D, Maani KD, et al. The Coping Strategies Patterns Based on Local Wisdom and Resilience Capital in Facing Natural Disaster Risk in Nagari Mandeh, Indonesia. Int J Sustain Dev Plan. 2023;18(1):315–25.
(14) Safira FA, Muryani C, Tjahjono GA. Tsunami Susceptibility Analysis in Coastal Area Petanahan District, Kebumen Regency. Jambura Geosci Rev. 2022;4(2):110–22.
(15) Abdul Latif RF, Muryani C, Yusup Y. Community Resilience in facing Tsunami Disaster in the Coastal Areas of Purworejo Regency. IOP Conf Ser Earth Environ Sci. 2021;884(1).
(16) Syamsidik, Rasyif TM, Suppasri A, Fahmi M, Al'ala M, Akmal W, et al. Challenges in increasing community preparedness against tsunami hazards in tsunami-prone small islands around Sumatra, Indonesia. Int J Disaster Risk Reduct [Internet]. 2020;47(November 2019):101572. Available from: https://doi.org/10.1016/j.ijdrr.2020.101572
(17) Masten AS. Resilience of children in disasters: A multisystem perspective. Int J Psychol. 2021;56(1):1–11.
(18) Ester Krisnawati. Communication Strategy for Tsunami Disaster Mitigation in Tourist Areas on the South Coast of Java Island (Case Study of the South Coast of Kebumen Regency). Proc Int Conf Commun Sci. 2022;2(1):512–9.
(19) Suwaryo PAW, Rahma DG, Waladani B, Safaroni A. Community Preparedness to Reduce Risk Disaster of Tsunami. Babali Nurs Res. 2021;2(1):40–8.
(20) Widyartini DS, Hidayah HA, Piranti AS, Warihjati A. Distribution pattern of macroalgae community in Pecaron Beach, Kebumen Regency. Proceeding ICMA-SURE. 2023;2(1):156.
(21) Fadlin, Waluyo G, Iwan Y, Ariyanti N, Ajeng Nurwantari N. Geodiversity of Menganti Paleo-Volcano: New Insight to Geotourism Potential in Kebumen, Central Java, Indonesia. J Geosci Eng Environ Technol. 2020;5(1):11–24.
(22) Hutchings SJ, Mooney WD. The Seismicity of Indonesia and Tectonic Implications. Geochemistry, Geophys Geosystems. 2021;22(9):1–42.
(23) Aldahlawi RY, Akbari V, Lawson G. A systematic review of methodologies for human behavior modelling and routing optimization in large-scale evacuation planning. Int J Disaster Risk Reduct [Internet]. 2024;110(May):104638. Available from: https://doi.org/10.1016/j.ijdrr.2024.104638
(24) Syahrial S, Badollahi MZ. Development of a Community-Based Marine Tourism Attraction in the Samboang Beach in Bulukumba Regency. J La Bisecoman. 2020;1(2):1–9.
(25) Mezmir EA. Qualitative Data Analysis: An Overview of Data Reduction, Data Display and Interpretation. Res Humanit Soc Sci. 2020;10(21):15–27.
(26) Donkoh S. Application of triangulation in qualitative research. J Appl Biotechnol Bioeng. 2023;10(1):6–9.
(27) Cugliari L, Cerase A, Amato A. Tsunami risk perception, a state-of-the-art review with a focus in the NEAM region. Front Earth Sci. 2022;10(December):1–25.
(28) Dzwigol H. Research Methodology in Management Science: Triangulation. Virtual Econ. 2022;5(1):78–93.

(29) Riviwanto M, Darwel D, Dwiyanti D, Juanda J. The Preparedness Level of Families with Disabilities Children in Facing the Earthquake and Tsunami Disaster in Padang, West Sumatra. Int J Disaster Manag. 2021;4(1):61–70.

(30) Yen E, Lin SC, Wu TR, Tsai YL, Chung MJ. Knowledge-Building Approach for Tsunami Impact Analysis Aided by Citizen Science. Front Earth Sci. 2020;8(October):1–12.

(31) Makinoshima F, Imamura F, Oishi Y. Tsunami evacuation processes based on human behaviour in past earthquakes and tsunamis: A literature review. Prog Disaster Sci [Internet]. 2020;7:100113. Available from: https://doi.org/10.1016/j.pdisas.2020.100113

(32) Räsänen A, Lein H, Bird D, Setten G. Conceptualizing community in disaster risk management. Int J Disaster Risk Reduct. 2020;45(January).

(33) Dias N, Haigh R, Amaratunga D, Rahayu H. A review of tsunami early warning at the local level - Key actors, dissemination pathways, and remaining challenges. Int J Disaster Risk Reduct [Internet]. 2024;101(June 2023):104195. Available from: https://doi.org/10.1016/j.ijdrr.2023.104195

(34) Salah P, Sasaki J. Knowledge, awareness, and attitudes toward tsunamis: A local survey in the southern coast of Iran. Sustain. 2021;13(2):1–19.

(35) Maly E, Suppasri A. The Sendai Framework for Disaster Risk Reduction at Five: Lessons from the 2011 Great East Japan Earthquake and Tsunami. Int J Disaster Risk Sci [Internet]. 2020;11(2):167–78. Available from: https://doi.org/10.1007/s13753-020-00268-9

(36) Berawi MA, Leviäkangas P, Siahaan SAO, Hafidza A, Sari M, Miraj P, et al. Increasing disaster victim survival rate: SaveMyLife Mobile Application development. Int J Disaster Risk Reduct. 2021;60(April).

(37) Tsai MT, Chang HW. Contribution of Accessibility to Urban Resilience and Evacuation Planning Using Spatial Analysis. Int J Environ Res Public Health. 2023;20(4).

(38) Putra AR, Dewi IN, Romli O, Wiguna W. Wisatawan Tanggap Bencana di Kawasan Wisata Anyer, Kabupaten Serang. JUKEMAS J Pengabdi Kpd Masy [Internet]. 2023;1(1):1–11. Available from: https://ejournal.lumbungpare.org/index.php/jukemas/article/view/13

(39) Senevirathne M, Amaratunga D, Haigh R, Clegg G. Enhancing systemic risk management to strengthen community resilience: Key recommendations from state-of-the-art practices of past complex incidents. Int J Disaster Risk Reduct. 2024;114(November).

(40) Behrens J, Løvholt F, Jalayer F, Lorito S, Salgado-Gálvez MA, Sørensen M, et al. Probabilistic Tsunami Hazard and Risk Analysis: A Review of Research Gaps. Front Earth Sci. 2021;9(April):1–28.

Analysis of the Distribution of Forest and Land Fire Severity Levels using the Normalized Burn Ratio (NBR) Algorithm on Landsat 8 Satellite Imagery in East Belitung Regency

A A N Sabrina[1], D N Sari[1*], B I Arijuddin[1], Taryono[1]

[1]Department of Geography, Faculty of Geography, Universitas Muhammadiyah Surakarta, Indonesia

*E-mail: dns104@ums.ac.id

Abstract. Deforestation is a major factor threatening Indonesia's forests every year. Indonesia deforested at an average rate of 2.54 million hectares per year between 2017 and 2021. A total of 131 forest and land fires were reported in Bangka Belitung Islands, one of the regions in Indonesia, between 2020 and 2024. The purpose of this study is to determine the severity of forest and land fires in East Belitung Regency. The Normalized Burn Ratio (NBR) technique was used on Landsat 8 imagery (path 122 and row 062) to analyze fire severity. The NBR value is obtained by comparing the Near Infrared (NIR) wavelength of Landsat 8 band 5 with the Shortwave Infrared (SWIR) wavelength of Landsat 8 band 7. Radiometric correction of Landsat 8 images aims to reduce errors in satellite images and is carried out using QGIS 3.32.3 software. The Difference Normalized Burn Ratio (ΔNBR) value is very important in assessing the level of damage. The ΔNBR formula used is (NBR pre fire - NBR post fire), with values that determine the level of damage in East Belitung being Min (-0.497122) and Max (0.819870). High ΔNBR values indicate that the area has suffered significant damage, while low ΔNBR values indicate high vegetation growth rates after the fire. Spatial analysis by sub-district in East Belitung shows that Dendang sub-district has 50.00% of the total area with high severity, indicating the most critical impact of the fire. This research is expected to be used to develop more effective policies for forest protection and sustainable land management.

1. Introduction

Forests are a vast and diverse ecosystem, encompassing a stretch of land rich in biological natural resources. In Indonesia, forests are one of the invaluable valuable assets, also known as one of the forests that hold the most natural wealth. The uniqueness of Indonesia's forests lies in the various levels of biodiversity of flora and fauna that abound within them. Indonesia is dubbed the lungs of the world. Forests function to produce the oxygen gas needed by humans because they absorb carbon dioxide gas. The area of Indonesian forests, according to data from the Ministry of Environment and Forestry (MOEF) in 2024, reached 125,664,549.9 hectares [1].

Deforestation is the main factor threatening Indonesia's forests every year. Significant forest and land fires in Indonesia transpired during the years 1982-1983, 1991, 1994, 1997-1998, 2006, and 2015. In 2015, the widespread and damaging effects of forest and land fires resulted in the destruction of approximately 2.61 million hectares of forested and terrestrial landscapes. In addition to the destruction of these environments, the fires also resulted in significant economic losses that had a considerable impact on various sectors. Furthermore, during the same year of 2015, the area subjected to combustion encompassed 2.61 million hectares of forests and land, which incurred staggering economic losses that amounted to a staggering total of approximately Rp 221 trillion, thereby highlighting the severe financial implications of such environmental catastrophes [2]. This situation shows that Indonesia's forests are not in a good condition. It is imperative that concerted efforts are made to overcome this issue. A prerequisite for effective countermeasures is a comprehensive understanding of the location of the fire.

Table 1. Burned area in Indonesia from 2019 - 2023.

Year	Area (Ha)
2019	1176884
2020	961587.9
2021	841391.4
2022	885238.4
2023	1395285

Source: Global Forest Watch.

The Bangka Belitung Islands, a region of Indonesia, are frequently impacted by forest and land fires. Data from the National Disaster Management Agency indicates that from 2020 to 2024, there were 131 cases of forest and land fires in this region. Of these, East Belitung Regency has the dubious distinction of being the location of 43 incidents, constituting approximately 32.82% of all forest and land fires in the Bangka Belitung Islands region.

Table 2. Burned area in East Belitung Regency from 2019 to 2023.

Year	Area (Ha)
2019	4.086.904
2020	2.826.155
2021	2781.67
2022	2.496.956
2023	4.528.838

Source: Global Forest Watch.

In the year 2024, East Belitung Regency has experienced several forest and land fire disasters that have destroyed many green areas that have long been a source of livelihood for the local community.

Table 3. Incidents of Forest and Land Fires in East Belitung Regency.

Date	Event
8/20/2024	Forest and Land Fire
8/18/2024	Forest and Land Fire
8/14/2024	Forest and Land Fire
8/12/2024	Forest and Land Fire

Source: bnpb.go.id

The collected data clearly shows that forest and land fires in East Belitung Regency occur frequently with a relatively high frequency and within a short period of time. Only in August 2024, four destructive fire incidents have already been recorded. This situation illustrates a highly dangerous condition, where forest and land fires have become a real threat to East Belitung Regency. East Belitung is characterized by critical land conditions, largely due to extensive tin mining activities that have left significant environmental impacts. Regional soils, including podsol, are affected by these activities, leading to various levels of land degradation. Podsol, usually acidic and nutrient-poor, is common in areas with high rainfall and sandy soil, conditions that are exacerbated by mining activities. A study using geographic information systems (GIS) identified that 12% of land in East Belitung is classified as critical, with an additional 43% as almost critical and 28% as potentially critical [3]. Podsol soil is usually sandy, which allows for rapid drying and increased flammability during dry conditions [4]. The aforementioned environmental conditions are the main factors that significantly contribute to the recurrent and increasingly prevalent forest and land fires in the East Belitung region, posing a serious threat to the local ecosystem and the socio-economic stability of the residents.

This research aims to evaluate the impact of forest and land fires in East Belitung Regency in 2024. The assessment of the intensity of both forest and land fires is facilitated through the integrated and synergistic utilisation of advanced Landsat 8 satellite imagery, in conjunction with the precise calculation of the Normalized Burn Ratio (NBR) index. The NBR index serves as a critical analytical tool in this research undertaking. The assessment of fire severity was conducted using Landsat 8 satellite imagery and the determination of the NBR index. The NBR index leverages the advanced capabilities of the Landsat 8 satellite, which is equipped with a state-of-the-art Thermal Infrared Sensor (TIRS) that boasts enhanced thermal imaging capabilities with exceptional spatial and temporal resolution.These images are made fully accessible to the open public at no charge. NBR is a quantitative approach that integrates near infrared (NIR) and shortwave infrared (SWIR) data from satellite images to derive pixel values. The NBR index has been identified as the most suitable metric for the estimation of fire severity from satellite imagery [5].

It is hoped that the analysis of hotspots and fire areas in the can be an illustration and a tool for policy-making efforts to mitigate forest and land fires quickly.Remote sensing and GIS technology can be used to monitor forest and land fires, and the process of analysing data using satellite data is a fast, precise and accurate method, so the process does not take a long time. However, this process is still constrained by several factors, especially cloud coverage.Activities involving forest and land fire monitoring are expected to provide information on a wide area.

2. Methods

2.1. Research areas and data used

The geographical location of the research sites is in East Belitung Regency, Bangka Belitung Islands Province. **Figure 1** is a visualization of the area to be studied. The study undertaken regarding the delineation of the intensity of forest and land fires through the application of the NBR Algorithm on Landsat 8 imagery within the East Belitung Regency utilizes remote sensing research methodologies. Remote sensing encompasses a suite of techniques that have been devised for the procurement and examination of data pertaining to the Earth's surface. At present, remote sensing is extensively employed as a principal data source for both observation and cartographic representation [6].

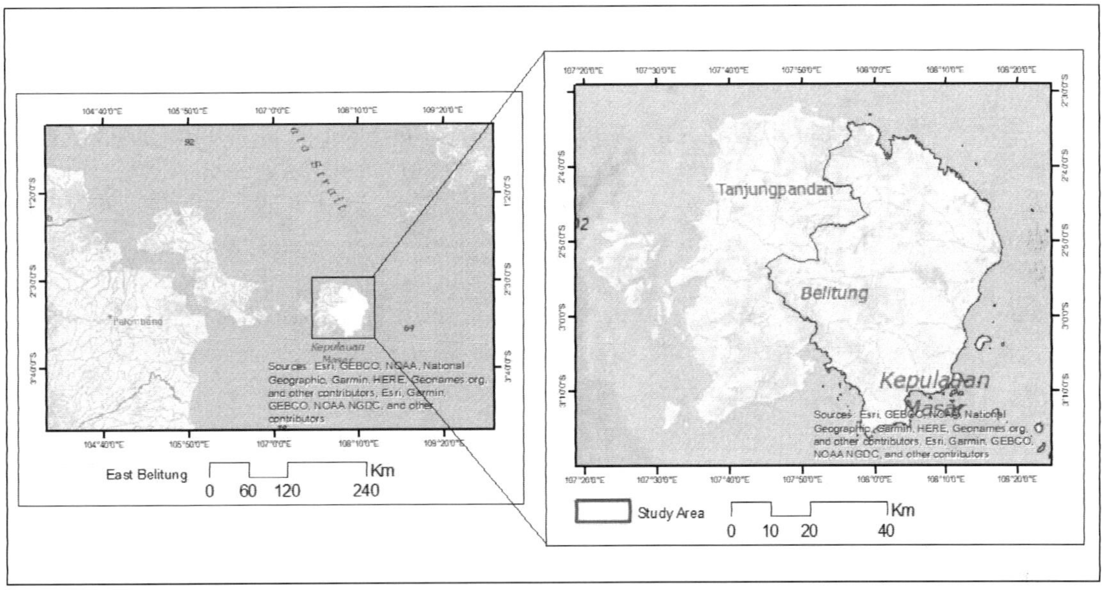

Figure 1. Research areas

The materials used in this research are:

1. Forest and Land Fire Data for East Belitung Regency. Source: (gis.bnpb.go.id)
2. Landsat 8 Satellite Imagery of East Belitung Regency 2024 Before the Forest Fire. Source: (earthexplorer.usgs.gov)
3. Landsat 8 Satellite Imagery of East Belitung Regency 2024 After the Forest Fire. Source: (earthexplorer.usgs.gov)
4. Hotspots in East Belitung Regency. Source: (firms.modaps.eosdis.nasa.gov)

The satellite image utilised in this study was Landsat-8 OLI/TIRS Level 2C path/row 122/62 pre fire (with acquisition date on Juny 4, 2024) and post fire (with acquisition date on August 23, 2024). Based on Table 3, the acquisition date was determined because the forest and land fires occurred between August 12 and 20, so satellite imagery data before and after that period is needed.

Landsat 8 is a satellite that has two sensors, the Thermal Infrared Sensor (TIRS) and the Operational Land Imager (OLI). The aim of Landsat 8 is to take a lot of detailed pictures of the Earth and collect lots of data about the environment and geography. Notably, the Operational

Land Imager encompasses a wide range of wavelengths, meticulously covering the visible spectrum, near-infrared, and shortwave infrared (commonly referred to as VNIR, NIR, and SWIR), which collectively enhance its observational effectiveness and versatility in capturing diverse environmental phenomena [7]. TIRS is cutting-edge technology that uses the concepts of quantum physics to analyze thermal radiation in order to quantify land surface temperature across two thermal bands. The spatial resolution of the Landsat 8 imagery is specified as 15 metres for panchromatic imaging and 30 metres for multispectral imaging, with these parameters applied to a 185-kilometre area.

2.2. Data Processing and Analysis

There are five parts to this research's processing step. The first step is called pre-processing, and it uses radiometric correction to adjust the picture for distortion. The NBR value is calculated from the 5 micrometre (5 μm) NIR and 7 micrometre (7 μm) SWIR wavelengths from Landsat-8 bands 5 and 7. Next, the ΔNBR (Difference Normalized Burn Ratio) is estimated from the Landsat-8 band 7 data. ΔNBR is used to calculate the degree of fire in the ground and vegetation fire areas. After that, the burnt areas are classified to detect which areas are classified as severe and which are classified as relatively safe. The final stage of the process is the testing of the accuracy of the research, which is used to verify the findings.

This study employed descriptive analytic techniques to give a thorough summary of the information gathered. Finding the NBR value, a crucial metric in land fire investigation, is the next step. The NBR value is obtained by comparing the Near Infrared (NIR) wavelength from Landsat 8 band 5 with the Shortwave Infrared (SWIR). Wavelength from Landsat 8 band 7. NBR is used to identify areas that have experienced fires by comparing the reflectance of the burned area with the reflectance of the unburned land. NBR is a method used to identify areas that have been burned. The range of NBR values is from -1 to 1, which can provide further insight into the vegetation condition of an area. When the NBR value approaches -1, it indicates that the area is a burned zone. Conversely, if the NBR value tends to be high towards 1, it is estimated that the area has good- quality vegetation. The equation used in the NBR method is as follows:

$$NBR = \frac{NIR - SWIR2}{NIR + SWIR2} \quad (1)$$

Where,
NBR: *Normalized Burn Ratio*
NIR: Band 5
SWIR 2: Band 7

The Differenced Normalized Burn Ratio (ΔNBR), a metric used to evaluate the intensity of land and forest fires in East Belitung Regency, is measured using data processing techniques to provide detailed information on the state of the environment.

$$\Delta \text{NBR} = NBR\ Pre\ Fire - NBR\ Post\ Fire \quad (2)$$

Where,
ΔNBR: Differences between pre fire and post fire NBR
NBRpre: NBR value before the fire
NBRpost: NBR value after a fire

In the visual representation shown in **Figure 2**, several key steps in this research are explained.

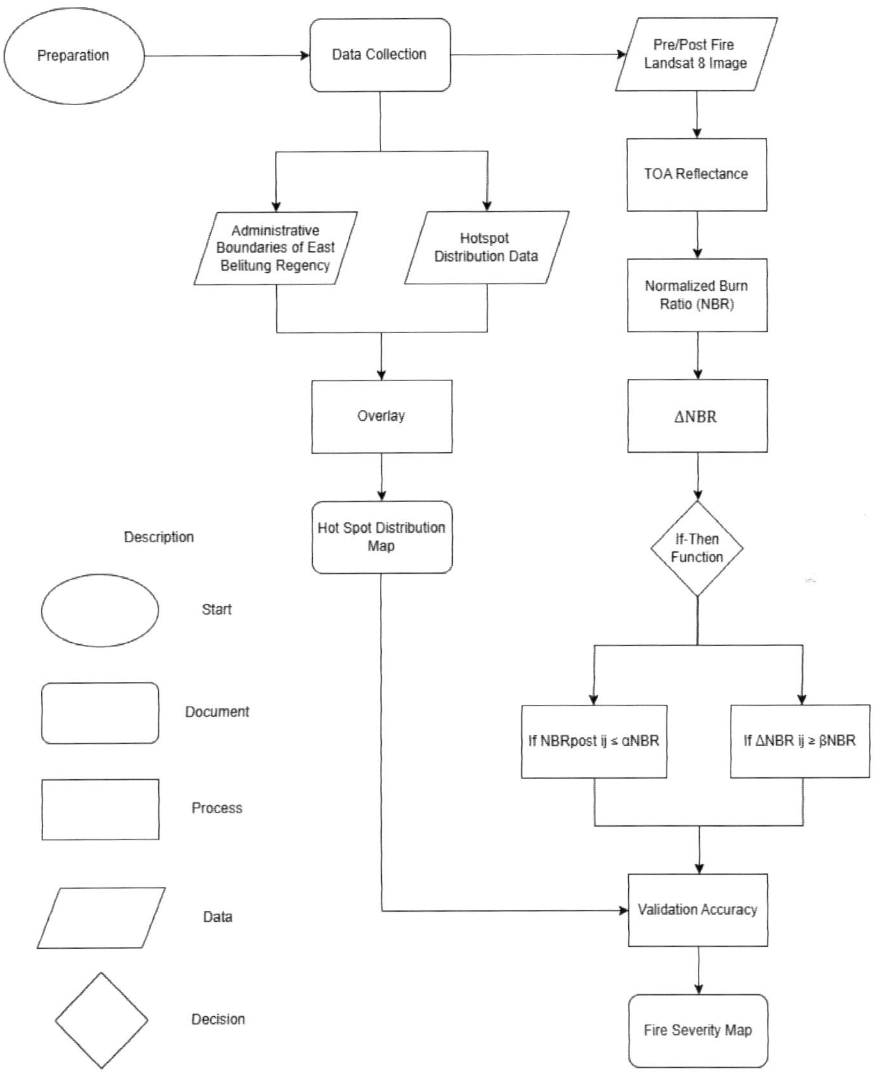

Figure 2. Research Flowchart

2.2.1. Pre-processing of Landsat data

Radiometric correction is a process that is employed to rectify disparities in image brightness that are attributable to atmospheric lighting, or to errors that occur during the acquisition of images [8]. Radiometric correction eliminates systematic errors in pixel intensity values of the

IOP Conf. Series: Earth and Environmental Science **1503** (2025) 012027 doi:10.1088/1755-1315/1503/1/012027

image to achieve accurate and consistent radiometry, converting spectral radiance into the true reflectance of the imaged scene [9]. The data is properly geometrically corrected so that the position of the pixels corresponds to how they appear on the Earth's surface [10]. The process was run using the Semi-Automatic Classification Plugin (SCP) in the QGIS software environment [11]. The reflectance rescaling factors included in the product metadata file, typically included in the mtl file, make the conversion procedure from Landsat to TOA Reflectance more effective.

2.3. Data Analysis Procedure

Detecting burned areas using the NBR algorithm requires a threshold calculation. The process of determining the threshold is executed through a meticulous computation that involves the determination of the mean, denoted as (μ), alongside the standard deviation, represented by the symbol (σ), derived from a series of area samples that effectively represent regions that have undergone combustion due to fire events. The threshold calculation used to determine burned areas in this study is $\mu + 2\sigma$ on the post-burn NBR image (NBR-Post), resulting in a ΔNBR image symbolized by the α and $\mu - 2\sigma$ values. The change (ΔNBR) is represented by the β value. The designation of a pixel as a burnt area is contingent upon the fulfillment of the following condition [12]:

$$\text{Condition 1 : if NBRpost ij} \leq \alpha\text{NBR} \quad (3)$$

$$\text{Condition 2 : if } \Delta\text{NBR ij} \geq \beta\text{NBR} \quad (4)$$

Where,
i : Pixel line number
j : Pixel column number
NBRpost ij : The NBR value following a fire at a specific pixel
ΔNBR ij : Changes in the NBR value for that particular pixel
αNBR : Mark threshold for NBR in the aftermath of a fire
βNBR : Mark threshold for NBR change

3 Results and Discussion

3.1 Analysis of Hotspot Distribution

Hotspot is a term for areas with higher surface temperatures detected by remote sensing, indicating the potential for forest fires [13]. The emerging hotspot provides initial indications of the possibility of a forest fire. However, the truth of these hotspot points needs to be examined more thoroughly by checking other fire data. This is important to identify whether the clustered hotspots indicate a significant forest fire. The recurrence of hotspot points in the same area also indicates that a fire is occurring in that location. The hotspot data was obtained from the NASA-FIRMS webpage, and the data set includes hotspot coordinates, as well as various supporting information such as satellite sensing, data acquisition time, and confidence level. The data can be downloaded in shp, csv and json formats.

Fire Information Resource Management System "FIRMS" provides timely access to data and imagery for active fire/hotspot data collection, mapping resources, and warning systems, allowing users to identify wildfire location, spatial extent, and intensity. NASA`s Land, Atmosphere Near real-time Capability for Earth observation (LANCE) provides critical support for users who want to observe a wide range of natural and human-induced events with near real-time (NRT) data and imagery delivered much faster than typical satellite imagery is possible with conventional processing techniques.

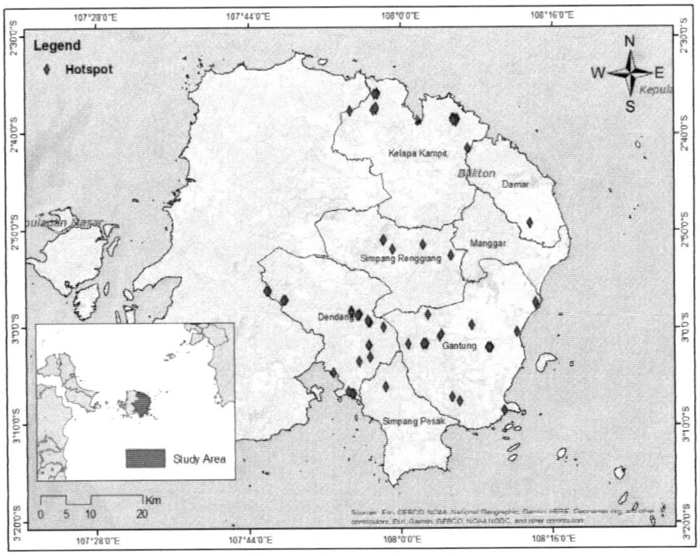

Figure 3. Map of Hotspot Distribution in East Belitung Regency in 2024

During the period from June 4, 2024 to August 23, 2024, the data shows that the most hotspots were identified in Dendang sub-district with 22 hotspots, followed by Gantung sub-district with 18 hotspots, then Kelapa Kampit with a total of 14 hotspots. Followed by Simpang Renggiang sub-district with 5 hotspots detected, Damar and Simpang Pesak sub-districts with 1 hotspot each. The total number of hotspots recorded during the specified period was 61.

Table 4. Hotspot Distribution per Sub-district.

Sub-District	Number of Hotspots
Damar	1
Dendang	22
Gantung	18
Kelapa Kampit	14
Manggar	-
Simpang Pesak	1
Simpang Renggiang	5

3.2 Normalized Burn Ratio (NBR)

In terms of the Normalized Burn Ratio (NBR) algorithm, it can be inferred that a numerical value close to 1 is a strong indicator that the particular area under consideration has good vegetation conditions, which may be due to factors such as sufficient water availability and the presence of healthy plant communities. Conversely, when the NBR value approaches -1, it indicates that the land in question is likely characterised by a significant lack of vegetation, potentially indicating a transition towards open conditions, which could result from various environmental pressures or disturbances that adversely affect plant growth and survival.

IOP Conf. Series: Earth and Environmental Science **1503** (2025) 012027 doi:10.1088/1755-1315/1503/1/012027

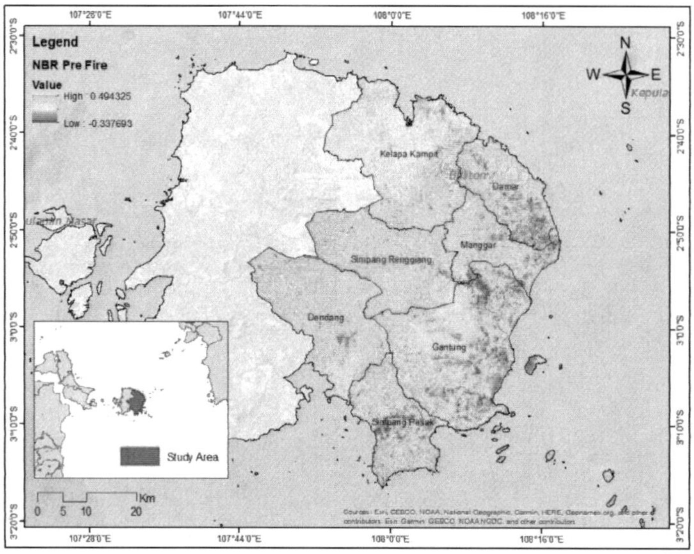

Figure 4. NBR Pre Fire Map.

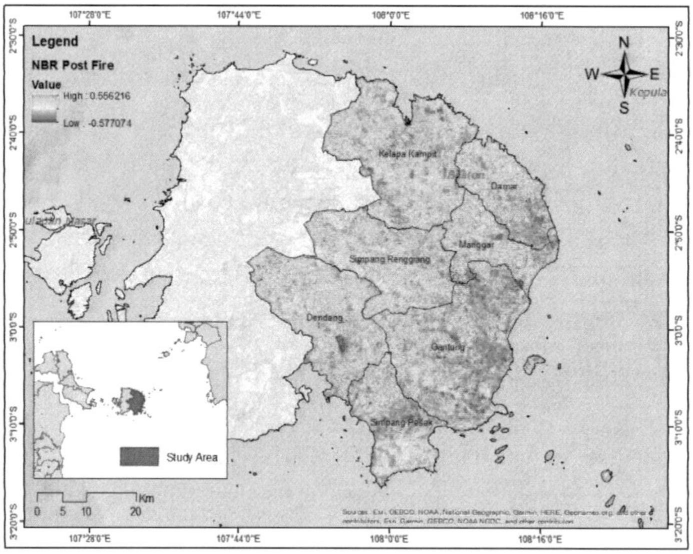

Figure 5. NBR Post Fire Map.

On the figure, the NBR values conventionally exhibit a spectrum of colors transitioning from deep red (indicative of high burn severity) to green (representative of low burn severity), thereby signifying a range of alterations in vegetation resulting from fire events [14]. With a commendable degree of precision, the NBR model demonstrates superior efficacy in delineating burned regions in comparison to alternative models.

The NBR methodology employs the NIR and SWIR spectral bands to analyze data and derive NBR values. In general, high NBR values show good vegetation health, whereas diminished NBR values suggest areas that have recently experienced burning or are devoid of vegetation [15]. The NBR methodology has emerged as a pivotal algorithm in the assessment of vegetation health and the identification of burned regions.

IOP Conf. Series: Earth and Environmental Science **1503** (2025) 012027 doi:10.1088/1755-1315/1503/1/012027

3.3 Delta/Difference Normalized Burn Ratio (ΔNBR)

In assessing the severity of forest and land fires, a series of steps must be followed to ensure accurate estimation. In a study conducted in East Belitung Regency, it was found that the use of ΔNBR values can provide a significant picture of the level of damage from fire events. ΔNBR serves as an important metric that describes the level of damage that occurs in a given area as a consequence of fire; an increasing ΔNBR value signifies a more severe and critical level of damage, while a decreasing value indicates a considerable level of vegetation recovery after burning. The determination of the ΔNBR is achieved through a thorough calculation that involves assessing the fluctuations in the NBR values observed during the temporal intervals that occur both prior to and subsequent to the occurrence of the fire event in question. This methodological approach facilitates comprehensive evaluation of the intensity of the fire's impact on the affected area and facilitates deeper understanding of the ecological consequences and recovery dynamics within the post-fire landscape. Table 5 shows the ΔNBR values and severity classifications. These classifications are derived from previous research [16]:

Table 5. Severity Based on ΔNBR Value.

ΔNBR	Fire Severity
< -0.25	High post-burn regrowth
-0.25 to -0.1	Low post-burn regrowth
-0.1 to 0.1	Unburned
0.1 to 0.27	Low severity burn
0.27 to 0.44	Moderate-low severity
0.44 to 0.66	Moderate-high severity
> 0,66	High severity burn

The comprehensive results obtained from careful calculation of the ΔNBR index yielded a very wide spectrum of values, with a min value of -0.497122 and a max of 0.819870. This wide range of values serves as an important indicator of the severity of fire events occurring within the defined study area. Based on this wide range of values, the intensity of fire impacts is systematically categorized into 7 different classifications. The implementation and subsequent establishment of a comprehensive fire severity classification system is of paramount importance for achieving a more profound and nuanced understanding of the extensive and often devastating damage inflicted upon the diverse landscapes that have been adversely impacted by the occurrence of wildfires, which have become increasingly prevalent in recent years. By systematically categorizing ΔNBR values into seven clearly defined severity classes, further analytical efforts can be made with the aim of assessing the various impacts of fire on the surrounding land and ecological environment. Furthermore, this valuable information on the spatial distribution of different severity levels is illustrated in more detail in **Figure 6**, which serves to provide a clear and in-depth visual picture of the overall distribution and intensity of fires within the study area boundaries.

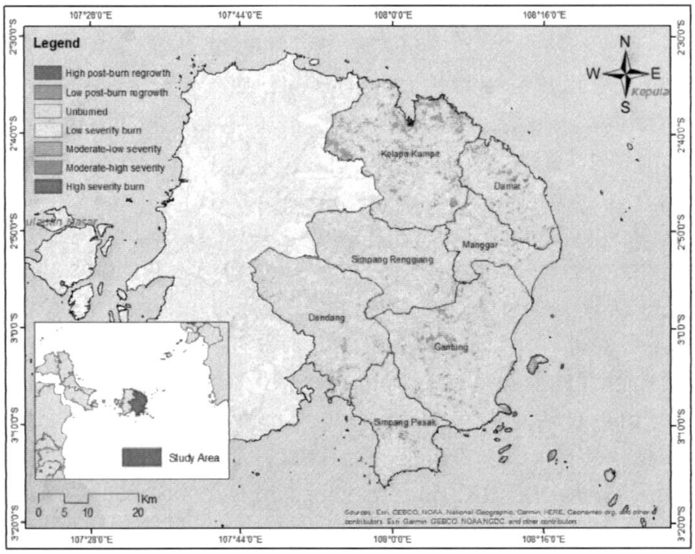

Figure 6. Map of Forest and Land Fire Severity Levels

The level of forest and land fires is greatly influenced by the interaction of various factors, including environmental variables, climatic conditions, and anthropogenic influences that collectively determine the overall intensity and impact of the fires. The conditions present before the outbreak of the fire, particularly those concerning the density of vegetation in the affected areas, have been proven to significantly influence the severity of burn injuries, where areas marked by denser vegetation tend to experience much more severe burns. In addition, the use of pre-fire vegetation indices, including the NBR, serves as a valuable predictive tool for assessing the potential severity of burns, it is imperative to acknowledge the pivotal role played by the formulation in this context and implementation of effective fire management strategies.

Table 6. Area Classification of Fire Severity by District

District	Moderate-Low		Moderate-High		High	
	Ha	%	Ha	%	Ha	%
Damar	40.86733	4.01%	0.18	0.88%	0	0.00%
Dendang	439.0203	43.12%	12.11082	59.52%	0.45	50.00%
Gantung	141.654	13.91%	2.086601	10.25%	0.18	20.00%
Kelapa Kampit	94.9157	9.32%	1.095469	5.38%	0	0.00%
Manggar	46.38585	4.56%	0.871622	4.28%	0.09	10.00%
Simpang Pesak	56.41302	5.54%	0.63	3.10%	0	0.00%
Simpang Renggiang	198.9519	19.54%	3.373822	16.58%	0.18	20.00%

This classification generates comprehensive areas associated with each severity classification across districts in East Belitung Regency, based on spatial analysis conducted at the district level to describe the severity of forest and land fires using ΔNBR. The prescribed fire severity category is classified as high severity, which is disseminated within the sub-district level. Dendang sub-district shows a contribution of 50.00%, followed by Gantung and Simpang Renggiang sub-districts, each contributing 20.00% Furthermore, Manggar sub-district contributes 10.00%.

3.4 Accuracy Test

In method validation, accuracy is often measured by combining truth and precision, although it is essentially a qualitative concept. Practitioners use the number of measurement errors to validate the method, comparing the calculated intervals with the acceptance criteria [17]. The total amount of data is the result of overlaying the total amount of hotspot data with the burned area data. The equation for the accuracy test is written as follows:

$$Accuracy = \frac{\sum Data\ Valid}{\sum Total\ Data} x\ 100\% \tag{5}$$

Where,
$\sum$ Valid Data = The number of burned area data that matches the hot spots
$\sum$ Total Data = The total spatial data of hot spots joined with the burned area

The accuracy assessment results show a value of 54.09%. Considering the 61 tested hotspot data samples and the 33 samples identified as overlapping data. Utilizing these values, pixel extraction corresponding to the burned area was carried out through the implementation of an if-then function algorithm. A pixel is classified as a burned area pixel when it meets two condition. Both condition are effective in distinguishing between burned and unburned land. Next, the burned areas undergo further examination to assess the fire's intensity.

Condition 1 : if NBRpost ij $\leq$ 0.456
Condition 2 : if ΔNBR ij $\geq$ -0.157

3.5 Confidence Class

The quality of a hotspot is determined by its confidence level. As the confidence interval increases, the probability that the area in question is a raging forest or land fire also increases. The MODIS Active Fire Product Manual presents the classification of confidence tiers as follows [18]:

Table 7. Confidence class of fire pixels.

Range	Confidence Class (C)
0%≤C <30%	Low
30% ≤C <80%	Moderate
80% ≤C ≤100%	High

The confidence class of the fire pixels for East Belitung was set at 54.09%. Based on this percentage, the value falls into the moderate category, so it is necessary to be aware of land and forest fires.

4 Conclusion

In order to comprehend and mitigate the consequences of forest and land fires, it is imperative to examine the NBR algorithm. This is a way to measure how intense a fire is using satellite images. The study was done in the East Belitung Regency area in 2024. In this study, 61 hotspots were identified on satellite images spread across six of the seven sub-districts in East Belitung Regency. The results of the analysis using the Differenced Normalized Burn Ratio (ΔNBR) index show a numerical range from -0.497122 to 0.819870. This range reflects the severity of fires that occur in the region. Based on the spatial analysis, Dendang sub-district has the most extensive area of severity, at 50.00% of all high levels of forest and land fire severity. The accuracy assessment results show a value of 54.09%. This value is based on 61 tested samples and 33 samples identified as overlapping data. People in East Belitung should be aware of forest and land fires because of this value.

Acknowledgement
Expressions of profound gratitude are extended to the esteemed Faculty of Geography of Universitas Muhammadiyah Surakarta (UMS), together with the UMS Center for Environmental Studies, for their exemplary efforts and unwavering commitment that has led to their appointment as the distinguished hosts of the highly anticipated International Conference on Geography and Disaster Management (ICGDM 2024), an integral part of the ISETH (2024). This landmark conference will serve as an important and influential platform for the widespread dissemination of groundbreaking research and innovative methodologies, with the overall goal of enhancing our understanding of the complex and diverse relationships that exist within the field of interdisciplinary geography, particularly as they relate to the interconnected themes of nature, technology, disaster management, and sustainable development. In this way, the conference seeks to foster an enriching dialogue among scholars, practitioners, and stakeholders, thereby contributing to the advancement of knowledge and practice in line with current challenges and future aspirations in these vital areas.

References

[1] MOEF, 2024, "Indonesia's total forest area" [Online]. Available https://www.menlhk.go.id/
[2] E, Endrawati, 2016, "Analisis data titik panas (Hotspot) dan areal kebakaran hutan dan lahan, Seminar Nasional Geomatika". *Direktorat Inventarisasi dan Pemantauan Sumber Daya Hutan, Ditjen Planologi Kehutanan dan Tata Lingkungan Kementerian Lingkungan Hidup dan Kehutanan*, pp. 273.
[3] Y. Suwarno, 2013, "Pemetaan Lahan Kritis Kabupaten Belitung Timur Menggunakan Sistem Informasi Geografis," *Majalah Ilmiah Globe*, vol. 15, no. 1
[4] E. Abakumov, 2020 "Effect of the wildfires on sandy podzol soils of Nadym region, Yamalo-Nenets Autonomous District, Russia," *Applied and Environmental Soil Science*, pp. 1–8.
[5] F. A. Kurnianto, E. A. Nurdin, E. I. Pangastuti, & S. Bella, 2024 "Monitoring Mining Impact for Geosites Using Time Series NDVI and Runoff in the East-ern Part of Southern Java Mountains, Indonesia" *Forum Geografi*, vol. 38, no. 2, pp. 291-301. doi: 10.23917/forgeo.v38i2.2996.
[6] H. Z. Hadibasyir, 2023 "Assessing performance of modified spectral indices as land surface temperature indicators in tropical urban areas" *IOP Conference Series Earth and Environmental Science*, vol. 1190, no. 1, pp. 012005.
[7] A. Kanav, & J. Kumar, 2024 "A GIS-Based Analysis of the Public Open Space Distribution and Accessibility in Gurugram City, India" *Forum Geografi*, vol. 38, no. 2, pp. 266-279. doi: 10.23917/forgeo.v38i2.4759.
[8] D. I. Rianti, & D. N. Sari, 2024 "Analysis of changes in vegetation density Index in Surakarta City 2018-2022 using LANDSAT-8 OLI/TIRS satellite imagery," *IOP Conference Series Earth and Environmental Science*, vol. 1357, no. 1, pp. 012033.
[9] A. Jenerowicz, D. Wierzbicki, & M. Kedzierski, 2023 "Radiometric Correction with Topography Influence of Multispectral Imagery Obtained from Unmanned Aerial Vehicles," *Remote Sensing*, vol. 15, no. 8, pp. 2059.
[10] H. Z. Hadibasyir, 2020, "Pemetaan Kondisi Ekologi Perkotaan Skala Mikro Menggunakan Citra Landsat 8 di Kota Semarang" *LaGeografia*, vol. 18, no. 3, pp. 209.
[11] H. Z. Hadibasyir, et al., 2025, "Assessing the performance of NDVI as a proxy for land surface temperature using linear and nonlinear models in different intensity of anthropogenic activities: a case study of Yogyakarta City, Indonesia," *IOP Conference Series Earth and Environmental Science*, vol. 1444, no. 1, pp. 012014.
[12] E. I. Putra, et al., 2022, "Assessing the severity of forest fire in Sungai Buluh Protected Peat Forest, Jambi," *IOP Conference Series Earth and Environmental Science*, vol. 959, no. 1, pp. 012059.
[13] A. Vatresia, A. et al., 2022 "Mining Fire Hotspots Over Nusa Tenggara and Bali Islands," *Indonesian Journal of Forestry Research*, vol. 9, no. 1, pp. 73–85.
[14] M. J. Van Gerrevink, & S. Veraverbeke, 2021 "Evaluating the hyperspectral sensitivity of the differenced normalized burn ratio for assessing fire severity," *Remote Sensing*, vol. 13, no. 22, pp. 4611.
[15] E. Alcaras, et al., 2022, "Normalized Burn Ratio Plus (NBR+): a new index for Sentinel-2 imagery," *Remote Sensing*, vol. 14, no. 7, pp. 1727.

[16] F. S. Afina, L. Syaufina, & I. S. Sitanggang, 2022, "Forest and Peatland Fire Severity Assessment at Siak Regency, Riau Province using Sentinel-2 Imagery," *Jurnal Pengelolaan Sumberdaya Alam Dan Lingkungan (Journal of Natural Resources and Environmental Management)*, vol. 11, no. 4, pp. 621–630.

[17] C. Yardin, C, 2015 "Exactitude et intervalles statistiques en validation de méthode," *16th International Congress of Metrology*, vol. 19, p. 02016.

[18] L. Giglio, 2015 "MODIS Collection 6 Active Fire Product User's Guide Revision A. Department of Geographical Sciences" University of Maryland.

IOP Conf. Series: Earth and Environmental Science **1503** (2025) 012028 doi:10.1088/1755-1315/1503/1/012028

Comparison of Landsat 8 and Landsat 9 Satellite Imagery for Land Cover Mapping in the Kendal - Pekalongan Coastal Area

Trida Ridho Fariz[1*], Amnan Haris[1], Nana Kariada Tri Martuti[2], Norma Eralita[3], Luthfi Hanum Saputri[1,3], Gilang Syahbananto[1], Cintiya Egi Purwadi[1], Zahra Rafidah[1], Sheeny Az-Zahra[1]

[1] Environmental science, Universitas Negeri Semarang, Indonesia
[2] Biology, Universitas Negeri Semarang, Indonesia
[3] Science Education, Universitas Negeri Semarang, Indonesia

[*]E-mail: trida.ridho.fariz@mail.unnes.ac.id

Abstract. This study evaluates land cover mapping using Landsat 8 and Landsat 9 satellite imagery, focusing on the coastal regions of Pekalongan Regency and Kendal Regency in Indonesia, which face significant environmental challenges due to climate change. Utilizing machine learning algorithms from Google Earth Engine (GEE), the research employs a supervised classification technique based on Random Forest (RF) to assess land cover. The analysis reveals overall accuracy values of 0.75 for Landsat 8 and 0.82 for Landsat 9, with Kappa values of 0.68 and 0.78, respectively. The accuracy of individual land cover types varies, with water bodies achieving the highest accuracy and agricultural land the lowest. The study suggests prioritizing Landsat 9 imagery for medium-scale mapping.

1. Introduction

Land cover is the physical appearance of the earth's surface from a biophysical perspective, such as buildings, vegetation, and so on. Understanding land cover conditions is essential to understanding the interaction between humans and the environment [1]. This is because changes in land cover can be caused by humans such as urbanization and policies, and their impacts will affect human activities [2]. Therefore, land cover data is very important not only for natural resource management and addressing environmental problems but also for the basis of spatial planning policies, coastal and watershed management and disaster risk reduction [3–5]. This further strengthens the urgency of representative mapping for sustainable development.

Remote sensing is a critical tool for land cover analysis, and since 1972, Landsat satellites have provided high-quality multispectral imagery with medium spatial and temporal resolution, making them a primary data source for environmental monitoring and land cover studies [6]. Landsat 8, launched in 2013, has been widely used in various ecosystems, including coastal, mountainous, urban, and rural areas [4,7,8]. On September 27, 2021, Landsat 9 was launched to complement Landsat 8, offering similar spatial and temporal resolutions but with a significant improvement in radiometric resolution—14 bits compared to Landsat 8's 12 bits [9,10]. This higher radiometric resolution allows Landsat 9 to distinguish 16,384 intensity levels, compared to Landsat 8's 4,096, enabling finer detection of subtle spectral variations. This capability is particularly crucial in coastal environments, where accurate differentiation of features such as water bodies, wetlands, and vegetation is essential for effective management and disaster risk reduction.

Accurate land cover mapping is vital for coastal management, as it supports sustainable spatial planning, ecosystem monitoring, and disaster mitigation. Coastal zones are dynamic environments where subtle spectral differences, such as those between turbid water, vegetation, and urban areas, can significantly impact decision-making. The higher radiometric resolution of Landsat 9 is expected to enhance the accuracy of land cover classification, particularly in challenging coastal areas where spectral overlap often leads to misclassification. Despite this potential, comparative studies between Landsat 8 and Landsat 9 remain limited [10], leaving a critical gap in understanding how these datasets can be best utilized collaboratively for improved environmental monitoring and policy formulation.

This study aims to compare land cover mapping using Landsat 8 and Landsat 9 imagery, employing machine learning algorithms on Google Earth Engine (GEE). The study focuses on the coastal area from Pekalongan Regency to Kendal Regency in Indonesia, a region facing significant environmental challenges, including land cover changes and coastal disasters exacerbated by climate change [11–14]. This area is also downstream of priority watersheds, such as the Bodri, Blorong, and Kupang watersheds, making accurate land cover mapping essential for integrated water resource management. By comparing the performance of Landsat 8 and Landsat 9, this study seeks to provide insights into their respective strengths and limitations, contributing to more effective land cover monitoring and sustainable spatial planning in coastal zones.

2. Methodology

2.1. Study location and data

This study is located in the coastal area of Central Java Province, Indonesia. This location includes several districts in Kendal Regency, Batang Regency, Pekalongan City and Pekalongan Regency (Figure 1). The data used are Landsat 8 and Landsat 9 satellite images accessed from GEE. The dataset is Surface Reflectance which is ready to use with an acquisition time of around August 2024. The selection of the acquisition time aims to obtain images with minimal cloud cover, because during that period the study location is in the dry season. Furthermore, both datasets are processed by cloud masking to obtain free cloud cover images in GEE [8]

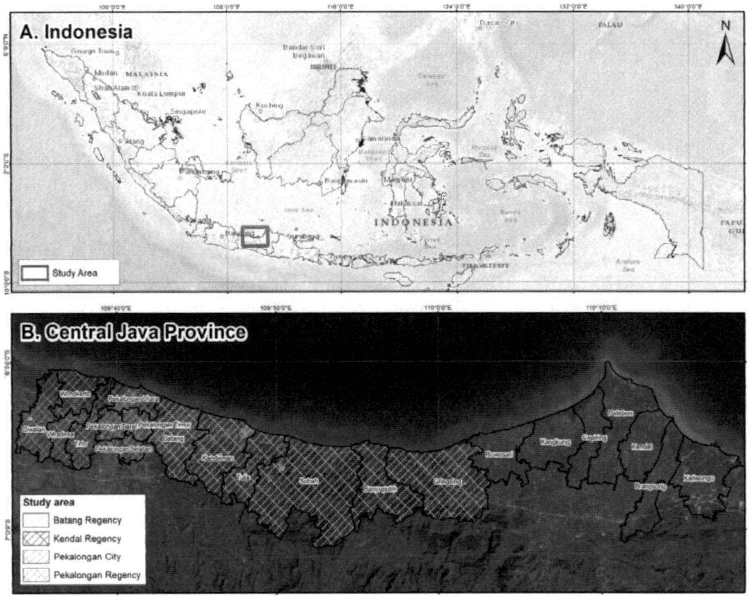

Figure 1. Study area

2.2. Land cover mapping and accuracy assesment

Land cover mapping uses a supervised classification technique based on RF (Random Forest) machine learning algorithms from GEE. RF was chosen because this machine learning algorithm has the best accuracy in several land cover mapping studies. RF is non-linear so it is suitable for application in areas with fragmented land cover spatial distribution [4,15,16].

Table 1. Description of land cover classes

Land cover classes	Code	Description
Water body	WB	Water body includes several land covers such as lakes, rivers, reservoirs, fishponds and flood areas
Built-up area	BA	Built-up areas include artificial impervirous surfaces such as urban settlements, rural settlements, roads and industrial areas
Agricultural land	AL	Agricultural land includes paddy fields and agricultural land mixed with bushes
Vegetation	VG	Vegetation includes land cover vegetation not agricultural land. The majority are vegetation in the form of trees such as forest plantations, mangrove forests, and mixed gardens
Bare soil	BS	Bare soil includes open land or land with little vegetation such as grass, bare soil also includes sand on the beach and land clearing

The mapping process begins with the collection of training samples, with a total of 320 training samples. The total training samples are divided into each land cover class. The land cover classes in this study are water body, built-up area, agricultural land vegetation, and bare soil (Table 1). After the training samples have been collected, the classification process can be carried out to produce a land cover map.

The next stage is collecting validation samples with a total of 230 samples. Validation samples are used for assessing the accuracy of the classification. The accuracy assessment is performed using

IOP Conf. Series: Earth and Environmental Science **1503** (2025) 012028 doi:10.1088/1755-1315/1503/1/012028

the kappa index, a widely accepted method for evaluating the accuracy of image classification [17,18]. The kappa index incorporates both Producer Accuracy (PA) and User Accuracy (UA), considering errors of omission and commission [19]. PA represents the proportion of correctly classified pixels relative to the total number of pixels in a particular class, as defined in the error matrix. This metric indicates the accuracy of classifying reference pixels and reflects omission errors. In contrast, UA measures the ratio of correctly classified pixels to the total number of pixels assigned to that class, providing an assessment of the map's reliability by showing how well it represents actual conditions [20]. The process of mapping, from training sample collection to accuracy assessment, was applied to both datasets.

3. Results and discussion

Landsat 8 and Landsat share the same spatial resolution and wavelength range but differ in their radiometric resolution. Landsat 9 provides a higher radiometric resolution of 14 bits, whereas Landsat 8 is limited to 12 bits. As a result, Landsat 9 can identify 16384 colors, compared to only 4096 colors for Landsat 8, allowing for a broader dynamic range in each spectral band of Landsat 9 [21].

Figure 2. Visual comparison of Landsat 8 and Landsat 9 satellite imagery in false red composite

Several studies have suggested that higher radiometric resolution helps avoid saturation in very bright or dark areas, allowing for more detail in features such as water [21]. This was also found in this study, where Landsat 9 imagery outperformed Landsat 8 in identifying dark water bodies (Figure 2). Likewise, the appearance of water bodies associated with agricultural land was clearly identified by Landsat 9 imagery.

Table 2. Accuracy assessment results

Sattelitte imagery	Assesment	Land cover classes					OA	KA
		WB	BA	AL	VG	BS		
Landsat 8	UA	0.92	0.74	0.56	0.72	1	0.75	0.68
	PA	1	0.96	0.66	0.56	0.47		
Landsat 9	UA	0.94	0.73	0.72	0.85	1	0.82	0.78
	PA	1	0.98	0.76	0.66	0.66		

The comparison between Landsat 8 and Landsat 9 satellite images in land cover mapping is by comparing the value of the accuracy assessment. Accuracy assessment (Table 2) shows good results on both maps. The overall accuracy values obtained were 0.75 (Landsat 8) and 0.82 (Landsat 9). Meanwhile, the Kappa accuracy values obtained were 0.68 (Landsat 8) and 0.78 (Landsat 9). The accuracy values of producers and users range from 0.56 to 1.00. The lowest accuracy assessment value is on agricultural land, and the highest is on water bodies.

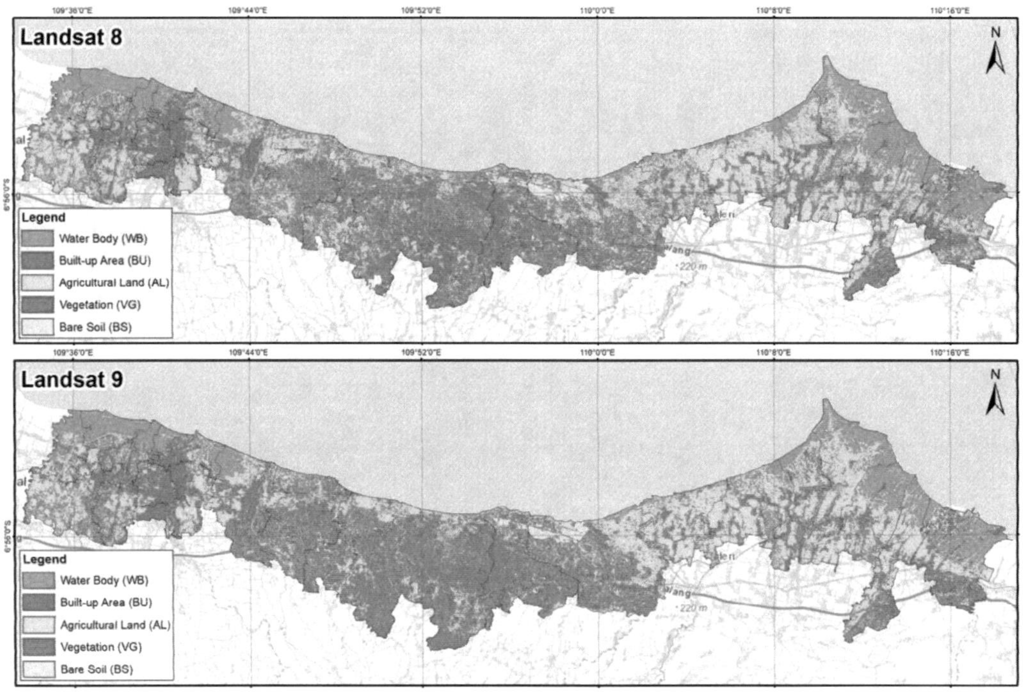

Figure 3. Land cover map from Landsat 8 and Landsat 9 imagery

The lowest accuracy in this study is in the agricultural land class. Misclassification is the entry of vegetation class into agricultural land, and vice versa. If seen on the land cover map, there is no significant difference between the two land cover maps (Figure 3). Land cover maps from Landsat 8 and Landsat 9 tend to have similar patterns and spatial distribution. However, if seen in more detail, misclassification on the land cover map from Landsat 8 tends to be more than from Landsat 9. An example is the entry of water bodies into the agricultural land class (Figure 4). In Figure 4, it is also seen that on the land cover map from Landsat 8 there are also many misclassifications where agricultural land is classified into vegetation. This proves that the difference in radiometric resolution between Landsat 8 and Landsat 9 provides a difference in the accuracy of land cover mapping results. This study can recommend that for medium-scale land cover mapping in recent years, Landsat 9 satellite imagery can be prioritized as the main data.

Figure 4. Comparison of land cover mapping results from Landsat satellite imagery with VHR (Very High-Resolution) imagery

This study has several limitations such as only mapping five types of land cover. Land cover types can be improved such as distinguishing rural settlements, urban settlements, industry, agircultural land types as well as dryland forests and mangrove forests. This is because the types of land cover that are not included may have different performance [21]. This study is also limited to the dry season, and there is no comparison for the rainy season or even land cover change studies. The richer the land cover mapping study on the Kendal - Pekalongan coast will add to the knowledge base as a basis for management policies in the area. This is considering the many developments in the area such as the expansion of Kendal Industrial Park and the construction of Batang Industrial Park which may be driving factors for land cover changes in the study location [22,23].

4. Conclusssion

The overall accuracy values for Landsat 8 and Landsat 9 were 0.75 and 0.82, respectively, with Kappa values of 0.68 and 0.78. These improvements highlight the potential of Landsat 9 for medium-scale land cover mapping. Given its higher radiometric resolution, Landsat 9 provides more accurate differentiation of land cover types, particularly water bodies, making it an ideal choice for recent land cover mapping. Coastal managers and policymakers can leverage Landsat 9's enhanced accuracy for more reliable environmental monitoring, land use planning, and habitat conservation. The data can support better tracking of coastal changes, including urban expansion and agricultural development, allowing for informed decision-making in vulnerable coastal regions. Additionally, it can improve disaster mitigation efforts by providing clearer assessments of tidal flood-prone areas and other environmental threats. However, future studies should expand land cover classes and incorporate multi-seasonal data to further enhance the utility of these datasets for long-term coastal management and disaster preparedness.

References

[1] Adipka A 2018 Analysis of Changes in Rice Field Land Use in Metro City Between 2000-2015

[2] Novianti E 2021 Impact of Land Use Changes From Agricultural To Non-Agricultural On Loss Of Agricultural Production And Food Availability In Urban And Suburban Areas In Bantul District

[3] Ojeda D F 2021 Flood and inundation hazard mapping of the town of Loreto, Corrientes Province *Reun. Comun. C ientíficas y Tecnológicas la UNNE*

[4] Fariz T R, Suhardono S and Fadhilla S 2024 Rapid mapping of land cover changes in tropical watershed in Indonesia using Google Earth Engine *IOP Conference Series: Earth and Environmental Science* vol 1414 (IOP Publishing) p 12077

[5] Ngana F R 2024 Mapping of Agricultural Land Use in Kupang City using Remote Sensing *AGRICA* **17** 96–104

[6] Susantl N I, Sanjoto T B and Tjahjono H 2012 Remote sensing application for land use change analysis 2002-2011 in Juana River Basin *Geo-Image J.* **1**

[7] Pramono C L, Meilani F, Reza A D and Foresty R S 2024 The impact of changes in land use on land surface temperature in Papringan Village, Kaliwungu District, Semarang Regency based on remote sensing *Crit. Issue Sustain. Futur.* **1** 1–16

[8] Sidiq W A B N, Fariz T R, Saputro P A and Sholeh M 2024 Risk Screening Environmental Indicators Model Change Based on Spectral Transformation Around New Yogyakarta International Airport *Ecol. Eng. Environ. Technol.* **25**

[9] Gross G, Helder D, Begeman C, Leigh L, Kaewmanee M and Shah R 2022 Initial cross-calibration of Landsat 8 and Landsat 9 using the simultaneous underfly event *Remote Sens.* **14** 2418

[10] You H, Tang X, Deng W, Song H, Wang Y and Chen J 2022 A study on the difference of LULC classification results based on Landsat 8 and Landsat 9 data *Sustainability* **14** 13730

[11] Khasanah E U 2023 Mapping of Coastline Changes and Mangrove Area in Kendal Regency, Central Java Province

[12] Nurhidayatullah D and Sugiyanto S 2023 Climate Change as Governmentality: A Practice Regime for Adaptation in Pekalongan *J. Gov. (Kajian Manaj. Pemerintah. dan Oton. Daerah)* **8**

[13] Purbani D and Borneo B B 2021 Perubahan Penggunaan Lahan dan Garis Pantai di Indramayu, Cirebon, Pekalongan, dan Semarang *Efektivitas Bangunan Pelindung Pantai Murah & Ramah Lingkungan* (IPB Press)

[14] Samudra F R and Jumadi S S 2024 Analysis of the Influence of Built-up Land Development on Surface Temperature Increases Related to the Urban Heat Island Phenomenon in Pekalongan City in 2013 and 2023

[15] Yusuf A R 2024 Land Cover Prediction In The Gowa-Makassar Border Area Using The Google Earth Engine Platform With Random Forest

[16] Fariz T R, Setyowati D L, Damayanti M, Rahmawati D and Nurhidayati E 2024 Population mapping in Kudus Regency using spectral built-up index with Google Earth Engine *Indones. J. Earth Hum.* **1** 64–75

[17] Oktaviah N D 2022 Utilization of SPOT 6 Imagery for Mapping Land Use in South Purwokerto District, Banyumas Regency

[18] Nurwendayanti L 2015 Mapping of Land Cover Changes in BKPH Rejuno FMU Saradan 1996-2014 Using Landsat Imagery

[19] Wijaya J and Tjahjono H 2024 Analysis of the Suitability of Land Use in Kendal Regency to the Regional Spatial Planning (RTRW) of Kendal Regency *Geo-Image J.* **13** 91–100

[20] Afasel D, Purnamasari R and Edwar E 2023 Land Cover Classification Using Supervised Machine Learning on Satellite Images Using Google Earth Engine *eProceedings Eng.* **9**

[21] Xu H, Zhang F, Li W, Shi J, Johnson B A and Tan M L 2024 Spatial-temporal pattern of change in production-living-ecological space of Nanchong City from 2000 to 2020 and underlying factors *Environ. Monit. Assess.* **196** 94

[22] Wijanaputri K O, Hamdani A A, Permata E A N, Magfiroh Y, Sihaloho E E, Suryaningrum I and Heriyanti A P 2024 Analysis of land changes on the coast of Batang Regency *Proceeding Semnas IPA* pp 199–206

[23] Prasetyo Y, Bashit N and Baihaqi H F 2021 Correlation Analysis of Kendal Industrial Park Development Area and Green Open Space *J. Kelaut. Trop.* **24** 102–12

Spatial Modeling for Sustainable Tourism Development: Evaluating Optimal Locations for Tourism Projects on the Southern Tetouan Coast, Morocco.

Mohammed ELAANZOULI[1*], Hamza AIT ZAMZAMI[1], Ayoub ZAHRANI[1], Oumaima BOUMEAZA[2], Abdelhamid MOULOUADE[3], Jamila SAIDI[1], Taieb BOUMEAZA[1]

[1] Laboratory of Dynamics of Spaces and Societies (LADES), Department of Geography, Faculty of Arts and Humanities. HASSAN II University of Casablanca, Morocco.
[2] LAGAGE Laoratory, Faculty of sciences Ben Msik, Hassan II University of Casablanca, Morocco.
[3] Laboratory of Recomposition of Space and Sustainable Development (REDD), Department of Geography, Faculty of Arts and Humanities, Chouaib Doukkali University of El Jadida, Morocco.

*E-mail : mohammed.elaanzouli-etu@etu.univh2c.ma

Abstract. This study examines the use of Geographic Information Systems (GIS) for spatial modeling to determine optimal locations for tourism projects along the southern Tetouan coast of Morocco. The methodology combines Multi-Criteria Decision Analysis (MCDA) with preference theory to assess natural, economic, and social factors influencing sustainable tourism development. Ten criteria, including proximity to the coast, tourist attractions, road infrastructure, elevation, forest areas, and social facilities, were evaluated. These factors were classified according to preference levels, ranging from "excellent" to "unsuitable," reflecting their importance in establishing viable tourism sites. Results reveal that areas closer to the coast, population centers, and main roadways are most suitable for tourism development, supporting both attractiveness and sustainability. This model serves as a strategic tool for stakeholders, guiding sustainable tourism initiatives while balancing regional economic and environmental considerations.

Keywords: Geographic Information Systems (GIS), Sustainable Tourism Development, Spatial Suitability Modeling, Multi-Criteria Decision Analysis (MCDA), Optimal Location Analysis.

1. Introduction

Spatial suitability modeling using Geographic Information Systems (GIS) is essential for identifying optimal locations for developments, particularly in tourism. These models combine geospatial data to highlight areas meeting specific criteria, aiding in site selection for tourism projects [1]- [2].

GIS enables effective planning by predicting suitable locations for tourism facilities while ensuring sustainability [3]. This study uses GIS to map tourism potential along the southern Tetouan coast, an area rich in natural and cultural attractions. GIS-based modeling ensures tourism development aligns with environmental and social sustainability [4], balancing economic benefits with resource preservation [5].

By applying GIS-based multi-criteria decision analysis (MCDA) and evaluating ten criteria—such as proximity to coastlines, elevation, and access to infrastructure—the research explores how spatial modeling enhances the strategic placement of sustainable tourism projects.

The findings offer a data-driven framework for decision-makers, supporting the creation of sustainable tourism that aligns with the region's ecological and economic goals [6].

1.1. Study area

The research area, situated in the northern part of Morocco in the Tangier-Tetouan-Al Hoceima region, includes the coastal area along the resources and ecosystems. From an administrative perspective [7], the region is within communal boundaries, which helps with governance but also presents difficulties in coordinating resource management, conservation, and Mediterranean Sea from Martil to Oued Laou. Various landscapes such as rocky cliffs and sandy coastlines, along with numerous waterways, define the distinctive ecological balance of this region. The Mediterranean climate in the area encourages diverse wildlife, making it an important location for environmental research. The closeness to the coast has led to an increase in urbanization and tourism growth, especially near Martil, putting strain on local urban planning.

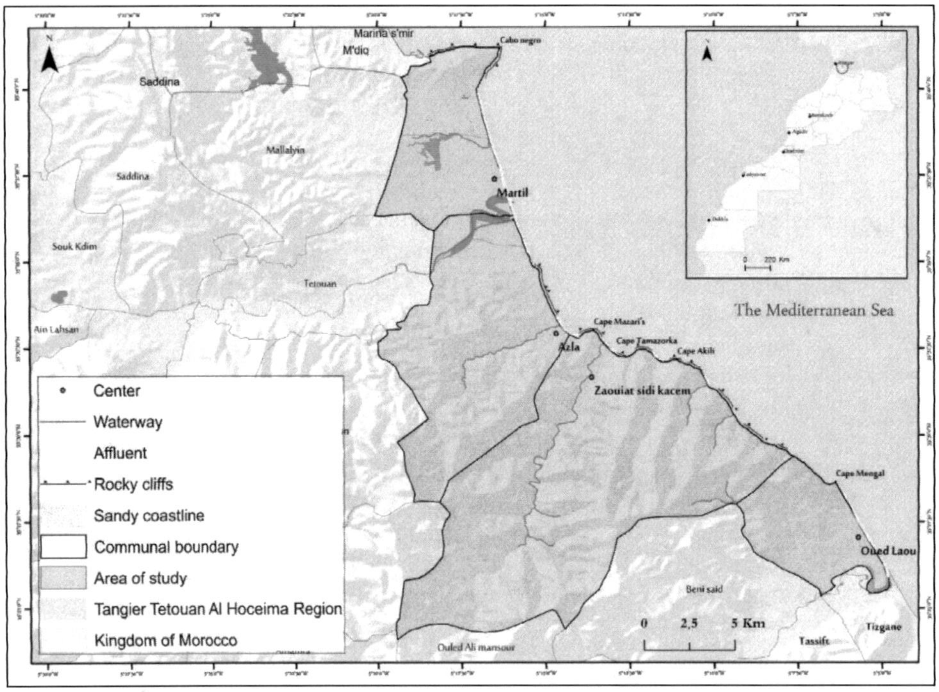

Figure 1: Map of the localization of the study area, (Source: Cartographic work based on the 2015 regional division).

2. Methodology

Multi-Criteria Decision Matrices are used to evaluate options based on several heterogeneous criteria in spatial modeling. These criteria can help identify the best locations for tourism projects based on a set of natural, economic, and social factors. This model is based on preference theory, which relies on analyzing relative preferences between options rather than absolute values. This means it focuses on how one option compares to another based on a set of criteria, involving a process of eliminating unacceptable options and selecting the best one through comparisons among different criteria.

This theory is founded on several principles, including the relative ranking of options based on a set of criteria, determining whether a specific option outperforms another based on multiple evaluations. Additionally, it involves criterion analysis, which assesses options based on how well they meet the established criteria, using a scale to preference one option over another (AHP).

2.1 Defining Criteria

Spatial modeling has traditionally focused on economic efficiency, often falling short in producing the best decisions until the emergence of multi-criteria analysis [8]. This approach allows for the consideration of a range of criteria and their combination to identify optimal options. We have identified ten different criteria that effectively contribute to regional tourism development, each with varying standard weights. Some of these criteria exceed the average effective standard weight of 6.88%, while others fall below this average. This assessment takes into account two main dimensions related to achieving sustainable tourism development goals, emphasizing the importance of natural resources and their utilization to foster economic and social development.

To explain these weights, the following reasoning can be provided, as shown in Table 1 (on the next page):

- Coastline proximity (20%): Key for eco-tourism and unique marine environments. The coastal law in Morocco (12-81) defines a distance of 100 meters from the highest tide mark towards the land as areas belonging to the maritime domain, where no planning is permitted.
- Proximity to tourist attractions (18%): Vital for attracting tourists by being close to hotspots.
- Proximity to roads, forests, and water bodies (12%, 10%, 12%): Important for accessibility and integration with nature.
- Elevation (8%): Adds value by offering a unique landscape and atmosphere.
- Proximity to basic services (6%) and population centers (4%): Important for logistics but less crucial in sustainable tourism.
- Topographic slope (5%): Less significant, mainly for safety and design.
- Proximity to recreational port (5%): Reduced significance due to distance, but may still be valuable for specific activities.

The spatial preference ratings for establishing tourism projects have been divided into five levels: Excellent (value of 9), Good (7), Medium (5), Low (3), and Unsuitable (1) [9]. These ratings vary based on the specific criterion being analyzed, as the importance of the value differs according to the criterion.

2.2 Calculating criteria Weight

In the beginning we Calculated the average values in each row of the normalized matrix (Table2), The result is the relative weight of each criterion, The consistency ratio (CR) must be calculated to ensure that the comparisons are logical:

Calculate the Consistency Index (CI) and the Consistency Ratio (CR) using using the formulas:

$$CI = \frac{\lambda\max - n}{n - 1} \qquad\qquad CR = \frac{CI}{RI}$$

Where:

λmax is the largest eigenvalue of the matrix.
n is the number of criteria.
RI is the random consistency index (depending on **n**
If **CR** < **0,1** the comparisons are acceptable

IOP Conf. Series: Earth and Environmental Science **1503** (2025) 012029 doi:10.1088/1755-1315/1503/1/012029

Table 1: The criteria used for selecting the best locations for establishing tourism projects

Order	Criterion	Proximity or Distance	Excellent (9)	Good (7)	Moderate (5)	Weak (3)	Not Suitable (1)
a	Proximity to coastline	Proximity (meter)	100 - 1500	3500 - 1500	3500 - 12500	> 12500	< 100
b	Proximity to Tourist Attractions	Proximity (meter)	< 100	100 - 1000	1000 - 4000	> 4000	-
c	Proximity to Waterways	Proximity (meter)	> 2500	1500 - 2500	1500 - 500	500 - 100	100
d	Proximity to Roads	Distance (meter)	< 500	1000 - 500	1000 - 2500	> 2500	-
e	Proximity to Forests	Distance (meter)	> 300	300 - 150	150 - 50	< 50	-
f	Elevation	Proximity (meter)	< 100	100 - 200	200 - 500	500 - 1000	-
g	Proximity to Services	Proximity (meter)	< 2500	2500 - 5000	5000 - 7500	> 75000	-
h	Slope	Proximity (degree°)	0° - 3°	3° - 6°	6° - 12°	12° - 14°	> 24°
i	Proximity to Marina	Proximity (meter)	< 1000	1000 - 5000	5000 - 10000	> 10000	-
j	Proximity to Centers	Proximity (meter)	< 150	150 - 750	750 - 1500	> 1500	-

Table 2*: Matrix for determining criteria weights By AHP method*

Order	Criteria	1	2	3	4	5	6	7	8	9	10	Weight (%)
a	Coastline	0.367	0.328	0.349	0.485	0.256	0.304	0.217	0.235	0.226	0.235	20%
b	Proximity to Tourist Attractions	0.122	0.162	0.263	0.209	0.154	0.13	0.157	0.161	0.157	0.161	18%
c	Proximity to Waterways	0.073	0.141	0.029	0.070	0.103	0.154	0.118	0.078	0.097	0.118	12%
d	Proximity to Roads	0.073	0.122	0.209	0.162	0.157	0.217	0.154	0.118	0.157	0.157	12%
e	Proximity to Forests	0.073	0.141	0.029	0.023	0.256	0.13	0.026	0.118	0.157	0.118	10%
f	Elevation	0.073	0.078	0.154	0.026	0.154	0.026	0.078	0.097	0.157	0.157	8%
g	Proximity to Services	0.061	0.078	0.026	0.078	0.065	0.154	0.039	0.118	0.157	0.078	6%
h	Slope	0.073	0.026	0.103	0.026	0.026	0.051	0.022	0.118	0.157	0.078	5%
i	Proximity to Marina	0.073	0.141	0.141	0.023	0.022	0.032	0.026	0.078	0.097	0.022	5%
j	Proximity to Centers	0.052	0.032	0.029	0.023	0.022	0.078	0.065	0.032	0.032	0.032	4%

2.3 Indicators of tourist arrivals increase opportunities for tourism investment

National and international investment constitutes an integral part of the tourism industry. It is also considered a productive process by providing tourism offerings, creating job opporunities for local population, and encouraging investment. The steady growth in tourist arrivals, along with the increasing number of overnight stays along the tetouan coast.

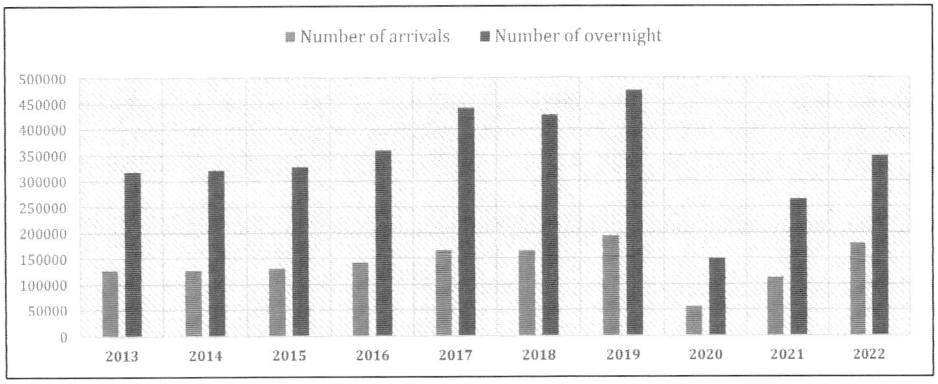

Figure2: Evolution of arrivals and overnight stays in classified hotels between 2013 and 2022

2.4 Motivations for Tourism Investment in the Study Area

Tourism investment plays an important role in revitalizing tourism and driving local development and sustainability, enhancing its economic status and elevating it. The study area offers a favorable environment for tourism investment due to several factors, as indicated by the chart.

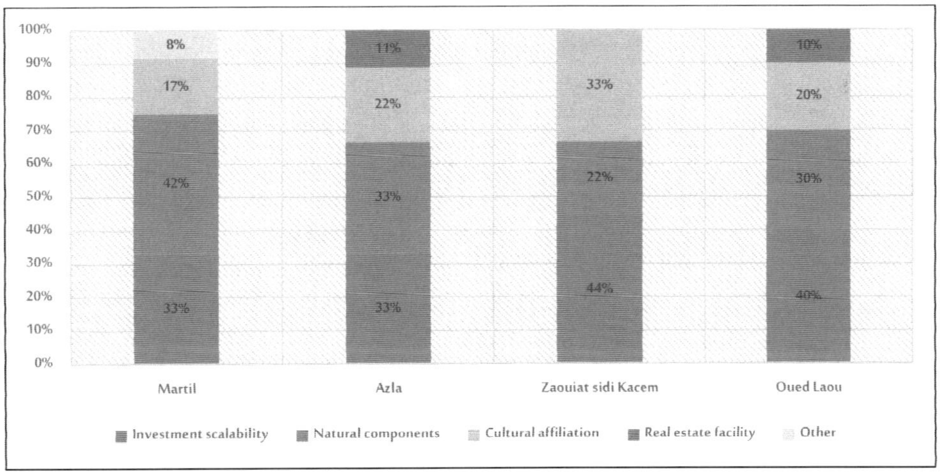

Figure3: Motivations for tourism investment (30 forms)

Based on Figure 3, the key motivation for investors in tourism projects is the perceived potential for growth, accounting for 37% of responses. Natural resources and attractions drive 32% of tourism investment, particularly factors like proximity to the sea or to popular tourist areas. Cultural affinity is another significant factor, with 23% of investors choosing locations where they have cultural ties, which also reduces land acquisition costs if they already own property.

Given the lack of land-use flexibility in some coastal areas due to regulations protecting the coast, coupled with real estate speculation that drives up land prices, some local investors opt to create projects in mountainous areas away from the coast [10]. The success of such projects in these regions relies on creating new tourist trails that incorporate rural areas and implementing tourist guides to help some of these villages benefit from tourism revenue.

The tourism growth and attraction of tourism projects rely on several factors, including the availability of basic services such as water and electricity, as well as the security and stability that Morocco offers (**Figure4**), especially in tourist areas, as confirmed by the sample of tourists we studied. Additionally, the national roads serve as a vital artery within the area of study. Furthermore, the potential to harness renewable energy sources like wind and solar power in mountainous regions, which remain distant from urban centers, presents an opportunity. This isolation, in fact, offers the chance to create sustainable eco-tourism projects that provide tourists with new, immersive experiences.

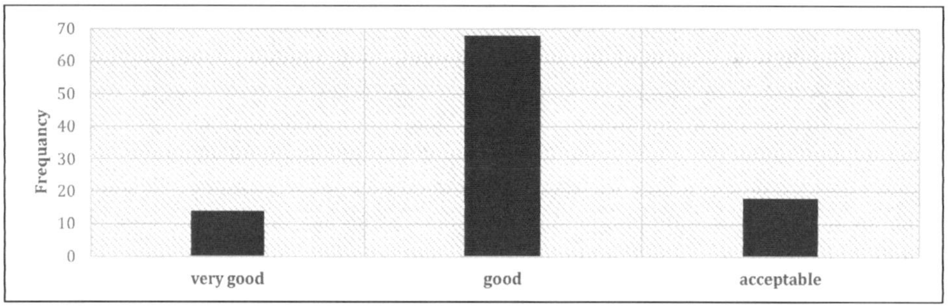

Figure4: The tourists views on the security aspect in the study area (300 forms)

It's clear that these factors can play a crucial role in enhancing the sustainability and attractiveness of tourism projects.

2.5 Modeling approach diagram

Given the growing interest in sustainable development and the increasing focus on tourism as an economic driver, particularly between 2013 and 2022, alongside the rising tourism investments, it has become critical to rethink how to achieve a balanced development that integrates both environmental and economic goals. As tourism planning gains prominence at various levels—central, regional, and local—throughout territorial communities, the focus must shift towards preserving natural resources while fostering economic growth.

To achieve this, sustainable planning must be rooted in scientific approaches, such as using geographic information systems (GIS), to identify the most suitable locations for tourism projects. This approach helps minimize environmental risks and ensures that the development of tourism is aligned with long-term sustainability goals, taking into account the needs of future generations.

After identifying the criteria to be used in constructing a suitable spatial model consisting of ten criteria, we subjected these criteria to cartographic analysis and logically reclassified them based on field data as well as legal and technical information available through Geographic Information Systems (GIS) programs, following the subsequent process.

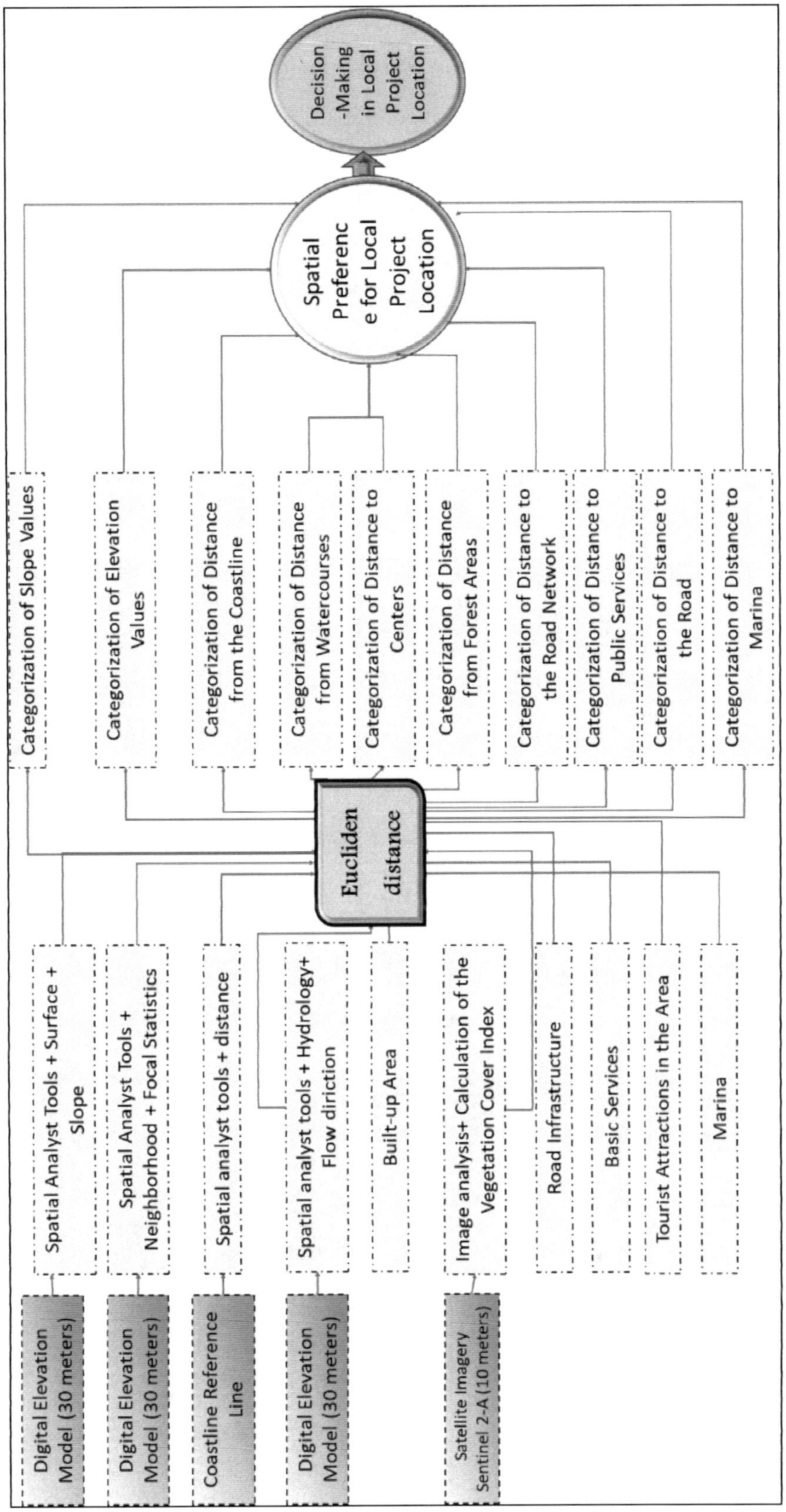

Figure5: The variables used in spatial modeling within geographic information systems for establishing a tourism project.

3. Results and Discussion

Creating job opportunities, reducing unemployment rates, improving individual incomes, and eliminating poverty or social vulnerability are key priorities in tourism development. If a destination experiences a large influx of tourists but fails to generate returns for its investment environment or improve the living conditions of local residents, it risks being far from achieving sustainability for its resources, and could eventually experience decline, in line with the tourism development theory.

Tourism investments directly contribute to job creation. Each tourism project within our area employs at least one local worker, with a total of 48% from the local community and 40% from outside the area where the project is located. Foreign workers represent only 12%.

Additionally, local residents engaged in freelance jobs related to tourism benefit, indirectly, from the tourism activity, leading to improvements in individual income. According to the survey, 60% of the participants reported that their daily income increases during the tourism season, while 25% noted only slight improvements. However, 15% indicated that they do not see any increase in their income. On the other hand, they face rising prices and pressure on social services.

Given the economic and social significance of the tourism sector, it is essential to ensure its sustainability by preserving the natural assets of the Tetouan coast and recommending the best locations for sustainable tourism projects.

3.1. Extracting Spatial Preference Results for Establishing Tourism Projects in the Study Area

In building an outstanding model, a set of elements, as previously identified, comes into play. However, the degree of influence of each element varies from one to another. Analyzing each element in isolation provides only preliminary results and cannot definitively determine the best locations for establishing tourism projects. Nevertheless, these analyses serve as a foundational basis for developing the model.

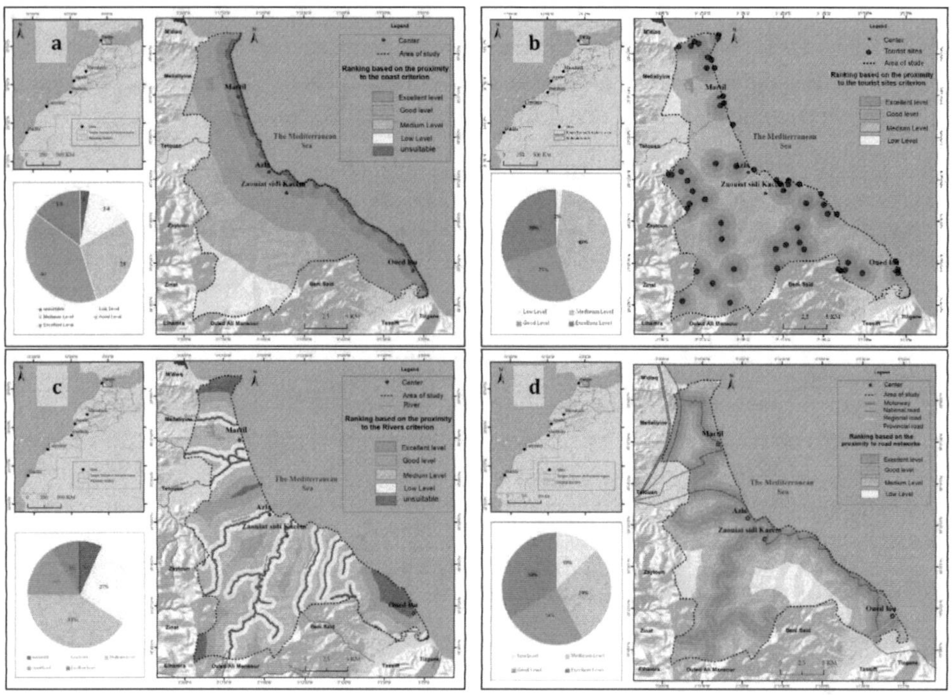

Figure 6: a-Coast proximity map, b-proximity to the tourist sites map, c- proximity to the rivers map, d-Roads proximity map.

a. the Proximity to Coast Criterion between Attractiveness and Sustainability

The criterion of proximity to the coast oscillates between attractiveness and sustainability, with both dimensions significantly influencing the planning and development of coastal areas in general, and tourism projects in particular [11]. Areas close to the coast are considered highly attractive; however, coastal regulations prohibit development within 100 meters, making these areas unsuitable for tourism projects, as illustrated on the map, representing 3% of the studied area.

In contrast, areas located less than 1500 meters from the baseline of the coast are deemed excellent in terms of their preference for attracting tourism projects, comprising 15% of the total area. Meanwhile, areas situated between 1500 and 3500 meters from the coast have a good preference for establishing tourism projects according to this criterion. Areas located between 3500 and 12500 meters are rated as having moderate suitability, accounting for 13.3% of the area. When the distance exceeds 12500 meters, these regions become less suitable for tourism projects, receiving a weak rating since the coast is a focal point for tourism activities and holds more weight than any other criteria influencing the establishment of tourism projects in our area.

b. Proximity Preference to Tourist Attractions in Establishing Tourism Projects

Proximity to tourist attractions in our field offers investment opportunities in the tourism sector and serves as an incentive for the sustainability of projects, which depend on the continuity and success of these tourist sites [12]. Additionally, the field of study provides a variety of tourism qualifications.

Based on this perspective and according to the results of this criterion, areas with excellent proximity preference are estimated to make up 30%, while areas with good proximity to tourist attractions represent about 25%. Areas with medium proximity account for 43%, and those with weak proximity are at 2%, according to the findings of this study.

c. The criterion of proximity or distance from Rivers in relation to tourism projects

The study area is intersected by several water bodies, primarily the Martil River and the Lau River, which create natural boundaries as they are perennial streams. In contrast, other rivers have seasonal flows, and their channels are considered unsuitable within a distance of 50 meters according to this criterion, representing 7% of the area.Areas located about 500 meters away are categorized as weak in terms of sustainable development, accounting for 27%, while medium areas represent 41%. Good and excellent areas, situated 1000 meters and more than 2500 meters away, represent 16% and 9%, respectively.

These percentages vary among the communities in the study area, with the Lau community offering the highest proportion of land suitable for establishing a tourism project, followed by Martil and Azla. Meanwhile, the mountainous character and river channels in the region of Sidi Qassem diminish the suitability of areas according to this criterion.

d. The criterion of proximity or distance from road axes and its impact on the sustainability of tourism projects.

The road network is considered one of the fundamental pillars of development in general, and particularly for encouraging tourism projects, as it is a key factor in attracting tourists [13]. A dense and high-quality road network ensures accessibility, making areas close to road axes more valuable and favorable for establishing tourism projects.

In the study area, approximately 18% of regions are within 100 meters of roads and are classified as excellent, while 52% are categorized as good, located between 100 to 500 meters from roads. Conversely, areas that experience isolation due to road distance—those more than 2000 meters away—constitute about 8% of the study area.

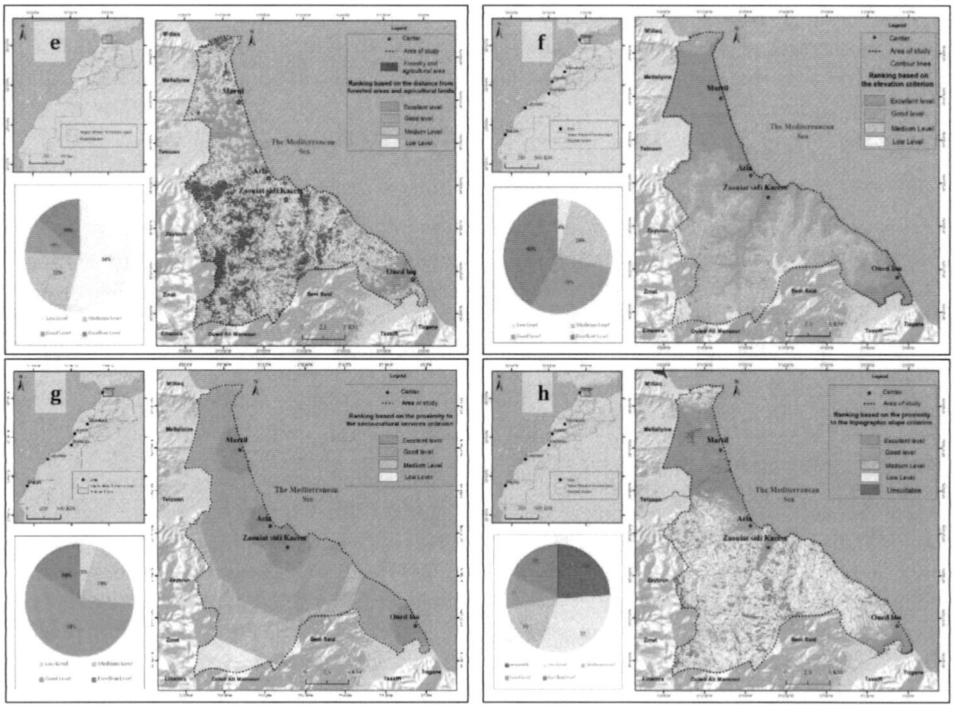

Figure 7 : e-Proximity to the forested area map, f-Elevation map, g-Proximity to Services map, h- Slope map.

e. The advantage of proximity or distance from forest areas and its role in establishing tourism projects.

The choice between proximity to or distance from forest areas depends on the nature of the targeted tourism project and the specific demographic of tourists [14]. Proximity to forests can enhance a project's appeal and provide diverse opportunities for eco-tourism activities. On the other hand, distance from forests may allow for easier spatial development [15]- [16].

Overall, it is essential to balance the benefits of natural resources with environmental protection and the needs of tourists. The degree of preference for proximity or distance from forests varies; only 7% of areas are considered excellent, while 41% are classified as medium. In contrast, 52% of areas are deemed unsuitable, reflecting the environmental considerations for sustainable development according to the established criteria.

f. The advantage of topographical elevation as a criterion for controlling the establishment of tourism project

Elevation plays an important role in selecting locations for tourism projects. Approximately 42% of areas in the study have an elevation of less than 150 meters, making them excellent for establishing tourism projects. Another 30% of areas, with elevations ranging from 150 to 300 meters, are considered good. In contrast, regions with elevations exceeding 300 meters have a moderate to low preference for tourism project development, accounting for 28%. This indicates that areas with lower elevations are generally more suitable for establishing tourism projects.

g. The criterion of proximity to social and cultural facilities.

The abundance of social and cultural facilities influences the concentration of tourism establishments. Areas located less than 2500 meters from these centers are considered excellent, representing 16% of our study area. The preference decreases as the distance from the studied centers increases, with areas 2500 to 7500 meters away classified as having good attractiveness, comprising the largest proportion at 58%. In contrast, areas further than 7500 meters are considered medium to weak in preference, representing 22% and 4%, respectively. However, these distant areas may be adjacent to other centers outside the study area.

h. The advantage of topographical slope as a criterion for establishing a tourism project.

Topographical slope is classified as a geographical factor that helps determine the location of a tourism project, either by providing panoramic views or facilitating water drainage [17]-[18]. However, it becomes unsuitable in areas prone to landslides[19]. In our study area, according to the data from Figure(4), the suitable area for establishing a tourism project varies in preference from excellent to good, accounting for 18% and 10% of the studied area, respectively. Areas with a medium preference represent 16%, while regions with low suitability due to steep slopes exceeding 12° account for approximately 32% and 24%, respectively.

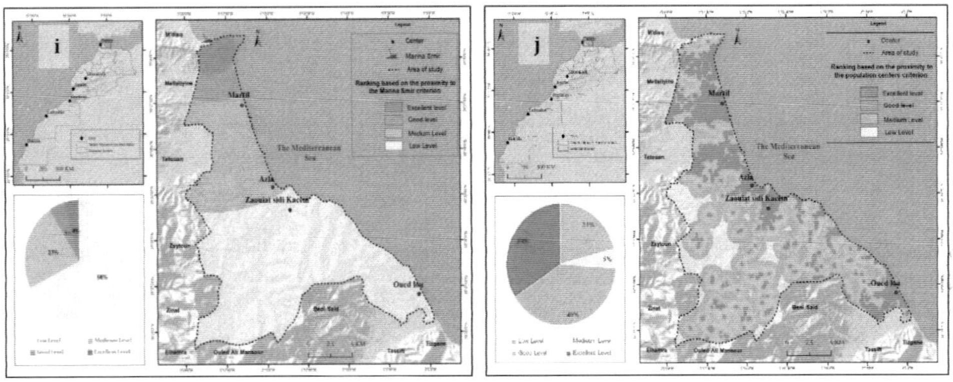

Figure 8: i- proximity to the marina map, j- proximity to the population centers map.

i. The criterion of proximity to the entertainment port and its importance in establishing a tourism project.

Proximity to a recreational port enhances the attractiveness of a tourism project by providing opportunities for various maritime activities. Therefore, areas located within 2500 meters from the port, as indicated on map, are considered to have excellent preference according to this criterion, although they represent a small percentage of the area, at just 4%.

Areas located between 2500 and 5000 meters away are classified as having good suitability, representing 5%. In contrast, areas more than 5000 meters away are rated as medium to weak, especially those exceeding 15 kilometers, which constitute 91% of the total area studied.

j. The criterion of proximity to population centers and its role in attracting tourism projects.

The study area exhibits variability in population density. While urban centers within the studied communities experience high population concentrations, making them initially suitable for attracting tourism projects, many rural population centers have densities exceeding 300 inhabitants/km^2. This average can serve as a benchmark for evaluating areas that are most suitable for establishing tourism projects, with 74% of the studied area classified as excellent or good, 21% as medium, and 5% as weak according to this criterion. Additionally, areas occupied by built environments are considered unavailable for tourism projects.

3.2. The result of spatial compatibility in choosing the most suitable areas for establishing tourism projects.

The spatial analysis of the established criteria revealed the extent of each criterion's impact on determining suitable areas for project establishment based on their priority in the study area. However, this was not sufficient for operationalizing these insights. Therefore, it was necessary to combine the spatial results of the ten adopted criteria through a cumulative matching process, utilizing the algebraic method (Algèbre de la carte), which considered the weight of each criterion. Ultimately, we arrived at the results presented in figure(9).

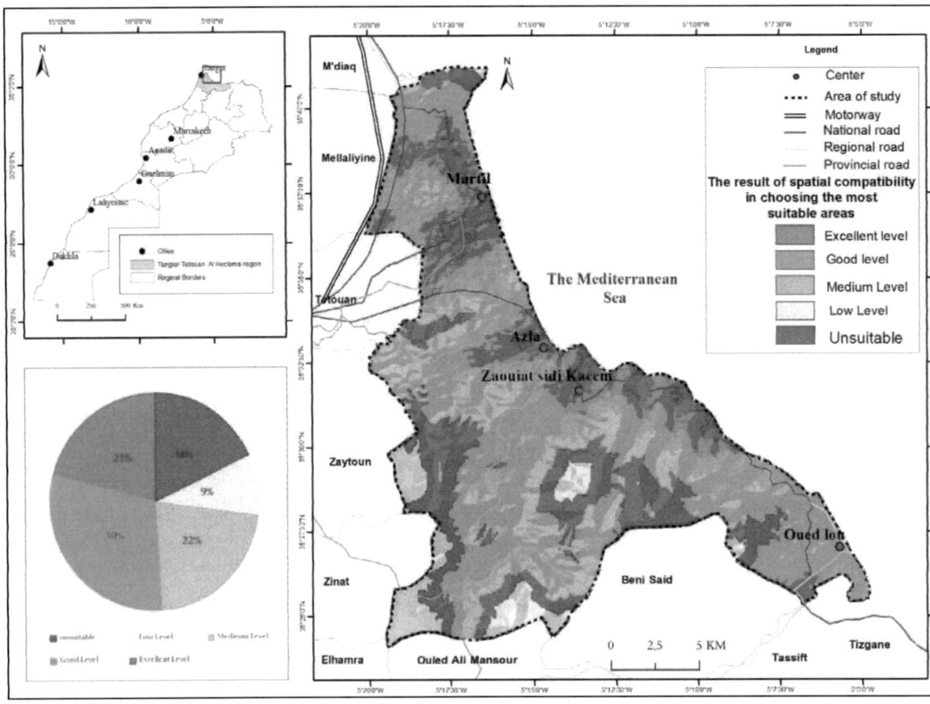

Figure 9 : Suitability Analysis Maps for Tourism Project Development

This model and the results obtained are primarily based on the role of Geographic Information Systems (GIS) in mitigating risks to natural resources such as forests, coastlines, waterways, and ecologically significant areas. With this in mind, our goal was to identify the most suitable locations for sustainable tourism development, ensuring better alignment with environmental sustainability principles [20]- [21]- [22].

This model has proven effective in planning designs at both the national and regional levels, where it can be utilized to identify the most suitable areas for sustainable tourism projects. This contributes to improving environmental planning on a large scale and promotes sustainable tourism development that balances the preservation of natural resources with the needs of tourism development across different regions.

a. Areas that are highly suitable and good for development.

Its area constitutes 60 km^2, accounting for 21% of the total studied area, with a clear concentration in the southern part of the Sidi kassm community and the western part of the Azla community.

These areas feature gentle slopes adjacent to the coast, extending up to 2500 meters, and offer sea views from moderate elevations. Additionally, their proximity to major roadways enhances their appeal. All these criteria, despite their varying weights, position these areas as the best locations for establishing sustainable tourism projects.

b. Areas suitable to a weak and moderate degree.

These areas represent a significant proportion of the studied region, with the weak areas accounting for 9% and the moderate areas approximately 22%. They include locations close to the coast and population centres within a distance of 100 to 1500 meters. These areas are known for their high attractiveness and are characterized by intense real estate speculation. On the other hand, this situation conflicts with the principles of sustainable development, which aim to preserve natural resources.

c. Areas unsuitable for establishing tourism projects.

The area of these regions is very small in our field, not exceeding 18% of its total area, equivalent to 50 km^2. However, they are of great importance for the sustainability of tourism activities in the region. They represent the coastline, riverbanks, areas with steep slopes, and locations that are very close to forested areas.

The integration of these results into planning processes is crucial due to their significant role in promoting sustainable spatial development. Spatial suitability analysis has proven effective in multiple domains, particularly in improving access to essential services such as schools and hospitals. Additionally, it has been instrumental in urban planning, supporting future expansion and infrastructure improvement.

For example, in the Netherlands, the Rijkswaterstaat project leveraged these techniques for strategic development, while in Canada, the Smart Growth initiative applied them to enhance sustainable urban expansion. Similarly, the MCDA (Multi-Criteria Decision Analysis) method has been successfully used in Germany and Denmark through the WindEurope project to identify optimal wind farm locations. In India, the National Solar Mission adopted this approach to determine the best sites for solar power plants.

3.3. Model validation

The results of spatial suitability analysis can help prioritize land-use planning designs, ensuring that land uses align with natural characteristics and the goals of sustainable development. However, our analysis revealed a lack of integration between our findings and the future directions of local communities in certain areas, where the reality contradicts the sustainable development goals aimed at balancing economic growth with the preservation of natural resources [23]- [24]- [25].

Some areas, according to the analysis results as shown on the map, are entirely unsuitable for expansion or the establishment of tourism projects due to their geographical characteristics. These areas consist of agricultural lands with complex real estate situations, or they are close to vital natural resources that could be damaged if these areas were used for residential or economic activities due to the pressure exerted on them.

This comparison is important to align institutional decisions that regulate spatial planning with the technical cartographic work based on legal frameworks, geographical specifics, and regional planning [26]. Doing so will help make better decisions and develop tourism projects in the most efficient and sustainable way, while also considering environmental and social sustainability.

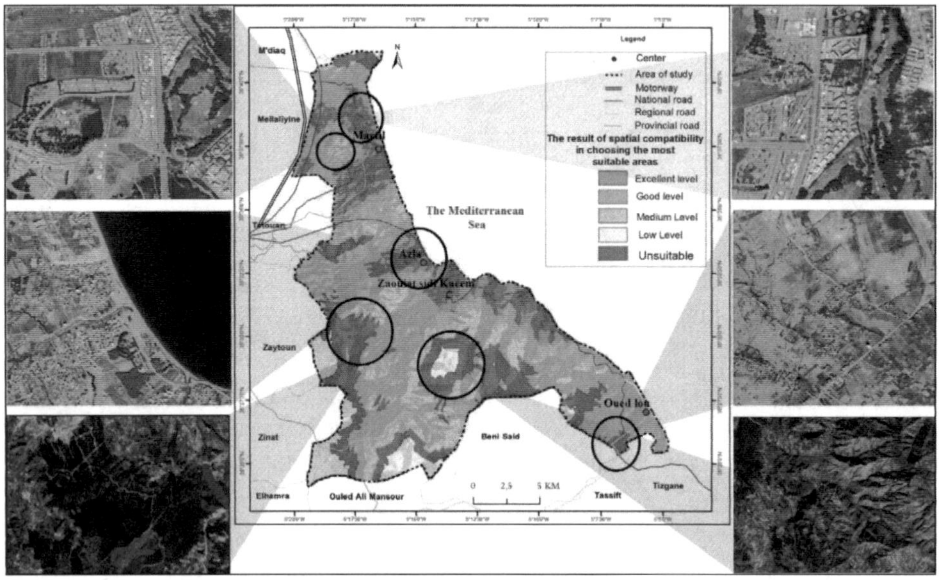

Figure 10 : The comparison between the results of spatial suitability analysis and field reality

Although applying this model in a small geographic area compared to its significance may seem limited, the similarity of spatial characteristics across Morocco's mountain and coastal tourist areas allows for its generalization to larger national regions. The same criteria used to build the model can be applied with different weights.

This shows how integrating spatial suitability with field reality requires ongoing collaboration between scientific analysis and field reviews to ensure projects fit the local environment and meet sustainable development goals[27].

3.4. The role of local residents in advancing the tourism sector and achieving sustainable development.

Local residents link the obstacles to tourism development to the lack of clear strategic visions and the weakness of the tourism infrastructure, as both are considered essential for advancing the sector. Among the suggestions put forward by local residents, the most prominent is the need to support local investments, encourage complementary tourism activities, establish diverse tourism projects, and create mountain tourism routes that would support the nature of coastal tourism, which is limited to the summer season. These initiatives could also provide regional benefits beyond the seasonal nature that characterizes the sector.

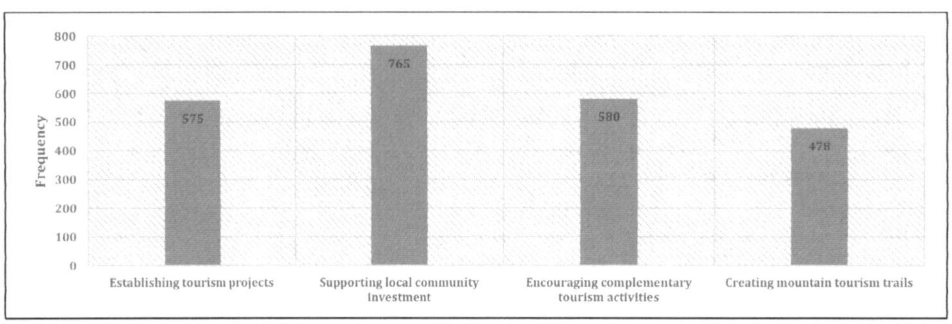

Figure 11 : Suggestions from locals for improving the tourism sector (more 1200 forms)

4. Conclusion

In the tourism sector, the use of multi-criteria decision-making techniques is especially important to optimize choices, as this field involves a wide range of factors to consider. Developing new tourist destinations requires analyzing multiple factors such as the attractiveness of the region, accessibility, infrastructure costs, and environmental impact. Multi-criteria analysis helps identify sites that offer the best balance among these elements.

With increasing awareness of environmental issues, multi-criteria analysis allows the inclusion of environmental factors (ecosystem preservation, waste reduction, water management) while balancing economic and social imperatives.

By optimizing the service offering (accommodation, transportation, activities) based on quality, cost, and satisfaction criteria, businesses can enhance their appeal and build customer loyalty.

Tourism projects often require significant investments. Multi-criteria analysis makes it possible to evaluate the potential profitability of projects by comparing costs and benefits, accounting for potential risks, and ensuring that funds are allocated to the most promising projects.

In tourism, various stakeholders (residents, local businesses, public authorities) may have different priorities. Multi-criteria analysis helps find an acceptable compromise by balancing each stakeholder's interests.

By identifying the most valued factors for tourists (e.g., safety, cultural authenticity, scenic beauty), regions can adjust their tourism offerings to better meet demand and stand out in the market.

By enabling more thoughtful and balanced decision-making, multi-criteria analysis is a significant asset for sustainable, profitable, and competitive tourism development.

It is important to enhance the use of multi-criteria analysis in the tourism industry to provide optimal solutions that balance economic, social, and environmental benefits, thus improving destination management.

Environmental values should be more deeply integrated at all stages of tourism project development, focusing on ecosystem preservation and minimizing the environmental impact of these projects.

Supporting sustainable tourism projects is essential, with governments and businesses investing in initiatives that provide local job opportunities and boost the local economy while preserving the environment.

It is crucial to involve all stakeholders (local communities, businesses, authorities) in tourism decision-making processes to ensure a balanced approach that meets diverse interests and achieves sustainable outcomes.

Further research and development should be encouraged in the field of tourism, with more studies conducted on the impact of tourism on the environment and local communities, ensuring decisions are based on accurate and realistic data.

Promoting sustainable destinations that adhere to environmental sustainability criteria should be a priority, helping attract environmentally conscious tourists and strengthening the global reputation of these destinations.

By implementing these recommendations, tourism can become more sustainable and effective, contributing to the development of tourism that aligns with the needs of both current and future generations.

References

[1] Dye, A. S., & Shaw, S. L. (2007). A GIS-based spatial decision support system for tourists of Great Smoky Mountains National Park. Journal of Retailing and Consumer Services, 14(4), 269-278. https://doi.org/10.1016/j.jretconser.2006.07.004

[2] Bahaire, T., & Elliott-White, M. (1999). The application of geographical information systems (GIS) in sustainable tourism planning: A review. Journal of Sustainable Tourism, 7(2), 159-174. https://doi.org/10.1080/09669589908667333

[3] Saaty, T. L. (1996). Decision making with dependence and feedback: The analytic network process. RWS Publications.

[4] Jovanović, V., & Njeguš, A. (2013). The application of GIS and its components in tourism. Yugoslav Journal of Operations Research, 18(2), 261-272. https://doi.org/10.2298/YJOR0802261J

[5] Saaty, T. L., & Vargas, L. G. (2012). Models, methods, concepts & applications of the analytic hierarchy process (Vol. 175). Springer Science & Business Media.

[6] Beedasy, J., & Whyatt, D. (1999). Diverting the tourists: A spatial decision-support system for tourism planning on a developing island. International Journal of Applied Earth Observation and Geoinformation, 1(3-4), 163-174. https://doi.org/10.1016/S0303-2434(99)85010-8

[7] Elaanzouli, M., Ait Zamzami, H., Moulouoade, A., Saidi, J., Ajaoun, M., & Boumeaza, T. (2024). The spatial implications of seasonal tourism on the southern Tetouan coast, Morocco. *A series of academic studies issued by the Mediterranean Center for Studies and Research*, 39, June 2024.

[8] Roy, B. (2013). Multicriteria methodology for decision aiding (Vol. 12). Springer Science & Business Media.

[9] Joerin, F., Thériault, M., & Musy, A. (2001). Using GIS and outranking multicriteria analysis for land-use suitability assessment. International Journal of Geographical Information Science, 15(2), 153-174. https://doi.org/10.1080/13658810051030487

[10] 6th World Congress on Snow and Mountain Tourism,www.unwto.org/fr/, consulté le 03/05/2024

[11] El Mrini, A., Anthony, E. J., Maanan, M., Taaouati, M., & Nachite, D. (2012). Beach-dune degradation in a Mediterranean context of strong development pressures: The Tetouan coast (Morocco). Ocean & Coastal Management, 63, 63-74. https://doi.org/10.1016/j.ocecoaman.2012.04.002

[12] Gharaibeh, A., Shaamala, A., Obeidat, R., & Al-Kofahi, S. (2011). The application of GIS in sustainable tourism development at Salt city, Jordan. Journal of Environmental Science and Engineering, 5(12), 1-11.

[13] Ait Zamzami, H., Elaanzouli, M., Zahrani, A., Saidi, J., Boumeaza, T., & Musiyam, M. (2024). Tourism investment and land use change in the northern piedmont of the western High Atlas (1989-2022), Morocco. *Forum Geografi, 38*(1), 83–93. https://doi.org/10.23917/forgeo.v38i1.4458

[14] Chen, Y., Yu, J., & Khan, S. (2010). Spatial sensitivity analysis of multi-criteria weights in GIS-based land suitability evaluation. Environmental Modelling & Software, 25(12), 1582-591. https://doi.org/10.1016/j.envsoft.2010.06.001

[15] Gigović, L., Pamučar, D., Lukić, D., & Marković, S. (2016). GIS-Fuzzy DEMATEL MCDA model for the evaluation of the sites for ecotourism development: A case study of "Dunavski ključ" region, Serbia. Land Use Policy, 58, 348-365. https://doi.org/10.1016/j.landusepol.2016.07.030

[16] Bunruamkaew, K., & Murayama, Y. (2011). Site suitability evaluation for ecotourism using GIS & AHP: A case study of Surat Thani province, Thailand. Procedia-Social and Behavioral Sciences, 21, 269-278. https://doi.org/10.1016/j.sbspro.2011.07.024

[17] Malczewski, J., & Rinner, C. (2015). Multicriteria decision analysis in geographic information science. Springer.

[18] Malczewski, J. (2006). GIS-based multicriteria decision analysis: A survey of the literature. International Journal of Geographical Information Science, 20(7), 703-726. https://doi.org/10.1080/13658810600661508

[19] Torres-Bejarano, F., González-Márquez, L. C., Díaz-Solano, B., Torregroza-Espinosa, A. C., & Cantero-Rodelo, R. (2018). Effects of beach tourists on bathing water and sand quality at Puerto Velero, Colombia. Environment, Development and Sustainability, 20(1), 255-269. https://doi.org/10.1007/s10668-016-9880-x

[20] Cybriwsky, R. Changing Patterns of Urban Public Space: Observations and Assessments from the Tokyo and New York Metropolitan Areas. Cities 1999,16, 223–231. [CrossRef]

[21] Pearce, D.G. An Integrative Framework for Urban Tourism Research. Ann. Tourism Res. 2001,28, 926–946. [CrossRef]

[22] Cope, D. Sustainable Development and the Trust. In The National Trust: The Next Hundred Years; Newby, H., Ed.; The National Trust: London, UK, 1995; pp. 53–69.

[23] T.H. Son, Z. Weedon, T. Yigitcanlar, *et al.* Algorithmic urban planning for smart and sustainable development: systematic review of the literature SCS, 2023 (Jul. 2023), Article 104562

[24] H. Xie, Y. Zhang, X. Zeng, *et al.* Sustainable land use and management research: a scientometric review Landsc. Ecol., 35 (Mar. 2020), pp. 2381-2411

[25] J. Li, Y. Wang Characteristic analysis and integration method of urban planning data based on GIS of internet of things SUSTAIN COMPUT-INFOR, 36 (Dec. 2022), Article 100801

[26] Taherdoost, H. 2017. Decision making using the analytic hierarchy process (AHP); A step by step approach. International Journal of Economics and Management Systems, 2.

[27] E.A. Moallemi, S. Malekpour, M. Hadjikakou, R. Raven, K. Szetey, M.M. Moghadam, R. Bandari, R. Lester, B.A. Bryan ocal Agenda 2030 for sustainable development Lancet Planet. Health, 3 (2019), pp. 240-241

IOP Conf. Series: Earth and Environmental Science **1503** (2025) 012030 doi:10.1088/1755-1315/1503/1/012030

Quantify Relation of SLR and Changes Along Karawang Coastal: GIS and Statistical Model Utilization

Romanna Friska Pricilia Manurung[1*], Nurul Khakhim[2] and Bowo Susilo[2]

[1] Master of Remote Sensing, Faculty of Geography, Universitas Gadjah Mada (UGM), Daerah Istimewa Yogyakarta, Indonesia
[2] Department of Geographic Information Science, Universitas Gadjah Mada (UGM), Daerah Istimewa Yogyakarta, Indonesia

*E-mail: romannafriskapriciliamanurung@mail.ugm.ac.id

Abstract. SLR has become a global and worldwide issue. However, its effects on shorelines, which depend on local variability and coastal geomorphology, make it necessary to monitor these impacts on a local scale. Not only observing shoreline shifting, this study also aims to evaluate the performance of Geographic Information Systems (GIS) in assessing shoreline change and its SLR in Karawang Regency. Using statistical models provided by QSCAT Plugin, we analyze SLR contribution to shoreline displacement activities along Karawang. Several models which are Net shoreline movement (NSM), End Point Rate (EPR), and Linear Regression Rate (LRR), are employed to measure the shoreline shifting and its change rates. The patterns of change rates are examined to determine the intercorrelation between SLR and dynamics of shoreline displacement using correlation coefficients. Results show AWEI SH is the most performed water index for extracting shoreline in Kawarang coastal. This area experiences accretion in over 51% of the area while erosion in about 40%, but the shoreline change's correlation with the SLR events is misaligned. The finding also highlights different spatial quality and number of data inputs don't have significant effects on LRR accuracy.

1. Introduction

Not only experiencing an annual global-trend rise of 18 cm for the past century [2], sea level is also projected to continuously increase. Despite the fact it occurs globally, changes in SLR are not uniform [2], therefore studies of its effects tend to be localized. Sea level rise impacts coastal areas, causing area loss of coastal low-lying lands, potential inundation by tidal floods, coastal erosion, seawater intrusion into aquifers [3], and shoreline displacement [1]. Thus, minimizing the potential damage to the coastal environment and supporting sustainable development need a comprehensive monitoring of the coastal activities.

Shorelines are generally defined as the instantaneous boundary between bodies of water and land [4]. Shoreline mapping requires proxies or substitute features as distinguishing references to indicate the boundary between land and water bodies [5]. Wetting limits and instant tidal level features are more commonly used for shoreline extraction with satellite imagery data sources than other proxies. Payo [6] explained that instantaneous waterlines are commonly used as shoreline indicators since they do not always necessitate images with the finest spatial resolution, unlike morphology-based proxies and vegetation boundaries.

Remotely sensed image utilization and its combination with Geographic Information System (GIS) is widely chosen for shoreline extraction, mapping, and monitoring approach [5] due to its cost-effectiveness and ease in the process of separating boundaries between the water bodies and lands including coastline detection. Delineation techniques applied vary from visual interpretation, and semi-automatic indexing with water index, eveloved to machine learning classification. Those shorelines should be evaluated by assessing the extracted shoreline accuracy, such as with a geometry accuracy test [9]. However, many coastal monitoring studies utilized and developed various shoreline extractions but didn't assess the accuracy of the shoreline [10].

Shoreline change includes quantification of the area and rate of change and analysis of influencing factors. Khakhim [7] found more than 20 GIS-based shoreline-change analysis studies spread across the world in the last 5 years. The most frequently used statistical model in shoreline change research is Net Shoreline Move supports the analysis of changes in areal extent and shifts in the position of the shoreline which allows erosion and accretion analysis [5]. To estimate change rate, both End Point Rate and Linear Regression Rate are common models used. Regression lines between accretion-erosion values and MSL trends yielded insights into SLR impact on shoreline change.

Based on previous studies, a comprehensive analysis of the change dynamic is needed. Karawang shoreline is expected to be affected by rising sea levels, as it is located directly near the Java Sea. According to Krismawardhani's research, it is found that the surface of the Java Sea has increased by about 37 to 39 mm/year. In addition, the coast along Karawang Regency experienced changes of up to 50% [12]. This study aims to observe shoreline changes in coastal Karawang which includes 1) analyzing and assessing the accuracy of the most suitable water index for shoreline extraction 2) mapping and calculating shoreline erosion and accretion by including quantitative analysis of LRR and EPR models 3) analyzing the relationship between SLR and shoreline change fluctuations in coastal Karawang district.

2. Method

2.1. Study Area

This research is performed along the coast of Karawang Regency, West Java, located at 107.098544 - 107.642983 E and 5.938903 - 6.215501 N. Karawang has a long coastline, which lies across nine sub-districts, with a length of 84.32 km directly facing the Java Sea (Figure 1). Its topography is relatively flat with a slope of about 0 - 2% in 80.44% of the area and coastal water depth ranges to a maximum of 5 meters, while offshore depths extend from 0 up to 20 meters [13].

Figure 1. Map of Karawang Coastal and study area boundaries, located in North of West Java

IOP Conf. Series: Earth and Environmental Science **1503** (2025) 012030 doi:10.1088/1755-1315/1503/1/012030

2.2. Datasets

Data used are long-term datasets from 28 years of Landsat satellite imagery (1996, 2000, 2004, 2008, 2010, 2016, 2020, 20240) and 8 years of Sentinel-2 imagery (2016 - 2024). The entire dataset has less than 20% cloud cover so it is eligible for use as the main data source in obtaining coastline data. The existing shoreline from the Indonesian Geospatial Information Agency (BIG) in 2022 and Ground Check Point (GCP) from Google Earth as reference data for the accuracy assessment. Mean Sea Level (MSL) records from tide gauges, which are widely distributed by the Permanent Service for Mean Sea Level (PSMSL), are utilized to analyze sea level rise trends [14].

2.3. Image Proccesing

Images recorded by sensors on satellites and aircraft generally have errors in both geometry and radiometric and must be corrected through pre-processing. In general, pre-processing consists of geometric and radiometric corrections to obtain an image that corresponds to the actual position on the Earth. Landsat satellite image products are available in Level 1TP correction, which is each image is corrected for terrain using DEM and for precision through an absolute adjustment process using GCP [11]. Radiometric correction is carried out to eliminate errors in pixels to obtain images with better quality. Radiometric correction consists of a radiometric calibration process to convert Digital Numbers (DN) to radians or reflectance and atmospheric correction to reduce object reflectance [15]. Landsat and Sentinel imagery products have passed the radiometric correction level. Besides those corrections, atmospheric correction is also important because the pixel transformation requires surface reflectance values. Landsat and Sentinel image products have undergone different levels of atmospheric correction with different algorithms.

2.4. Shoreline Extraction and Accuracy Assessment

Shorelines are obtained from two resources, Landsat as main data and complemented by Sentinel-2 imagery. A semi-automatic shoreline extraction technique will be applied to both images, namely water indexing transformation. The water index was selected as it meets several criteria, including the study's spatial coverage and the chosen shoreline indicator, which is instantaneous waterlines. This research was conducted at a location with a coastline length of 84 km so that it falls into the regional scale of 20 - 100km [10]. Four selected water indices are determined according to the summary of shoreline indicators extracted from multispectral satellite images and the corresponding extraction method [6], namely NDWI McFeeter version [21], Modified-NDWI or known as MNDWI developed by Xu [22], and both of AWEI by Feyisa [16]. With different formulas, the representation of water bodies obtained from each index has a distinct visualization (Figure 2).

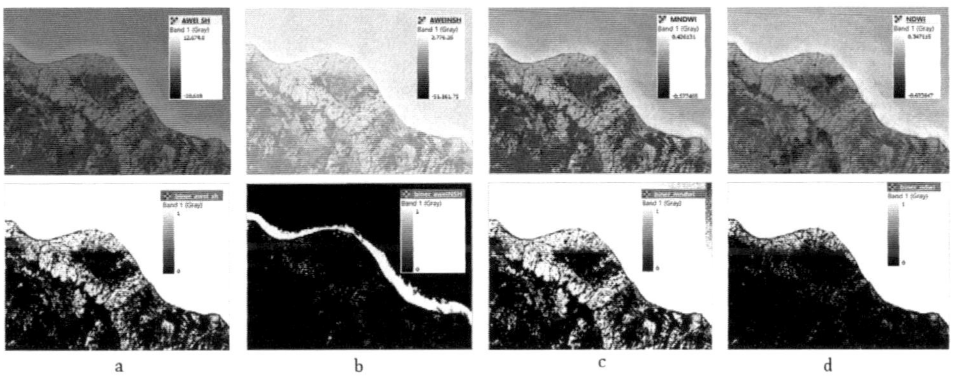

Figure 2. Pixel transformation with water indices a) AWEI SH b) AWEI NSH c) MNDWI d) NDWI

IOP Conf. Series: Earth and Environmental Science **1503** (2025) 012030 doi:10.1088/1755-1315/1503/1/012030

Water indices rely on spectral bands that are highly responsive to water bodies, enabling the separation between land and water objects. NDWI by McFeeter [21] is able to detect water boundaries by emphasizing water bodies through the calculation of pixel values with the following equation:

$$NDWI = \frac{Green - NIR}{Green + NIR}$$

The calculation of MNDWI by Xu [22] is carried out with the following equation:

$$MNDWI = \frac{Green - SWIR1}{Green + SWIR1}$$

Feyisa [16] offered AWEI-Sh dan AWEI-Nsh formula to improve the effectiveness and increased accuracy in detecting and analyzing water bodies. Four sensitive-to-water bands are utilized in the AWEI formula. The calculation is done with the following equation [16]:

$$AWEI\ SH = Blue + 2.5 \times Green - 1.5 \times (NIR + SWIR1) - 0.25 \times SWIR2$$

$$AWEI\ NSH = 4 \times (Green - SWIR1) - (0.25 \times NIR + 2.75 \times SWIR2)$$

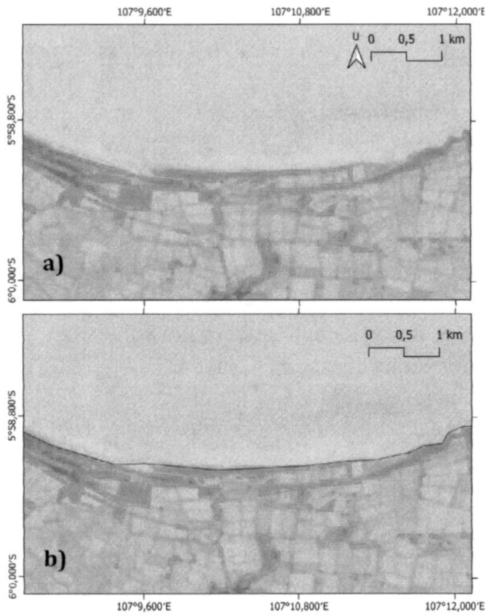

Figure 3. Hybrid technique combination to refine the shoreline delineation a) shoreline with water index extraction b) after refining with visual interpretation

Digital interpretation with water indices transformation applied on the pixel has limitations in distinguishing the target water body, which is the beach, from water bodies from ponds in coastal Karawang. A hybrid technique, a combination of digital and visual interpretation, is needed to obtain a more accurate coastline as applied in the coastal district of Batang which has the same land cover conditions as this study [7]. The shoreline delineation in areas not covered by the digital interpretation was improved following the shoreline seen in the same image where the index transformation was applied (Figure 3).

IOP Conf. Series: Earth and Environmental Science **1503** (2025) 012030 doi:10.1088/1755-1315/1503/1/012030

Accuracy assessment is an important part of data processing to find out the quality of the data used7by knowing the level of accuracy. Spatial accuracy indicates how precisely the positions of classified objects in the data align with their actual locations in the real world [17], encompassing both positional and thematic accuracy components. Here we calculate the positional accuracy using Root Mean Square Error (RMSE), to measure the closeness of extracted objects to their reference counterparts, with smaller distance values indicating higher accuracy. Shoreline accuracy tests can use actual shoreline data from visual interpretation of high-resolution imagery, Ground Control Points, and existing shorelines.

2.5. Shifting Analysis and its uncertainty

The extracted shoreline will be calculated for the amount of change (areal) and the rate of change using QSCAT (QGIS Shoreline Change Analysis Tool), which is additional software contained in the open-source QGIS software. Calculation of shoreline change requires baseline parameters as a measurement reference, the shoreline to be measured, and transects that intersect perpendicular to the baseline and form a point of intersection with each shoreline. NSM model is employed to define area changes. Negative area change indicates erosion, while positive change means accretion on the observation transect. Change-rate estimation is executed with 2 models, EPR and LRR. EPR only calculates the pace between two shorelines in the starting and ending points model. The LRR model calculates shoreline change by placing regression lines at all shoreline points in each transect. Several comparison scenarios will be carried out in the process of calculating the change rate, quantitatively defining the factors that affect value of LRR model.

The confidence interval (CI) is one way to measure uncertainty or also called the degree or level of uncertainty. The most commonly used confidence interval for shoreline analysis is the 95% CI. Statistical model in the QSCAT plugin in QGIS uses the same confidence interval of CI95% (symbolized as LCI95) as it is adapted from the DSAS toolbox (due to the plugin version update, the latest CI is 99%). LCI is derived by scaling the standard error of the slope (LSE) using a two-sided test statistic corresponding to a 95% confidence level set by the user. If the LRR is 1.34 meters per year with an LCI of approximately ±0,5 meters, the LCI95% suggests a 95% probability that the actual LRR ranges from 0,84 meter to 1,84 meters per year [18].

2.6. Sea Level Data

Sea level rise values were obtained from trend analysis of mean sea level (MSL) values. Monthly MSL (1996 - 2009) was aggregated into annual data. MSL data trend is analyzed to get SLR values with the least squares regression method. The result is the equation Y= ax+b and the value of MSL changes per year (SLR). Correlation between SLR value and shoreline changes which are classified as accretion and erosion are demonstrated. The relation is represented by coefficient determination, where X data represents the average MSL value and Y represents value of shoreline shifting over the same period as the MSL measurements. The test allows for quantifying the impact of SLR on shoreline displacement history.

3. Result and Discussion

3.1. Shoreline Mapping

Shorelines were estimated based on several water indices that enhance the water body and separate it from the land by involving several bands that are highly sensitive to water. A pixel value which is greater than 1 represents the water and less than 1 is land. Combining semi-automatic detection with visual interpretation, shorelines were refined. Each water index yields a different shoreline. Table A shows accuracy comparisons from shorelines demarcated by each water index in two different periods. Least RMSE value indicates the most performed index for shoreline demarcating in Karawang Coastal. RMSE values from the accuracy test obtained by comparing with the existing shoreline are greater than 100m due to the different tidal datum references. The existing shoreline downloaded from the Indonesian Geospatial Information Agency (BIG) is shoreline data measured in 2022 with Mean High-Water Spring (MHWS) and shorelines from the satellite use an instant tidal level based on real conditions captured by the satellite sensor. Images from Landsat and Sentinel-2 were captured around 9 to 10 am (GMT+7, Western Indonesia Time) with a water level is 0.1, below the High-Water Level. The highest water level at Karawang Coastal is approximately 0,44 m with the lowest water level is 0,34 below Mean Sea Level [13].

Table 1. RMSE value of shoreline with different indices transformation

Index	RMSE with Baseline 2022 (m)		RMSE with Google Earth GCP (m)	
	Landsat 2022	Sentinel 2022	Landsat 2024	Sentinel 2024
MNDWI	114.76	126.14	47.41286249	19.41320073
NDWI	151.36	159.98	58.68398614	38.32549427
AWEI NSH	-	141.51	-	22.03327708
AWEI SH	114.68	122.49	38.01988443	17.91591173

AWEI SH performed the best in representing the shoreline both in Landsat and Sentinel as shown in Table 1, as the minimum overlap distance was from AWEI SH and followed by MNDWI. AWEI and MNDWI were good at detecting mixed pixels that represent small ponds and rivers [19], AWEI SH performed the best for extracting water bodies in non-volcanic land cover class [9]. This also can be qualitatively explained by the findings of Novico et al. (2021) which identified some locations of Karawang Coastal have sediment composition consisting of organic matters such as organic shells. Based on this, all shorelines used in the next analysis are obtained with AWEI SH index.

3.2. Shoreline Shifting Measurement on Karawang Coastal

Transects are employed from a reference line and overlaid on all multi-temporal shorelines (Figure 4) from Landsat and Sentinel-2 data separately. With approximately 800 transects, shoreline changes are analyzed with different statistical models, NSM, EPR, and LRR.

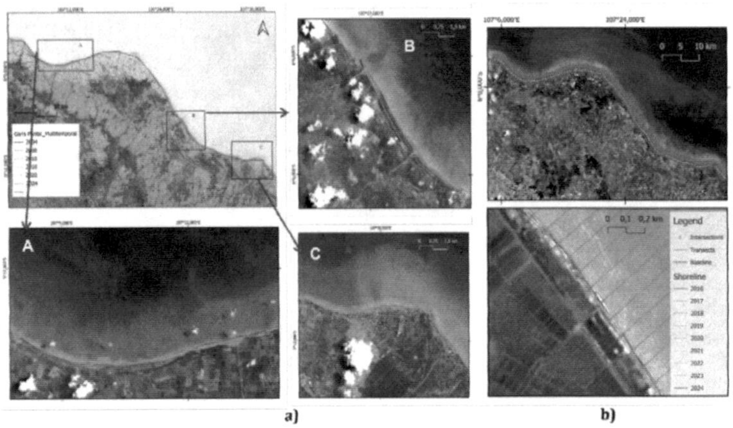

Figure 4. Multi-temporal shoreline a) overlaying along Karawang Coastal zone b) hypothetical baseline and transect generated on the shoreline

NSM quantifies the length that lies between the current and oldest shorelines and disregards any intermediate data. [18]. Supposing that the analysis concerns short-term area change, NSM

IOP Conf. Series: Earth and Environmental Science **1503** (2025) 012030 doi:10.1088/1755-1315/1503/1/012030

calculation should be carried out separately such as 1996 – 2000, 2000 – 2004, 2004 – 2008, etc. Dividing the distance by the time elapsed of 2 shoreline positions, results in a rate of change, EPR. Another model utilized is LRR, a common statistical model to calculate the pace of shoreline movement.

According to the statistical calculation with model uncertainty set as half of pixel resolution size, 15 m and 5 m for Landsat and Sentinel-2, respectively, landward shifting known as erosion shown in negative value and accretion manifested by positive shifting is evident here. The result indicates a spatial interconnection of all models used (Figure 5). It visualizes several levels of shifting rate in the Karawang Coastal, a greater rate of erosion/accretion occurs at the same place as greater shifting (area changes).

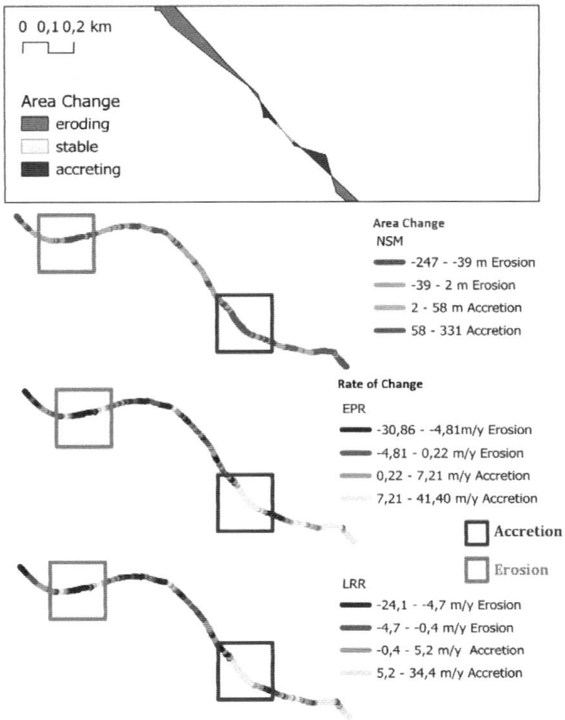

Figure 5. Shoreline movement in Karawang Coastal from different statistical model calculations, NSM estimates total area shifting while EPR and LRR quantify the change rates

This map highlights areas with the extensive and highest rates of erosion and accretion are the parts experiencing the fastest changes in the shoreline compared to others. Changes calculated from long-term data start from 2 meters, reaching 247 meters for erosion and 331 meters for accretion indicating that the movement is not uniform. The highest accretion mostly occurs in eastern regions, B and C (based on Figure 4) ranging from 5,2 m/year up to 34,4 m/year according to the LRR model. Similar results are also found in a previous study from Fauzie [23] which was conducted in the same regency and divided the eastern zone as the same as this study. It shows that from 1997 – 2006 accretion dominated and mostly occurred in the eastern with change rates value between 5,48 m/year to 10,78 m/year. Over the next 9 years (2006 - 2015), the area continued to shift seaward, but the shifting rates decreased to 3,1 m/year. Inversely, the western area eroded more in 2006-2015.

Table 2. Summary of area change of Karawang Coastal derived from NSM calculation

Shoreline Shifting (m)	Landsat 1996-2024	Sentinel 2016-2024
Average Erosion	-113.11	-55.67
Average Accretion	217.92	90.4
Average	68,99	17,49
Eroding		
Total transects:	319	356
(%) transects:	40.23%	**46.23%**
Accreting		
Total transects:	417	368
(%) transects:	**52.59%**	47.79%
Stable (%) transects	7.19%	5.97%

A comparison of changing areas from different time series demonstrates the changes are not constant from time to time. Table 2 shows average net changes calculated from NSM in long-term and short-term observation. Karawang shoreline changed up to 93% of the area within 28 years, slightly increased in the last 8 years and only less than 10% remains stable. Approximately 319 transects (40,23%) experience eroding and more than 52% are accreting from 1996 through 2024. Erosion is greater, reaching up to 46,23% of all transects while accretion is lower based on short-term data. This goes with Novico [20] research which shows erosion is greater than accretion from 2000 – 2018 and Fauzie [23] shows erosion is even more from 2006 to 2015. Those indicate the coast has experienced great erosion in the past few years. Average rate of landward shifting reduced from 113.11 meters (long-term observation) to 55.67 meters over the past 8 years, yet the number of transects is continuously increasing. Average shifting denotes the mean distance of shoreline shift from all transects.

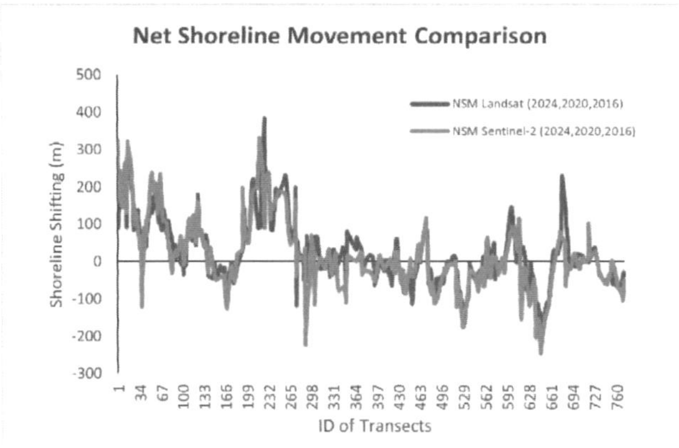

Figure 6. Comparison of NSM value through the same transects measurement on shoreline data obtained from different medium-spatial resolution satellite images

Anticipating the effects of different spatial resolutions, the NSM values of 2 medium-spatial resolution data sources are also compared, as shown in Figure 6. It displays NSM from Landsat and Sentinel-2 with the same number of inputs, 3 shorelines, for the two sets of data that is 2016, 2020, and 2024. The result is not the same, but the changes in overall transect ID are quite similar. This comparison is only valid if the same transect is used in the calculation.

IOP Conf. Series: Earth and Environmental Science **1503** (2025) 012030 doi:10.1088/1755-1315/1503/1/012030

3.3. Quantification of shifting rate based on EPR and LRR

Table 3 presents a summary of the change analysis results obtained using the EPR and LRR models for different periods of observation. In the long-term analysis, EPR values along the transects indicate a maximum landward shoreline movement of 12,6 meters/year, and mean rate is 4,19 meters/year observed in 38.4% of total transects applied. Meanwhile, the average seaward movement is 7,92 meters/year, affecting 51.5% of the transects, with a maximum value of 22,4 meters per year. According to EPR, accretion is the dominant process, as reflected in the average shoreline shift of 2,46 m/year (a positive value indicating accretion). The uncertainty associated with this measurement is ±0.76 m/year. This finding is further supported by the NSM calculation, which confirms that accretion is the prevailing trend in the long-term dataset. Based on the LRR model, approximately 56,4% of the transects in this study indicate land accretion, with mean value is around 7,4 m/year, while erosion is lower at 3.5 m/year. Overall, the results highlight that accretion dominates, with a shoreline displacement of around 2,6 m/year. To account for uncertainty, LRR applies a 99% confidence interval (LCI), which is ±7 m/year.

Comparing long-term and short-term observation, the average shifting rate from EPR calculation is quite stable ranging from 2,1 to 2,4 m/year. On the other hand, LRR shows very distinct results where shoreline average movement estimation is 2,6 m/year within 28 years but drops to only 0,9 m/year in the past 8 years (2016-2024). Several factors that may contribute to this discrepancy include differences in spatial resolution between Landsat (30 meters) and Sentinel-2 (10 meters) as well as variations in the timing of observations. Herein, a comprehensive analysis of LRR models is carried out. This study provides an overview of the factors that are crucial in LRR accuracy. An important requirement of this comparison is that the transect used for the entire calculation is the same and that all aspects of the datasets other than the factor being compared are the same.

Table 3. Value of shifting rates computed using EPR and LRR model in meter per year

Shoreline Change rate	Landsat 1996-2024		Sentinel 2016-2024	
	EPR	**LRR**	**EPR**	**LRR**
Highest Erosion	-12.6	-13.33	-30.86	-24.08
Highest Accretion	22.46	22.29	41.4	34.4
Average Shifting	2.46	**2.66**	2.19	**0.98**
Eroding:				
Total transects	305	345	341	399
(%) transects	38.46%	43.51%	44.29%	51.82%
Average Erosion	-4,19	-3,55	-7,23	-5,99
Accreting:				
Total transects	409	448	355	371
(%) transects	51.58%	56.49%	46.10%	48.18%
Average Accretion	7,92	7,45	11,69	8,49

One of the issues in LRR model calculation is the effect of data input. It is stated by Ciritci & Türk [8] accuracy of LRR relies on the quality of shoreline data and the duration of the observation period, and also improves as the amount of data increases. Figure 7 displays 3 comparisons of those factors that are mostly assumed to affect the LRR value. In this study, long-term data consists of 8 shoreline data at 4-year intervals and short-term data 9 at 1-year intervals. Datasets from Sentinel-2 with 3 shorelines data (2016, 2020, 2024) and Sentinel-2 with 9 data (from 2016 to 2024) are compared and show coefficient of determination (R^2) is 0,8. It explains that although

IOP Conf. Series: Earth and Environmental Science **1503** (2025) 012030 doi:10.1088/1755-1315/1503/1/012030

the data inputs are different, it has less effect on the LRR calculation. The second comparison evaluates spatial resolution impact (Figure 7b). Two datasets from Landsat and Sentinel-2 with the same data input and time range of observation (2016, 2020, 2024) are compared. That is also not the main factor as LRR values are fitted each other with R^2 value of 0,76. It is shown in Figure 7C that the temporal range of data observed is the major factor affecting the LRR R^2 value of 0,36. LRR values from both datasets are not dependent on each other.

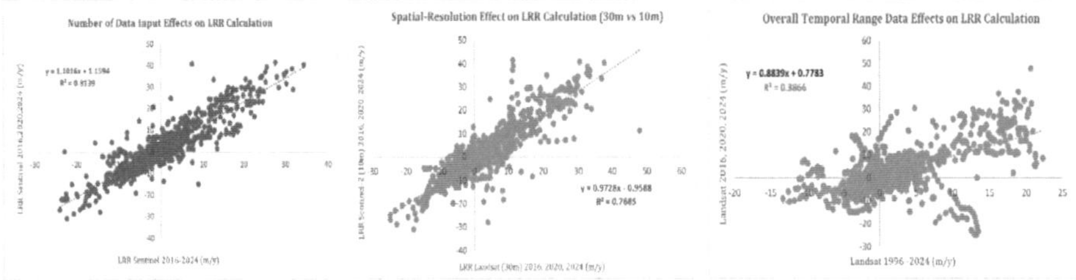

a) b) c)

Figure 7. Comparison of key factors a) number of data inputs b) spatial resolution of shoreline data source c) temporal range of data observed which are estimated affecting the accuracy of LRR calculation

Compared to the EPR model which works by dividing the total distance with time range between two shorelines, the LRR model calculation is based on an accepted statistical concept, fitting a least-square, so all available data is used in the calculation. However, the temporal range of data observed has more influence on LRR calculation than the number of data inputs. It is indicated by the first comparison above, that both datasets have the same temporal range at 8 years of observation. As long as shoreline data between the oldest and newest is available, it is considered sufficient to represent the fluctuation of the datasets. This analysis is evidence that LRR is more sensitive to time range than other aspects. According to the analysis, LRR is preferable for long-term monitoring due to its sensitivity to time span, as long as there are more than 2 shorelines in the datasets.

3.4. SLR and Shoreline Changes Intercorrelation

Understanding how the SLR trend affects shoreline changes requires analyzing the relationship between sea-level variations and shoreline movement patterns. Figure 8 is the monthly Mean Sea Level (MSL) height aggregated into yearly MSL. MSL value fluctuated, decreasing from 2000 to 2004 and starting to increase in 2005. A regression line is drawn along the MSL and shows an increasing trend with a positive coefficient correlation is 0,66. This highlights an upward trend in sea level across the study area during the observed period.

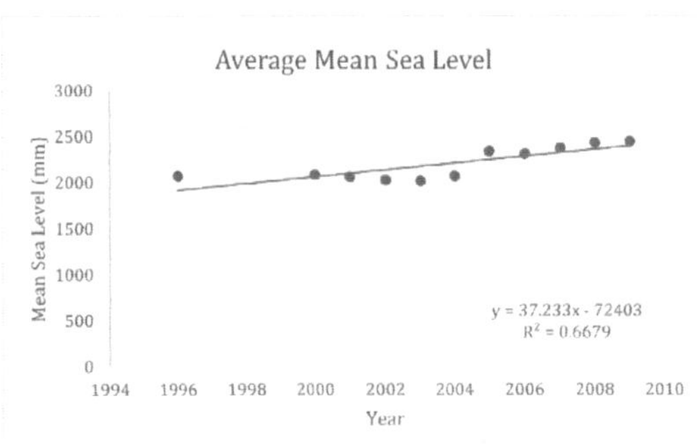

Figure 8. MSL height trend line signifies a rise in sea level.

IOP Conf. Series: Earth and Environmental Science **1503** (2025) 012030 doi:10.1088/1755-1315/1503/1/012030

Figure 9 is a regression chart of accretion and erosion with SLR trend where y is the shoreline shift value as the dependent variable and x is the average MSL value. The linear regression line and coefficient correlation value describe the relation strength of those variables. It is assumed that if SLR is the primary driver of shoreline change in a region, its correlation with erosion should be quite high. Erosion and accretion accumulation is calculated with NSM on shoreline data from Landsat with 4-year intervals (1996 - 2000, 2000 - 2004, 2004 - 2008). Both accretion and erosion on Karawang Coastal have a strong relation with SLR trend displayed by $R^2 = 0.9$ for accretion and $R^2 = 0.8$ for erosion. Technically, the accuracy of the results can be higher if the MSL availability is longer.

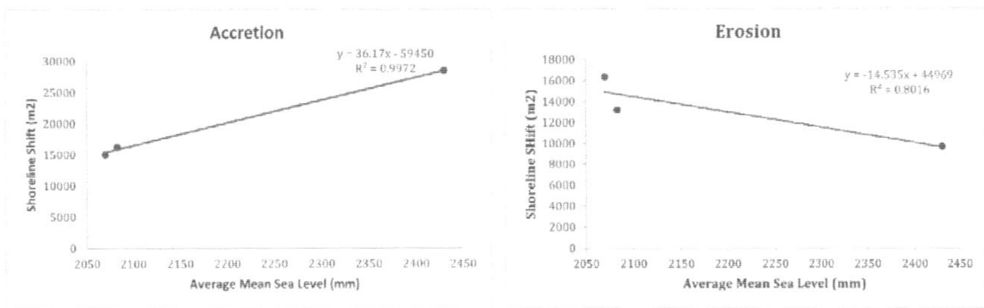

Figure 9. Linear regression between shoreline landward shifting (accretion) and recession (erosion) with sea-level change

The strong but negative correlation of SLR with shoreline recession indicates that SLR has less impact on shoreline shifting. Contrary to the previous relation, SLR is positively correlated with accretion which does not explain that SLR has a primary role in Karawang shoreline changes. The fundamental concept linking SLR to shoreline change is that rising sea levels typically lead to increased beach erosion, which serves as a key indicator of SLR effects [24] on account of greater displacement by the wave action on the beach [1]. Therefore, other physical factors may affect the changes such as sediment deposition from the river run-off due to multiple river outlets of the Citarum watershed.

4. Conclusion

In this study, long-term and short-term shoreline change analysis is performed by combining remote sensing data and techniques with GIS analysis along the Karawang Coast. The most suitable water index for extracting shoreline in this study is AWEI-SH with the smallest value of RMSE compared to the rest water indices. Based on 28 years of Landsat data, Karawang Regency has experienced a higher rate of accretion (>51%) than erosion (40%). according to the NSM model. Yet, the shifting changed where erosion is growing greater while accretion is lower in the last 8 years. EPR model quantifies a similar average rate of change in both long-term and short-term monitoring, starting at 2.1 meters per year and slightly rising to 2.4 meters per year. On the contrary, LRR value is significantly different compared to the two time-range observations. Therefore, LRR is a better model for long-term data because of its sensitivity to the time or temporal range. Quantitative results show a mismatched correlation between shoreline shifting and sea level rise. This finding eliminated SLR, as the factor which is assumed as shoreline change causion. However, SLR may not affect the shoreline shifting but still can possibly threaten Karawang Coastal zone by inundation.

Acknowledgments

This research is part of projects titled "Korea-Indonesia Marine Technology Cooperation Research Center (20220512)", and "Establishment of the Integrated Ocean Fisheries Technology Training Center and the Enhancement of Capacity Building in Indonesia (PG54670)" which are funded by the Ministry of Oceans and Fisheries, Korea.

References

[1] Narayana, AC 2016 Encyclopedia of Estuaries: Shoreline Change *Springer Netherlands*

[2] IPCC 2007a UN Intergovernmental Panel on Climate Change, Fourth Assessment Report – climate change

[3] Nicholls, R J, & Leatherman, S P 1995 "Global sea-level rise" As climate changes: international impacts & Implications in Strzepek, K M &Smith, J B eds *Cambridge: Cambridge University Press*, pp 92–123

[4] Kafrawy, S Basiouny, M Ghanem, E & Taha, A 2017 Performance Evaluation of Shoreline Extraction Methods Based on Remote Sensing Data Journal of Geography, *Environment and Earth Science International*, 114 1–18

[5] Kasim, F 2012 Pendekatan Beberapa Metode dalam Monitoring Perubahan Garis Pantai Menggunakan Dataset Penginderaan Jauh Landsat dan SIG Some Approaching Methods

[6] Payo, A Hennen, M Martinez, J Monteys, X Jaegler, T Martin-Lauzer, F-R Jacobs, C & Ellis, M A 2020 Monitoring Coastal Change from Space; What End Users Need and What Is Feasible p 213–228

[7] Khakhim, N Kurniawan, A Pranowo, W S Khasanah, E U & Halilintar, P 2024 Shoreline Morphological Change Prognostic Model based on spatiotemporal framework imagery data on the northern coast of Java, Indonesia *Kuwait Journal of Science*, 514

[8] Ciritci, D & Türk, T 2019 Automatic Detection of Shoreline Change by Geographical Information System GIS and Remote Sensing in the Göksu Delta, Turkey *Journal of the Indian Society of Remote Sensing*, 472 p 233–243

[9] Wicaksono, A & Wicaksono, P 2019 Akurasi Geometri Garis Pantai Hasil Transformasi Indeks Air Pada Berbagai Penutup Lahan di Kabupaten *Jepara Majalah Geografi Indonesia* 331 (86)

[10] McAllister, E Payo, A Novellino, A Dolphin, T & Medina-Lopez, E 2022 Multispectral Satellite Imagery and Machine Learning for The Extraction of Shoreline Indicators in *Coastal Engineering* 174

[11] Basiouny, M Kafrawy, S Ghanem, E & Taha, A 2017a Shoreline Change Rate Detection and Future Prediction Using Remote Sensing and GIS Techniques: A Case Study of Ras EL-Hekma, North Western Coast, Egypt *Journal of Geography, Environment and Earth Science International*, 93 p 1–14

[12] Setiabudi, A R & Maryanto, T I 2018 Deteksi Perubahan Garis Pantai di Pesisir Kabupaten Karawang dengan Aplikasi Digital Shoreline Analysis System DSAS *Reka Geomatika* 2 p 42 – 50

[13] Agus, F Liliek, S RobertoPP 2020 Kajian Hidro – Oseanografi di Perairan Kabupaten Karawang PELAGICUS: *Jurnal IPTEK Terapan Perikanan dan Kelautan* 11 p 39-51

[14] Permanent Service for Mean Sea Level PSMSL, 2024, "Tide Gauge Data", Retrieved 16 Dec 2024 from http://wwwpsmslorg/data/obtaining/

[15] Sukojo, B M & Mahmudi, A R 2018 Studi Analisis Ketelitian Geometrik Citra Satelit Resolusi Tinggi Sebagai Peta Dasar Rencana Detail Tata Ruang Perindustrian Studi Kasus: Kawasan Pt Sier Surabaya

[16] Feyisa, GL Meilby, H Fensholt, R Proud, SR 2014 Automated Water ExtractionIndex: A New Technique for Surface Water Mapping Using Landsat Imagery *Remote Sensing of Environment* 140 p 23–35

[17] Whiteside, T G S W Maier, G S Bogss 2014 Area-based and Location Based validation of classified image Objects International *Journal of Applied Earth Observation and Geoinformation*, 28 p 117–130

[18] Himmelstoss, E A, Rachel E H M G K and Amy S F 2018 Digital shoreline analysis system dsas version 50 user Guide Technical Report, *US Geological Survey*

[19] Acharya, T D Subedi, A & Lee, D H 2018 Evaluation of Water Indices for Surface Water Extraction in a Landsat 8 Scene of Nepal Sensors Switzerland, 188

[20] Novico, F D A Siddik, Lufiandi, A Albab, A Mulia, D Kusnida, R A Komarudin, D Ranawijaya, I Kamariah, C Endyana, H Bachtiar and Hendarmawan 2021 Interdisciplinary Approach for Qualitatively Monitoring Coastline Dynamics in North Java Coast, Case study: Karawang Regency Indonesia *IOP Conf Series: Earth and Environmental Science* 944

[21] McFeeters, S.K., 1996. The use of the Normalized Difference Water Index (NDWI) in The Delineation of Open Water Features. *Int. J. Rem. Sens.* 17 (7) p 1425–1432

[22] Xu, H., 2006. Modification ff Normalized Difference Water Index (NDWI) to Enhance Open Water Features in Remotely Sensed Imagery. *Int. J. Rem. Sens.* 27 (14) p 3025–3033

[23] Fauzie K 2017 Analysis of Short and Medium-Term Coastal Abrasion and Accretion Rates Using Gis in Karawang, West Java

[24] Cazenave, A., & Llovel, W. (2010). Contemporary sea level rise. Annual Review of Marine Science, 2(1), 145–173. https://doi.org/10.1146/annurev-marine-120308-081105

AUTHOR INDEX